D0942749

About IFPRI and the 2020 Vision Initiative

The International Food Policy Research Institute (IFPRI®) was established in 1975 to identify and analyze alternative national and international strategies and policies for meeting food needs of the developing world on a sustainable basis, with particular emphasis on low-income countries and on the poorer groups in those countries. Research results are disseminated to policymakers, opinion formers, administrators, policy analysts, researchers, and others concerned with national and international food and agricultural policy. IFPRI also contributes to capacity strengthening of people and institutions in developing countries that conduct research on food, agriculture, and nutrition policies.

"2020 Vision for Food, Agriculture, and the Environment" is an IFPRI initiative to develop a shared vision and consensus for action for meeting future world food needs while reducing poverty and protecting the environment. Through the 2020 Vision Initiative, IFPRI brings together divergent schools of thought on these issues, generates research, and develops policy recommendations.

IFPRI's research and capacity-strengthening and communications activities are made possible by its financial contributors and partners. IFPRI receives its principal funding from governments, private foundations, and international and regional organizations, most of which are members of the Consultative Group on International Agricultural Research (CGIAR). IFPRI gratefully acknowledges the generous unrestricted funding from Australia, Canada, China, Finland, France, Germany, India, Ireland, Italy, Japan, the Netherlands, Norway, South Africa, Sweden, Switzerland, the United Kingdom, the United States, and the World Bank.

Proven Successes in Agricultural Development

Proven Successes in Agricultural Development
A Technical Compendium to *Millions Fed*

An IFPRI 2020 Book

Edited by David J. Spielman and Rajul Pandya-Lorch

International Food Policy Research Institute
Washington, D.C.

International Food Policy Research Institute
2033 K Street, NW
Washington, D.C. 20006-1002, U.S.A.
Telephone +1-202-862-5600
www.ifpri.org

DOI: 10.2499/9780896296695

Library of Congress Cataloging-in-Publication Data

Proven successes in agricultural development : a technical compendium to
Millions Fed / edited by David J. Spielman and Rajul Pandya-Lorch.
    p.   cm.
Includes bibliographical references and index.
ISBN 978-0-89629-669-5 (alk. paper)
  1. Agricultural development projects—Developing countries—Case
studies.   2. Agriculture and state—Developing countries.   I. Spielman,
David J.   II. Pandya-Lorch, Rajul.
HD1417.P768   2010
338.109172'4—dc22   2010035433

Contents

Foreword

Since the late 1950s, the share of the world's population suffering from hunger and undernutrition has dropped from about one-third to one-seventh. Successes in agricultural development contributed substantially to these gains by feeding billions of people—by increasing food supplies, reducing food prices, and creating new income and employment opportunities for the rural poor.

Progress has slowed, however, and efforts to meet future food needs are continually challenged by factors such as climate change, environmental degradation, and population growth. By drawing on lessons from the past, we can once again put agriculture to work in ending hunger and undernutrition. But to do this, something more is required—evidence on where, why, and how past interventions in agricultural development have succeeded.

To identify and examine successes in agricultural development and draw out the lessons they offer, the Bill & Melinda Gates Foundation (BMGF) called upon the International Food Policy Research Institute (IFPRI) to assess the evidence on what works in agriculture—what sorts of policies, programs, and investments in agricultural development have actually reduced hunger and poverty. The project builds on a similar effort supported by BMGF and led by the Center for Global Development (CGD) called Millions Saved: Proven Successes in Global Health, summarized in a CGD book of the same name, and it stands out as a unique analytical contribution to both global and local discussions on agriculture's role in our future.

The successes highlighted in the Millions Fed project were chosen through a rigorous selection process that was supported by insights and advice from a committee of internationally recognized experts on agricultural development. The selected successes are described in a series of in-depth case studies that synthesize the evidence on a given intervention's impact on hunger and undernutrition. Each case study was carefully evaluated by the Millions Fed project team, several anonymous peer reviewers, and IFPRI's independent Publications Review Committee.

Overviews of these studies, written in a style accessible to a diverse global audience, were released in 2009 by IFPRI in a book titled *Millions Fed: Proven Successes in Agricultural Development*. The book then became the centerpiece for policy dialogues held in Washington, London, Addis Ababa, Seattle, Beijing, New Delhi, Dhaka, and Nairobi in late 2009 and early 2010. The book, along with supplementary materials on the project, is available at www.ifpri.org/millionsfed.

This compendium is yet another product of the Millions Fed project. The compendium reaches out to a more technical audience by bringing together the in-depth case studies published in the *Millions Fed* book. Each chapter compiles the evidence of what works, examines the rigor with which the evidence was generated, and assesses the pros and cons of each success. Like all analytical materials prepared for the Millions Fed project, this compendium was carefully peer reviewed by the project team, anonymous reviewers, and IFPRI's independent Publications Review Committee.

The technical studies in this compendium, along with the introductory analysis and the final chapter on trends in impact assessment, should be of great interest to the research community. It is our hope that the book will stimulate vigorous debate and dialogue—among not only researchers working strictly in the field of agricultural development but also those studying the linkages between health and agriculture, the relationship between urban and rural development, the tools of impact assessment, and many other topics.

Our ultimate hope is that the Millions Fed project will direct more attention to the need for well-designed policies and investments in agricultural development that help end hunger and undernutrition. Without greater attention to these issues, we stand to lose the impressive gains achieved in improving global food security since the 1950s.

Shenggen Fan
Director General, IFPRI

Prabhu Pingali
Deputy Director, BMGF

Acknowledgments

We gratefully acknowledge the support of the Bill & Melinda Gates Foundation for this project, and we appreciate the early encouragement received from Prabhu Pingali and Ellen McCullough on the need to bring together the technical analysis prepared for the Millions Fed project.

We sincerely thank the authors for their valuable contributions and insights. The time, effort, and dedication they gave to the Millions Fed project since early 2009—from the very first draft papers to this compendium—are greatly appreciated.

We express our warm appreciation to our colleague Sivan Yosef for her tremendous commitment and enthusiasm. Without her superb coordination skills and careful attention to details, the publication of this compendium would not have been possible.

We thank Klaus von Grebmer for his leadership on the Millions Fed communications strategy, as well as Evelyn Banda, Gwendolyn Stansbury, and John Whitehead for their support in editing and production. We also thank Heidi Fritschel, Clare Wolfowitz, and Kathryn Bertram for providing superb editorial insights in the preparation of these chapters, along with Djhoanna Cruz and Etenesh Yitna for their administrative assistance.

Finally, we extend our appreciation to everyone who contributed to the entire Millions Fed project. Special appreciation is extended to the Millions Fed Advisory Committee for their guidance and insights, to IFPRI's Publications Review Committee for managing an intensive peer review process, and to the many anonymous reviewers who provided thoughtful comments on the chapters presented here.

Chapter 1

Fifty Years of Progress

David J. Spielman and Rajul Pandya-Lorch

In the late 1950s around a billion people—about one-third of the world's population—were estimated to go hungry every day. Famines were threatening millions in Asia and Africa in particular, and prospects for feeding the world's booming population looked bleak. In response to this alarming picture, scientists, policymakers, farmers, and concerned individuals initiated a concerted push to boost agricultural production and productivity in developing countries. Developing and industrialized countries, together with development agencies and civil society organizations, pursued a range of interventions in agriculture: they applied modern science to crop and livestock production, constructed irrigation systems, developed new cultivation practices to conserve natural resources, introduced policies to encourage farmers to grow and sell more food, and launched many other programs in agricultural development.

The result? About a billion people now go hungry every day, according to the United Nations Food and Agriculture Organization (FAO 2009). This result may look like failure, and in one sense it is. The fact that 1 billion people remain hungry and malnourished is a tragedy on a grand scale. Looked at another way, however, the present situation reflects astounding success. Although the *absolute* number of people who are hungry has remained the same, the *relative* figure—the proportion of the world's population that has remained hungry—has declined dramatically. In the mid-1960s, when the global population was about 3.3 billion, only about 2 billion people were getting enough to eat. Today's population has burgeoned to more than 6 billion—and some 5 billion people now have enough food to live a healthy and productive life.

Clearly, progress has been made. China and India, once viewed as dire cases, have experienced agricultural booms. China slashed the number of its hungry people from 303 million in 1979–81 to 122 million in 2003–05, singlehandedly making a significant dent in world hunger because of its sheer size. In the face of rapid population growth, India also cut the number of people suffering from chronic hunger, from 262 million in 1979–81 to 231 million in 2003–05 (FAO 2002, 2008). Efforts to increase the global availability of food have led to enormous gains in agricultural productivity and food production, with yields of many staple crops multiplying severalfold. Great strides also have been made in improving the quality of food so that it contributes to good nutrition and in improving the ability of the most vulnerable groups—most significantly, women and children—to access the food they need for survival.

Importantly, these efforts have done more than just feed millions. The interventions of the past half century have also demonstrated that agriculture can be a key driver of growth and development for many of the world's poorest countries. Whereas scholars and leaders once viewed the agricultural sector with disdain—as a drag on their attempts to promote growth and modernization—they now recognize that agriculture can be a leader in the process of economic and social development (de Janvry 2009).

Yet 1 billion remain hungry. In fact, progress in overcoming hunger has reversed in the face of the recent food price crisis and the global economic downturn. In Latin America and the Caribbean, 45 million people still go hungry. In Sub-Saharan Africa, the number of chronically hungry people has almost doubled, from 125 million in 1979–81 to 212 million in 2003–05. In South Asia, although the declines in hunger are commendable, the sheer size of the population that remains hungry—314 million—is overwhelming (FAO 2008). In short, more remains to be done.

Although the causes of chronic hunger and persistent malnutrition are complex, the experiences of the past 50 years show that the solutions are by no means beyond our reach. But what do we really know about what works in agricultural development and about where, when, and why some interventions succeed? Which policies, programs, and investments in agricultural development can substantially reduce hunger and malnutrition? And which of these interventions can do so within a changing global landscape characterized by growing natural resource scarcities, climate change, global market volatility, and major health and demographic changes?

The purpose of this book is to identify and describe successes in agricultural development that have made substantial contributions to reducing hunger and malnutrition. It is not an exhaustive compilation of all successes that have occurred during the past 50 years. Rather, it is an in-depth analysis based on 20 case studies that aims to give readers a better understanding of what worked and why. By drawing key insights and lessons from past successes, this book intends to inform future policymaking and leverage future investments in ways that will contribute to overcoming hunger and malnutrition.

Methodology

A study such as this one cannot claim to be an exhaustive review of all policies, investments, and programs in agricultural development; five decades of rich and diverse experiences simply cannot be summed up in a single volume. So instead this book focuses on relatively large-scale and long-term successes that were backed by strong evidence of positive impact.

The methodology used to identify and analyze these successes is detailed in this section. This methodology draws on several previous efforts to document successes in development, including studies by Gabre-Madhin and Haggblade (2004) and Levine (2004). Additional insights were extracted from studies by Krishna, Uphoff, and Esman (1997); Uphoff, Esman, and Krishna (1998); World Bank (2006, 2008); and the Asia Pacific Association of Agricultural Research Institutions (2009).

A first step in identifying these successes was to seek input from those who participate in or study agricultural policies, investments, and programs that aim to improve food security and reduce hunger—practitioners, scholars, policymakers, and many others. To this end, the project team circulated a global "call for nominations" of successes in agricultural development in late 2008 while it simultaneously compiled a list of potential successes from experts in the field (including the project's own Advisory Committee, listed on pp. 601–602) and from information garnered from scholarly literature, project documents, websites, and other sources. The call for nominations was distributed by e-mail, posted at the project's website, linked to the websites of several organizations, and circulated in English, Spanish, and French. Conservative estimates of total circulation suggest that the call for nominations reached 58,000 individuals globally, although it is difficult to arrive at truly accurate figures given the nature of its transmission. By early 2009, 326 nominations had been received. This final figure reflects all submissions, suggestions, and recommendations received from all sources, inclusive of incomplete online submissions, similar nominations provided by different sources, and suggestions that required further information gathering by the project team.[1] Accounting for incomplete and repeated submissions, the total number of unique nominations was slightly greater than 250. Table 1.1 provides a breakdown of all 326 nominations by source.

The project team then evaluated these nominations. The sorting process was based on the application of two qualifying criteria that had to be met in order for a nominated intervention to be considered further and five evaluative criteria that focused on the specific attributes and impacts of the intervention.

The first qualifying criterion was that the intervention must have been operational in at least one developing country. For the purposes of this project, developing countries are those classified as low-income, lower-middle-income, or higher-

Table 1.1 Nominations, by source

Source	Percent
Open call for nominations	56
Advisory Committee recommendations	21
Expert opinion poll / key informant interviews	15
Project team research and literature review	8
Total	100

Source: Authors.

middle-income countries according to the income groups defined by the World Bank (2009) or according to the equivalent classification that was current when the intervention was being implemented. Note that this criterion does not imply that interventions were chosen because they benefited only developing countries: if an intervention generates benefits that also accrue to high-income industrialized countries, it was still considered.

The second qualifying criterion was that the intervention must have engaged agriculture directly—that is, it must have operated on constraints that are specific to agriculture. This criterion excluded certain types of interventions that operated on agriculture indirectly. For example, although there is strong evidence indicating that basic education, health, and sanitation programs targeting the rural poor contribute to increasing their labor productivity, and thus their incomes and nutritional status, these interventions were not considered here because their point of entry is not directly related to the production, distribution, marketing, or consumption of agricultural goods and services. Similarly, although rural school feeding programs, rural conditional cash transfer programs, rural safety net programs, and food aid are often viewed as important to increasing rural incomes, building rural assets, and improving nutrition, their generally indirect impact on agriculture means that they were ruled out from consideration here.[2]

Once these two criteria were met, the intervention was to meet five evaluative criteria:

- importance—the intervention should have tackled an important food security problem by addressing the needs of a vulnerable group;

- scale—the intervention should have operated at scale, measured in terms of whether the number of beneficiaries exceeded several hundred thousand individuals or whether the intervention was, at a minimum, national in coverage;

- time and duration—the intervention should have been (1) fully operational at scale long enough to generate significant reductions in hunger or improvements in food security and (2) implemented in the past 50 years;

- proven impact—the intervention should have been supported by documented and rigorous evidence of a clear and measurable impact on individual or household hunger or nutritional status; and

- sustainability—the intervention should have been sustainable, whether in financial terms (cost-effectiveness) or in broader social, political, or environmental terms.

Although efforts were initially made to apply each criterion as a necessary (but not sufficient) condition for selection, it was recognized that few cases would meet all criteria. Therefore, the criteria were relaxed somewhat, although the qualifying criteria were maintained as necessary conditions, along with the following evaluative criteria: importance, scale, and time and duration.

With respect to the proven impact and sustainability criteria, it became apparent early on that very few interventions were supported by what may be termed "state-of-the-art" impact assessments that documented the effects of the intervention through randomized experiments that established attribution by combining before-and-after comparisons with treatment-and-control comparisons. In many cases, such experimental methods did not exist when the intervention was in operation; in other cases, neither the resources nor the expertise were available to undertake such data-intensive assessments. Relaxing these criteria meant that alternative forms of evidence were accepted. These alternative forms of evidence included geospatial imagery documenting changes in agroecological landscapes over time, quantitative evidence using estimation techniques that were not necessarily the most up-to-date methods, and qualitative evidence gleaned from policy analyses and from surveys conducted among direct beneficiaries.

In most cases, insufficient evidence was available in refereed academic publications to provide sufficient evidence of success. Thus, many of these alternative forms of evidence had to be drawn from the "gray literature"—documents produced and published by governmental agencies, academic institutions, and other groups that are not distributed or indexed by commercial publishers. That evidence appeared in the gray literature implies that it was of a lower standard or quality than evidence published in the refereed academic literature, raising concerns about whether gray literature could contribute to meaningful evaluation of the success of an intervention. This was of particular concern in cases in which the gray literature was authored by individuals or organizations that were principal parties to the intervention itself. If gray literature was more readily available than academic literature, extensive efforts were made to cross-check and verify claims from different and independent sources wherever possible.

As a result of this rigorous review process, this book does not feature several types of successes. First, the book does not cover nonagricultural interventions, such as rural health, rural education, or rural social protection programs, for example. These programs undoubtedly comprise an important class of intervention, but they do not *directly* address agriculture in its strictest sense, as discussed earlier. However, one success case that was included—improving micronutrient consumption in Bangladesh, discussed in Chapter 20—does feature a rural nutrition program that promotes home-based production of fruits, vegetables, and livestock to improve nutrition and health among the poor.

Second, this book does not examine programs that integrate agriculture with health, education, microfinance, microenterprise, governance, and other development priorities. These represent an important class of intervention, but because of the complex synergies between these many activities, it is hard to disentangle the evidence. However, several successes featured in this book, while primarily defined as agricultural development programs, do examine the importance of integrated approaches. The study of community forestry in Nepal discussed in Chapter 5, which highlights the importance of integrating agricultural development with local governance, is one such success.

Third, this book does not cover cases of failure. Often learning lessons is as much about observing the failures as it is about observing the successes. But although there are many failures in agricultural development from which to learn—and many studies that highlight the causes of these failures—in this book we have chosen to focus on the successes only, primarily because it is the absence of successes in agricultural development that has marginalized its importance in discussions of how to improve food security and reduce hunger in developing countries.

What this book does highlight is a set of successes that are based on syntheses of evidence from multiple sources that range from first-hand accounts by individual participants and beneficiaries to large-scale impact-evaluation studies that combine both quantitative and qualitative evidence at the highest levels of academic rigor. This implies that the successes highlighted in this book are not evidenced by a common set of impact-assessment methodologies, indicators, or conclusions; the book does not assess success on a single set of indicators that describe the numbers of millions fed—or the quantitative improvement in food security—resulting from an intervention.

This may not be a disadvantage to the present analysis when considered more closely. A glance at the chapters in this book indicates that the interventions vary so greatly—in terms of what they aim to achieve, how they do so, and what they actually end up accomplishing—that a single indicator runs the risk of reducing an intervention's impacts to something entirely impractical.

Moreover, the casual reader should not conclude that only those interventions that are backed by rigorous impact-assessment materials and definitive indicators are successes. For example, a program that was not rigorously evaluated by teams of independent scholars conducting lengthy household surveys may nonetheless be a success. Or a program that was initially seen as a success may nonetheless fail in the long term.

In fact, there are many successes in agricultural development that are not covered by this volume but that may have equal merit. Examples include smallholder cultivation of high-value export crops in Kenya and Guatemala; systems of rice intensification that have become popular in several countries during the past two decades; New Rice for Africa, which is being developed and disseminated for farmers in several Sub-Saharan African countries; or the Greenbelt Movement in Kenya, which has encouraged community-based tree planting on a massive scale.

But even with these caveats in mind, we know that there are clearly discernible pathways—interventions that seek to improve crops, livestock, forestry, and fisheries; conserve natural resources; and strengthen the markets, institutions, and policies that relate to these social and economic activities—that link agricultural development with improvements in food security. These pathways and the stepping-stones along them are the main focus of impact assessments and thus the main focus in proving success.

Results

The entire exercise of identifying potential successes in agricultural development yielded several interesting results. Note, however, that the results are not meant to be representative of trends in agricultural development. At best, they characterize the outcomes of a purposive research process and allow readers to reflect on how the process might have influenced the study's outcomes, both positively and negatively.

In terms of geographic representation, the largest number of nominations were for interventions in Sub-Saharan Africa, followed by South Asia and then by East and Southeast Asia and the Pacific (Table 1.2). A further breakdown of the data indicates that the largest numbers of interventions were identified in India (19 percent), Kenya (12 percent), and Nigeria (10 percent), with other countries accounting for under 10 percent each.

Efforts to categorize interventions by their primary function provide further insights into the nomination process. Although there is a nearly infinite number of ways to classify interventions in agricultural development and extensive overlaps in

Table 1.2 Nominations, by region

Region	Percent
Latin America and the Caribbean	10
East/Southeast Asia and Pacific	17
South Asia	25
Middle East and North Africa	7
Eastern Europe and Central Asia	2
Sub-Saharan Africa	40
Total[a]	100

Source: Authors.
[a]Because some nominations covered more than one region, the percentages given above are based on a total number of nominations that exceeds 326, that is, the total number of nominations given in Table 1.1. The total given here may not sum to 100 percent due to rounding.

doing so, the following themes reflect the major categories that emerged from the nominations:

- intensifying staple food production,

- integrating people and the environment,

- expanding the role of markets,

- diversifying out of major cereals,

- reforming economywide policies, and

- improving food quality and human nutrition.

Table 1.3 shows that almost one-third of all nominations were identified as agricultural production interventions, a category that most commonly describes the development and dissemination of technologies such as improved cultivars, chemical fertilizers, irrigation, and equipment used in agricultural production activities. Diversification out of food staples and into other crops such as horticulture and livestock account for 23 percent of nominations, followed by interventions that seek to expand the role of markets in agricultural development, which account for 21 percent. Interventions that seek to integrate people and the environment—primarily programs focused on natural resource management—account for an additional

Table 1.3 Nominations, by theme

Theme	Percent
Intensifying staple food production	29
Integrating people and the environment	15
Expanding the role of markets	21
Diversifying out of major cereals	23
Reforming economywide policies	6
Improving food quality and human nutrition	6
Total	100

Source: Authors.
Note: Although some nominations covered more than one theme, the percentages given above are based on efforts to identify a single or dominant theme for each intervention.

15 percent of nominations. Policy interventions that seek to promote economywide reforms account for just 6 percent of all nominations, as do interventions in food quality and human nutrition.

At first glance, these results seem to reflect the prevalent historical focus on interventions to intensify agricultural production through the application of modern inputs and technologies, followed by interventions to diversify rural livelihoods and integrate farmers into markets—points that are discussed in detail later. But recalling that the nominations were provided by self-selecting respondents, it remains difficult to interpret the results in any particular direction. Only two points are worth bearing in mind. First, the large number of programmatic interventions in food staples, diversification, and market development may reflect a tendency among respondents to nominate specific projects with which they are associated or familiar as opposed to nominating large policy changes, which are far less attributable to individuals. Second, closer examination of these interventions might suggest that many nominations cut across several categories at once and are driven by a combination of production, natural resource, market, and policy factors.

Of further note is the temporal nature of the interventions identified, as shown in Table 1.4. Approximately 75 percent of the interventions began in the 1990s, whereas 68 percent were of less than 15 years' duration. This may suggest that the identification process itself—the nomination, expert opinions, and literature reviews—tended to draw out examples from fairly recent time periods. However (and keeping in mind the nonrepresentativeness of this sample), it may also suggest something about the short time horizons that tend to characterize government and donor commitments to agriculture, a point also discussed in detail later.

Finally, it is worth examining the nominations in terms of the project's criteria (Table 1.5). Only 86 nominations (26 percent of the total) met the minimum

Table 1.4 Nominations, by time and duration

Start date of the intervention	Percent
1950s or earlier	2
1960s	3
1970s	8
1980s	17
1990s	31
2000s	39
Total[a]	100

Duration of the intervention	Percent
1–4 years	17
5–9 years	22
10–14 years	24
15–19 years	12
20 years or more	25
Total[a]	100
Ongoing interventions (as of 2009)	**54**

Source: Authors.

[a]Because some nominations provided incomplete information on the duration and time period of the intervention, the percentages given above are based on a total number of nominations that is less than 326, that is, the total number of nominations given in Table 1.1. The totals given here may not sum to 100 percent due to rounding.

Table 1.5 Nominations, by criteria

Indicator	Percent
Interventions that met the criterion for	
1. Developing-country relevance	100
2. Agricultural relevance	97
3. Importance	94
4. Scale	67
5. Time and duration	40
6. Proven impact	51
7. Sustainability	57
Interventions that met criteria 1–5	26
Interventions that met criteria 1–7	20
Interventions that met at least	
1 criterion	100
2 criteria	100
3 criteria	97
4 criteria	83
5 criteria	64
6 criteria	43
7 criteria	20

Source: Authors.

Note: The totals given here may not sum to 100 percent due to rounding.

requirements, that is, the two qualifying criteria and the three most basic evaluative criteria (importance, scale, and time and duration). From among these, only 64 nominations (20 percent) met all seven criteria. Subsequent review and validation of the published material on these interventions resulted in the elimination of an additional 39. For the remaining 25 interventions, studies were commissioned to comprehensively document the intervention and its supporting evidence. Of these, the studies on 20 were completed, peer reviewed, and accepted for publication.

What Worked?

Successes in developing-country agriculture are extremely rich and diverse in nature, varying in time, space, and character. Some successes have emerged for just a few short years to trigger long episodes of growth and development. Some have resulted from years of dogged persistence that yielded returns despite substantial risks, uncertainties, and doubts. Others were achieved because communities took action to ensure their own survival under difficult environmental conditions. Still others were inspired by leaders and organizations who marshaled the resources needed to contain the spread of crop and livestock diseases that know no boundaries.

The pathways to success are also extremely varied. Some cases demonstrate how an improved crop variety or cultivation practice contributed to improving food security by increasing crop output per hectare of land, lowering production costs, or reducing crop losses caused by pests, diseases, drought, or soil erosion. Others demonstrate how new agricultural technologies improved the sustainable use of scarce resources such as fertile soil and water or enhanced the nutritional quality of food that people both cultivate and consume. Still others illustrate how changes in incentives—whether public policies, commercial regulations, or socioeconomic norms—encouraged farmers to produce more food, pursue more sustainable cultivation practices, and participate more actively in the marketplace.

But these pathways to success are not simply about increasing the *physical* supply of food. Rather, they are about reductions in hunger that result not only from an improvement in the physical availability of food but also from a change in an individual's ability to secure quality food (Sen 1981; Dreze and Sen 1991). This change may result from any number of situations: an improvement in an individual's ability to produce food within the farm household, an increase in income that provides a consumer with greater purchasing power in the market, or a shift in norms that reduces the impact of practices and behaviors that limit an individual's entitlement to food within the household, community, or society. Here we examine these pathways by looking at successes in the six different thematic areas described earlier: (1) intensifying staple food production, (2) integrating people and the environment, (3) expanding the role of markets, (4) diversifying out of major cereals, (5) reforming economywide policies, and (6) improving food quality and human nutrition.

Intensifying Staple Food Production

A loose timeline of recent successes in agricultural development begins somewhere in the mid-20th century, when the menace of war, hunger, and disease loomed large for many developing countries that had just gained independence from colonial control or influence. Crisis—whether the result of human actions such as conflict, oppression, or complacency or of natural causes such as drought or pests—was a key driver of these early successes in agriculture.

During the late 1940s and early 1950s, astute political leaders keenly recognized that hunger was a threat to long-term security, development, and prosperity. In India, for example, then–Prime Minister Jawaharlal Nehru put agriculture at the forefront of the national agenda following the country's independence in 1947 by allocating substantial attention and public resources to building rural roads, expanding irrigation systems, generating rural power, establishing state agricultural universities and research centers, constructing fertilizer plants, and promoting land reform. The real and perceived threat of famine ushered in an era in which policymakers' key priority was to increase the output (greater production) and yields (greater production from a given area of land) of staple foods.

One of the first major successes, described by Dubin and Brennan in Chapter 2, came from a global effort to fight wheat rusts—a plague that has been known to humanity for thousands of years but had never been effectively contained. Wheat rusts are actually fungi that can rapidly decimate wheat as it matures in the field, and they are thus a threat to food security in industrialized and developing countries alike. The late Nobel Prize Laureate Norman Borlaug, with the eventual backing of policymakers, scientists, and philanthropists, catalyzed a global effort to combat the scourge by bringing modern science to bear on the problem—by breeding rust-resistant wheat varieties in Mexico with the help of innovative research methods. As a result of this global effort, about 117 million hectares of land under wheat cultivation were protected from wheat rusts, directly ensuring the food security of 60 to 120 million rural households and many more millions of consumers. Importantly, it also secured a place for science and technology in developing-country agriculture and gave rise to a global agricultural research system, including the Consultative Group on International Agricultural Research, dedicated to finding scientific solutions to end hunger and food insecurity.

The wheat rust success evolved into a much larger and more multidimensional series of successes that began in the 1960s and came to be known as the Green Revolution. As Hazell explains in Chapter 3, Asia's experience during this revolution started with the introduction of improved rice and wheat varieties for irrigated land that could be cultivated twice a year instead of once. The process continued into the 1990s as successes expanded to lesser-known staple crops such as millet and sorghum and to more marginal areas dependent on rain rather than irrigation. The investments in science and technology—along with complementary investments in irrigation systems, road networks, fertilizer production, and food price stabilization policies—that underwrote the Green Revolution paid off handsomely. Farmers rapidly adopted the new farming practices and technologies to such a massive extent that between 1965 and 1990 cereal output and yields doubled, pulling India and other Asian countries back from the brink of famine. Between 1970 and 1990, an estimated 1 billion people benefited from the Green Revolution in terms of improved access to food, increased earnings from agriculture, or both. Estimates by Evenson et al. (2006) suggest that without a Green Revolution, about 30 million children would have died in the developing world between 1970 and 2000, with more than two-thirds of these children in Asia alone.

Successes in Sub-Saharan Africa were smaller in magnitude but no less important in addressing the persistent threat of hunger in the region. In East and Southern Africa, applications of modern science to maize improvement led to growth in both maize output and yields among the region's primarily small-scale, resource-poor farmers (Smale and Jayne 2010).[3] Between 1965 and 1990, maize yields in Kenya, Malawi, Zambia, and Zimbabwe increased annually between 1 and 5 percent—rates that compare respectably with yield and production growth rates in countries such as the United States—while annual maize production increases ranged from 1.8 to 3.3 percent in these same countries.

In West Africa between 1971 and 1989, the application of modern science similarly helped contain the spread of a cassava mosaic virus (a disease) and mealybug (an insect), as explained by Nweke in Chapter 4. Both threats can generate major losses in cassava, a crop that is central to the sustenance and incomes of the region's poorest farmers, particularly in times of drought or crisis. By breeding cassava varieties that were resistant to the mosaic disease and by introducing a parasitic wasp to destroy mealybug in countries such as Ghana and Nigeria, the potential damage posed by these two threats was effectively contained. The introduction of disease-resistant cassava varieties is estimated to have contributed to making an additional 1.4 million tons of *gari* (a granular, fermented cassava flour commonly used in cooking) available per year, enough to feed 29 million people in the region.[4] Similarly, the mealybug control program is estimated to have reduced losses from infestations by an estimated 2.5 tons per hectare.

Integrating People and the Environment

By the 1970s, concerns emerged about the equity and environmental implications of rapid agricultural development (Staatz and Eicher 1998). These new concerns encouraged a move away from a strictly yield-increasing outlook on food staple productivity to a more complex perspective of agriculture and rural development. Sustainable development issues came to the forefront of the development discourse, partly in response to issues that accumulated during the Green Revolution, such as the overuse of agricultural chemicals, the depletion of scarce water resources, and the neglect of farmers' input into policymaking. New policies, programs, and investments were specifically designed to integrate rural communities into decisionmaking processes about their own development as a way of addressing sustainability along with equity issues. The idea that agricultural development could work if driven by direct community participation, environmentally sustainable cultivation practices, and supportive public policies gained a global following.

Experiences in Nepal that began in the 1970s illustrate this change in perspective, as demonstrated by Ojha, Persha, and Chhatre in Chapter 5. During this period, a series of prescient legislative reforms and innovative forestry programs contributed to a transformation of the country's strictly conservation-focused approach to its natural forests into a more broadly based strategy that encompassed forest use, enterprise development, and livelihood improvement with direct benefits for the rural poor. Partly as a result of these reforms and programs, an estimated one-third of Nepal's population is participating in community forestry activities and directly managing over one-fourth of Nepal's forest area as a means of improving household food security and livelihoods.

In Burkina Faso and Niger during the 1980s, the rediscovery of community-based knowledge in the form of traditional agricultural management practices helped to transform the Sahelian region's arid landscape into productive agricultural land. In Chapter 6, Reij, Tappan, and Smale explain how, in the wake of repeated droughts, farmers began innovating on simple practices: protecting and managing indigenous trees and shrubs among crops to provide fodder and firewood and to improve soil fertility; digging pits on barren, degraded land to concentrate organic manure and rainwater for planting; and constructing stone contour bunds to control rainfall and runoff and to combat erosion. With technical support from charismatic community leaders and nongovernmental organizations, the dissemination of these practices helped Sahelian households to intensify and expand their cultivation of essential food staples such as sorghum and millet and to earn more income from the sale of crop surpluses, fodder, firewood, fruit, and other products. In Burkina Faso's Central Plateau, the rehabilitation of between 200,000 and 300,000 hectares of land translated into roughly 80,000 tons of additional food per year, or enough to sustain about one-half million people in the region. In southern Niger, similar efforts are estimated to have transformed approximately 5 million hectares of land, improving the food security of at least 2.5 million people.

In Argentina, large-scale farmers adopted a different set of resource-conserving cultivation techniques, resulting in a significant increase in the global production of soybeans in particular. In Chapter 7, Trigo, Cap, Malach, and Villarreal explain how during the 1980s Argentinian farmers, researchers, extension workers, and private companies worked together to promote zero-tillage cultivation—a crop management technique in which farmers essentially plant seeds in unplowed fields to maximize the gains from intensive double cropping and to lower production costs, with the added benefits of reducing land degradation, conserving soil fertility, and economizing on scarce water resources. By 2008, the area of land under zero tillage reached nearly 18 million hectares. The use of zero tillage, along with the introduction of herbicide-resistant soybean varieties and other factors, improved soil fertility by reversing decades of erosion, created an estimated 200,000 new agricultural jobs, and provided the international market with new supplies of soybeans that contributed to keeping global food prices low.

During roughly the same period in the 1980s, small-scale farmers in the Indo-Gangetic Plains—a vast region that encompasses parts of Bangladesh, India, Nepal, and Pakistan—began experimenting with similar zero-tillage techniques. In Chapter 8, Erenstein shows that an estimated 620,000 wheat farmers have adopted some form of zero-tillage cultivation since these experiments began, accounting for about 1.8 million hectares of land in the region and generating average income gains of US\$180–340 per household, particularly in the Indian states of Haryana and Punjab.[5]

Expanding the Role of Markets

In spite of these successes, many developing countries still suffered from slow growth, general economic malaise, and persistent food insecurity through the 1980s. A shift to more market-driven development took hold in many countries during this period. In some countries, this shift came in the form of structural adjustment programs that sought to rein in public deficits, improve national balances of payments, liberalize markets, and encourage private investment in the economy (Staatz and Eicher 1998). In other countries, this shift occurred after the recognition that efficient supply chains played an important role in improving the production incentives for farmers, increasing incomes from farming, and improving food security. Market forces were expected to contribute to agricultural development, for example, by freeing up seed and fertilizer markets from state-owned monopolies, removing price-setting policies in agricultural commodity markets to encourage more vibrant trading, and closing the supply chain gaps that link farmers to markets through traders, processors, distributors, wholesalers, and retailers.

In Bangladesh, government moves to liberalize agricultural input markets in the 1980s led to an easing of restrictions on the importation and sale of irrigation equipment, such as low-lift power pumps and shallow tubewells. As Hossain explains in Chapter 9, these seemingly minor reforms stimulated the rapid growth of irrigated dry-season rice farming, which subsequently grew to account for 90 percent of the increase in rice production in Bangladesh between 1988 and 2007. And with this growth in rice production came a decline in the real rice prices facing food-insecure households and, ultimately, significant reductions in poverty in the country.

In China, policy reforms that promoted private investment in agriculture, along with breakthroughs in rice research, fostered the growth of a vibrant seed industry for hybrid rice, as shown by Li, Xin, and Yuan in Chapter 10. Hybridization, first demonstrated in maize by scientists in the United States in 1918, is a process in which inbred parent lines of a crop are crossed to create seeds that are characterized by greater yield potential than either parent, an outcome known as "hybrid vigor." This vigor tends to decline with each generation of seed that is saved and replanted, so farmers need to purchase new seed each season to realize the full yield gains of these hybrids. In China, as in the United States and other industrialized countries, this characteristic of hybrid seed supported the rapid emergence of an entirely new industry—one that distributes good-quality hybrid seed to farmers. Hybrid rice in China spread so quickly that between 1978 and 2008 it had grown to account for 63 percent of all land under rice cultivation. Importantly, its yield advantages helped China to feed an additional 60 million people per year during this period.

In India, as Pray and Nagarajan discuss in Chapter 11, similar policy reforms and scientific advances in the mid-1990s encouraged the growth of private investment in the marketing of improved seeds for pearl millet and sorghum, including hybrids. These two crops are essential sources of sustenance and income for some 14 million poor households in India. Although together they account for just 10 percent of the total cropped area in India, they are cultivated in the country's arid and semiarid regions, where nearly 60 percent of the rural population lives. The emergence of private seed companies, combined with good public research, has not only provided an estimated 6–9 million farmers with access to improved seeds that have increased yields by 60–75 percent in recent decades but also served as the foundation for an important industry in India's growing economy.

Reforms in Burkina Faso's cotton sector that began in 1992 brought together experiences from both market liberalization and cash crop development as drivers of success in agricultural development, according to Kaminski, Headey, and Bernard in Chapter 12. Saddled with a state-led cotton development strategy that was branded as inefficient, inequitable, and destabilizing to the national economy by the late 1980s, Burkina Faso pursued a reform path that combined efforts to strengthen the role of cotton farmers' groups before partially liberalizing input and output markets. Partly as a result of these reforms, and even despite consistently low world prices for cotton, Burkina Faso has emerged as the leading African exporter of cotton based on a threefold increase in production since the early 1990s. The cotton sector's growth has absorbed more than 200,000 new farmers who either were engaged in the cultivation of other crops or were return migrants from neighboring countries experiencing civil strife.

In Kenya, as Ariga and Jayne explain in Chapter 13, policy reforms in the early 1990s contributed to the rapid growth of private investment in fertilizer and maize marketing, the outcome of which has been a dramatic reduction in the time, effort, and costs associated with purchasing fertilizer and selling surplus maize production. The average distance that small farmers had to travel to purchase fertilizer decreased by half between 1997 and 2007, with similar decreases observed in the distances traveled to sell maize. The application of fertilizer to maize increased by 70 percent between the mid-1990s and 2007, contributing to an increase in both yields and availability of this vitally important staple crop for Kenyan consumers.

Diversifying out of Major Cereals

The emphasis on markets also opened up new opportunities for cultivating and marketing nonstaple crops—commodities such as legumes, fruits, and vegetables as well as dairy, livestock, and fish—as a means of increasing farm incomes and improving food security among the poor. Each success offers a different angle on how small-scale farmers, entrepreneurs, and policymakers responded to growth in market opportunities.

Across a range of Asian countries—Bangladesh, Bhutan, China, India, Myanmar, Nepal, Pakistan, Sri Lanka, and Thailand—the move away from food staples was exemplified by the diffusion of improved mungbeans, a little-known pulse crop that is high in protein, iron, and other micronutrients and is particularly useful in maintaining soil fertility. Thanks to an international research program and active farmer participation in the research process—as explained in Chapter 14 by Shanmugasundaram, Keatinge, and Hughes—beginning in the mid-1980s a wide range of mungbean varieties was released with traits such as higher yields, shorter maturity times, and other qualities that targeted a variety of agroecological conditions in the region. These improvements contributed to yield gains of 28 to 55 percent among an estimated 1.5 million farmers and were key factors in the 35 percent increase in global mungbean production between 1984 and 2006.

Closely related to these successes are advances made in the areas of livestock and fisheries, which produce commodities that sometimes receive much less consideration than do food staple and high-value crops but are no less important to millions of small-scale, resource-poor farmers and pastoralists and to consumers who depend on milk, meat, and fish as key sources of sustenance and nutrition.

Global efforts to control and eradicate rinderpest—a livestock disease that, in its severest form, is capable of killing 95 percent or more of the animals it infects—

reiterate the importance of livestock to rural livelihoods and food security. Roeder and Rich describe in Chapter 15 the concerted global, regional, and national efforts made in recent decades to control the spread of rinderpest through cattle vaccination, quarantine measures, and disease surveillance, measures that have played an important role in securing the livelihoods of small-scale farmers who keep livestock, as well as pastoralists whose livelihoods depend primarily on the health of their herds. Programs operating in Asia and Africa have helped to avoid potentially massive financial losses in terms of milk, meat, animal traction, and the main livelihood assets of many pastoralists and have brought rinderpest to the edge of eradication, the first time a disease has been eradicated since smallpox in humans.

In India, Operation Flood, an innovative national program that ran from 1970 to 1996, helped create a national dairy industry that integrated small-scale farmers—many of them women—with village-level dairy cooperatives, commercial dairy processors and distributors, and new technologies to modernize the industry. Cunningham explains in Chapter 16 how India went from being a net importer of dairy products to a major player in the global dairy market with the backing of a supportive policy environment that ensured the dairy industry's steady growth and development. Between 1970 and 2001, dairy production in India increased at the respectable rate of about 4.5 percent per year, with estimates during 2007–08 indicating that dairy production has exceeded 100 million tons per year. As a result, millions of consumers now have better access to milk and other dairy products while India has become a top global producer of buffalo and goat milk, the sixth largest producer of cow milk, and an exporter of milk powder.

In the Philippines, the Genetic Improvement of Farmed Tilapia project that ran from 1988 to 1997 played an important part in enhancing the role of fish as a source of income and protein for many farmers and consumers, according to Yosef in Chapter 17. By breeding a tilapia strain that originated in Africa, the project developed a new strain that is faster growing and more resistant to environmental stresses than other strains. These improvements significantly boosted fish yields and output, thus increasing the availability of fish for consumers, reducing market prices, and providing a less expensive source of protein for the country's poor.

Reforming Economywide Policies

As the emphasis of agricultural development has evolved and diversified over recent decades, the role of the agricultural sector in the wider economy has similarly changed. Economic policy reforms have contributed significantly to changing the traditional urban biases that historically discriminated against farmers and, ultimately, against the poor (Lipton 1977, 1984). In some cases, trade and fiscal policy reforms have changed how both trade and aid are leveraged for development, transforming dependencies on food aid into more effective, long-term opportunities for development financing. In other cases, monetary policy reforms have reduced the distorting effects of exchange rates and lending policies on the agricultural sector, allowing for more rapid growth and development.

The most dramatic case in point comes from China, as explained in Chapter 18 by Bruce and Li. Between 1978 and 1984, China undertook a series of policy reforms that transformed the country's food and agricultural sectors and reduced hunger on a scale unrivaled in history. The reforms effectively reintroduced household farming after more than 30 years of collective agriculture. This new approach to agriculture—the Household Responsibility System—gave farmers the incentive to sell their surplus farm production to the market. By returning more than 95 percent of China's farmland to some 160 million farm households, the reforms directly contributed to an increase in rural incomes by 137 percent, a reduction in rural poverty by 22 percent, and an increase in grain production by 34 percent. Gains in on-farm efficiency also led to a 47 percent increase in the rural labor force available for nonagricultural employment, a shift that fueled rapid industrial growth in rural China and, more broadly, China's remarkable march to industrialization during the past three decades.

In Vietnam, a series of similar reforms between 1987 and 1993 fundamentally shifted the country's economy to a greater market orientation, immediately transforming the agricultural sector, according to Kirk and Nguyen in Chapter 19. During 1989–92, the agricultural sector emerged from its stagnation and grew at a rate of 3.8 percent per year while the country shifted from being a net food-importing country to the world's third-largest exporter of rice in 1989. Within a decade, more than 10 million households—representing about 87 percent of peasant households—had received land use certificates for about 78 percent of Vietnam's agricultural land. These reforms, together with other market liberalization policies, encouraged farmers to produce food staples, livestock, and high-value crops far more productively, and for substantially greater market gain, than in previous eras. The reforms contributed substantially to Vietnam's dramatic reductions in poverty and contributed to both economic growth and industrialization.

Improving Food Quality and Human Nutrition

Although massive gains in improving the availability of and access to food have been achieved in China, India, and many other developing countries as a result of these successes, far less has been achieved in improving the quality of food. Scholars have argued that the decades-old effort to raise people's incomes to boost their calorie consumption and protein intake should be refocused to include improvements in people's micronutrient intake and dietary diversity (Bouis 2000; Allen 2003; Berti,

Krasevec, and FitzGerald 2004). With this shift in focus comes the recognition that the pathways through which agricultural development affects hunger and food security are more complex than previously understood.

Taking aim at this challenge is an innovative program in Bangladesh that promotes home gardening, small livestock production, and nutrition education. Iannotti, Cunningham, and Ruel explain in Chapter 20 how Helen Keller International, a nongovernmental organization, worked in partnership with more than 70 local organizations and the Government of Bangladesh to encourage food-insecure households to grow their own micronutrient-rich foods for both home consumption and the market. These homestead food production programs have reached an estimated 5 million individuals and contributed to combating micronutrient deficiencies that can be major causes of diseases such as night blindness, particularly among women and children.

Caveats

Five decades of investment in agricultural development have contributed significantly to feeding billions of people. Early interventions were critical to improving the availability of food by bringing modern science, rural infrastructure, public policy, and international collaboration to bear on the challenge of enhancing yields and output to feed millions. Many later interventions sought to integrate community participation and environmental sustainability into agricultural development, with important repercussions for the use of local knowledge resources and natural resources in combating hunger. Other interventions worked to strengthen the role of markets and agricultural incomes by encouraging the commercialization of small-scale farmers' production, improving supply chain efficiencies, and loosening the state regulation of both input and commodity markets. Still other interventions have focused on addressing the nutritional aspects of hunger, particularly the issues of micronutrient deficiency, dietary diversity, and food quality. But a few caveats are in order.

First, successes are rarely stand-alone events; rather, many are closely related in scope and are intertwined over time. In China, the impact of the Household Responsibility System (1978–84) on rural food security was partly driven by the introduction of hybrid rice and other agricultural technologies. In Kenya, early successes in breeding improved maize provided the productivity gains needed to leverage the subsequent liberalization of both fertilizer and maize markets. In the Indo-Gangetic Plains, the promotion of zero-tillage cultivation techniques has been partly an attempt to reduce the dependence on chemical inputs that were heavily promoted during Asia's Green Revolution in previous decades. In short, the interrelatedness of these episodes of success demonstrates how interventions in agricultural development are solidly based on what went before.

Second, successes have not occurred or accumulated at a consistent pace over the past five decades. Instead, the historical record has been peppered with starts and stops. In Malawi, Zambia, and Zimbabwe, for example, the gains that came with the dissemination of improved maize (1965–90) came to a halt in the 1990s due to unsustainable fiscal burdens, erratic reforms, and bad weather. The benefits of community forestry in Nepal from 1978 to the present have similarly been disrupted by civil strife and political crises in the country. And the global effort to combat wheat rusts has been renewed by the emergence of UG99, a rust race that can overcome the resistance conferred by breeders during the past 50 years.

Finally, many successes that are important in reducing hunger and malnutrition have been experienced beyond the ones presented in detail here. Some are briefly mentioned in the chapters that follow: the spread of community forestry programs in Cameroon, India, Mexico, and Tanzania; the development of Brazil's Cerrado region; the diffusion of insect-resistant cotton in China and India; and the introduction of pro-poor regulatory reforms in Kenya's dairy industry. Others have not accumulated sufficient evidence to prove their impact but may intuitively be seen as successes. Still others may be smaller in size and scale or have yet to capture the world's attention but are no less important.

Why Did It Work?

These successes in agricultural development provide valuable insights for the future—insights that are important to those directly involved in policy, programs, and investments in agriculture, including policymakers designing progressive legislation, donors investing in projects and programs, nongovernmental organizations working with vulnerable farm communities, scholars studying growth and development, scientists breeding new crops in labs and fields, farmers' associations promoting their members' voices and interests, and people wanting to help ensure that agricultural development translates into reduced hunger. In all, eight key elements emerge that appear to have driven these successes.

Science and Technology

Sustained investment in agricultural research and development is vital to developing-country agriculture. The application of science and technology to agricultural development—whether by developing advanced techniques for crop breeding or updating farmers' traditional soil and water management practices—is a common determinant of success. The critical role of long-term public investment in science and technology plays out across the entire developing world, from Asia to Latin America and Sub-Saharan Africa, and across a range of successes, from major foodcrops such as rice, wheat, and maize to lesser-known crops such as millet, sorghum, cassava, and mungbeans and also to livestock and fisheries.

These successes also demonstrate just how difficult it is to sustain public investment in agricultural science and technology in the face of competing demands for public resources. Often policymakers and donors are put off by the long lead times required to develop new technologies for small-scale farmers or by the unending need to continuously develop new technologies in an effort to stay ahead of coevolving pests and diseases, changes in market preferences, and new environmental stresses such as global climate change. Nonetheless, sustained significant public investment is vital to ensure and maintain successes in agricultural development that address chronic hunger and persistent malnutrition.

Complementary Investments

Still, science and technology are not enough: hunger and malnutrition are complex phenomena, and there are no silver bullets in the struggle against them. To improve the chances that science and technology will make a real contribution to improving food security, sustained public investment in the hardware and software of agricultural development is also critical. This includes public investment in irrigation schemes, rural road networks, rural education, market infrastructure, and regulatory systems. The private sector tends not to invest in these directly because they seemingly offer little opportunity for profit, despite their value to society as a whole. Long-term public investment in the building blocks of agricultural development is a necessary condition for success and is evident in each and every success case—from building and improving irrigation systems in India and Pakistan to providing extension and education services on zero-tillage cultivation techniques for Argentinean farmers. Conversely, the absence of sustained investment in infrastructure, supportive policies, and robust markets after 1993 in many East and Southern African countries stalled the gains in agricultural productivity growth and hunger reduction that had been achieved with the introduction of improved maize in the region. Clearly, sustained public investment can pay high dividends in terms of addressing chronic hunger and persistent malnutrition.

Private Incentives

But even with sustained public investment in science, technology, and complementary investment areas, little can be achieved without the right incentives. Putting policies in place that encourage farmers, entrepreneurs, and companies to invest in agriculture and ensuring that markets provide accurate and timely price signals to these private-sector actors increase the likelihood of success in agricultural development.

China demonstrated this with the return to household farming, in which a change in incentives encouraged farmers to invest in their land's productivity and grow more food. Kenya demonstrated something similar by loosening state control over fertilizer and maize markets, dramatically reducing smallholders' costs of purchasing inputs and marketing surplus production. And, as Burkina Faso demonstrated by strengthening the role of farmers' organizations as cotton sector reforms heightened the competitiveness of the cotton sector, private incentives work best when market participants can respond effectively to these incentives, collectively or individually.

Cooperation and Collaboration

Many successes are built around the notion of cooperation and collaboration. Partnerships among diverse actors in the agricultural sector—research institutes, community-based organizations, private companies, government agencies, and international bodies—are evident in almost all successes. But collaborative interventions are a tricky business and require know-how in effectively managing public and private resources, orchestrating foreign assistance and community resources, and managing relationships among sometimes disparate interest groups.

Still, partnerships underscore many successes in agricultural development. Examples of successful partnerships include the scientific collaborations that developed rust-resistant wheat strains and semidwarf rice and wheat varieties, the grassroots-level partnerships that improved forestry management in Nepal and land management in the Sahel, the global and regional cooperation that helped eradicate rinderpest, and the public–private partnerships that brought improved sorghum and millet hybrids to Indian farmers.

Timing and Planning

Many successes result from good timing, whether by chance or by design. In some cases, the time was simply right for the intervention: the technological, economic, social, and political elements were all in place. In other cases, the intervention was adjusted to ensure that the timing was right: gradual reforms were undertaken step by step, calculated measurements of the potential gains and losses were undertaken, and strong support was provided to those affected by the reforms. For example, Burkina Faso's experience with the development of its cotton sector has proceeded relatively smoothly partly because of its staged effort to strengthen farmers' organizations *before* liberalizing the cotton sector. China's experience with a return to household farming has generated such significant gains in food security partly because of the carefully sequenced introduction of localized experiments in land tenure rights to the country as a whole.

Experimentation and Evolution

Often successes emerge from localized experiments that allow participants to learn from their mistakes, adapt to changes in the landscape, evolve as the playing field becomes more complex, and pursue incremental, step-by-step approaches to scaling up. Creating space for local experimentation and innovation is a critical means of generating big bangs from incremental changes. For example, China's land tenure reforms began as a local experiment undertaken by administrators in just one poverty-ridden county but rapidly evolved into a national-level reform program. The homestead food production program in Bangladesh integrated a learning process into its activities to ensure that the intervention could be improved with the accumulation of new evidence from the sciences and new experiences at the grassroots level. By encouraging farmers to improve on their traditional soil and water management practices in the Sahel or by involving farmers in the breeding of improved mungbeans in Asia, long-term investments in agricultural development paid off handsomely.

Community Involvement
Similarly, by vesting communities with a stake in ownership of a development process, grassroots participation contributes much to the long-term sustainability of a success. Involving communities and smaller groups in local consultations, policy deliberations, scientific research, and experimentation is part of building from the bottom up to achieve success. Similarly, local practices, customs, and knowledge are the seeds of big successes. The successes in community forestry in Nepal, intensification of dryland cultivation in the Sahel, and homestead food production in Bangladesh, among many others, are all testaments to the value of community involvement and engagement.

Leadership and Dedication
Often the solutions needed to address agricultural development challenges require dedicated individuals to make the difference—champions to push an issue to the forefront of the public's consciousness, demonstrate what can be done in the face of seemingly insurmountable challenges, or mobilize the political and financial capital to overcome inertia. Some of these individuals are well known, such as Norman Borlaug, who received the Nobel Prize for his contributions to agricultural development. Others may be less well known, such as Yacouba Sawadogo, a farmer from Gourga in Burkina Faso who contributed significantly to the spread of soil fertility management techniques in the barren fields of the Sahel. Still others—the unsung heroes whose efforts have fed millions—include the extension agents with good ideas about how to improve local crop production and marketing, the credit officers who disburse and collect on small loans to small farmers, or the community organizers who help their fellow farmers find new ways of combating crop pests and diseases. These champions, both renowned and anonymous, are essential ingredients of success. Creating an environment that encourages leadership on such issues and rewards individuals based on their merit is important in creating success.

What Can We Learn from It?

Looking to the future, the changing realities of the global food and agriculture system and the persistence of hunger in the developing world indicate that more and more frequent successes are needed. Agriculture is increasingly driven by market demand, consumer preferences, regulatory scrutiny, and ethical considerations. Agriculture is far more commercial and far more globalized through domestic market growth, international trade, and global finance than ever before. Emerging information, communications, and biological technologies are providing new opportunities for farmers and consumers, while climate change is imposing new constraints on agricultural practices, rural livelihoods, and the resilience of agro-ecological systems. New demographic concerns are emerging with the continuing HIV/AIDS pandemic, changing age structures in some developing countries, rapid urbanization and rural flight, and growing regional and global migration.

The tools needed to address these evolving realities have changed during the past five decades. But how can the successes of the past help inform and influence agricultural investments that will contribute to substantially reducing hunger in the future? A few reflections are offered here.

Success Is Not a Substitute for Strategy

Individual successes of any size or scale must stimulate broader and more sustained processes of national and global success-building. But these processes are feasible only if countries pursue good strategies, create supporting policies, and encourage appropriate levels of investment and experimentation needed to accumulate successes that eventually add up to a sustained success. Without these necessary conditions, successes will likely be scattered, occasional events—outcomes of an unexpected scientific breakthrough or a one-off policy correction. Rather than generating these types of short-lived successes, decisionmakers should design and implement strategies that take a comprehensive approach to raising agricultural productivity, increasing incomes, and reducing poverty. Comprehensive strategies along these lines can encourage many intertwined successes in agriculture to emerge with a frequency that adds up to a national or regional success story rather than a fleeting, intermittent, or serendipitous set of successes.

Success Is a Process

Agricultural development must address a range of ever-changing priorities and challenges—containing the transboundary movement of new diseases and pests, strengthening ecosystem resilience in the face of climate change, improving the governance of global trade in food and agricultural products, encouraging both public and private investment in developing-country agriculture, and more effectively articulating grassroots voices on global and local issues. Therefore, successes are generated and sustained through experiential processes. This means discovering by doing, learning from mistakes, and adapting to change. Designing an intervention that allows for learning and adaptation can increase the likelihood of success.

Success Is Recognizable

Sometimes successes emerge only in retrospect once a substantial amount of time has passed to allow for reflection. But for successes in agricultural development to be recognized as such, they need to be sufficiently supported by strong evidence, as shown by Maredia in Chapter 21. Such evidence comes in many shapes and sizes, ranging from first-hand accounts of individual participants to large-scale impact evaluation studies that combine both quantitative and qualitative evidence at the highest levels of academic rigor. Regardless of the type and level of evidence, the key point is that successes in agricultural development—and failures, too—need to be systematically documented, examined, and shared so that others can learn lessons, adapt them to different circumstances and contexts, and avoid similar pitfalls.

Success Can Be Ambiguous

In many cases, it is immediately obvious that there is no such thing as an "unequivocal success." Many successes are accompanied by some type of trade-off. Increases in food production may depend on the intensive use of harmful chemicals. Productivity gains may generate price collapses that hurt farmers but benefit consumers. Scarce public resources allocated to rural infrastructure may be funds that could be used for other investments. In fact, many successes are characterized by a mix of pros and cons. In Burkina Faso, the continued central role of the state in the country's cotton sector is criticized because of the public resources required to sustain the system. In Nepal, difficulties in extending the gains from community forestry to underrepresented social groups are cited as a source of concern for the policy's long-term outlook. And in countries that have benefited from the Green Revolution, the excessive and inappropriate use of fertilizers and pesticides that damage waterways and natural ecosystems, irrigation practices that lead to salt buildup on good farmland, and water scarcities resulting from the overuse of groundwater and water from major river basins are similarly noted as major shortcomings of the Revolution.

Unfortunately, this ambiguity may be one reason that agricultural development became such an unpopular topic among both governments and donors in the 1980s, along with other reasons, such as fatigue over the extensive lag times needed to demonstrate an impact, frustrations with moving large government bureaucracies into action, and tendencies among investors and entrepreneurs to steer clear of all but the most low-risk investments in the agricultural sector. As a result, public investment and donor assistance declined precipitously during this period: agricultural research spending stagnated, while rural infrastructure development came to a halt in many developing countries (Pardey et al. 2006; Beintema and Stads 2008).

But investments in agricultural development have generated sizable dividends for society, demonstrating that agriculture is not only an important means of reducing poverty but also a worthwhile part of an investment portfolio (Alston et al. 2000; Fan 2008). For example, the research that culminated in the development and release of disease-resistant, high-yielding cassava varieties in Nigeria during the 1970s is estimated to have achieved an annual economic rate of return of 55 percent over a 31-year period. Similar estimates of the returns on investments in research indicate similarly high figures: 19–66 percent for research on wheat rust resistance, 43–64 percent for hybrid maize research in East and Southern Africa, 50 percent and higher for research on the modern crop varieties introduced during Asia's Green Revolution; and 70 percent for research on tilapia improvement in the Philippines.

Returns to agricultural development projects are comparable. In Burkina Faso and Niger, projects designed to bring degraded or new land under cultivation yielded returns ranging from 20 to 147 percent depending on the location and the natural resource management technique being applied. The global control and eradication of rinderpest is estimated to have generated returns ranging from 11 percent in Côte d'Ivoire to 118 percent in Burkina Faso, while Operation Flood in India generated an estimated return of 45 percent.

These pros and cons mean that success in agricultural development requires careful consideration of difficult trade-offs. While interventions that increase the availability, access, and quality of food are all desirable, the resources available to undertake these interventions are limited, suggesting the need to weigh the benefits against the costs in terms of economic and financial gains, environmental impacts, and sociopolitical importance. But the repercussions of failing to invest in agricultural development are clear: continued and persistent hunger among the rural poor and food-insecure households. And the precipitous rise in global food prices and hunger in 2008 makes these repercussions even more visible and urgent.

Looking Ahead

Decisions have to be made on where, when, and how to invest in agricultural development. Without the Green Revolution, millions of households in Asia would still be facing the threat of hunger and famine today. Without hybrid maize in East and Southern Africa, hybrid millet and sorghum in the arid and semiarid tropics of India, and improved cassava varieties in Ghana and Nigeria, food-insecure farm households and consumers would have far fewer opportunities—and less food and income—than they have today. Similarly, without the policy changes that strengthened the market incentives facing Chinese farmers or those that improved the availability of seed and fertilizer to Kenyan farmers, agriculture in these countries might have been far less productive and far less sustainable as a livelihood than it is today.

Progress in feeding the world's millions has slowed, while the challenge of feeding its future millions remains enormous and subject to new uncertainties in the global food and agricultural system. Rapid degradation of the world's natural resource base, changes in rainfall and moisture availability due to global climate change, and volatility associated with closely integrated international markets suggest that learning from successes in agricultural development is now more urgent than ever.

Ultimately, the essentials remain unchanged: we need to increase the production of, access to, and quality of food to end hunger and feed millions. All of the lessons learned here must be applied and adapted for the future but with a greater sense of urgency and commitment than ever before.

Notes

1. A complete list of the nominations is available at <http://www.ifpri.org/sites/default/files/milfednominations.pdf>.

2. It is worth noting here that conditional cash transfer programs such as the Programa de Educación, Salud y Alimentación (Progresa) in Mexico were, in fact, treated as health (rather than agricultural) interventions by Levine (2004) in the predecessor to this project, titled Millions Saved: Proven Successes in Global Health and summarized in a book of the same name published by the Center for Global Development.

3. A summary of the study by Smale and Jayne (2010) is featured in a predecessor publication to this volume, Spielman and Pandya-Lorch (2009), *Millions Fed: Proven Successes in Agricultural Development.* The summary is based on an in-depth study prepared by these same authors and published in Haggblade and Hazell (2010), *Successes in African Agriculture: Lessons for the Future.* The contribution of this material by these authors and editors is greatly appreciated.

4. Throughout this book, all tons referred to are metric tons.

5. Unless otherwise noted, throughout this book all dollars are U.S. dollars.

References

Allen, L. H. 2003. Interventions for micronutrient deficiency control in developing countries: Past, present, and future. *Journal of Nutrition* 133 (11S-II): 3875S–3878S.

Alston, J. M., C. Chan-Kang, M. C. Marra, P. G. Pardey, and T. J. Wyatt. 2000. *A meta-analysis of rates of return to agricultural R&D: Ex pede herculem?* Washington, D.C.: International Food Policy Research Institute.

APAARI (Asia-Pacific Association of Agricultural Research Institutions). 2009. APAARI success stories. <www.apaari.org/publications/apaari-success-stories>. Accessed July 15, 2009.

Beintema, N. M., and G. Stads. 2008. *Measuring agricultural research investments: A revised global picture.* Agricultural Science and Technology Indicators (ASTI) Initiative background note. Washington, D.C.: International Food Policy Research Institute.

Berti, P. R., J. Krasevec, and S. FitzGerald. 2004. A review of the effectiveness of agriculture interventions in improving nutrition outcomes. *Public Health Nutrition* 7 (5): 599–609.

Bouis, H. 2000. Improving human nutrition through agriculture: The role of international agricultural research. *Food and Nutrition Bulletin* 21 (4): 550–566.

de Janvry, A. 2009. Agriculture for development: New paradigm and options for success. Elmhirst lecture presented at the 27th Conference of the International Association of Agricultural Economists, August 16–22, in Beijing, China.

Dreze, J., and A. Sen. 1991. *Hunger and public action.* New York: Oxford University Press.

Evenson, R. E., S. Msangi, T. B. Sulser, and M. W. Rosegrant. 2006. Green Revolution counterfactuals. Paper presented at the annual meeting of the American Agricultural Economics Association, July 23–26, in Long Beach, Calif., U.S.A.

Fan, S. 2008. *Public expenditures, growth and rural poverty: Lessons from developing countries.* Baltimore: Johns Hopkins University Press.

FAO (Food and Agriculture Organization of the United Nations). 2002. *The state of food insecurity in the world 2002.* Rome.

———. 2008. *The state of food insecurity in the world 2008.* Rome.

———. 2009. *More people than ever are victims of hunger.* Rome.

Gabre-Madhin, E. Z., and S. Haggblade. 2004. Successes in African agriculture: Results of an expert survey. *World Development* 32 (5): 745–766.

Haggblade, S., and P. B. R. Hazell, eds. 2010. *Successes in African agriculture: Lessons for the future.* Baltimore: Johns Hopkins University for the International Food Policy Research Institute.

Krishna, A., N. Uphoff, and M. J. Esman. 1997. *Reasons for hope: Instructive experiences in rural development.* West Hartford, Conn., U.S.A.: Kumarian Press.

Levine, R. 2004. *Millions saved: Proven successes in global health.* Washington, D.C.: Center for Global Development.

Lipton, M. 1977. *Why poor people stay poor: Urban bias in world development.* Cambridge, Mass., U.S.A.: Harvard University Press.

———. 1984. Urban bias revisited. *Journal of Development Studies* 20 (3): 139–166.

Pardey, P. G., N. M. Beintema, S. Dehmer, and S. Wood. 2006. *Agricultural research: A growing global divide?* IFPRI Food Policy Report. Washington, D.C.: International Food Policy Research Institute.

Sen, A. 1981. *Poverty and famines: An essay on entitlement and deprivation.* New York: Oxford University Press.

Smale, M., and T. Jayne. 2010. Maize in Eastern and Southern Africa: "Seeds of success" in retrospect: Hybrid maize in Eastern and Southern Africa. In *Successes in African agriculture: Lessons for the future,* ed. S. Haggblade and P. B. R. Hazell. Baltimore: Johns Hopkins University for the International Food Policy Research Institute.

Spielman, D. J., and R. Pandya-Lorch, eds. 2009. *Millions fed: Proven successes in agricultural development*. Washington, D.C.: International Food Policy Research Institute.

Staatz, J. M., and C. K. Eicher. 1998. Agricultural development ideas in historical perspective. In *International agricultural development*, 3rd ed., ed. C. K. Eicher and J. M. Staatz. Baltimore: Johns Hopkins University Press.

Uphoff, N., M. J. Esman, and A. Krishna. 1998. *Reasons for success: Learning from instructive experiences in rural development*. West Hartford, Conn., U.S.A.: Kumarian Press.

World Bank. 2006. *Agriculture investment sourcebook*. Washington, D.C.

———. 2008. World development report 2008: Agriculture for development. Washington, D.C.

———. 2009. *Data and statistics: Country groups*. <go.worldbank.org/K2CKM78CC0>. Accessed July 15, 2009.

Chapter 2

Combating Stem and Leaf Rust of Wheat: Historical Perspective, Impacts, and Lessons Learned

H. J. Dubin and John P. Brennan

Wheat Rust: A History of Losses

Wheat, one of the world's staple foodcrops, is under constant attack in farmers' fields from diseases, insects, and weeds. Rusts are a group of fungi that have plagued wheat since ancient times. They are among the most destructive plant pathogens in the world, noted for their severe attacks on cereal crops. It is estimated that cereal rusts reduce total grain yields by about 10 percent annually (Agrios 2005). Under epidemic conditions, they can cause crop losses of 60–100 percent, trigger famines, and even ruin whole economies (Park, Bariana, and Wellings 2007). Three types of rusts infect wheat: stem (black) rust (*Puccinia graminis* Pers. f. sp. *tritici* Eriks. & E. Henn), leaf (brown) rust (*P. triticina* Eriks.), and stripe (yellow) rust (*P. striiformis* Westend f. sp. *tritici*).

Infection may occur on any of the plant parts above ground, but generally cereal rusts attack the stem and leaves. During the rust's uredinial stage, the disease appears as rusty spots or pustules that contain millions of urediniospores or infective propagules. These spores develop in the spring and summer in all three rusts and can continually reinfect wheat crops, causing epidemics. Billions of wind-borne spores can be carried thousands of kilometers (Agrios 2005).

Rusts may debilitate or kill young wheat plants, but more typically they reduce foliage, root growth, and yield by decreasing photosynthesis, increasing respiration rate, and decreasing translocation of carbohydrates. They move the carbohydrates to the infected areas and use them for growth (Agrios 2005).

Rust pathogens have an excellent ability to vary via mutation. Some may also vary through sexual reproduction and thus overcome resistance genes. In the case of stem rust of wheat, the alternate host, barberry (*Berberis vulgaris,* an ornamental shrub originally from Europe), historically played an important role in variability in the United States and other countries. The successful eradication of barberry has, however, reduced the influence of the sexual cycle on the disease (Roelfs, Singh, and Saari 1992).

Epidemics of wheat rust have decimated crops for millennia. Kislev (1982) published the first note about the existence of stem rust in prebiblical times in Israel. He found germinating urediniospores, uredinia, and hyphae on fragments of lemmas in a storage jar from the late Bronze Age (ca. 3300 BCE). In Roman times, various authors noted the importance of rust in the production of wheat and barley. Likewise, rust has been recorded on wheat in India for centuries (Nagarajan and Joshi 1975). Howard and Howard (1909) estimated that losses from wheat rusts in parts of India on occasions reached 50 percent or more, and they argued that the losses from wheat rusts in India each year exceeded the losses from all other pests combined. In the past century or so, epidemics affecting traditional tall varieties of wheat have been recorded in Europe, North America, South America, and Australia. In 1953 and 1954, a rust epidemic in the U.S. Midwest caused the loss of 35 percent of the spring bread wheat crop and 80 percent of the durum wheat crop (Stakman and Harrar 1957). In 1951, about 40 percent of the wheat crop in Chile, mainly durum, was destroyed by a stem rust epidemic. Stakman and Harrar (1957) quote sources saying that once resistance in the main durum variety broke down, durum was no longer grown in Chile because of the subsequent low yields.

In 1973 (before any adoption of semidwarf wheats in Australia), a stem rust epidemic in southeastern Australia was rated the most severe in the history of the Australian wheat industry (Park 2007). The losses in southeastern Australia were estimated at $200–300 million, or 25–35 percent of the value of production in that part of Australia.

During the past 50 years, an international effort to identify and breed wheat varieties that not only have high yields but also are resistant to rust has helped protect and improve wheat yields and feed millions. The success of the effort depended on a remarkable network of people, national and international agricultural research centers, gene banks, and nursery programs, underpinned by a free and open worldwide system of exchanging information and plant genetic materials. Although the adoption of modern semidwarf varieties of wheat has reduced the frequency and severity of rust epidemics, Saari and Prescott (1985) list 33 rust epidemics that occurred in Africa and Asia between 1970 and 1985 (Table 2.1). Notable epidemics occurred in Ethiopia, India, and Pakistan.

Table 2.1 Developing-country losses due to wheat rust diseases

Rust type	Yield losses in susceptible varieties		Endemic areas as a proportion of total wheat areas (%)	Number of recent epidemics (1970–85)	Hot spots (areas where disease is most severe)
	Average in endemic area (%)	Average in epidemic (%)			
Stem rust	40	Up to 100	50	7	Parana State, Brazil; Highlands of Kenya and Ethiopia; South India
Leaf rust	15–20	Up to 50	90	8	Bangladesh, China, India, Mexico, Pakistan
Stripe rust	40	Up to 100	33	18	Highlands of South America and East Africa; North Africa; Middle East; Indo-Gangetic Plains of India and Pakistan

Source: Hanson, Borlaug, and Anderson (1982).

In 1976–77, a leaf rust epidemic developed in northwestern Mexico, where more than 70 percent of the country's wheat crop was produced (Hanson, Borlaug, and Anderson 1982). Seasonal conditions favoring disease development coincided with drought, land tenure problems, and difficulties with seed production that reduced the seed supply of resistant replacement varieties. To head off what appeared to be a wheat leaf rust epidemic of "catastrophic proportions" (Dubin and Torres 1981, 45), Mexican authorities advised farmers who had not yet planted wheat to plant safflower instead. The remaining crops in the region were treated with an aerial application of fungicides through a government-sponsored program. Although the fungicides did not eradicate the disease, they reduced the losses. The yields in 1977 were reduced by only 15 percent from 1976 levels in the Yaqui and Mayo valleys, compared with more than 40 percent reductions in the nearby Carrizo Valley, where no spraying was undertaken (Dubin and Torres 1981).

International Collaboration and the Development of Rust-Resistant Wheat

The long-term international control of stem and leaf rust through durable resistance has to be studied in the context of international cooperation both before the creation of the International Maize and Wheat Improvement Center (CIMMYT) and after.[1] Incorporation of rust resistance was, and continues to be, a key aspect of the international wheat-breeding programs, and the development of disease-resistant varieties is considered one of the major impacts resulting from these wheat-breeding activities.

The Birth of the International Nursery System

The international germplasm nursery system—a global network in which scientists work together for the common good and freely test and exchange germplasm and information—was born out of the epidemic of stem rust race 15B that began in North America in the early 1950s. This rust race had been seen for some years before but only in low amounts (Stakman and Harrar 1957). In the summer of 1950, however, 15B increased in the U.S. spring wheat areas and then appeared in Mexico in 1950–51. It culminated in the severe epidemics of 1953–54 throughout North America (Borlaug 2007).

The U.S. Department of Agriculture (USDA) and similar agencies in Mexico,[2] Canada, and across Latin America became so concerned that they held a stem rust conference in St. Paul, Minnesota, in November 1950. Although many countries had exchanged wheat breeding lines informally for years, at that meeting participants agreed to formalize the screening of wheat germplasm—and thus was born the International Spring Wheat Rust Nursery (ISWRN) Program.[3] Seven countries—Argentina, Canada, Chile, Colombia, Ecuador, Mexico, and the United States—participated in the program initially. More than 1,000 wheat lines were tested annually for rust resistance in each location (Plucknett, Ozgediz, and Smith 1990; Kolmer 2001). By the mid-1950s this nursery and its associated breeding programs were successful in bringing stem rust under control. All germplasm and information were freely shared and made available to the cooperators and others who were interested. The ISWRN was the flagship of rust-screening nurseries worldwide until its demise in 1987 due to lack of funding (Kolmer 2001).

In the 1950s, the Mexico/Rockefeller wheat program established a new effort under the leadership of Norman E. Borlaug, applying the principles of cooperation in Latin America. This effort culminated in the Inter-American Nursery Trials initiated in 1960 and the Near East and North Africa Spring Wheat Yield Nursery initiated in 1962. In 1964 the two nurseries merged into the International Spring Wheat Yield Nursery (ISWYN). The nursery's objectives had evolved to include evaluation for yield and additional diseases, as well as exchange of materials among breeding programs worldwide. The Borlaug program truly began the opening of the commons, a free germplasm exchange system, and worldwide collaboration. The international nurseries and the concomitant training helped to standardize data collection and produce reliable information that could be analyzed over time and space (Byerlee and Dubin 2008).

One of the collateral benefits of the nursery program was the expansion of the genetic base of the Mexican program through new germplasm. By the late 1950s, the nurseries grown in Mexico included about 50,000 entries. In the meantime, U.S. and Canadian wheat research programs asked to plant off-season nurseries in Mexico. This practice reduced breeding time and added to the germplasm exchanges and networking among programs. In the 1960s and 1970s, scientists in other countries asked to plant off-season nurseries in Mexico. In this way, Mexico became an informal center of international germplasm exchange and information on spring wheat (Byerlee and Dubin 2008).

Shuttle Breeding and Semidwarf Genes

In the 1940s, the Borlaug breeding program in Mexico focused on increasing production and stem rust resistance in the tall wheat varieties. The program achieved these goals with the wheat varieties Supremo and Frontera until the emergence of rust race 15B (Bickel 1974). Fortuitously, Borlaug had bred resistance to 15B into Kentana 48 with crosses from Kenya wheat and later with Chapingo 52, Chapingo 53, Bajio 53, and others (Kolmer 2001). But how did he achieve this in just a few short years? In the mid-1940s, Borlaug realized that Mexico could not be self-sufficient in wheat by depending solely on the traditional highland areas. He wanted to produce wheat in areas of northern Mexico that had started irrigation, such as the Yaqui Valley in Sonora state (Borlaug 2007). This move marked the beginning of what is known as "shuttle breeding"—a methodology that would affect the lives of millions of people.

How was it done? Borlaug would plant his nurseries in May in the highlands of the Mexico City area (Chapingo/Toluca) and harvest them in October. Then he would plant in the Yaqui Valley in November and harvest in April. By producing two harvests a year, he halved the time required to release a rust-resistant variety (from 10–12 years to 5–6 years). The two locations were 1,600 kilometers and 10° latitude apart. The summer elevations were 2,200 to 2,600 meters above sea level in a cool, wet climate, and the winter elevation was at sea level in a desert climate. This breeding method resulted in rust-resistant wheat and also bred out day-length sensitivity, allowing the varieties to be planted over wide latitudes. Furthermore, Borlaug bred resistance to the prevalent diseases—including stem rust, leaf rust, stripe rust, and a series of other pathogens—in both areas by selecting only highly resistant materials. In four years, he had promising varieties available.

Once the new varieties such as Yaqui 50 and others that were resistant to stem and leaf rust were released, Borlaug continued to look for better yields. He found that the tall wheat varieties fell down (lodged) under conditions of good soil fertility and irrigation. In 1949 Orville Vogel of the USDA had started developing dwarf wheat at Washington State University using Japanese materials, and Borlaug

thought dwarf varieties would work well in the excellent irrigated growing conditions in northern Mexico. In 1953 Vogel sent Borlaug some breeding material
called Norin 10–Brevor. The attempt to incorporate its dwarfness was successful
on the second attempt, and it became clear that a higher-yield type of wheat was
forthcoming (Dalrymple 1986).

In 1960 the Mexican wheat program released the last tall wheat, Nainari 60.
Soon after, new high-yielding semidwarf wheat varieties (modern varieties) were
released, including Pitic 62, Penjamo 62, Sonora 64, Lerma Rojo 64, Siete Cerros
66, Inia 66, and others. Several of these wheat varieties were sent to India and
Pakistan, where they launched the Green Revolution (Bickel 1974).

Formalization of the International Research System

By 1964, Borlaug's program had instituted the ISWYN, from which germplasm
was being sent to collaborators in developing and other countries. In the early
1960s, the training programs and germplasm exchange became more formalized.
The Mexico/Rockefeller program was superseded by the Inter-American Food
Crop Improvement Program in 1960 and expanded to include maize, wheat, and
potatoes.

In 1963, at the urging of President Adolfo López Mateos of Mexico, the
Rockefeller Foundation pledged support to the creation of a wheat and maize
institute that other donors, such as the U.S. Agency for International Development
(USAID), the Ford Foundation, and the Inter-American Development Bank
(IADB), quickly signed on to. CIMMYT came into existence soon after, with the
construction of its facilities completed by 1966.

It was during this period that international germplasm exchanges, information
sharing, and human resource development took place more formally.[4] The system
evolved over time, as described in the following sections.

Germplasm Development, Breeding, and Crop Management

The germplasm used by the original Mexico/Rockefeller program and later by
CIMMYT came from many countries around the world. At that time, free exchange
of germplasm was the rule, and this approach maximized diversity. Rasmusson
(1996) estimated that nearly half of the progress made by breeders was due to
exchange of germplasm.

Crosses between crop varieties are made using the shuttle breeding technique
developed in Mexico. With this technique, it takes only five to six years from the
initial crossing to the international distribution of advanced spring wheat lines to
national programs. The CIMMYT wheat program makes up to about 8,000 crosses
a year, focusing on diverse megaenvironments and their needs. Since 1945 the program has made more than 200,000 crosses, and it annually tests more than 4,000
advanced spring wheat lines at sites worldwide (Rajaram and Van Ginkel 2001).

This large-scale crossing program of genetically diverse germplasm and multi-location testing has yielded widely adapted germplasm.

Crosses are made based on the international testing program's results from around the world. In this way the process may break undesirable genetic linkages and pyramid desired genes for important characters. The recycling of the best geno-types based on the international data is like a large "recurrent selection program." In this way the breeding incorporates tolerance or resistance even to unidentified stresses. As genetic information is obtained through molecular genetics and other technology, more precise crosses are being made. Furthermore, changes in breeding methodology also decreased the number of crosses needed. Thus the efficiency of the breeding program has been significantly increased.

Disease screening focuses on disease "hot spots," such as Kenya for stem rust and Ecuador for stripe rust. These are areas of high variability for virulence genes of the rusts. The lines that have the lowest levels of rust infection (a low average coefficient of infection) at these and all other sites are selected as the most likely to have more durable resistance. This is corroborated with genetic studies wherever possible (Rajaram, Singh, and Torres 1988). In addition, breeding nurseries are heavily inoculated with selected races of rusts in Mexico, and only the most resistant are used for crossing.

The main traits selected for in the spring wheat program are high and stable yields, resistance to diseases and insects, tolerance to stresses such as heat and drought, and grain quality. Input efficiency has been successfully selected for some years with respect to phosphorus, nitrogen, and water. Some new nutritional characteristics being studied are zinc and iron, both commonly deficient in the diets of developing-country populations.

Crop management in the early Borlaug program and later at CIMMYT was, and still is, critically important. The best germplasm is useless if the appropriate technology is not used. The Green Revolution in the 1960s would not have been successful without the proper land preparation, weed control, and fertilizer and water application. Thus Borlaug and his team had to work closely with the Indian and Pakistani programs to get these practices in place; training in agronomy went hand in hand with that in breeding and pathology (Bickel 1974).

The International Nurseries Network

In 1970, the CIMMYT annual report stated that the role of international nurseries was to give collaborators

- basic information about adaptability of varieties, yield potential, and disease and pest resistance;

- parental materials for accelerating their breeding programs;

- indications of which varieties might serve as immediate introductions into potentially high-production areas; and

- a means of evaluating promising breeding materials on a worldwide basis and fostering international cooperation.

Although the basic objectives have remained essentially the same, much more is now included. The International Wheat Improvement Network (IWIN), as it is now called, is the intermediary between CIMMYT's wheat program and an international network of wheat researchers. Researchers receive improved germplasm through nurseries and return data from trials to CIMMYT, which catalogues, analyzes, and distributes them. The ultimate beneficiaries are farmers (Payne 2004). IWIS, the International Wheat Information System that CIMMYT uses for analysis and data storage, is freely available to all collaborators (Payne et al. 2002).

The range of nurseries has grown and become more sophisticated over the years to serve the needs of the national programs. Their activities range from segregating materials to providing advanced materials and special nurseries for biotic or abiotic stresses such as rust or heat tolerance.

By 1992 CIMMYT had developed the concept of megaenvironments, or zones, areas with similar production constraints and climatic conditions. There are now 12 megaenvironments (Trethowan et al. 2005), which allow for better targeting of germplasm. Nonetheless, the germplasm still has a high degree of broad adaptation for diverse targeted environments (Trethowan and Crossa 2006).

From 1994 to 2000, CIMMYT shipped 1.2 million samples of wheat seed to more than 100 countries (Fowler, Smale, and Gaiji 2001). The total number of diverse sets of nurseries shipped to developing countries from 1973 to 2006 reached more than 2,000 per year in the late 1980s. This was due to a large increase in the number of young scientists being trained and of countries joining the system.[5] Due to funding shortfalls after 1988, the number of nurseries sent decreased.

Seed produced and shared by CIMMYT has always been considered international public goods. It was a common objective to help increase food production in the emerging nations. Because CIMMYT does not name varieties, any nation could release the same germplasm under different names, and they did. It was important that cooperators have a sense of ownership and that CIMMYT be seen as an honest broker regarding the sharing of germplasm and information.

Human Resource Development

At CIMMYT, human resource development was understood to be of equal importance to germplasm development. Without trained scientists willing to work long hours in the field, Borlaug knew there would not be the favorable yield and resistance outcomes he knew were possible with the modern wheat (Bickel 1974). CIMMYT offered practical, hands-on field training courses with ample theory as well. They were for B.Sc.-level scientists. The six-month training courses produced the cadre of young scientists who continue to produce the wheat varieties and quality data that are shared by all today. This shared commitment to the common goal of increasing food production and to working in the field, very often under difficult conditions, was paramount in establishing an esprit de corps. This spirit inspired the scientists to share germplasm and information with their colleagues worldwide, irrespective of nationalities, politics, or other issues.

More than 1,360 young scientists from 90 countries have attended these courses. Much as in the international nurseries, training increased strongly from 1967 to the mid-1980s, with a peak of 69 trainees from 32 countries in 1986. The numbers then dropped, largely due to funding constraints but also as some programs matured. There was a modest recovery in the late 1990s to 2002 with an increased focus on Afghanistan and Central Asia.

The Scientific Impacts of Rust Resistance Breeding

Wheat scientists have been quite successful in the past 40 to 50 years in controlling stem and leaf rust. This section presents the scientific bases and background of the resistance used to achieve this, albeit with some setbacks, as we shall later see.

Types of Rust Resistance Used in Wheat Breeding

The genetic resistance to cereal rusts is often presented as of two basic types:

- *Race-specific resistance,* also known as specific, major gene, or seedling resistance, among others. This type is clearly conditioned by the interaction of specific genes in the host with those in the pathogen. There is an obvious differential reaction, and races can be determined (Dyck and Kerber 1985).

- *Non-race-specific resistance,* also known as partial, general, minor gene, nonspecific, adult plant, and slow rusting, among others. This type is characterized by a non-differential interaction. It is not possible to discern races, and it generally allows some sporulation of the rust (Parlevliet 1985).

Some kinds of rust resistance tend to last for only a few years, whereas others are longer lasting, or durable. Even durable-type resistance, however, does not last forever. Parlevliet (1985) noted that no resistance is truly durable in an evolutionary sense. The need to evolve is a natural imperative for survival in biology.

Once resistance to a disease has been developed, strenuous efforts are needed to maintain that resistance. Maintenance research is designed to ensure that productivity gains are maintained and do not deteriorate over time. Maintenance breeding aims to maintain current yields (based on past productivity increases) in the face of evolving pathogens and changing abiotic constraints. The development of durable resistances rather than race-specific resistance is a significant form of maintenance breeding that has proven very effective for rust resistance.

CIMMYT Methodology for Rust Resistance Breeding

In the Mexico/Rockefeller program, the search for stem rust and leaf rust resistance used materials from Kenya, South America, and the United States, where resistance was known to exist. The breakdown of specific resistance, however—causing boom and bust cycles in wheat production—accelerated the search for more durable, long-lasting types of resistance. During the 1970s, scientists were able to release new varieties with non-race-specific resistance in Mexico that soon spread to other countries.

The strategy with the cereal rusts has been, and continues to be, to breed for slow rusting (nonspecific) resistance based on historically proven durable resistance genes (Singh et al. 2008). This durable resistance would be provided by a gene or genes that function in the adult plant stage with intermediate effects in combination with genes of minor effects and additive action. Additional minor genes for resistance, and combinations of them with different specific genes, provide a degree of diversity to further strengthen the nonspecific resistance (Rajaram and Van Ginkel 2001).

Breeding for nonspecific rust resistance proved successful, but scientists still sought a better genetic understanding of the genes being dealt with. In the early 1980s, R. P. Singh joined the CIMMYT program as a rust geneticist, and he made significant progress in understanding the genes involved with slow rusting and how to use them more efficiently (Singh et al. 2008).

Deployment of Resistant Varieties

The adoption of modern varieties that yielded improved rust resistances was rapid, widespread, and well documented (Evenson and Gollin 2003). Although the first wave of improved varieties focused on maximizing yield gain, developers of the second wave attempted not only to increase yield but also to maintain those higher yields as wheat faced evolving attacks from rusts and other diseases and pests. By 2002, nearly 95 percent of the developing world's wheat consisted of modern varieties (Lantican, Dubin, and Morris 2005).

Over time, the nature of the resistance incorporated into those modern varieties changed. The initial deployment of race-specific resistance in the 1960s was replaced during the 1970s by nonspecific resistance (Smale et al. 1998). Despite a resurgence of race-specific resistance in the 1980s, by 1995 the resistance deployed in bread wheats was almost entirely nonspecific resistance.

The widespread use of rust-resistant cultivars worldwide has reduced all three rusts as significant factors in wheat production in recent decades. Despite the evolution of pathogen virulence, these cultivars have generally been released before the rusts could inflict major damage (Expert Panel 2005). Prolonged use of resistance then led to a decline in levels of inoculum both locally and particularly in areas that were considered hot spots and had previously been sources of wind-blown inoculum for other regions.

Large stem rust epidemics have not been a feature of global wheat production for many years. In each of 12 main global wheat production regions, stem rust was listed as of major importance historically (that is, there would be severe losses without the cultivation of resistant varieties) (Expert Panel 2005). By 2005, in only 1 of these 12 regions (East Africa) was stem rust of major importance (because of the new Ug99 strain that was identified in 1999). In a further 8 regions, it was listed as minor (that is, it often occurs but is of little significance), while in the remaining 3 regions it was listed as "local" (that is, it occurs in only a small part of the region, although losses can occasionally be severe if susceptible cultivars are grown).

Experimental Evidence of Benefits of Rust Resistance on Yield

Experiments in Mexico have shown that increased rust resistance can do more to protect yield progress than increases in yield potential per se (Sayre et al. 1998). A study of 15 CIMMYT-derived varieties released between 1966 and 1988 in northwest Mexico showed that good genetic yield progress was made over 20 years, but that the benefits from protecting the yield progress with leaf rust resistance were three times greater than the benefits from genetic yield progress. Similar results have been obtained with different planting dates and other localities (Sayre et al. 1998).

A major component of the measured yield gains is thus the development and maintenance of rust resistance (Dixon et al. 2006). Byerlee and Moya (1993) classified genetic gains in wheat into three distinct categories: (1) gains in yield potential, (2) improvement in disease resistance, and (3) maintenance of disease resistance. Byerlee and Moya (1993) suggest that the greatest impact from wheat breeding in the period 1960–90 was maintenance of disease resistance, in the sense that breeders developed newer varieties that incorporated newer sources of resistance against evolving races of the three rust pathogens.

The Economic and Social Impacts of Rust Resistance Breeding

To assess the impacts of rust resistance, the rate of return on investments in rust resistance can be calculated by estimating the cost of developing and delivering resistance to farmers, the benefits of that resistance to farmers, and the impact on global wheat prices. Resistance also has wider social impacts on poverty, nutrition, and food security, which are also explored.

Costs of Rust Resistance to Farmers

Rust resistance, because it is embedded in the wheat seed, has no specific additional cost for the individual farmer. Instead, the costs are absorbed by the funders of the research and breeding programs that developed the varieties with the resistance, although some costs are passed on to farmers in developed countries through the price of seed or through a levy on production. For wheat growers in developing countries, however, those direct costs have mostly been zero or close to zero.

The true economic cost to farmers of effective resistance to rust is the yield reduction they might have to incur in years of no disease to grow the varieties with the highest levels of resistance. Where the resistances have been incorporated into well-adapted high-yielding varieties, that resistance is effectively free to farmers. The widespread success of the international research system in incorporating the highest levels of resistance into well-adapted, high-yielding backgrounds in the Green Revolution and post–Green Revolution periods has meant that the economic costs to farmers of using the resistances have been minimal.

If varieties become susceptible, national programs must replace those cultivars as quickly as possible. During the period of replacement, the economic cost of resistance to farmers can be substantial, because they may have to grow lower-yielding varieties to achieve effective resistance. This can lead to increased use of fungicides to control the disease in higher-yielding but susceptible varieties or to increased risk-taking by farmers where fungicide use is not a realistic economic option.

Costs of Rust Resistance to Research Organizations

For most developing countries, the cost of the development of rust resistance has been met by public sector agencies, whether international agricultural research centers (IARCs) or national agricultural research systems (NARSs). In developed countries, private-sector breeders increasingly meet the cost. Because incorporating rust resistance is an inherent part of the breeding operation, it is difficult to define the cost of resistance as distinct from the rest of the breeding activities. Also, because of the high level of international cooperation and collaboration on disease resistance and the generally public availability of resistance genes and parental materials developed, the costs of developing resistance in particular production environments are difficult to identify separately.

Because of this difficulty, Marasas, Smale, and Singh (2004) included the full cost of CIMMYT's wheat genetic improvement, thus overestimating the costs of the resistance activities. An alternative is to take a proportion of the total investment in breeding as the cost of rust resistance, in which case total global costs of wheat improvement research must be estimated first.

Heisey, Lantican, and Dubin (2002) provide detailed estimates of total expenditure on investments in wheat genetic improvement for developing countries from both NARSs and Consultative Group on International Agricultural Research (CGIAR) centers (CIMMYT and ICARDA, the International Center for Agricultural Research in the Dry Areas) from 1965 to 1990 (Table 2.2). In 1990 that estimate was $112 million (1990 PPP). Since that time, NARS investment is estimated to have increased steadily in real terms. Heisey, Lantican, and Dubin (2002) indicated that CIMMYT's investment in wheat improvement research declined between 1990 and 2000, while ICARDA's was estimated to have remained at $1 million per year. Projecting those same levels from 2000 to 2005 and converting the result to 2006 dollars, for consistency with the benefit estimates, gives annual costs of investment in wheat genetic improvement research of approximately $196 million in 2005.

A significant proportion of breeders' efforts is related to disease resistance. For example, Adusei and Norton (1990) found that 41 percent of wheat research at U.S. research stations was dedicated to maintenance research; most of that maintenance research for wheat was strongly related to wheat disease resistance, particularly rust resistance. Given that 30–50 percent of breeders' efforts is related to disease resistance, the total annual cost of rust resistance for developing countries is estimated at $59–98 million in 2006 dollars (equivalent to $0.50–0.84 per hectare of wheat across all developing countries).

Table 2.2 Total investment in wheat genetic improvement, 1965–2005 (US$ millions)

Investment	1965	1970	1975	1980	1985	1990	1995	2000	2005
Expenditures (1990 dollars)									
By developing-country NARSs	30	41	56	74	87	98	105[a]	113[a]	118[a]
By CGIAR centers	1	7	9	13	14	14	11[a]	11[a]	11[a]
Total for developing countries	31	48	65	87	101	112	116	124	129
Total expenditure (2006 dollars)	47	73	99	132	153	169	176	189	196
Cost of developing rust resistance (2006 dollars)									
30% of genetic improvement	14	22	30	40	46	51	53	57	59
50% of genetic improvement	23	37	49	66	77	85	88	94	98

Source: Derived from Heisey, Lantican, and Dubin (2002).
Notes: NARSs means national agricultural research systems; CGIAR means Consultative Group on International Agricultural Research.
[a]Estimated; includes estimated expenditure on wheat research at both the International Maize and Wheat Improvement Center and the International Center for Agricultural Research in the Dry Areas.

Costs of Making Resistance Available

One cost of having effective resistance available in farmers' fields is the cost of infrastructure to enhance seed increase and the seed distribution networks where varieties need to be replaced to maintain resistance. However, estimates of those costs are not available.

Valuing the Benefits of Rust Resistance

In assessing the economic benefits of resistance, it is appropriate first to consider the alternatives to resistance and the options available to farmers if resistance were not available. When farmers are faced with wheat diseases for which no effective genetic resistance is available, they can decide to accept the losses that will occur, in some cases looking to other crops that can be grown in place of wheat. They can also use fungicides to help control rusts, although for some small-scale farmers fungicides may be unavailable or too costly (Expert Panel 2005).

Assessment of the economic and social impacts of wheat rusts, and therefore of the value of their control through resistance, is complex. It is difficult to distinguish the impact of the development of disease resistance from the impact of modern semidwarf varieties and the Green Revolution; all are intrinsically intertwined. As modern semidwarf varieties were spread and taken up by farmers, they incorporated different levels of rust resistance. Losses are accurately measured only when disease development is severe, and even then it is difficult to disaggregate losses due to rust from those due to other biotic and abiotic stresses (Marasas, Smale, and Singh 2004).

However, a number of positive impacts of rust resistance at the local level can be identified:

- Varieties with improved rust resistance have resulted in higher yields over time than would have occurred without that resistance.

- Farmers have experienced increased yield stability with rust resistance as the epidemics that would have occurred in seasons that favored the rusts were reduced or prevented.

- The quality of the grain would have been lower without resistance, because rust epidemics can cause smaller, pinched grains, resulting in lower prices for marketed grain.

- With increased yields and possibly increased local prices, farmers operating in a market economy would receive higher incomes from disease-resistant varieties than from varieties without that resistance.

There are no direct estimates of the global value of wheat rust resistance. However, a number of studies have estimated the benefits of particular resistances to particular countries or groups of countries. Three sources of estimates of the value of rust resistance can be examined:

1. Studies of the value of leaf rust resistance breeding for developing countries

2. Estimates of the costs of the new strain of stem rust Ug99 that provide an equivalent estimate of the value of maintaining the resistance before it was overcome

3. Direct estimates of the value of wheat rust resistance for India and Pakistan

In addition, several studies have estimated the benefits of the development of modern semidwarf varieties and the Green Revolution. Given that rust resistance is part of that genetic improvement, those estimates provide a context for the value of rust resistance and allow separate estimates of that value, depending on the proportion of total breeding benefits that can be attributed to rust resistance.

Studies of the Value of Leaf Rust Resistance in Developing Countries
The first study to measure the impact of durable rust resistance was that of Smale et al. (1998), who assessed the benefits of nonspecific rust resistance in the Yaqui Valley in Mexico. They estimated the benefits over that period at $17 million (in 1994 dollars), equivalent to $0.85 million per year for the average of 150,000 hectares of wheat in the Yaqui Valley at that time (or $5.67 per hectare, equivalent to $7.71 per hectare in 2006 dollars). If the same value per hectare were projected to the total wheat area in developing countries, the benefits would be approximately $902 million per year in 2006 values (Table 2.3).

Marasas, Smale, and Singh (2004) assessed the economic impact of CIMMYT's efforts to breed leaf rust–resistant spring bread wheat varieties since 1973. They estimated the yield losses avoided by having resistant rather than susceptible varieties across developing countries and valued those benefits across megaenvironments. Over the 25-year period, total gross benefits were estimated at $7.46 billion (in 1990 dollars). The annual benefits over the period 1973–97 averaged $298 million per year (in 1990 dollars), or $5.16 per hectare. Converting to 2006 values, the benefits were $7.83 per hectare. Applying that benefit to all wheat in developing countries in 2006 gives estimated total benefits valued at $917 million per year in 2006 prices (see Table 2.3).

Estimates of the Losses from Ug99
Estimates of the losses likely to occur from the new stem rust strain Ug99 help us estimate the value of the resistance that was in place prior to the development of

Table 2.3 Estimated benefits of resistance in developing countries (2006 US$)

Study	Country/region	Rust	Value of resistance ($/hectare)	All developing countries[a] ($ millions)
Smale et al. (1998)	Yaqui Valley, Mexico	Leaf rust	7.71	902
Marasas, Smale, and	Spring bread wheat,			
Singh (2004)	developing countries	Leaf rust	7.83	917
Hodson, Singh, and	East Africa, Middle East,			
Dixon (2005)	South Asia	Stem rust	21.58	2,527
Brennan and Quade	India, Pakistan	Stem rust	1.53	180
(2004)	India, Pakistan	Leaf rust	2.53	297
	Mean	Stem rust	11.56	1,353
	Mean	Leaf rust	6.03	705

Source: Authors.
[a]Based on 117 million hectares in developing countries in 2006.

that strain. Hodson, Singh, and Dixon (2005) examined the potential impact of the Ug99 race if it were to spread across East Africa, the Middle East, and the entire Indo-Gangetic Plains. Based on 10 percent production losses of susceptible varieties, the potential aggregate losses were estimated as 8.3 million tons across 57 million hectares. Using the 2006 weighted average price of $148 per ton, those losses would be valued at $1.2 billion, or $21.58 per hectare across the potentially affected regions. This provides another estimate of the value of stem rust resistance in these regions, which produce 19 percent of the world's wheat. If the same value per hectare were applied to all wheat in developing countries, the value of the resistance would be $2.5 billion (see Table 2.3). Cornell University (2008) also estimated that if Ug99 and its derivatives were to establish themselves in North Africa, the Middle East, and South Asia, annual losses could reach $3 billion in any given year.

The Estimated Value of Wheat Disease Resistance in India and Pakistan
Brennan and Quade (2004), in a study of research capacity building, estimated that the value of rust resistance for India and Pakistan, respectively, was $35 million and $18 million for stem rust and $32 million and $55 million for leaf rust (in 2006 dollars). For both countries, this is equivalent to $1.53 per hectare for stem rust and $2.53 per hectare for leaf rust, which is lower than what was found in other country studies (see Table 2.3). These values for resistance are lower because the potential losses if there were no resistance were assessed as considerably lower than in other studies.[6]

Estimates of the Value of Modern Semidwarf Wheat Varieties
Other studies of the value of wheat breeding achievements have generally focused on the overall benefits from the development of modern semidwarf varieties or the

contribution of CIMMYT to those improvements (for example, see Byerlee and Moya 1993; Heisey, Lantican, and Dubin 2002; Lantican, Dubin, and Morris 2005). The findings of a selection of these studies are shown in Table 2.4.

Only a proportion of the benefits of international wheat breeding improvement can be attributed to rust resistance research. As previously discussed (see "Costs of Rust Resistance to Research Organizations"), some 30 to 50 percent of the wheat improvement effort could be attributed to the development and maintenance of rust resistance. The benefits of rust resistance measured in this way would range from $0.67 to $2.87 billion per year (in 2006 dollars) using 30 percent of total semidwarf benefits (see Table 2.4).

These estimates are broadly consistent with the range of the more direct estimates developed earlier from the studies of rust resistance in Table 2.3.

Evenson and Gollin (2003) reported on a detailed study on the impacts of crop genetic improvement across all crops on world production, prices, and food intake from 1960 to 2000. In one scenario, drawn from Evenson and Rosegrant (2003), they found that the modern semidwarf varieties led to approximately 17 percent higher production in developing countries, a subsequent significant reduction in average prices, and an overall impact on nutrition (Table 2.5). Although these benefits cannot be attributed to rust resistance in wheat, if the same benefits for all crops were ascribed to wheat and 30 percent of wheat benefits were attributed to rust resistance, wheat rust resistance would provide significant achievements (see Table 2.5).

Table 2.4 Summary of estimated benefits from international wheat improvement research

Source	Region/level	Total benefits of modern semidwarf wheat varieties		Value at 2006 prices (US$ billions)	Benefits attributable to rust resistance[a] (US$ billions)
Byerlee and Moya (1993)	Spring wheat, developing countries, 1977–90	15.34	Million metric tons	2.27	0.68
Byerlee and Traxler (1995)		2.5 billion	1990 US$	3.79	1.14
Heisey, Lantican, and Dubin (1998)	Wheat, developing countries	1.6–6.0 billion	1990 US$	2.43–9.15	0.73–2.73
Evenson (2000)		3.4–6.3 billion	1990 US$	5.16–9.56	1.55–2.87
Lantican, Dubin, and Morris (2005)	Wheat, developing countries	2.0–6.1 billion	2002 US$	2.24–6.84	0.67–2.05
All studies				2.24–9.56	0.67–2.87

Source: Authors.
[a]Based on 30 percent of total benefits from modern semidwarf wheats.

Table 2.5 Impacts of modern semidwarf varieties on key global parameters in
developing countries, 1960–2000 (percent)

Parameter	Increase/decrease due to semidwarf varieties	Increase/decrease attributable to rust resistance[a]
Crop production in developing countries	15.9 to 18.6	5.2
Crop prices, all countries	–35 to –66	–15.2
Imports by developing countries	–27 to –30	–8.6
Children malnourished in developing countries	–6.1 to –7.9	–2.1
Calorie consumption per capita in developing countries	13.3 to 14.4	4.2

Source: Evenson and Gollin (2003).
[a]Based on 30 percent of midpoint of range.

Impact on Global Wheat Prices

Diseases can affect grain quality, so a reduction in diseases can lead to higher local prices for individual farmers. However, at the broader market level, large changes in production can change the market prices for all wheat. For example, large crop losses from disease may imply price increases that are passed on to consumers or unforeseen imports purchased at world market prices that may be unfavorable (Marasas, Smale, and Singh 2004). Whenever large quantities of grain are not available, whether through disease losses or through diversion to other uses such as ethanol production, prices increase significantly.

Conversely, an increase in production resulting from improved rust resistance is likely to lead to lower prices over time (Alston, Norton, and Pardey 1995). For example, the real price of wheat in India paid by consumers decreased by approximately 2 percent per year from 1970 to 1995 (Dixon et al. 2006). If 30 percent of the production increases reported by Evenson and Gollin (2003) as a result of modern semidwarf wheat varieties in developing countries were the result of improved rust resistance, a price reduction on the order of 10 to 20 percent would be attributable to the increased production resulting from the improved rust resistance.

Estimated Returns on Investment in Rust Resistance

Although the data reported here do not enable a precise estimate of the returns on investment in wheat rust resistance, some broadly indicative estimates (in 2006 dollars) can be developed from the previous analysis and discussion:

- The global investment in the development, deployment, and maintenance of rust resistance in wheat in developing countries is likely to be $60–100 million per year.

- The value of individual epidemics avoided or reduced by the resistance is likely to be up to $300 million per year.

- The estimated benefits of rust resistance in developing countries is between $0.2 and $2.5 billion per year for the individual rusts.

- Based on a partition of the benefits from modern semidwarf wheat varieties and taking 30 percent of the total benefits for rust resistance gives a separate estimate of the benefits of rust resistance of $0.7–2.9 billion per year.

- On that basis, the estimated total benefits of rust resistance globally are likely to be in the range of $0.4–2.0 billion per year.

Using conservative estimates of benefits of $0.4, $0.8, and $1.2 billion per year and lags between investment and returns of five, seven, and ten years, the internal rate of return on the expenditure on rust resistance between 1960 and 2006 has been assessed (Table 2.6). With benefits of $0.8 billion per year and seven-year lags, the internal rate of return is estimated at 38 percent per year. This indicates a successful outcome from the development of improved rust resistance.

Impacts on Poverty, Nutrition, and Food Security

Agricultural research that generates broad-based productivity increases is an effective means of reducing poverty through income generation and rural employment. Developing-country populations can realize significant benefits from having effective resistance to wheat rust. Many of these farmers and societies rely extensively on the wheat crop for their livelihood and nutrition, and because large-scale fungicide treatment is not feasible for them, improved rust resistance has a direct influence on both poverty alleviation and nutrition.

Table 2.6 Estimated internal rate of return from investments in rust resistance, 2006

Lags (years)	Benefits (US$ billions per year)		
	0.4	0.8	1.2
	Rate of return (percent)		
5	35	54	66
7	26	38	46
10	19	27	32

Source: Authors.

Both large-scale and small-scale farmers have adopted the improved varieties of wheat (Lipton and Longhurst 1989), although in many regions smaller-scale (and poorer) farmers may not have achieved the same level of productivity improvement as larger-scale farmers (Dixon et al. 2006). Byerlee and Moya (1993) and Marasas, Smale, and Singh (2004) found that farmers in irrigated areas received the largest share of benefits from wheat breeding, and Dixon et al. (2006) noted that about half of the world's population living in poverty is located in the large irrigated areas of South Asia alone.

Though farmers are the primary beneficiaries of improved wheat varieties, for poor producers lower prices can offset the advantages of higher yields, and the benefits of increased yields and reduced variability can be further offset if accompanied by increased levels and prices of inputs (Dixon et al. 2006). Whether producers are better off will depend on their environment, the level of yield gains in that environment, and the extent to which they market their production. Studies such as that of Evenson and Gollin (2003) show that the overall net effect has been a significant gain for producers. Thus it is likely that most producers are better off, while only a minority is worse off.

On the other hand, all consumers will benefit from any price reduction. Such price changes resulting from rust resistance research would lead to an overall shift in welfare from producers to consumers (Alston, Norton, and Pardey 1995). There is evidence that urban poor as well as rural poor have made significant gains through the modern semidwarf varieties (Harris, Hunter, and Lewis 1995; Ravallion and Datt 1995). In addition, higher levels of production can reduce absolute poverty in rural areas through increased demand for harvest and postharvest labor. There have also been downstream benefits to the economy where that increased production found its way into a marketing chain (Lipton and Longhurst 1989).

Where production was consumed on-farm rather than traded, the families could expect to have improved nutrition from the larger harvests. Benefits of modern semidwarf varieties incorporating rust resistance are the gain in calorie consumption per capita in developing countries, and the reduction in the percentage of malnourished children (see Table 2.5). Calorie consumption per capita in developing countries significantly increased between 1960 and 2000, by approximately 13–14 percent, and the number of children malnourished fell by about 7 percent, or more than 30 million children (Evenson and Gollin 2003). If rust resistance was responsible for 30 percent of the increase in global wheat production, it has made a significant impact on nutrition.

The increased availability of and lower prices for food are especially beneficial to the poor, who spend a large share of their income on food. An Expert Panel (2005) notes that in countries where wheat is a staple, low-income households tend to spend a larger proportion of their income on it than do higher-income households.

Price falls are beneficial to both urban and rural consumers, who have more disposable income for other goods (Dixon et al. 2006).

The World Bank (2005) found that the empirical evidence suggested that for every 1.0 percent increase in the productivity of wheat the extent of poverty has been reduced by 0.5–1.0 percent. If this ratio were extrapolated to the figures from Table 2.5, the 5.2 percent increase in productivity attributable to rust resistance would have reduced poverty by between 2.6 and 5.2 percent across all developing countries.

Nagarajan and Joshi (1975) discussed the extent to which rust epidemics in India have led to famines in the past. They concluded that the failure of the monsoons and the associated "Kharif" summer-season crops was the main reason for past famines. As a result, they argued that rust epidemics worsen famine conditions if they occur before or after a monsoon.

Therefore, improving rust resistance is an effective way to provide benefits to the poor, though they receive benefits only in proportion to their production and consumption. While rust resistance is an effective (and low-cost) way to benefit poor farmers, it is not a targeted way to direct benefits to poor farmers rather than to wealthier farmers.

Rust resistance has clearly enhanced food security in many developing countries by eliminating, or at least reducing the frequency of, serious epidemics. The availability of effective rust resistance in developing countries, especially those with food deficits, has precluded the need for a number of strategies that would have been needed to improve food security, such as reduced emphasis on food aid imports for countries or alternative crop systems for farmers. Farmers especially have experienced increased yield stability with improved rust resistance. Gollin (2006) has shown that the absolute magnitude of yield variability for wheat in developing countries declined with the spread of modern semidwarf wheat varieties. The value of the increased yield stability is equivalent to yield increases worth $143 million per year (Gollin 2006). Rust resistance is clearly responsible for a significant proportion of these benefits through its contribution to that improvement in stability of yields.

The increased stability of yields leads to a more stable and cohesive society with greater food security than if the harvests were subject to regular, or even occasional, destructive epidemics inducing famine. At the household level, for example, Bunch (1982) found that the social consequences of a crop failure are high, which is especially important when considering that farmers (particularly poor farmers) are generally risk-averse (Dixon et al. 2006).

Overview of Impacts

In summary, a wide range of economic and social indicators show that the development and deployment of resistance to stem and leaf rusts in developing countries

Table 2.7 Summary of outcomes from rust resistance, 2006

Measure of impact	Size of impact
Total wheat area affected[a]	117 million hectares
Estimated number of households affected[b]	60–120 million
Changes in wheat yields	
Value of benefits per hectare of wheat[c]	US$6–$12
Equivalent average annual yield increase[d]	4–8%
Equivalent average annual yield increase[e]	108–216 kilograms per hectare
Estimated increase in wheat production in developing countries[f]	5.2%
Estimated reduction in children malnourished in developing countries[f]	2.1%
Estimated increase in calorie consumption in developing countries[f]	4.2%

Sources: FAOSTAT (2009); authors' calculations.
[a]From FAOSTAT data for 2006.
[b]Assuming an average farm size of approximately 1–2 hectares in developing countries.
[c]See Table 2.3.
[d]Valuing per hectare benefits at a 2006 price of US$148 per metric ton.
[e]Assuming average yields in developing countries of 2.7 metric tons per hectare.
[f]See Table 2.5.

have brought about positive outcomes. Across some 60–120 million households, wheat yields and returns have increased, bringing about significant increases in aggregate wheat production (Table 2.7). Those increased wheat supplies have also resulted in improvements in nutrition, such as a 2.1 percent reduction in child malnourishment and a 4.2 percent increase in calorie consumption for consumers across developing countries.

Sustainability: Factors That Influenced the Long-Term Success of Breeding for Durable Wheat Rust Resistance

The overarching action that shaped the success of the international wheat breeding effort was a true cooperation or collaboration among the parties. In addition to CIMMYT, the NARS and several governmental or university programs also have played major roles in the development of the durable type of stem and leaf rust resistance and its use in developing countries. Principal among them were the USDA, the University of Minnesota, Agriculture and Agri-Food Canada, the University of Saskatchewan, PBI–Cambridge, IPO–Wageningen, and the University of Sydney. Most of the other factors that helped produce the success of breeding for durable rust resistance fall under this heading of cooperation and collaboration, while others were related to social situations or individual and institutional policy. Key success factors, as well as several negative factors and controversial issues, are examined in the following sections.

Free Exchange of Germplasm

The informal, free exchange of germplasm before the advent of the ISWRN and thereafter with IWIN adhered to the classic definition of "open-source" collaboration, whereby all cooperated and all benefited. This collaboration, coupled with the CIMMYT training of young scientists from the NARS, produced the basic tools needed to achieve the goals of the programs. Because CIMMYT did not release varieties directly, countries were able to select and give their own names to new releases and thus have a sense of ownership in the breeding process and the releases as well.

International Nursery System and Information Sharing

The birth of the international nursery system out of the stem rust epidemic of the early 1950s started the collaboration and set the tone for the Borlaug program in Mexico, then Latin America, and finally South Asia and the rest of the world. Without the free exchange of germplasm and the nursery system as the vehicle, the incorporation and distribution of durable rust resistance worldwide would likely not have happened. Information sharing based on the data generated and reported via the IWIS and its predecessors over the years was also a major factor in helping disseminate the information on rust resistance and in the ultimate release of the resistant varieties.

Multilocation Testing System

This system, starting with shuttle breeding in Mexico and then used at more than 100 sites around the world, has continually broadened the gene pool of the wheat program. The sites are those where wheat may be a major or minor crop and represent many different environments with unique biotic and abiotic stresses. The key is the cycling of the best germplasm based on the multisite/multiyear testing results. This method is of paramount importance, in conjunction with "hot spot" testing, for the rapid incorporation of the best disease resistance. In this way, advanced lines are made available to the NARS for crossing, selection of segregating populations, advanced lines, and possible rapid release after local yield trials.

Human Resource Development

Trained scientific personnel were as necessary as the germplasm for the national programs over the years. The practical field and classroom training provided by CIMMYT helped to develop the young scientists' commitment and dedication to food production. This commitment contributed to producing reliable data that were returned for analysis and then used in the broad breeding program.

Regional Programs and Networking

In a peripheral way, regional programs had a significant effect on the development of a durable type of resistance. The fact that international staff lived in the wheat-growing regions and interacted often with national staff allowed them to participate in the process of selection of national and international germplasm. In areas of South America and East Africa (Kenya and Ethiopia), regional programs distributed nurseries monitored for the rusts. Lamentably, these nurseries were discontinued in the mid-1980s.

Food Shortages

The international breeding initiative in the 1940s was born out of necessity. Mexico had severe food shortages, and the government requested help from the United States. The success of the Rockefeller program into the late 1950s and early 1960s gave birth to the larger international effort in wheat and rice that arose in response to the severe food shortages on the Indian subcontinent (Paddock and Paddock 1967). Food production could not keep up with the population increase at that time, and the international community responded effectively. The Green Revolution was the result.

Clear Focus on Food Production

The Borlaug program had a clear focus on food production in Mexico and its major limiting factor in wheat stem rust. Borlaug himself had to battle to keep that focus throughout his career, especially in the early years (Bickel 1974). A persistent theme in CIMMYT was to remember that the mission was to produce food for the poor and, as Borlaug recently said, "a will to win" (Borlaug 2009).

Long-Term Commitments

The Rockefeller Foundation's support of the Mexican Oficina de Estudios Especiales, headed by J. G. Harrar, and the Borlaug wheat research effort, among others, allowed the program to thrive. Donors such as USAID, the World Bank, the IADB, the Canadian International Development Agency, Official Development Assistance, and others continued to support the work that allowed the dissemination of durable resistance throughout the wheat world. The long-term commitment was and continues to be essential for agriculture research and development. An additional factor in the success of these efforts was the dedication of the international staff involved over the years in the CIMMYT Wheat Program. The fact that many were able to make a career of breeding, pathology, agronomy, and related disciplines meant that a cadre of experienced and knowledgeable scientists was available to produce the resistance and train the national staff necessary to get the product into the farmers' fields.

Impact Assessment

Economic assessments of wheat breeding efforts have been powerful tools to support the breeding programs over the years. Critical appraisal has created an awareness of key data in assessing contributions. This has helped focus the research effort at CIMMYT and make the center's intervention more effective. The ongoing economic assessment of how well the program was meeting its targets meant that the scientists involved were continually challenged to justify their progress and achievements.

Negative Factors that Affected the Process

Long-term support is needed for a good return on investments in agriculture research and development. But between 1980 and 2005, agriculture's share of official international development support had dropped from about 20 percent to 4 percent due to factors such as decreases in international commodity prices, a shifting of support to debt relief and social development, opposition from environmental groups that accused agriculture of contributing to natural resource degradation, and a shift within the support given to agriculture from productivity improvement to natural resources management and policy research (World Bank 2007).

As the overall international agricultural development budget decreased, so did the CIMMYT wheat budget, which, in real terms, was cut by more than half from 1980 to 2002 (Byerlee and Dubin 2008). Also, the share of budget allocated to unrestricted core funding fell from over 80 percent in 1990 to around 45 percent in 2006 and even further by 2008. Donors restricted funds to specific projects, very often short term in nature, to preserve the identity of the funds. As described earlier, the main components of breeding and international germplasm exchange need long-term core funding.

The international development community's reduction of its support to commodity research programs had repercussions throughout all programs at CIMMYT. In 1988 the CIMMYT wheat program had 16 pathologists, with 5 working at least part time on rusts. In 1999 there were 6 pathologists, with 2 working on rusts (CIMMYT 1988, 1999). In 1987 the ISWRN was discontinued due to funding issues. From the mid-1980s to about 1990, CIMMYT decreased its pathology support in East Africa. The last CIMMYT regional pathologist/breeder left in 1989. In addition, the International Disease Trap Nursery and the Latin American Rust Nursery were discontinued due to changes in the program. Finally, in the mid- to late 1990s, two key international rust programs that collaborated closely with CIMMYT were curtailed due to staff retirements. All these changes culminated in the elimination of the wheat program as such, at least until it was reestablished in 2006 as the result of another restructuring within CIMMYT.

Controversial Aspects of the International Wheat Breeding Efforts

There has been considerable controversy on the environmental, socioeconomic, and political impacts of modern varieties and the Green Revolution. However, the role of wheat rust resistance is less controversial. The increased yields as a result of improved rust resistance mean that production can take place on smaller cropped areas, decreasing demand on marginal or stressed land for crop production to meet global food demands.

Still, critics of the Green Revolution (e.g., Griffin 1974) have questioned the sustainability of intensive cultivation, noting concerns such as the environmental consequences of soil degradation, chemical pollution, aquifer depletion, and soil salinity. Evenson and Gollin (2003) note that, although these are valid criticisms, it is unclear that alternative scenarios would have allowed developing countries to meet the human needs of their expanding population with less environmental impact.

In the initial period of intensive cultivation up to 1980, the modern wheat varieties (and the associated rust resistance) introduced in irrigated areas during the Green Revolution required greater use of inputs by small farmers, including fertilizers and herbicides. These issues were partly addressed beginning around 1980, when the expansion in the area under modern semidwarf varieties of wheat occurred in rainfed areas, where inputs were used less intensively (Byerlee 1996). By the mid-1980s, more than 50 percent of the area sown to wheat in rainfed areas of developing countries was planted to semidwarf varieties. Thus the association between modern semidwarf varieties and high inputs diminished in the post–Green Revolution period.

Furthermore, the CIMMYT wheat breeding program has been selecting wheat germplasm efficient in input use for many years. Compared to that for tall varieties, CIMMYT germplasm requires less nitrogen and phosphorus and concomitantly less land to produce the same amount of wheat (Ortiz-Monasterio et al. 1997; Ortiz-Monasterio 2009). To the extent that the rust resistance was an integral part of the modern varieties, resistance was associated with the initial move to higher-input agriculture. It was not directly related because, by its nature, it obviated the use of fungicides as a regular input to wheat in modern farming systems in developing countries.

The reduction in fungicide use from improved rust resistance has had a positive impact on environmental and human health. Hundreds of millions of liters of fungicides would need to be applied to wheat crops around the world if the rust resistance had not been developed and deployed. Rust resistance also precludes the misuse of fungicides in the farming environment, which would be a constant threat if fungicide use were widespread globally.

An indirectly related issue with durable resistance is the genotypic background into which it is placed. While there have been claims that the Green Revolution reduced genetic diversity, Byerlee (1996) argued that the evidence is mixed. Although diversity was reduced in the early stages of the Green Revolution, recent works (Smale and McBride 1996; Smale et al. 2002; Lantican, Dubin, and Morris 2005; Warburton et al. 2006) have shown that the genetic base of the CIMMYT spring bread wheat germplasm continues to broaden with the continued introgression of new sources of wheat germplasm.

However, although the genetic diversity of varieties generally increased in the 1980s and 1990s, the diversity of rust resistance was not so broad based. In relation to the spectrum of durable stem and leaf rust resistance genes, the arsenal at this stage is not as great as desired, although the group of resistance genes has lasted for 30 to 50 years with only limited failures. Furthermore, the resistance imparted by Sr31 was such that it precluded detection of other genes (a masking effect) without morphological or molecular markers once the Sr31 was present. However, the reliance on Sr31 for varieties in East Africa, the Middle East, and South Asia led to vulnerability to the Ug99 strain of stem rust.[7]

On the issue of economic sustainability, Byerlee (1996) points out that the development of durable resistance can have an economic cost if farmers replace varieties less frequently, thereby failing to realize genetic gains in yield potential in new varieties. Only when they replace varieties rapidly are farmers able to achieve rust resistance with the highest-yielding varieties.

Finally, it is worth commenting on the social and political sustainability of global efforts to breed for durable wheat rust resistance. Although there were some political issues relating to the Green Revolution and the change to higher-input wheat production associated with it, there is nothing intrinsically political in the development of durable rust resistance. At times, the push from CIMMYT scientists and their colleagues was seen by some countries' leaders as unnecessary interference in their seed production and distribution, as in Pakistan in the period immediately prior to the leaf rust epidemic of 1977–78. However, the free provision by CIMMYT of the seed for release by the NARS meant that the NARS had ownership of the varieties. The future impediments to continuing successful deployment of rust resistance appear likely to be related to funding rather than any inherent political issues associated with the process

Lessons Learned and Conclusions

The development of rust resistance and its deployment in developing countries has provided a high rate of return on the investment involved. The experience from the efforts to develop and maintain rust resistance has provided several lessons for anyone attempting to replicate that success in other situations. Several key factors need to be put into place to ensure success for any future efforts to bring about international agricultural programs.

Clear Focus and Adequate Resources

One lesson is that strong leadership is needed to ensure that a clear focus on the objective is maintained. That involves long-term funding and staffing to ensure that goals are met and achievements are maintained. As a result, donors and administrators must carefully analyze where the payoffs have been and maintain support to those programs; continuous assessment of progress toward the objective is required. New technology should be supported when it can provide advantages but not at the cost of giving up a successful approach. For example, biotechnology provides a good opportunity to repeat and improve progress so far, though its role needs to be assessed carefully.

International Collaboration and Training

The success in rust resistance clearly demonstrates that a collegial approach is required, involving close collaboration between the NARS, CGIAR, and other international scientists. One key component of that collaborative approach is the ready exchange of genetic materials among those involved. A continuous stream of diverse germplasm is needed if initial successes are to be maintained over time. If the technology loses its effectiveness (as when the resistance breaks down), scientists must have the replacement technology available and ready to adopt. The rust resistance experience also clearly shows that human resource development is as important as the germplasm effort.

Infrastructure to Maintain Progress

When facing an evolving threat such as plant diseases, continual vigilance and surveillance of the pathogen and host are needed. This might include, among other things, the use of an international early warning system, trap nurseries, and rust population monitoring. As discussed earlier, no form of resistance lasts forever. Therefore, there can be no complacency when working with nature, and the successful role of the IARCs in developing the global strategy for rust resistances must not allow the NARS to become complacent about the threats posed. To ensure that new threats do not present significant problems, anticipatory thinking is needed.

Emphasis on Generating Effective Impacts

The technology had maximum impact where it addressed a significant issue for the maximum number of farmers and where it was broadly adaptable and durable. The low or zero cost to poor farmers meant that those who most needed it could readily adopt the new technology. In particular, technology embedded in seed is likely to lead to effective outcomes, because it requires little investment by farmers to receive the benefits. It is also clear that programs to enhance the adoption of the improved technology produced significant benefits.

Free Exchange of Germplasm and Accommodation of Intellectual Property Rights Issues

Despite its significant contribution to the success of rust resistance, the free exchange of germplasm has come under pressure within the CGIAR since the early 1990s.

The declining core funding that reduced operating funds for the germplasm networks and the increased role of private-sector breeding and biotechnology programs in developed countries worked toward reducing the free exchange of germplasm. In addition, the Agreement on Trade-Related Aspects of Intellectual Property Rights and Convention on Biological Diversity were both signed in 1993. These all led to uncertainties and higher transaction costs with respect to international germplasm exchanges. In some cases, it has resulted in diminished germplasm movement (Byerlee and Dubin 2008). The International Treaty on Plant Genetic Resources for Food and Agriculture and the Standardized Material Transfer Agreement for all germplasm exchange are responses to the possible consequences of the agreements. Nevertheless, it will be essential to mitigate the effects of intellectual property rights to ensure free exchange of materials if similar programs are to succeed in the future.

Practical Technologies

In the future, regardless of the technologies and programs involved, it is imperative that scientists who work in the field and not only the laboratory produce the varieties and seed for farmers. This is to ensure that the technology is directly relevant to farmers and so that the scientists can respond immediately and practically to future problems that arise.

The Recent Breakdown of Durable Stem Rust Resistance: A Reality Check

The success of the intervention described here has allowed donor fatigue and a realignment of immediate priorities for funding away from the maintenance of rust resistance in some areas over time. Thus the world wheat crop became more vulnerable to a new threat as IARC funding declined over time. The financial sustainability of the development of rust resistance has been uncertain because of the shortage of funds. At the farm level, the financial returns have come at close to zero cost and will be readily sustained if the resistance can be maintained in well-adapted varieties.

In 1998 a new stem rust race, Ug99, was identified in Uganda that rendered previously resistant lines susceptible to stem rust (Pretorius et al. 2000). Now major genes and slow-rusting genes have been identified and are being incorporated into germplasm. The germplasm is being evaluated through multilocation testing in many key areas, including Ethiopia and Kenya, where Ug99 and some of its descendants are now endemic (Singh et al. 2008). However, the new race is mutating and spreading as feared. It was observed in Yemen in 2006 and in Iran in 2007 (Nazari et al. 2009).

An expert panel report (Expert Panel 2005) declared that Ug99 was a threat to world wheat production because it was predicted to migrate across the Red Sea to Yemen, then to the Middle East, and subsequently to Central and South Asia. Those areas, with a population of 1 billion people, produce 19 percent of the world's wheat. The report predicted that either wind currents or inadvertent transport would eventually carry Ug99 to North Africa, Europe, West Asia, China, Australia, and the Americas (Cornell University 2008). Once Ug99 and its derivatives have established themselves in North Africa, the Middle East, and South Asia, annual losses could reach $3 billion in any given year (Cornell University 2008).

The success in developing global rust resistances allowed some NARSs to become complacent about the threats posed, which is now evident in countries threatened by Ug99. Resources will need to be increased across a range of research skills and infrastructure to manage the rust infections and the response to them (Expert Panel 2005). Although countries such as India have the resources and infrastructure to prepare for and respond to the Ug99 threat (ICAR 2008), many poorer countries need considerable external resources to meet those needs now without encountering bottlenecks (Expert Panel 2005). Substantial efforts are under way to provide the countries under threat with varieties that will overcome the problems caused by Ug99.

Conclusions

The successful efforts of the international wheat stem and leaf rust resistance programs over the past half century have had significant economic returns as well as positive impacts on poverty reduction, nutrition, food security, and the environment. Nevertheless, decreased donor support for the programs in recent years has had negative effects, and the recent occurrence of a new strain of stem rust that defeated key durable resistance genes has put large wheat areas in developing countries at risk. It is clear that there is a critical need for continuous research and vigilance to keep ahead of the ever-changing pathogens to maintain the progress that has been made.

Notes

1. Durable rust resistance broadly means resistance that has been tested over time and space and lasted a long time (Johnson 1981). However, today the term *durable* in plant pathology is in transition and would commonly be used in conjunction with adult-plant, non-race-specific resistance, as noted in a subsequent section. See Dubin and Brennan (2009). The International Maize and Wheat Improvement Center, known by its Spanish acronym (CIMMYT), is an internationally funded, not-for-profit organization that conducts research and training related to maize and wheat throughout the developing world (www.cimmyt.org).

2. In Mexico, this included the Mexico/Rockefeller program, an international agricultural development program founded in 1943 to help Mexico increase food production.

3. The international rust nursery program later had spring wheat and winter wheat screening nurseries, but we focus primarily on the spring bread wheat here as well as throughout this chapter. Spring bread wheat is the major wheat type grown in the developing world.

4. For a more complete discussion of the international agricultural research system see Dubin and Brennan (2009).

5. In particular, a program to develop wheat for the more tropical environments brought in a number of nontraditional wheat-producing countries.

6. Murray and Brennan (2009) found that the value of rust resistance in wheat in Australia was $272 million for stem rust and $94 million for leaf rust (in 2006 dollars), equivalent to $22.81 and $7.89 per hectare, respectively. These estimates are more consistent with the other estimates for developing countries in Table 2.4 than with those of Brennan and Quade (2004).

7. It is significant that Ug99 and recent mutations of this strain attack a broad array of resistance genes. Scientists have worked to develop diverse sets of genes and forms of resistance (for example, see McIntosh, Wellings, and Park 1995). Research is ongoing to obtain molecular markers for other stem and leaf rust resistance genes of both the major and durable slow rusters with additive minor genes. New candidate durable-type resistance genes are being studied, as is strengthening of the Sr2 complex for stem rust and more minor genes of additive effects for both rusts (Singh et al. 2008). Critical to this work will be continued funding over the long term.

References

Adusei, E. O., and G. W. Norton. 1990. The magnitude of agricultural maintenance research in the USA. *Journal of Production Agriculture* 3: 1–6.

Agrios, G. 2005. *Plant pathology.* 5th ed. Boston: Elsevier Academic Press.

Alston, J. M., G. W. Norton, and P. G. Pardey. 1995. *Science under scarcity: Principles and practice for agricultural research evaluation and priority setting.* Ithaca, N.Y., U.S.A.: Cornell University Press.

Bickel, L. 1974. *Facing starvation: Norman Borlaug and the fight against hunger.* New York: Readers Digest Press.

Borlaug, N. E. 1966. Basic concepts which influence the choice of methods for use in breeding for diverse resistance in cross pollinated and self pollinated crop plants. In *Breeding pest-resistant trees,* ed. H. D. Gerhold, E. J. Schreiner, R. E. McDermott, and J. A. Winieski. Oxford, U.K.: Pergamon Press. Quoted in S. Rajaram and M. van Ginkel, Mexico: 50 years of wheat breeding, in *The world book of wheat,* ed. A. P. Bonjean and W. J. Angus. Paris: Lavoisier, 2001.

———. 2007. Sixty-two years of fighting hunger: Personal recollections. *Euphytica* 157: 287–297.

Brennan, J. P., and K. J. Quade. 2004. *Genetics of and breeding for rust resistance in wheat in India and Pakistan: ACIAR Projects CS1/1983/037 and CS1/1988/014.* ACIAR Impact Assessment Series Report 25. Canberra: Australian Centre for International Agricultural Research.

Bunch, R. 1982. *Two ears of corn: A guide to people-centered agriculture.* Oklahoma City, Okla., U.S.A.: World Neighbors.

Byerlee, D. 1996. Modern varieties, productivity, and sustainability: Recent experience and emerging challenges. *World Development* 24 (4): 697–718.

Byerlee, D., and H. J. Dubin. 2008. Crop improvement in the CGIAR as a global success story of open access and international collaboration. *International Journal of the Commons* 4: 452–480.

Byerlee, D., and P. Moya. 1993. *Impacts of international wheat breeding research in the developing world, 1966–90.* Mexico City: International Maize and Wheat Improvement Center.

Byerlee, D., and G. Traxler. 1995. National and international wheat improvement research in the post–Green Revolution period: Evolution and impacts. *American Journal of Agricultural Economics* 77: 268–278.

CIMMYT (International Maize and Wheat Improvement Center). Various years. *CIMMYT annual report.* Mexico City.

———. 1988. *CIMMYT annual report.* Mexico City.

———. 1999. *CIMMYT annual report.* Mexico City.

Cornell University. 2008. Durable rust resistance in wheat project. <http://www.wheatrust.cornell.edu/about/backgroundandrationale.cfm>. Accessed March 16, 2009.

Dalrymple, D. G. 1986. *Development and spread of high-yielding wheat varieties in developing countries.* Washington, D.C.: U.S. Agency for International Development.

Dixon, J., L. Nalley, P. Kosina, R. La Rovere, J. Hellin, and P. Aquino. 2006. Adoption and economic impact of improved wheat varieties in the developing world. *Journal of Agricultural Science* 144: 489–502.

Dubin, H. J., and J. P. Brennan. 2009. *Combating stem and leaf rust of wheat: Historical perspective, impacts, and lessons learned.* IFPRI Discussion Paper 910. Washington, D.C.: International Food Policy Research Institute.

Dubin, H. J., and E. Torres. 1981. Causes and consequences of the 1976–77 wheat leaf rust epidemic in northwest Mexico. *Annual Review of Phytopathology* 19: 41–49.

Dyck, P., and E. Kerber. 1985. Resistance of the race-specific type. In *The cereal rusts,* vol. 2: *Diseases, distribution, epidemiology, and control,* ed. A. P. Roelfs and W. Bushnell. New York: Academic Press.

Evenson, R. E. 2000. Crop germplasm improvement: A general perspective. Paper presented at the annual meeting of the American Association for the Advancement of Science, February 21, in Washington, D.C.

Evenson, R. E., and D. Gollin. 2003. Assessing the impact of the green revolution, 1960 to 2000. *Science* 300 (5620): 758–762.

Evenson, R. E., and M. Rosegrant. 2003. The economic consequences of crop genetic improvement programmes. In *Crop variety improvement and its effect on productivity: The impact of international agricultural research,* ed. R. E. Evenson and D. Gollin. Wallingford, U.K.: CABI Publishing.

Expert Panel. 2005. *Sounding the alarm on global stem rust: An assessment of race Ug99 in Kenya and Ethiopia and the potential for impact on neighboring regions and beyond.* Expert Panel on the Stem Rust Outbreak in Eastern Africa. Mexico City: International Maize and Wheat Improvement Center.

FAOSTAT (Food and Agriculture Organization of the United Nations Statistical Database). 2009. <http://faostat.fao.org>. Accessed March 2009.

Fowler, C., M. Smale, and S. Gaiji. 2001. Unequal exchange? Recent transfers of agricultural resources and their implications for developing countries. *Development Policy Review* 19 (2): 181–204.

Gollin, D. 2006. *Impacts of international research on intertemporal yield stability in wheat and maize: An economic assessment.* Mexico City: International Maize and Wheat Improvement Center.

Griffin, K. B. 1974. *Political economy of agrarian change: An essay on the green revolution.* Cambridge, Mass., U.S.A.: Harvard University Press.

Hanson, H., N. E. Borlaug, and R. G. Anderson. 1982. *Wheat in the Third World.* Boulder, Colo., U.S.A.: Westview Press.

Harris, J., J. Hunter, and C. M. Lewis. 1995. *The new institutional economics and Third World development.* London: Routledge.

Heisey, P. W., M. A. Lantican, and H. J. Dubin. 2002. Impacts of international wheat breeding research in developing countries, 1966–1997. Mexico City: International Maize and Wheat Improvement Center.

Hodson, D. P., R. P. Singh, and J. M. Dixon. 2005. An initial assessment of the potential impact of stem rust (race Ug99) on wheat producing regions of Africa and Asia using GIS. Proceedings of the 7th International Wheat Conference, November 27–December 2, in Mar del Plata, Argentina.

Howard, A., and G. L. C. Howard. 1909. *Wheat in India: Its production, varieties, and improvement.* Calcutta, India: Thacker, Spink and Company.

ICAR (Indian Council of Agricultural Research). 2008. New race of stem rust Ug99—Be aware and do not be scared. <http://www.icar.org.in/news/New-Race-of-Wheat-Stem-Rust.htm>. Accessed March 16, 2009.

Johnson, R. 1981. Durable resistance: Definition of, genetic control, and attainment in plant breeding. *Phytopathology* 71: 567–568.

Kislev, M. E. 1982. Stem rust of wheat 3300 years old found in Israel. *Science* 216: 993–994.

Kolmer, J. A. 2001. Early research on the genetics of *Puccinia graminis* and stem rust resistance in wheat in Canada and the United States. In *Stem rust of wheat: From ancient enemy to modern foe,* ed. P. D. Peterson. St. Paul, Minn., U.S.A.: APS Press.

Lantican, M. A., H. J. Dubin, and M. L. Morris. 2005. *Impacts of international wheat breeding research in the developing world, 1988–2002.* Mexico City: International Maize and Wheat Improvement Center.

Lipton, M., and R. Longhurst. 1989. *New seeds and poor people.* London: Unwin Hymen.

Marasas, C. N., M. Smale, and R. P. Singh. 2004. *The economic impact in developing countries of leaf rust resistance breeding in CIMMYT-related spring bread wheat.* Economics Program Paper 04-01. Mexico City: International Maize and Wheat Improvement Center.

McIntosh, R. A., C. R. Wellings, and R. F. Park. 1995. *Wheat rusts: An atlas of resistance genes.* Melbourne, Australia: Australian Commonwealth Scientific and Research Organization.

Murray, G. M., and J. P. Brennan. 2009. *The current and potential costs from diseases of wheat in Australia.* Canberra, Australia: Grains Research and Development Corporation.

Nagarajan, S., and L. M. Joshi. 1975. An historical account of wheat rust epidemics in India and their significance. *Cereal Rust Bulletin* 3 (2): 29–33.

Nazari, K., M. Mafi, A. Yahyaoui, R. P. Singh, and R. F. Park. 2009. Detection of wheat stem rust (*Puccinia graminis* f. sp. *tritici*) race TTKSK in Iran. *Plant Disease* 93: 317.

Ortiz-Monasterio, J. I. 2009. E-mail to author (HJD), May 8.

Ortiz-Monasterio, J. I., K. D. Sayre, S. Rajaram, and M. McMahon. 1997. Genetic progress in wheat yield and nitrogen use efficiency under four N rates. *Crop Science* 37: 898–904.

Paddock, W., and P. Paddock. 1967. *Famine 1975! America's decision: Who will survive.* Boston: Little, Brown, and Company. Quoted in L. Hesser, *The man who fed the world: Nobel Laureate Norman Borlaug and his battle to end world hunger.* Dallas, Tex., U.S.A.: Durban House, 2009.

Park, R. F. 2007. Stem rust of wheat in Australia. *Australian Journal of Agricultural Research* 58: 558–566.

Park, R. F., H. S. Bariana, and C. S. Wellings. 2007. Preface. *Australian Journal of Agricultural Research* 58: 469.

Parlevliet, J. E. 1985. Resistance of the non-race specific type. In *The cereal rusts,* vol. 2: *Diseases, distribution, epidemiology, and control,* ed. A. P. Roelfs and W. Bushnell. New York: Academic Press.

Payne, T. S. 2004. The international wheat improvement network at CIMMYT. <http://www.cimmyt.org/Research/wheat/IWISFOL/IWIN.htm>. Accessed April 19, 2009.

Payne, T. S., B. Skovmand, C. G. Lopez, E. Brandon, and A. McNab, eds. 2002. The international wheat information system (IWIS)TM, version 4, 2001. Mexico: CIMMYT. Compact disk.

Plucknett, D. L., and N. J. H. Smith. 1986. Sustaining agricultural yields. *BioScience* 36 (1): 40–45.

Plucknett, D. L., S. Ozgediz, and N. J. H. Smith. 1990. *Networking in international agricultural research.* Ithaca, N.Y., U.S.A.: Cornell University Press.

Pretorius, Z. A., R. P. Singh, W. W. Wagoire, and T. S. Payne. 2000. Detection of virulence to wheat stem rust resistance gene Sr31 in *Puccinia graminis* f. sp. *tritici* in Uganda. *Plant Disease* 84: 203.

Rajaram, S., and M. Van Ginkel. 2001. Mexico: 50 years of wheat breeding. In *The world book of wheat,* ed. A. P. Bonjean and W. J. Angus. Paris: Lavoisier.

Rajaram, S., R. P. Singh, and E. Torres. 1988. Current CIMMYT approaches to breeding wheat for rust resistance. In *Breeding strategies for resistance to the rusts of wheat,* ed. N. W. Simmonds and S. Rajaram. Mexico City: International Maize and Wheat Improvement Center.

Rasmusson, D. C. 1996. Germplasm is paramount. In *Increasing yield potential in wheat: Breaking the barriers,* ed. M. P. Reynolds, S. Rajaram, and A. McNab. Mexico City: International Maize and Wheat Improvement Center.

Ravallion, M., and G. Datt. 1995. *Growth and poverty in rural India.* Policy Research Working Paper 1405. Washington, D.C.: World Bank.

Roelfs, A. P., R. P. Singh, and E. E. Saari. 1992. *Rust diseases of wheat: Concepts and methods of disease management.* Mexico City: International Maize and Wheat Improvement Center.

Saari, E. E., and J. M. Prescott. 1985. World distribution in relation to economic losses. In *The cereal rusts*, vol. 2: *Diseases, distribution, epidemiology, and control*, ed. P. Roelfs and W. R. Bushnell. Orlando, Fla., U.S.A.: Academic Press.

Sayre, K. D., R. P. Singh, J. Huerta-Espino, and S. Rajaram. 1998. Genetic progress in reducing losses to leaf rust in CIMMYT derived Mexican spring wheat cultivars. *Crop Science* 38 (3): 654–659.

Singh, R. P., D. P. Hodson, J. Huerta-Espino, Y. Jin, P. Njau, R. Wanyera, S. Herrera-Foessel, and R. W. Ward. 2008. Will stem rust destroy the world's wheat crop? *Advances in Agronomy* 98: 271–310.

Smale, M., and T. McBride. 1996. Understanding global trends in the use of wheat diversity and international flows of wheat genetic resources. Part 1 of *CIMMYT 1995/96 World Wheat Facts and Trends: Understanding global trends in the use of wheat diversity and international flows of wheat genetic resources*. Mexico City: International Maize and Wheat Improvement Center.

Smale, M., R. P. Singh, K. Sayre, P. Pingali, S. Rajaram, and H. J. Dubin. 1998. Estimating the economic impact of breeding nonspecific resistance to leaf rust in modern bread wheats. *Plant Disease* 82: 1055–1061.

Smale, M., M. P. Reynolds, M. Warburton, B. Skovmand, R. Trethowan, R. P. Singh, I. Ortiz-Monasterio, and J. Crossa. 2002. Dimensions of diversity in modern spring bread wheat in developing countries from 1965. *Crop Science* 42: 1766–1779.

Stakman, E. C., and J. G. Harrar. 1957. *Principles of plant pathology.* New York: Ronald Press.

Trethowan, R., and J. Crossa. 2006. Forty years of international bread wheat trials: What have we learned? In *Extended abstracts of the international symposium on wheat yield potential, Challenges to International Wheat Breeding,* March 20–24, in Ciudad Obregon, Mexico, ed. M. P. Reynolds and D. Godinez. Mexico City: International Maize and Wheat Improvement Center.

Trethowan, R., D. Hodson, H.-J. Braun, W. Pfeiffer, and M. van Ginkel. 2005. Wheat breeding environments. In *Impacts of wheat breeding research in the developing world 1988–2002,* ed. M. Lantican, H. J. Dubin, and M. Morris. Mexico City: International Maize and Wheat Improvement Center.

Warburton, M. L., J. Crossa, J. Franco, M. Kazi, R. Trethowan, S. Rajaram, W. Pfeiffer, P. Zhang, S. Dreisigacker, and M. van Ginkel. 2006. Bringing wild relatives back into the family: Recovering genetic diversity in CIMMYT improved wheat germplasm. *Euphytica* 149 (3): 289–301.

World Bank. 2005. *Agricultural growth for the poor: An agenda for development.* Directions in Development Series. Washington, D.C.

———. 2007. *World development report 2008: Agriculture for development.* Washington, D.C.

Chapter 3

The Asian Green Revolution

Peter B. R. Hazell

D riven by rapid advances in the sciences and substantial public investments and policy support for agriculture, the Green Revolution was just one aspect of a much larger transformation of global agriculture during the 20th century. The story of English wheat is typical. It took nearly 1,000 years for wheat yields to increase from 0.5 to 2.0 tons per hectare, but then wheat yields climbed to over 7 tons per hectare during the 20th century. These advances were fueled by modern plant breeding, improved agronomy, and the development of inorganic fertilizers and modern pesticides. Most industrial countries had achieved sustained food surpluses by the second half of the 20th century and abolished the threat of food shortages.

These advances were much slower in reaching today's developing countries. Although the colonial powers had invested in improving the production of tropical export crops, they had invested relatively little in the food production systems of their colonies. This neglect together with rapidly growing populations led to widespread hunger and malnutrition by the 1960s, especially in developing Asia, which had a growing dependence on food aid from the rich countries. Sequential droughts in India during the mid-1960s highlighted the precarious nature of the situation, and a 1967 report of the U.S. President's Science Advisory Committee concluded: "The scale, severity and duration of the world food problems are so great that a massive, long-range innovative effort unprecedented in human history will be required to master it" (quoted in Evans 1998).

In response, the Rockefeller and Ford Foundations took the lead in establishing an international agricultural research program to help transfer and adapt scientific advances already available around the world to the conditions of the developing countries (Tribe 1994). The first investments were made in research on rice and wheat, two of the most important foodcrops for developing countries. The breeding of improved varieties combined with the expanded use of fertilizers, other chemical inputs, and irrigation led to dramatic yield increases for these two crops in Asia and Latin America, beginning in the late 1960s. This development was coined the "Green Revolution" by USAID Administrator W. S. Gaud, hoping it would help contain the communistic "Red Revolution" that was capitalizing on poverty in developing countries at the time.

Although the term *Green Revolution* originally described developments related to rice and wheat, the term has since been used to refer to the development of high-yielding varieties of a number of other major foodcrops important to developing countries. These include sorghum, millet, maize, cassava, and beans. Moreover, there is now a full-fledged system of international agricultural research centers, the Consultative Group on International Agricultural Research (CGIAR), that works on many aspects of developing-country agriculture.

The Green Revolution was a continuing process of change rather than a single event, and even today, continuing improvements of cereal varieties and management practices help support and advance the high levels of productivity that were initially attained. Although the main thrust of the Asian Green Revolution occurred during the period 1965–90, it had many technology and policy antecedents in the rice revolution that began in Japan in the latter part of the 19th century and spread to Taiwan and Korea during the late 1920s and 1930s (Jirström 2005). This chapter focuses on the more narrowly defined Asian Green Revolution of 1965–90.

The Green Revolution spread rapidly across developing Asia, and the resultant increases in food production pulled the region back from the edge of an abyss of famine and led to regional food surpluses within 25 years. It lifted many people out of poverty, made important contributions to economic growth, and saved large areas of forest, wetlands, and other fragile lands from conversion to cropping. The investments and policies that underpinned the Green Revolution were highly successful in achieving the objectives of the time, and produced a high rate of economic return. There are lingering social and environmental problems that still need to be resolved. Additionally, it will be important to (1) continue to increase the productivity of the Green Revolution areas to meet the growing demands for cereals for food, feed, and fuel, and (2) spread intensive farming methods more widely, particularly to Sub-Saharan Africa, which has failed so far to adequately intensify its food production systems and as a result suffers from worsening poverty and hunger.

The Intervention

The Green Revolution was driven by a technology revolution, comprising a package of modern inputs—irrigation, improved seeds, fertilizers, and pesticides—that together dramatically increased crop production. But its implementation also depended on strong public support for developing the technologies; building up the required infrastructure; ensuring that markets, finance, and input systems worked; and ensuring that farmers had adequate knowledge and economic incentive to adopt the technology package. Public interventions were especially crucial for ensuring that small farmers were included, without which the Green Revolution would not have been as pro-poor as it was. Attempts have been made to separate the contributions of the different components of the Green Revolution package, but in practice it was the combined impact of interventions and their powerful interactions that made the difference.

Irrigation

Asia had already been investing heavily in irrigation prior to the Green Revolution, and by 1970 around 25 percent of the agricultural land was already irrigated (Table 3.1). In India, 10.4 million hectares of land were irrigated by canals in 1961 and 4.6 million hectares by tanks (Evenson, Pray, and Rosegrant 1999). Significant additional investments were made across Asia during the Green Revolution era, and the irrigated area grew from 25 to 33 percent of the agricultural area between 1970 and 1995 (see Table 3.1).

Table 3.1 Indicators of input use during the Green Revolution in Asia, 1967–90

Country	Irrigated area (percentage of agricultural area) 1970	1995	Fertilizer (kilograms per hectare) 1970	1995	Annual growth rate in agricultural work force, 1967–82	Annual growth rate in agricultural land area, 1967–82
Bangladesh	11.6	37.6	15.7	135.5	1.07	0.05
China	37.2	37.0	43.0	346.1	1.92	0.03
India	18.4	31.8	13.7	81.9	1.59	0.19
Indonesia	15.0	15.2	9.2	84.7	1.41	0.00
Malaysia	5.9	4.5	43.6	148.6	0.57	1.03
Myanmar	8.0	15.4	2.1	16.9	1.93	−0.21
Nepal	5.9	29.8	2.7	31.6	1.82	1.56
Pakistan	67.0	79.6	14.6	116.1	2.41	0.33
Philippines	11.0	16.6	28.9	63.4	1.90	1.72
South Korea	51.5	60.8	251.7	486.7	−0.07	−0.38
Sri Lanka	24.6	29.2	55.5	106.0	1.69	−0.05
Thailand	14.2	22.7	5.9	76.5	2.17	2.52
Vietnam	16.0	29.6	50.7	214.3	1.58	0.54
Total	25.2	33.2	23.9	171.1	1.76	0.28

Source: Rosegrant and Hazell (2000).

Fertilizer

Like irrigation use, fertilizer use across Asia was also growing prior to the Green Revolution. In 1970, 23.9 kilograms of plant nutrients were applied per hectare of agricultural land and the average use rapidly grew to 102.0 kilograms per hectare by 1995 (see Table 3.1).

Improved Seeds

Irrigation and fertilizer use helped increase cereal yields, but their full impact was realized only after the development of high-yielding varieties. Scientists sought to develop cereal varieties that were more responsive to plant nutrients and had shorter and stiffer straw that would not fall over under the weight of heavier heads of grains. They also wanted tropical rice varieties that could mature more quickly and grow at any time of the year, thereby permitting more crops to be grown each year on the same land. Varieties also needed to be resistant to the major pests and diseases that flourish under intensive farming conditions and to retain desirable cooking and consumption traits.

Borrowing from the rice breeding work undertaken in China, Japan, and Taiwan, the fledgling International Rice Research Institute in the Philippines developed semidwarf varieties that met most of these requirements and could be grown under a wide range of conditions. Similar achievements were made for wheat after Norman Borlaug (later awarded a Nobel Prize for his work) crossed Japanese semidwarf varieties with Mexican varieties at what is now known as the International Center for Maize and Wheat Improvement in Mexico.

The adoption of high-yielding varieties occurred quickly (Table 3.2), and by 1980 about 40 percent of the total cereal area in Asia was planted to modern varieties (World Bank 2007). This had increased to about 80 percent of the cropped area by 2000.

It should be noted that the high-yielding varieties that powered the Green Revolution were not developed overnight but were the product of a long and sustained research effort. The initial varieties that were released also had to be adapted to counter evolving pest and environmental problems and be better suited to local conditions and consumer needs. This required a continuing process of agricultural research and development (R&D).

Public Investment and Policy Support

The Green Revolution was more than a technology fix. It also required a supporting economic and policy environment. The need to educate farmers about the new technology, rapidly expand input delivery and credit systems so they could adopt the new inputs, and increase processing, storage, trade, and marketing capacities to handle the surge in production was considered too great a challenge for the

Table 3.2 Percentage of harvested area planted with modern varieties in Asia, 1965–2000

Region, year	Rice	Wheat	Maize
South Asia			
1965	0.0	1.7	0.0
1970	10.2	39.6	17.1
1975	26.6	72.5	26.3
1980	36.3	78.2	34.4
1985	44.2	82.9	42.5
1990	52.6	87.3	47.1
1995	59.0	90.1	48.8
2000	71.0	94.5	53.5
East and Southeast Asia and the Pacific			
1965	0.3	0.0	0.0
1970	9.7	0.0	16.2
1975	27.0	14.8	39.5
1980	40.9	27.5	61.7
1985	54.1	34.3	65.9
1990	63.5	58.7	73.0
1995	71.1	78.8	83.2
2000	80.5	89.1	89.6

Source: Gollin, Morris, and Byerlee (2005); used by permission of Oxford University Press.

private sector on its own at the time, especially if small farmers were to participate (Johnson, Hazell, and Gulati 2003). It was also necessary to ensure that adoption of the technology package was profitable for farmers. To achieve these ends, governments across Asia actively intervened in launching and implementing the Green Revolution. Some but not all public interventions were market mediated, and all were backed by substantial public investments in agricultural development (Djurfeldt and Jirström 2005).

The levels of public investment needed to launch and sustain the Green Revolution were impressive. Asian countries not only invested heavily to launch their Green Revolution but also continued to invest in agriculture to sustain the gains that were achieved. On average, Asian countries were spending 15.4 percent of their total government spending on agriculture by 1972 and they doubled the real value of their agricultural expenditures by 1985 (Table 3.3). The need to sustain investment levels is especially great in the case of agricultural R&D, because there are long lead times in developing new products and farmers continually need new crop varieties and natural resource management practices to stay ahead of evolving pests, environmental problems, and changing market demands.

Table 3.3 Government expenditures on agriculture in Asia, 1972–90 (1985 U.S. dollars) (purchasing power parity)

Country	1972	1975	1980	1985	1990	1972	1975	1980	1985	1990
	(million dollars)					(percentage of total government expenditure)				
Bangladesh	2,358	528	1,187	1,749	1,269	23.2	11.0	12.3	15.7	5.4
China	11,595	17,843	24,542	21,113	28,229	8.5	12.1	12.4	8.3	8.9
India	15,491	13,680	22,877	30,549	39,109	22.1	9.7	14.6	12.6	11.5
Indonesia	1,436	3,020	5,026	4,351	6,157	7.6	9.8	9.6	6.8	7.6
Malaysia	348	458	1,264	1,581	1,830	4.0	4.2	7.2	7.9	5.7
Myanmar	272	219	655	874	296	12.5	13.3	23.6	24.5	9.3
Nepal	107	136	257	541	254	13.7	15.5	16.4	22.0	8.5
Pakistan	740	1,031	1,168	971	1,312	5.7	6.7	5.4	2.9	2.6
Philippines	416	1,145	729	604	1,409	4.5	9.0	5.3	5.7	6.0
South Korea	537	993	1,129	2,244	4,332	3.8	6.3	4.1	5.8	6.9
Sri Lanka	627	449	589	2,124	614	12.3	9.0	5.7	20.0	5.8
Thailand	902	767	1,850	3,181	3,190	7.8	5.9	8.1	11.7	10.4
Total	34,828	40,269	61,273	70,151	88,001	15.4	10.5	12.4	10.9	9.6

Source: Rosegrant and Hazell (2000); used by permission of Asian Productivity Organization.

Governments also shored up their farm credit systems; subsidized key inputs, especially fertilizer, power, and water; and intervened in markets to ensure that farmers received adequate prices each year to make the technologies profitable. Many governments used their interventions to make sure that small farmers participated in the Green Revolution and did not get left behind. Substantial empirical evidence at the time showed that small farms were the more efficient producers in Asia, and land reform and small-farm development programs were implemented to create and support large numbers of small farms. Agricultural growth led by small farms proved not only more efficient, but also more pro-poor—a win–win proposition for growth and poverty reduction.

The Government of India, for example, placed a top priority on agricultural development right after independence. Prime Minister Jawaharlal Nehru realized the importance of physical and scientific infrastructure for modern agriculture. From 1947 to 1952, the government allocated about 30 percent of its budget to agriculture and irrigation, and this spending led to an impressive buildup of rural roads, irrigation, rural power, state agricultural universities, and a national agricultural research system. Fertilizer plants were also set up. Another important policy intervention was land reform, characterized by the abolition of landowners and their intermediaries, increased tenancy security, ceilings on the size of landholdings, and use of cooperatives and community development programs.

These efforts were intensified in the mid-1960s as India became more dependent on U.S. food aid despite its own agricultural potential. The Indian government set up the Intensive Agricultural District Program, which invested heavily in agricultural extension and distribution of subsidized inputs. It also established the Food Corporation of India, which bought excess production at a guaranteed price in order to guarantee stability to farmers. The government took an active role in coordinating interventions from donors and development partners, who financed agricultural extension as well as R&D. All of this came on top of India's considerable existing infrastructure, including roads and irrigation systems. In total, the Indian government coordinated interventions all along the market chain to enable the entire agricultural system to function.

As a result, most rural small farms could profitably obtain inputs such as scale-neutral high-yield seeds, fertilizers, irrigation, and credit. Despite the heavy government presence, the private sector was also permitted a key role. The dual presence of a private and a public marketing system actually helped complement and improve the efficient distribution of inputs to farmers. The success of the Green Revolution was also catalyzed by combining subsidies to factor inputs with public investments in infrastructure (roads, power, and irrigation), as well as research and extension, later followed by marketing policy interventions to ensure that farmers had access to market outlets at stable prices (Johnson, Hazell, and Gulati 2003).

The Impacts

Average cereal yields grew impressively in Asia; over the period 1965–82, wheat yields grew by 4.1 percent per year and rice yields by 2.5 percent per year (Table 3.4). Higher yields and profitability also led farmers to increase the area of rice and wheat they grew at the expense of other crops. And, with faster growing varieties and irrigation, they grew more crops on their land each year. This led to even faster growth in cereal production (see Table 3.4). All these gains were achieved with negligible growth (0.42 percent per year) in the total area planted to cereals.

The Impact on Food Production

Asia-wide, total cereal production grew by 3.57 percent per year over 1967–82, with average annual growth rates of 5.43 percent, 3.25 percent, and 4.62 percent for wheat, rice, and maize, respectively (see Table 3.4). The growth rates were considerably higher in the bread basket areas (for example, Punjab and Haryana in India and Central Luzon in the Philippines) where the Green Revolution was launched. Cereal production in Asia virtually doubled between 1970 and 1995, from 313 to 650 million tons per year. Although the population increased by 60 percent, the increase in food production was sufficient that, instead of experiencing widespread famine, Asia saw an increase in calorie availability per person of nearly 30 percent, and wheat and rice became less expensive (Table 3.5 and ADB 2000).

Table 3.4 Annual growth rates in cereal production
in Asia, 1967–82 (percent)

Crop	Area	Yield	Production
Wheat	1.30	4.07	5.43
Maize	1.09	3.48	4.62
Rice	0.70	2.54	3.25
Other grains	−1.76	1.63	−0.15
All cereals	0.42	3.13	3.57

Source: Rosegrant and Hazell (2000).

Table 3.5 Indicators of change in Asia, 1970–95

Indicator, year	India	Other South Asia	People's Republic of China	Southeast Asia	Developing Asia
Total population affected (millions)					
1970	554.9	156.2	834.6	204.4	1,750.2
1995	929.0	293.9	1,226.3	343.7	2,792.9
Percent change	67.4	88.2	46.9	68.2	59.6
Agricultural population affected (millions)					
1970	370.4	140.2	650.9	191.7	1,353.2
1990	500.7	190.7	830.2	244.1	1,765.7
Percent change	35.2	36.0	27.5	27.3	30.5
Cereal yield (metric tons per hectare)					
1970	0.925	1.197	1.769	1.352	1.317
1995	1.743	1.846	4.007	2.237	2.627
Percent change	88.4	54.2	126.5	65.6	99.5
Cereal production (millions of metric tons)					
1970	92.8	25.4	161.1	33.8	313.2
1995	174.6	48.1	353.3	73.6	649.6
Percent change	88.1	89.3	119.3	117.8	107.4
Per capita income (US$/year)					
1970	241.0	187.0	91.0	351.0	177.0
1995	439.0	299.0	473.0	1027.0	512.0
Percent change	82.2	60.0	419.8	192.6	189.3
Calorie consumption (kcal/person/day)					
1970	2,083	2,184	2,019	1,945	2,045
1995	2,388	2,274	2,697	2,596	2,537
Percent change	14.6	4.1	33.5	33.5	24.1
Poverty (millions)					
1975	472.2	568.9	108.1	1,149.2	
1990s	514.7	269.3	40.2	824.2	
Percent change	9.0	−52.7	−62.8	−28.3	
Poverty (percent of population)					
1975	59.1	59.5	52.9	58.7	
1990s	43.1	22.2	11.5	29.9	

Sources: Asian Development Bank (2000), Tables 1 and 2; Rosegrant and Hazell (2000), Table V.8; FAO (2009).

The Impact on Production Fluctuations

The high-yielding cereal varieties were developed to give higher yields in favorable environments, such as irrigated areas with high fertilizer usage. This led to some initial concern that they would be more vulnerable to pest and weather stresses than traditional varieties, increasing the risk of major yield and food production shortfalls in unfavorable years. Early work by Mehra (1981) among others suggested that yield variability for cereals in India was increasing relative to increases in average yield (higher coefficients of variation) at the national level, raising the specter of a growing risk of national food shortages and high prices in some years. Subsequent analysis showed that, at the plot level, many modern varieties were no more risky than traditional varieties in terms of downside risk[1] and that also, although some crop yields measured at regional and national levels were becoming more variable (a greater problem for maize and other rainfed cereals than for wheat or rice[2]), this was largely the result of more correlated or synchronized patterns of spatial yield variation across space (Hazell 1982, 1989). Several scholars suggested that these changes might be attributable to the widespread adoption of input-intensive production methods that led to larger and more synchronized yield responses to changes in market signals and weather events, shorter planting periods with mechanization, and the planting of large areas to the same or genetically similar crop varieties (Hazell 1982; Ray 1983; Rao, Ray, and Subbarao 1988). Later studies showed that rice and wheat yields generally became more stable in Asia in the 1990s, but the patterns for maize and coarse grains were more mixed, especially at country and subregional levels (Singh and Byerlee 1990; Deb and Bantilan 2003; Larson et al. 2004; Gollin 2006; Sharma, Singh, and Kumari 2006; Chand and Raju 2008).

National yield and production variability became less of a policy issue after market liberalization policies were implemented in the late 1980s and early 1990s, enabling international trade to play a greater role in stabilizing market supplies and prices. But because large areas of major cereals are still planted to relatively few modern varieties, concern remains about the risk that genetic uniformity might make crops vulnerable to catastrophic yield losses from changes in pests, diseases, and the climate. The absence of any catastrophic crop failures despite the Green Revolution is due in large part to extensive behind-the-scenes scientific work to prevent such disasters. Crop genetic uniformity has been counteracted by spending more on conserving genetic resources and making them accessible for breeding purposes, using breeding approaches that broaden the genetic base of varieties supplied to farmers, and changing varieties more frequently over time in order to stay ahead of evolving pest, disease, and climate risks (Smale et al. 2009). The international agricultural research system also spends significant shares of its budget on "maintenance" research in order to provide national systems with new germplasm on a timely basis in response to emerging new pest, disease, and climate risks.

Indirect Income and Employment Impacts

Productivity growth in agriculture can have far-reaching impacts on the productivity and growth of regional and national economies. There are several growth linkages that drive this relationship: lower food prices for workers, more abundant raw materials for agro-industry and export, release of labor and capital (in the form of rural savings and taxes) to the nonfarm sector, and increased rural demands for nonfood consumer goods and services, which, in turn, support growth in the service and manufacturing sectors.

The powerful economywide benefits emanating from the Green Revolution were amply demonstrated during the Green Revolution era in Asia (Mellor 1976). In India, the lack of change in the share of nonagricultural employment in total national employment for over a century, until the full force of the Green Revolution was under way in the 1970s, provided strong circumstantial evidence of the importance of agricultural growth as a motor for the Indian economy. This was also confirmed by Rangarajan (1982), who estimated that an increase of 1.0 percentage point in the agricultural growth rate stimulated an increase of 0.5 percent in the growth rate of industrial output and an increase of 0.7 percent in the growth rate of national income.

Regional growth linkage studies have also shown strong multiplier impacts from agricultural growth to the rural nonfarm economy (Bell, Hazell, and Slade 1982; Hazell and Haggblade 1991; Hazell and Ramasamy 1991). The size of the multipliers varies depending on the method of analysis chosen, and for Asia they vary between $0.30 and $0.85; that is, each dollar increase in agricultural income leads to an additional increase in rural nonfarm earnings of between $0.30 and $0.85 (Haggblade, Hazell, and Dorosh 2007). The multipliers tend to be larger in Green Revolution regions because of better infrastructure and market town development, greater use of purchased farm inputs, and higher per capita incomes and hence consumer spending power (Hazell and Haggblade 1991).

The Impact on Poverty

Although the primary goal of the investments underlying the Green Revolution was to increase food production, they also helped slash poverty. Reliable poverty data are not available for the early Green Revolution period, but in 1975 nearly three out of every five Asians still lived on $1 a day. This number declined to fewer than one in three by 1995 (Rosegrant and Hazell 2000). The absolute number of poor people declined by 28 percent, from 1,150 million in 1975 to 825 million in 1995. These reductions in poverty would have been even more impressive if the total population had not grown by 60 percent over the same period. The vast majority of the poor who were lifted out of poverty were rural and obtained important shares of their livelihood from agriculture and allied activities.

Given the complex causes underlying poverty and the diversity of livelihoods found among poor people, the relationship between the Green Revolution and poverty alleviation is necessarily complex, and this has led to considerable and contentious debate in the literature. Before turning to that literature, it is useful to first consider the conceptual basis for these complexities.

There are a number of pathways through which the Green Revolution might have benefited the poor (Hazell and Haddad 2001). Within adopting regions, the Green Revolution could have helped poor farmers directly by increasing their production, providing more food and nutrients for their own consumption, and increasing the surplus of products available for them to sell for cash income. Small farmers and landless laborers could have gained additional agricultural employment opportunities and higher wages within adopting regions.

The Green Revolution could also have benefited the poor in less direct ways. Growth in adopting regions could have created employment opportunities for migrant workers from other less dynamic regions. It could also have stimulated growth in the rural and urban nonfarm economy with benefits for a wide range of rural and urban poor people. Greater food production could also have led to lower food prices for all types of poor people. It might also have improved poor people's access to foods that are high in nutrients and crucial to their well-being—particularly for poor women.

But the Green Revolution could also have worked against the poor. Although its technology was in principle scale neutral it may nevertheless have favored large farms because of their better access to irrigation water, fertilizers, seeds, and credit. The accompanying use of machines and herbicides might have displaced labor, leading to lower wage earnings for agricultural workers. By favoring some regions or farmers over others, the Green Revolution might have harmed non-adopting farmers by lowering their product prices even though only the adopting farmers benefited from cost-reducing technologies.

Given that many of the rural poor are simultaneously farmers, paid agricultural workers, and net buyers of food and have nonfarm sources of income, the net impact of the Green Revolution on their poverty status can be complex, with households experiencing gains in some dimensions and losses in others. For example, the same household might have gained from reduced food prices and from higher nonfarm wage earnings but lost from lower farmgate prices and agricultural wages. Measuring the net benefits to the poor requires a full household income analysis of direct and indirect impacts, as well as consideration of the impacts on poor households that are not engaged in agriculture and/or live outside adopting regions. Much of the controversy seen in the literature about how the impact of the Green Revolution on the poor has arisen because too many studies have taken only a partial view of the problem.

Much of the literature available on the Green Revolution's impact focuses on the direct poverty impacts within adopting regions, while a smaller body of literature assesses the broader and indirect poverty impact arising through food price changes and intersectoral linkages.

Impacts within Adopting Regions.   A number of village and household studies conducted soon after the release of Green Revolution technologies raised concern that large farms were the main beneficiaries of the technology and poor farmers were either unaffected or made worse off. Later evidence showed mixed outcomes. Small farmers did lag behind large farmers in adopting Green Revolution technologies, but many of them eventually did adopt them (Pinstrup-Andersen and Hazell 1987). Many of these adopters of small farms benefited from increased production, greater employment opportunities, and higher wages in the agricultural and nonfarm sectors (Lipton with Longhurst 1989). In some cases, small farmers and landless laborers actually ended up gaining proportionally more income than did larger farmers, resulting in a net improvement in the distribution of village income (Hazell and Ramasamy 1991; Thapa, Otsuka, and Barker 1992; Maheshwari 1998).

Freebairn (1995) performed a meta-analysis of 307 published studies on the Green Revolution. The primary concern of nearly all the studies that he reviewed was changes in inequality and income distribution rather than absolute poverty; the latter emerged as a more important issue in the 1990s. Freebairn found that 40 percent of the studies he reviewed reported that income became more concentrated within adopting regions, 12 percent reported that it remained unchanged or improved, and 48 percent offered no conclusion. He found more favorable outcomes of the revolution in the literature on Asia than elsewhere and found that the authors of the Asian studies gave more favorable conclusions than did those of the non-Asian studies. Freebairn also found that later studies reported equity outcomes similar to those reported in earlier studies, thereby casting some doubt on the proposition that because small farmers adopted later than large farmers, equity improved over time. However, it should be noted that his analysis did not include repeat studies undertaken at the same sites over a longer period of time, as did that of Hazell and Ramasamy (1991), Hayami and Kikuchi (2000), and Jewitt and Baker (2007), all of whom found favorable longer-term impacts on inequality. Freebairn also found that microbased case studies reported the most favorable outcomes, while macro-based essays reported the worst outcomes.

Walker (2000) argues that reducing inequality is not the same thing as reducing poverty and may be much more difficult to achieve through technologically driven agricultural growth. More recent studies focusing directly on poverty confirm that improved technologies do have a favorable impact on many small farmers, but the gains for the smallest farms and landless agricultural workers can be too small to raise them above poverty thresholds (Hossain et al. 2007; Mendola 2007).

However, the poor can benefit in other ways, too. Hossain et al. (2007) found that in Bangladesh, the spread of high-yielding varieties of rice helped reduce the vulnerability of the poor by stabilizing employment earnings, reducing food prices and their seasonal fluctuations, and enhancing their ability to cope with natural disasters. Use of participatory research methods in the selection of improved rice varieties in Uttar Pradesh, India, has been shown to empower women as decisionmakers in their farming and family roles as well as to lead to the greater adoption of improved varieties (Paris et al. 2008).

Indirect Impacts on Poverty. A number of econometric studies have used cross-country and/or time-series data to estimate the relationship between agricultural productivity growth and poverty and they cover the Green Revolution era. These studies have generally found that agricultural productivity growth has high poverty reduction elasticities. Thirtle, Lin, and Piesse (2003) estimate that each 1.00 percent increase in crop productivity reduces the number of poor people by 0.48 percent in Asia. For India, Ravallion and Datt (1996) estimate that a 1.00 percent increase in agricultural value added per hectare leads to a 0.40 percent reduction in poverty in the short run and a 1.90 percent increase in the long run, the latter arising through the indirect effects of lower food prices and higher wages. Fan, Hazell and Thorat (1999) estimate that each 1.00 percent increase in agricultural production in India reduces the number of rural poor by 0.24 percent. For Asia, these poverty elasticities are still higher for agriculture than for other sectors of the economy (Hasan and Quibria 2004; World Bank 2007).

There is some evidence that the poverty elasticity of agricultural growth may be diminishing because the rural poor are becoming less dependent on agriculture. In Pakistan, for example, agricultural growth was associated with rapid reductions in rural poverty in the 1970s and 1980s, but the incidence of rural poverty hardly changed in the 1990s despite continuing agricultural growth (Dorosh, Khan, and Nazli 2003). This was partly because a growing share of the rural poor households (46 percent by 2002) had become disengaged from agriculture; even small farm households and landless agricultural worker households received about half their income from nonfarm sources (Dorosh, Khan, and Nazli 2003).

Lower food prices and growth linkages to the nonfarm economy play an important role in most of the results just cited, and these benefit the urban as well as the rural poor. These indirect impacts have sometimes proved more powerful and positive than the direct impacts of R&D on the poor within adopting regions (Hazell and Haddad 2001). A question arises as to whether the power of these indirect benefits has diminished over time with market liberalization and greater diversification of Asian economies.

Interregional Disparities. The Green Revolution began in regions with assured irrigation, and although it subsequently spread to areas that depended more on rainfed crops, it did not benefit many of the poorest regions (Prahladachar 1983). The widening regional income gaps that resulted have been buffered to some extent by interregional migration. In India the Green Revolution led to the seasonal migration of more than 1 million agricultural workers each year from the eastern states to Punjab and Haryana (Oberai and Singh 1980; Westley 1986). These numbers were tempered in later years as the Green Revolution technology eventually spilled over into eastern India in conjunction with the spread of tubewells. In a study of the impact of the Green Revolution in a sample of Asian villages, David and Otsuka (1994) asked whether regional labor markets were able to spread the benefits between adopting and non-adopting villages and found that seasonal migration did go some way toward fulfilling that role. But although migration can buffer widening income differentials between regions, it is rarely sufficient to avoid them. In India, for example, regional inequalities widened during the Green Revolution era (Galwani, Kanbur, and Zhang 2007), and the incidence of poverty remains high in many less-favored areas (Fan and Hazell 2000).

The Impact on Nutrition

The Green Revolution was successful in increasing the per capita supply of food and reducing the prices of food staples in Asia. Making food staples more available and less costly has proved an important way to benefit poor people (Fan, Hazell, and Thorat 1999; Rosegrant and Hazell 2000; Fan 2007). Several micro-level studies from the Green Revolution era in Asia found that higher yields typically led to greater calorie and protein intake among rural households in adopting regions. For example, Pinstrup-Andersen and Jaramillo (1991) found that the spread of high-yielding varieties of rice in North Arcot District, South India, led to substantial increases over a 10-year period in the energy and protein consumption of farmers and landless workers. Their analysis showed that, after controlling for changes in nonfarm sources of income and food prices, about one-third of the calorie increase could be attributed to increased rice production. Ryan and Asokan (1977) also found complementary net increases in protein and calorie availability as a result of Green Revolution wheat in the six major producing states of India, despite some reduction in the area of pulses grown.

More aggregate analysis of the impacts of rising incomes on diets and nutrient intake has proved more complex, particularly as concern has shifted from calorie and protein deficiencies to micro nutrients and broader nutritional well-being. Food price declines are, in general, good for households that purchase more food than they sell, for they amount to an increase in their real income. Real income increases can be used to increase the consumption of important staples and to purchase more diverse and nutritionally rich diets. However, a study of Bangladesh showed that a downward trend in the price of rice from 1973–75 to 1994–96 was accompanied by upward trends in the real prices of other foods that are richer in micronutrients, making these less accessible to the poor (Bouis 2000). Similar patterns were observed in India during the 1970s and 1980s when farmers diverted land away from pulses to wheat and rice, leading to sharp increases in the price of pulses and a drop in their per capita consumption (Kennedy and Bouis 1993; Kataki 2002).

Since then there have been substantial changes in food intake patterns in rural Asia. In India, for example, the share of cereals in total food expenditure has declined while that of milk, meat, vegetables, and fruits has increased. Per capita consumption of cereals has also fallen in absolute terms (Nasurudeen et al. 2006) and this is true for all income groups. However, because deficiencies in iron and the B vitamins are common among the poor, the increases in micronutrient-rich foods must not always have been high enough to offset the decline from cereals. Deficiencies are found in other micronutrients (for example, vitamins C and D), but these are not related to reductions in cereal consumption.

As the Green Revolution unfolded, strategies were implemented to enhance the nutritional quality of the diets of the poor. These included

1. improvements in the productivity of fruits, vegetables, livestock, and fish, in home gardens, on farms, and in ponds, for on-farm consumption and more generally to increase the marketed supplies of these nutrient-rich foods;

2. promotion of biodiversity in foodcrops, especially traditional crops and cultivars that are rich in nutrients; and

3. biofortification of major food staples.

Although improved technologies have helped enhance the nutritional value of diets, studies show that the most effective results are obtained when technology interventions are complemented by investments in nutrition education and health services and are targeted in ways that give women additional spending power (Ali and Hau 2001; Berti, Krasevec, and FitzGerald 2004).

Returns to Public Investments

Given the key role that the public sector played in launching and sustaining the Green Revolution, it is important to ask if the returns to its investments were justified.

Fan, Gulati, and Thorat (2008) have estimated the returns to different types of public investments in agriculture in India over a four-decade period, beginning in the 1960s (Table 3.6). The marginal returns to these investments in terms of growth and poverty alleviation were very favorable in the early stages of the Green Revolution, and many additional investments, especially those in rural roads and agricultural R&D, continued to yield high returns through the 1990s (see Table 3.6).

Table 3.6 Returns to agricultural growth and poverty reduction from investments in public goods and subsidies in different phases of India's Green Revolution, 1960s–90s

Returns	1960s	1970s	1980s	1990s
Returns in agricultural GDP (rupees in return / rupees spent)				
Road investment	8.79	3.8	3.03	3.17
Educational investment	5.97	7.8	3.88	1.53
Irrigation investment	2.65	2.1	3.61	1.41
Irrigation subsidies	2.24	1.22	2.38	n.s.
Fertilizer subsidies	2.41	3.03	0.88	0.53
Power subsidies	1.18	0.95	1.66	0.58
Credit subsidies	3.86	1.68	5.2	0.89
Agricultural R&D	3.12	5.9	6.95	6.93
Decrease in the number of poor people per million rupees spent				
Road investment	1,272	1,346	295	335
Educational investment	411	469	447	109
Irrigation investment	182	125	197	67
Irrigation subsidies	149	68	113	n.s.
Fertilizer subsidies	166	181	48	24
Power subsidies	79	52	83	27
Credit subsidies	257	93	259	42
Agricultural R&D	207	326	345	323

Source: Fan, Gulati, and Thorat (2007).
Note: n.s. means not significant.

Table 3.7 Rates of return to agricultural research, 2000

Region	Number of estimates	Median rate of return
Africa	188	34
Asia	222	50
Latin America	262	43
Middle East / North Africa	36	11
All developing countries	683	43
All developed countries	990	46

Source: Alston et al. (2000).

Although the returns to most input subsidies were initially high and have declined sharply over time, suggesting the need for an exit strategy, this does not negate the important role they played in the early years of the Green Revolution in helping to promote small-farm-led growth.

Other studies provide a more focused analysis of the returns to public investments in agricultural research during the Green Revolution era. Evenson, Pray, and Rosegrant (1999) reviewed several impact studies from South Asia and found that the returns to national agricultural R&D investments exceeded 60 percent in all cases. At commodity levels, Alston et al. (2000) reviewed 222 impact studies from Asia and found a median rate of return of 50 percent, higher than in other developing country regions (Table 3.7).

Sustainability

Declining Growth in Yields and Total Factor Productivity

The Green Revolution was built on rapid growth in cereal yields and an associated increase in total factor productivity (TFP) that enabled food prices to decline. A weak form of sustainability requires that these higher levels be sustained into the future. But given growing populations and demands for cereals for food and feed, a stronger form of sustainability means that those yields and TFP need to continue to grow.

In reality, cereal yields have continued to rise on average across Asia since the Green Revolution era, but annual growth rates are slowing (Rosegrant and Hazell 2000; Hazell 2008). This is confirmed by more careful, microbased studies of wheat and rice yields in the Indo-Gangetic Plain (Cassman and Pingali 1993; Murgai, Ali, and Byerlee 2001; Bhandari et al. 2003; Ladha et al. 2003), in India's major irrigated-rice growing states (Janaiah, Otsuka, and Hossain 2005), and in East Asia's rice bowls (Pingali, Hossain, and Gerpacio 1997).

There are several possible reasons for this slowdown: displacement of cereals on better lands by more profitable crops such as groundnuts (Maheshwari 1998); diminishing returns to modern varieties when irrigation and fertilizer use are already at high levels; and low cereal prices relative to input costs until recently, which have made additional intensification less profitable. But there are concerns that the slowdown also reflects a deteriorating crop-growing environment in intensive monocrop systems. Murgai, Ali, and Byerlee (2001) and Ali and Byerlee (2002), for example, report deteriorating soil and water quality in the rice–wheat system of the Indo-Gangetic Plain, and Pingali, Hossain, and Gerpacio (1997) report degradation of soils and buildup of toxins in intensive paddy systems in a number of Asian countries.

These problems are reflected in the growing evidence of stagnating or even declining levels of TFP in some of these farming systems (Janaiah, Otsuka, and Hossain 2005). Ali and Byerlee (2002) have shown that degradation of soil and water are directly implicated in the slowing of TFP growth in the wheat–rice system of the Pakistan Punjab. Ladha et al. (2003) examined data from long-term yield trials at multiple sites across South Asia and found stagnating or declining yield trends when input use is held constant. One consequence has been that farmers have had to use increasing amounts of fertilizers to maintain the same yields over time (Pingali, Hossain, and Gerpacio 1997). There is also concern that pest and disease resistance to modern pesticides now slows yield growth and that breeders have largely exploited the yield potentials of major Green Revolution crops—though sizable gaps still remain between experiment-plot and average farmer yields.

Environmental Problems

The concerns about the environmental stresses that may underlie the decline in growth rates of yields and TFP also link to broader worries about the environmental sustainability of the Green Revolution due to issues including excessive and inappropriate use of fertilizers and pesticides that pollute waterways and kill beneficial insects and other wildlife, irrigation practices that lead to salt buildup and eventual abandonment of some of the best farming lands, increasing water scarcities in major river basins, and retreating groundwater levels in areas where more water is being pumped for irrigation than can be replenished (Hazell and Wood 2008). Some of these outcomes were inevitable as millions of largely illiterate farmers began to use modern inputs for the first time, but the problem was exacerbated by inadequate extension and training, an absence of effective regulation of water use and quality, and input pricing and subsidy policies that made modern inputs too inexpensive and encouraged excessive use.

Just how serious are the environmental problems associated with the Green Revolution and are they likely to undermine future food production and Asia's ability to feed itself? Measuring environmental impacts is difficult, and as a result good empirical evidence is fragmentary, often subjective, and sometimes in direct contradiction with the overall trends in agricultural productivity. The best evidence relates to the degradation of irrigated land, increasing water scarcities, and the consequences of poor pest management practices.

Degradation of Irrigated Land.   There is growing evidence that poor irrigation practices have led to significant waterlogging and salinization of irrigated land. The Comprehensive Assessment of Water Management in Agriculture (2007) estimates that nearly 40 percent of irrigated land in dry areas of Asia is affected by salinization. For Pakistan, Ghassemi, Jakeman, and Nix (1995) estimate that 4.2 million hectares of irrigated lands (26 percent total) are affected by salinization, while Chakravorty (1998) claims that one-third of the irrigated area is subject to waterlogging and 14 percent is saline. For India, Dogra (1986) estimates that nearly 4.5 million hectares of irrigated land are affected by salinization and a further 6 million hectares by waterlogging, while Umali (1993) claims that 7 million hectares of arable land has been abandoned because of excessive salts.

Water Scarcity.   Even more worrying for irrigated agriculture is the threat from the growing scarcity of fresh water in much of Asia. Many countries are approaching the point at which they can no longer afford to allocate two-thirds or more of their fresh water supplies to agriculture (Comprehensive Assessment of Water Management in Agriculture 2007). Most of the major river systems in Asia are already fully exploited at least part of the year, and the massive expansion of tubewell irrigation in South Asia has led to serious overdrawing of groundwater and to falling water tables. On the Indian subcontinent, groundwater withdrawals have surged from less than 20 cubic kilometers to more than 250 cubic kilometers per year since the 1950s (Shah et al. 2003). More than a fifth of groundwater aquifers are overexploited in Punjab, Haryana, Rajasthan, and Tamil Nadu, and groundwater levels are falling (Postel 1993; World Bank 2007). Even as current water supplies are stretched, the demands for industry, urban household use, and environmental purposes are growing (Rosegrant and Hazell 2000; Comprehensive Assessment of Water Management in Agriculture 2007). It would seem that either many Asian farmers must learn to use irrigation water more sparingly and more sustainably or the irrigated area will have to be reduced.

Pest Management. Pest problems emerged as an important problem early in the Green Revolution era because many of the first high-yielding varieties released had poor resistance to some important pests. The problem was compounded by a shift to higher cropping intensities, monocropping, high fertilizer use (which creates dense, lush canopies in which pests can thrive), and the planting of large adjacent areas to similar varieties with a common susceptibility. Control was initially based on prophylactic chemical applications, driven by the calendar rather than the incidence of pest attacks. This approach disrupted the natural pest–predator balance, and led to a resurgence of pest populations that required even more pesticide applications to control. Problems were compounded by the buildup of pest resistance to the commonly used pesticides. As pesticide use increased so did environmental and health problems. Rola and Pingali (1993) found that the health costs of pesticide use in rice reached the point at which they more than offset the economic benefits from pest control.

Efforts to Achieve Environmental Sustainability

A growing awareness of these environmental problems has led a few Green Revolution critics to argue for a drastic reversion to the traditional technologies that dominated Asia before the Green Revolution (Shiva 1991; Nellithanam, Nellithanam, and Samati 1998). Such authors claim that yield growth rates were already high before the Green Revolution but ignore the fact that this growth was largely the result of the spread of irrigation and fertilizers prior to the introduction of high-yielding varieties (Evenson, Pray, and Rosegrant 1999).

More generally, environmental concerns have led to a significant research response and a wider array of more sustainable technologies and farming practices. Some of these have been spearheaded by environmentally oriented nongovernmental organizations (NGOs) that have contested the Green Revolution approach and undertaken research and extension activities of their own. Others have been developed by national and international R&D systems for improving water, pest, and soil fertility management within intensive Green Revolution systems.

One of the outcomes of greater NGO involvement has been a lively debate about competing farming paradigms, and "alternative" farming has been offered as a more sustainable and environmentally friendly alternative to the modern input-based approach associated with the Green Revolution.[3] The alternative farming approach includes extremes that eschew the use of any modern inputs as a matter of principle (for example, organic farming), but also includes more eclectic whole-farming systems approaches such as low external input farming (Tripp 2006) and eco-agriculture (McNeely and Scherr 2003). In practice, most alternative farming approaches cannot match the high productivity levels achieved by modern farming methods in Green Revolution areas (Pretty et al. 2007). The yield gap is even greater if the extra land needed to produce plant nutrients and organic matter without the use of fertilizer is factored in (Hazell 2008).

Other efforts to improve sustainability involve improving farming practices to use inputs more efficiently rather than replacing them. For example, fertilizer efficiency can be improved through more precise matching of nutrients with plant needs during the growing season and by switching to improved fertilizers such as controlled-release fertilizers and deep placement technologies. Examples of these approaches are site-specific nutrient management, developed by the International Rice Research Institute and its partners, and urea deep placement, pioneered by the International Fertilizer Development Center. The Rice–Wheat Consortium, a partnership of CGIAR centers and the National Agricultural Research System from Bangladesh, India, Nepal, and Pakistan, has promoted the adoption of zero-tillage farming, which involves the direct planting of wheat after rice without any land preparation. This technology saves labor, fertilizer, and energy; minimizes planting delays between crops; conserves soil; reduces irrigation water needs; increases tolerance to drought; and reduces greenhouse gas emissions (Erenstein et al. 2007; World Bank 2007). Technical research has also shown the potential to increase yields in irrigated farming with substantial savings in water use (Mondal et al. 1993; Guerra et al. 1998). Finally, integrated pest management (IPM) combines pest-resistant varieties, natural control mechanisms, and the judicious use of some pesticides (Waibel 1999). Studies show that IPM farmers save significantly on costs (for labor and pesticides), with no impact on productivity (Rasul and Thapa 2003; Susmita, Meisner, and Wheeler 2007).

Despite the development of more sustainable technologies and farming practices for Asia's Green Revolution areas, their uptake and spread remains inadequate. There are several possible reasons, including the high levels of knowledge required for their practice; perverse incentives caused by input subsidies, labor constraints, and insecure property rights; difficulties of organizing collective action; and externality problems. These constraints require more calibrated policy responses, and developing these remains a major challenge for the future management of the Green Revolution areas.

Lessons Learned and Conclusions

Several general lessons can be drawn from this study of the Asian Green Revolution. First, technological barriers to expanded food production among small and large farmers in developing countries can be alleviated and at a cost that is far less than the resulting gains in growth and poverty reduction. Second, Green Revolutions do not just happen; they require considerable and sustained nurturing by the state. Third, Green Revolutions are not necessarily pro-poor or environmentally benign, and achieving favorable outcomes requires appropriate and supporting government policies. We discuss each in turn and then consider the implications for realizing a comparable Green Revolution in Africa.

How to Overcome Technological Barriers to Food Production

Asia was able to break out of its food production constraint by bringing the force of the 20th-century scientific revolution in agriculture to its farmers. Governments and their international partners invested heavily in agricultural R&D, extension, irrigation, and fertilizer supplies, and farmers made major changes in their traditional and well-honed farming systems. The switch from low-input, low-output farming to high-input, high-output farming was not without its problems, but it sufficed to provide the needed productivity breakthroughs that had otherwise failed to materialize.

The initial Green Revolution technology package worked best for wheat and rice in the best-irrigated areas, but within 10 to 15 years the technologies had evolved to accommodate the challenges of many poorer regions growing a wider range of foodcrops under rainfed or less-assured irrigation conditions. Continuing advances in the agricultural sciences have increased the range of areas that can benefit from Green Revolution technologies, and often the major bottleneck to their uptake is public policies and investments rather than lack of suitable technologies.

How to Make Green Revolutions Happen

Market forces alone are insufficient for launching Green Revolutions in poor developing countries, where market chains for food staples are typically characterized by numerous failures and coordination problems (Dorward, Kydd, and Poulton 1998). More than a single technology fix occurred in Asia; a set of policy initiatives and preconditions came together to create an enabling and sustained economic environment that ensured that farms of all sizes could participate in a fully functional market chain for food staples. These initiatives and preconditions included access to a game-changing technology package, threshold levels of infrastructure and market and institutional development, and an enabling policy environment.

The Preconditions Needed for a Green Revolution. If farmers are to adopt Green Revolution technologies, they need access to a package of affordable inputs (fertilizer, improved seed, pesticides, and irrigation water), seasonal credit to buy them each season, extension to provide the knowledge they need to use them, and assured access to markets at profitable and stable prices. A Green Revolution takes off only if all these things come together in an integrated way and under conditions that enable significant productivity gains. Achieving these preconditions requires critical accumulated levels of investment in agricultural R&D, extension, roads, irrigation, power, and other infrastructure, as well as effective public and private institutions that serve agriculture. In Asia, these things were built up over several decades in an integrated way—guided by national agricultural development plans—and were already advanced before the Green Revolution.

An Enabling Economic Environment.   Although many Asian countries initially discriminated against agriculture in their macroeconomic, taxation, and industrial-sector policies (Krueger, Schiff, and Valdés 1991),[4] they offset many of these biases by subsidizing key inputs such as fertilizers, water, and power; shoring up farm credit systems; and intervening in markets to ensure that farmers received fair and stable prices. The net result was to ensure that Green Revolution technologies were profitable for farmers. Moreover, although the Asian Green Revolution was initiated and led by governments, the private sector was given an important mediating role, which helped reduce marketing inefficiencies and corruption.

Sustained Investment and Support.   Asian countries not only invested heavily to launch the Green Revolution but continued to invest in agriculture to sustain the gains that were achieved. On average, Asian countries were spending more than 15 percent of their total government spending on agriculture by 1972, and they doubled the real value of their agricultural expenditures by 1985 (see Table 3.3). The need to sustain investment levels is especially great with regard to agricultural R&D, because there are long lead times in developing new products and farmers continually need new crop varieties and natural resource management practices to stay ahead of evolving pest and environmental problems.

The investments made by Asian governments were driven by concerns about food insecurity and poverty, and these investments made impressive contributions toward overcoming those problems. But they also paid off handsomely in terms of their economic rates of return. Even input subsidies generated favorable benefit–cost ratios in the early years as they helped kick-start fledgling markets. A key lesson is that a wide range of investments are needed to launch a Green Revolution, but the mix of expenditures needs to be adjusted as a Green Revolution matures. For input subsidies, this means it is necessary to have an adequate exit strategy from an early stage.

How to Make Green Revolutions Pro-Poor

The small-farm sector needs to lead Green Revolutions to be pro-poor, but this does not automatically happen without supportive government policies. In Asia the conditions under which the Green Revolution proved pro-poor included (1) a scale-neutral technology package that could be profitably adopted on farms of all sizes, (2) an equitable distribution of land with secure ownership or tenancy rights, (3) modern input and credit systems that served small farms at prices they could afford, (4) public extension systems that prioritized small farms, and (5) product markets and price support policies that ensured that small farms received stable and profitable prices. Meeting these conditions typically required proactive efforts by governments in the form of land reforms, small-farm development programs, and input and credit subsidies. Some Asian countries—particularly those that began with inequitable land distribution—were unable to meet these conditions.

How to Make Green Revolutions Environmentally Sustainable

The Green Revolution made important environmental contributions by saving large areas of forest and woodland from conversion to agriculture. But it also generated some environmental problems of its own, imposing high off-site externality costs on populations at large and undermining the long-term sustainability of some intensive farming systems. Inappropriate management of modern inputs by farmers was the primary cause, and the problem was exacerbated by inadequate extension and training, ineffective regulation of water quality, and input pricing and subsidy policies that made modern inputs too inexpensive and encouraged excessive use.

Policy and institutional reforms that correct inappropriate incentives can make an important difference. Improved technologies, such as precision farming, integrated pest management, and improved water management practices, can even increase yields while reducing chemical use, implying that intensification does not have to be inconsistent with good management of the environment. More calibrated policy responses are needed to accelerate farmers' adoption of these kinds of improved practices, and developing these policies remains a major challenge for the future management of the Green Revolution areas.

Implications for Achieving an African Green Revolution Today

The Green Revolution was not confined to Asia; it successfully spread to large parts of Latin America, the Middle East, and North Africa. But despite several attempts to bring the Green Revolution to Africa, it has not yet happened at the scale that is needed. By missing out on the Green Revolution, Africa has seen its average cereal yields stagnate since 1960 while those in Asia and other developing regions have nearly tripled. This has been a major factor underlying the trend decline in per capita food availability in Africa, as well as worsening poverty and malnutrition (World Bank 2007).

An important reason for the failure of the Green Revolution in Africa is the nature of Africa's farming systems. Irrigation as well as rice and wheat play much smaller roles in Africa than in Asia; hence Africa simply could not benefit much from the first round of Green Revolution technologies and had to wait for improvements in crops such as maize, sorghum, millet, and cassava grown under rainfed conditions. A more fundamental problem is that Africa has invested relatively little in developing its rural infrastructure, leading to unusually high transport and marketing costs for African farmers.[5] Import prices for fertilizer are also high because many African countries are landlocked and buy too little fertilizer to secure significant price discounts (Morris et al. 2007). The net result is that it is simply not profitable for most African farmers to shift to high-input, high-output farming systems. A related problem has been the lack of government and donor commitment to agricultural development in Africa. African governments have also been far less effective than those in other regions in creating a supportive policy environment for their farmers.

How can Africa realize its Green Revolution potential? Simply copying the Asian Green Revolution model is not enough. Although a successful Green Revolution must be built around rapid increases in the use of improved seeds, fertilizer, and water, Africa's relatively low irrigation potential, diverse rainfed farming systems, and degraded soils mean that these must be developed and promoted in flexible ways that enable farmers to combine them and adapt them to local conditions (InterAcademy Council 2004). As in Asia, the state will need to play a leading role in ensuring that farmers have affordable and sustained access to critical packages of key inputs, credit, and marketing services (Dorward et al. 2004). But rather than following the kind of top-down approach that worked in Green Revolution Asia, African countries will need to take a more flexible and opportunistic approach that builds strategic partnerships between key actors at local, meso, and macro levels.

Two promising Africa-led initiatives that may yet catalyze the development of the continent's Green Revolution potential are the Alliance for a Green Revolution in Africa and the Comprehensive African Agricultural Development Program of the New Economic Partnership for Africa. But if they are to succeed in mobilizing significant political support and funding for a Green Revolution, they will need to demonstrate early and significant successes.

Notes

1. See relevant case study material in Anderson and Hazell (1989).

2. In contrast to the situation in India, Tisdell (1988) found that the relative yield and production variability of foodgrain fell at district and national levels in Bangladesh over a similar time period.

3. Alternative farming is sometimes also called "sustainable" or "ecological" farming.

4. The worst of these biases were removed in the late 1980s and early 1990s as part of structural adjustment programs, but they prevailed during the early years of the Green Revolution.

5. Africa still has less infrastructure than Asia had at the beginning of its Green Revolution. The road density in mainland Africa from 2000 to 2005 ranged from 3 to 4 square kilometers per 1,000

in Ethiopia and Mali to 50 to 70 square kilometers per 1,000 in Namibia and Botswana (World Bank 2008). In contrast, India's average road density at the start of the Green Revolution was 388 square kilometers per 1,000 (Spencer 1994).

References

Ali, M., and D. Byerlee. 2002. Productivity growth and resource degradation in Pakistan's Punjab: A decomposition analysis. *Economic Development and Cultural Change* 50 (4): 839–863.

Ali, M., and V. T. B. Hau. 2001. Vegetables. In *Bangladesh: Economic and nutritional impact of new varieties and technologies.* Technical Bulletin 25. Tainan, Taiwan: The World Vegetable Center.

Alston, J. M., C. Chan-Kang, M. C. Marra, P. G. Pardey, and T. J. Wyatt. 2000. *A meta-analysis of rates of return to agricultural R&D, ex pede Herculem?* Research Report 113. Washington, D.C.: International Food Policy Research Institute.

Anderson, J. R., and P. B. R. Hazell, eds. 1989. *Variability in grain yields: Implications for agricultural research and policy in developing countries.* Baltimore: Johns Hopkins University Press.

Asian Development Bank. 2000. *Rural Asia: Beyond the Green Revolution.* Manila: Asian Development Bank.

Bell, C., P. B. R. Hazell, and R. Slade. 1982. *Project evaluation in regional perspective: A study of an irrigation project in northwest Malaysia.* Baltimore: Johns Hopkins University Press.

Berti,·P. R., J. Krasevec, and S. FitzGerald. 2004. A review of the effectiveness of agriculture interventions in improving nutrition. *Public Health Nutrition* 7 (5): 599–607.

Bhandari, A. L., R. Amin, C. R. Yadav, E. M. Bhattarai, S. Das, H. P. Aggarwal, R. K. Gupta, and P. R. Hobbs. 2003. How extensive are yield declines in long-term rice–wheat experiments in Asia? *Field Crops Research* 81: 159–180.

Bouis, H. E., ed. 2000. Special issue on improving nutrition through agriculture. *Food and Nutrition Bulletin* 21 (4).

Cassman, K. G., and P. Pingali. 1993. Extrapolating trends from long-term experiments to farmers' fields: The case of irrigated rice systems in Asia. In *Measuring sustainability using long-term experimenters.* Proceedings of the working conference at Rothamsted Experimental Station, U.K., April 28–30, 1993. Funded by the Agricultural Science Division of The Rockefeller Foundation.

Chakravorty, U. 1998. The economic and environmental impacts of irrigation and drainage in developing countries. In *Agriculture and the environment: Perspectives on sustainable rural development,* ed. E. Lutz. Washington, D.C.: World Bank.

Chand, R., and S. S. Raju. 2008. *Instability in Indian agriculture during different phases of technology and policy.* Discussion Paper NPP 01/2008. National Centre for Agricultural Economics and Policy Research. New Delhi: Indian Council of Agricultural Research.

Comprehensive Assessment of Water Management in Agriculture. 2007. *Water for food, water for life:*

A comprehensive assessment of water management in agriculture. London and Colombo, Sri Lanka: Earthscan and International Water Management Institute.

David, C. C., and K. Otsuka. 1994. *Modern rice technology and income distribution in Asia.* Boulder, Colo., U.S.A.: Lynne Rienner.

Deb, U. K., and M. C. S. Bantilan. 2003. Impacts of genetic improvement in sorghum. In *Crop variety improvement and its effects on productivity,* ed. R. E. Evenson and D. Gollin. Wallingford, U.K.: CABI.

Djurfeldt, G., and M. Jirström. 2005. The puzzle of the policy shift—The early green revolution in India, Indonesia, and the Philippines. In *The African food crisis: Lessons from the Asian green revolution,* ed. G. Djurfeldt, H. Holmén, M. Jirström, and R. Larsson. Wallingford, U.K.: CABI.

Dogra, B. 1986. The Indian experience with large dams. In *The social and environmental effects of large dams,* vol. 2, ed. E. Goldsmith and N. Hildyard. London: Wadebridge Ecological Centre.

Dorosh, P., M. Khan, and H. Nazli. 2003. Distributional impacts of agricultural growth in Pakistan: A multiplier analysis. *Pakistan Development Review* 42 (3): 249–275.

Dorward, A., J. Kydd, and C. Poulton, eds. 1998. *Smallholder cash crop production under market liberalisation: A new institutional economics perspective.* Wallingford, U.K.: CABI.

Dorward, A., J. Kydd, J. Morrison, and I. Urey. 2004. A policy agenda for pro-poor agricultural growth. *World Development* 32 (1): 73–89.

Erenstein, O., U. Farook, R. K. Malik, and M. Sharif. 2007. *Adoption and impacts of zero tillage as a resource conserving technology in the irrigated plains of south Asia.* Comprehensive Assessment Research Report 19. Colombo, Sri Lanka: International Water Management Institute.

Evans, L. T. 1998. *Feeding the ten billion: Plants and population growth.* Cambridge, U.K.: Cambridge University Press.

Evenson, R. E., C. E. Pray, and M. W. Rosegrant. 1999. *Agricultural research and productivity growth in India.* Research Report 109. Washington, D.C.: International Food Policy Research Institute.

Fan, S. 2007. Agricultural research and urban poverty in China and India. In *Agricultural research, livelihoods, and poverty: Studies of economic and social impacts in six countries,* ed. M. Adato and R. Meinzen-Dick. Baltimore: Johns Hopkins University Press.

Fan, S., A. Gulati, and S. Thorat. 2007. *Investment, subsidies, and pro-poor growth in rural India.* Discussion Paper 716. Washington, D.C.: International Food Policy Research Institute.

———. 2008. Investment, subsidies, and pro-poor growth in rural India. *Agricultural Economics* 39 (2): 163–170.

Fan, S., P. Hazell, and S. Thorat. 1999. *Linkages between government spending, growth, and poverty in rural India.* Research Report 110. Washington, D.C.: International Food Policy Research Institute.

FAO (Food and Agriculture Organization of the United Nations). 2009. FAOSTAT. <http://faostat .fao.org>. Accessed August 20, 2010.

Freebairn, D. K. 1995. Did the green revolution concentrate incomes? A quantitative study of research reports. *World Development* 23 (2): 265–279.

Galwani, K., R. Kanbur, and X. Zhang. 2007. *Comparing the evolution of spatial inequality in China and India: A fifty-year perspective.* Development, Strategy, and Governance Division Discussion Paper 44. Washington, D.C.: International Food Policy Research Institute.

Ghassemi F., A. J. Jakeman, and H. A. Nix. 1995. *Salinization of land and water resources: Human causes, extent, management, and case studies.* Canberra, Australia: Centre for Resource and Environmental Studies, Australian National University.

Gollin, D. 2006. *Impacts of international research on intertemporal yield stability in wheat and maize: An economic assessment.* Mexico City: International Maize and Wheat Improvement Center.

Gollin, D., M. Morris, and D. Byerlee. 2005. Technology adoption in intensive post-green revolution systems. *American Journal of Agricultural Economics* 87 (5): 1310–1316.

Guerra, L. C., S. I. Bhuiyan, T. P. Tuong, and R. Barker. 1998. *Producing more rice with less water from irrigated systems.* SWIM Paper 5. Colombo, Sri Lanka: International Water Management Institute.

Haggblade, S., P. B. Hazell, and P. A. Dorosh. 2007. Sectoral growth linkages between agriculture and the rural nonfarm economy. In *Transforming the rural nonfarm economy,* ed. S. Haggblade, P. B. Hazell, and T. Reardon. Baltimore: Johns Hopkins University Press.

Hasan, R., and M. G. Quibria. 2004. Industry matters for poverty: A critique of agricultural fundamentalism. *Kyklos* 57 (2): 253–264.

Hayami, Y., and M. Kikuchi. 2000. *A rice village saga: Three decades of green revolution in the Philippines.* London: Macmillan.

Hazell, P. B. R. 1982. *Instability in Indian foodgrain production.* Research Report 30. Washington, D.C.: International Food Policy Research Institute.

———. 1989. Changing patterns of variability in world cereal production. In *Variability in grain yields: Implications for agricultural research and policy in developing countries,* ed. J. R. Anderson and P. B. R. Hazell. Baltimore: Johns Hopkins University Press.

———. 2008. *An assessment of the impact of agricultural research in South Asia since the green revolution.* Rome: Consultative Group on International Agricultural Research Science Council Secretariat.

Hazell, P. B. R., and L. Haddad. 2001. *Agricultural research and poverty reduction.* 2020 Vision for Food, Agriculture, and the Environment Discussion Paper 34. Washington, D.C.: International Food Policy Research Institute.

Hazell, P. B. R., and S. Haggblade. 1991. Rural–urban growth linkages in India. *Indian Journal of Agricultural Economics* 46 (4): 515–529.

Hazell, P. B. R., and C. Ramasamy. 1991. *Green revolution reconsidered: The impact of the high yielding rice varieties in South India.* Baltimore and New Delhi: Johns Hopkins University Press and Oxford University Press.

Hazell, P. B. R., and S. Wood. 2008. Drivers of change in global agriculture. *Philosophical Transactions of the Royal Society B* 363 (1491): 495–515.

Hossain, M., D. Lewis, M. L. Bose, and A. Chowdhury. 2007. Rice research, technological progress, and poverty: The Bangladesh case. In *Agricultural research, livelihoods, and poverty: Studies of economic and social impacts in six countries,* ed. M. Adato and R. Meinzen-Dick. Baltimore: Johns Hopkins University Press.

InterAcademy Council. 2004. *Realizing the promise and potential of African agriculture: Science and technology strategies for improving agricultural productivity and food security in Africa.* Amsterdam.

Janaiah, A., K. Otsuka, and M. Hossain. 2005. Is the productivity impact of the green revolution in rice vanishing? *Economic and Political Weekly* 40 (53): 5596–5600.

Jewitt, S., and K. Baker. 2007. The green revolution re-assessed: Insider perspectives on agrarian change in Bulandshahr district, Western Uttar Pradesh, India. *Geoforum* 38 (1): 73–89.

Jirström, M. 2005. The state and green revolutions in East Asia. In *The African food crisis: Lessons from the Asian green revolution,* ed. G. Djurfeldt, H. Holmén, M. Jirström, and R. Larsson. Wallingford, U.K.: CABI.

Johnson, M., P. B. R. Hazell, and A. Gulati. 2003. The role of intermediate factor markets in Asia's green revolution: Lessons for Africa? *American Journal of Agricultural Economics* 85 (5): 1211–1216.

Kataki, P. K. 2002. Shifts in cropping system and its effect on human nutrition: Case study from India. *Journal of Crop Production* 6 (1/2): 119–144.

Kennedy, E., and H. Bouis. 1993. *Linkages between agriculture and nutrition: Implications for policy and research.* Washington, D.C.: International Food Policy Research Institute.

Krueger, A. O., M. Schiff, and A. Valdés. 1991. *The political economy of agricultural pricing policy.* Baltimore: Johns Hopkins University Press.

Ladha, J. K., D. Dawe, H. Pathak, A. T. Padre, R. L. Yadav, B. Singh, Y. Singh, Y. Singh, P. Singh, A. L. Kundu, R. Sakal, N. Ram, A. P. Regmi, S. K. Gami, A. L. Bhandari, R. Amin, C. R. Yadav, E. M. Bhattarai, S. Das, H. P. Aggarwal, R. K. Gupta, and P. R. Hobbs. 2003. How extensive are yield declines in long-term rice: Wheat experiments in Asia? *Field Crops Research* 81: 159–180.

Larson, D. W., E. Jones, R. S. Pannu, and R. S. Sheokand. 2004. Instability in Indian agriculture: A challenge to the green revolution technology. *Food Policy* 29 (3): 257–273.

Lipton, M., with R. Longhurst. 1989. *New seeds and poor people.* Baltimore: Johns Hopkins University Press.

Maheshwari, A. 1998. Green revolution, market access of small farmers and stagnation of cereals' yield in Karnataka. *Indian Journal of Agricultural Economics* 53 (1): 27–40.

McNeely, J. A., and S. J. Scherr. 2003. *Ecoagriculture: Strategies to feed the world and save wild biodiversity.* Washington, D.C.: Island Press.

Mehra, S. 1981. *Instability in Indian agriculture in the context of the new technology.* Research Report 25. Washington, D.C.: International Food Policy Research Institute.

Mellor, J. W. 1976. *The new economics of growth: A strategy for India and the developing world.* Ithaca, N.Y., U.S.A.: Cornell University Press.

Mendola, M. 2007. Agricultural technology adoption and poverty reduction: A propensity-score matching analysis for rural Bangladesh. *Food Policy* 32 (3): 372–393.

Mondal, M. K., M. N. Islam, G. Mowla, M. T. Islam, and M. A. Ghani. 1993. Impact of on-farm water management research on the performance of a gravity irrigation system in Bangladesh. *Agricultural Water Management* 23 (1): 11–22.

Morris, M., V. Kelly, R. Kopicki, and D. Byerlee. 2007. *Promoting increased fertilizer use in Africa.* Washington, D.C.: World Bank.

Murgai, R., M. Ali, and D. Byerlee. 2001. Productivity growth and sustainability in post-green revolution agriculture: The case of the Indian and Pakistan Punjabs. *World Bank Research Observer* 16 (2): 199–218.

Nasurudeen, P., A. Kuruvila, R. Sendhil, and V. Chandresekar. 2006. The dynamics and inequality of nutrient consumption in India. *Indian Journal of Agricultural Economics* 61 (3): 363–373.

Nellithanam, R., J. Nellithanam, and S. S. Samati. 1998. Return of the native seeds. *The Ecologist* 28 (1): 29.

Oberai, A., and H. Singh. 1980. Migration flows in Punjab's green revolution belt. *Economic and Political Weekly* 15: A2–A12.

Paris, T. R., A. Singh, A. D. Cueno, and V. N. Singh. 2008. Assessing the impact of participatory research in rice breeding on women farmers: A case study in Eastern Uttar Pradesh, India. *Experimental Agriculture* 44: 97–112.

Pingali, P. L., M. Hossain, and R. V. Gerpacio. 1997. *Asian rice bowls: The returning crisis.* Wallingford, U.K.: CABI.

Pinstrup-Andersen, P., and P. Hazell. 1987. The impact of the green revolution and prospects for the future. *Food Reviews International* 1: 1–25.

Pinstrup-Andersen, P., and M. Jaramillo. 1991. The impact of technological change in rice production on food consumption and nutrition. In *The Green revolution reconsidered: The impact of the high-yielding rice varieties in South India,* ed. P. B. R. Hazell and C. Ramasamy. Baltimore and New Delhi: Johns Hopkins University Press and Oxford University Press.

Postel, S. 1993. Water and agriculture. In *Water in crisis: A guide to the world's fresh water resources,* ed. P. H. Gleick. New York: Oxford University Press.

Prahladachar, M. 1983. Income distribution effects of the green revolution in India: A review of empirical evidence. *World Development* 11 (11): 927–944.

Pretty, J. N., A. D. Noble, D. Bossio, J. Dixon, R. E. Hine, F. W. T. Penning de Vries, and J. I. L. Morrison. 2007. Resource conserving agriculture increases yields in developing countries. *Environmental Science and Technology* 40 (4): 1114–1119.

Rangarajan, C. 1982. *Agricultural growth and industrial performance in India.* Research Report 33. Washington, D.C.: International Food Policy Research Institute.

Rao, C. H. H., S. K. Ray, and K. Subbarao. 1988. *Unstable agriculture and droughts: Implications for policy.* New Delhi: Vikas Publishing.

Rasul, G., and G. Thapa. 2003. Sustainability analysis of ecological and conventional agricultural systems in Bangladesh. *World Development* 31 (10): 1721–1741.

Ravallion, M., and G. Datt. 1996. How important to India's poor is the sectoral composition of economic growth? *World Bank Economic Review* 10 (1): 1–26.

Ray, S. K. 1983. An empirical investigation of the nature and causes for growth and instability in Indian agriculture: 1950–80. *Indian Journal of Agricultural Economics* 38 (4): 459–474.

Rola, A. C., and P. L. Pingali. 1993. *Pesticide, rice productivity, and farmer's health: An economic assessment.* Washington, D.C.: World Resources Institute and the International Rice Research Institute.

Rosegrant, M. W., and P. B. R. Hazell. 2000. *Transforming the rural Asia economy: The unfinished revolution.* Hong Kong: Oxford University Press.

Ryan, J. G., and M. Asokan. 1977. Effects of green revolution in wheat on production of pulses and nutrients in India. *Indian Journal of Agricultural Economics* 32 (3): 8–15.

Shah, T., A. D. Roy, A. Qureshi, and J. Wang. 2003. Sustaining Asia's groundwater boom: An overview of issues and evidence. *Natural Resources Forum* 27 (2): 130–141.

Sharma, H. R., K. Singh, and S. Kumari. 2006. Extent and source of instability in foodgrains production in India. *Indian Journal of Agricultural Economics* 61 (4): 647–666.

Shiva, V. 1991. The green revolution in the Punjab. *The Ecologist* 21 (2): 57–60.

Singh, A. J., and D. Byerlee. 1990. Relative variability in wheat yields across countries and over time. *Journal of Agricultural Economics* 41 (1): 21–32.

Smale, M., P. Hazell, T. Hodgkin, and C. Fowler. 2009. Do we have an adequate global strategy for securing the biodiversity of major food crops? In *Agrobiodiversity and economic development,* ed. A. Kontoleon, U. Pascual, and M. Smale. Oxon, U.K.: Routledge.

Spencer, D. 1994. *Infrastructure and technology constraints to agricultural development in the humid and sub-humid tropics of Africa.* Environment and Production Technology Division Discussion Paper 3. Washington, D.C.: International Food Policy Research Institute.

Susmita, D., C. Meisner, and D. Wheeler. 2007. Is environmentally friendly agriculture less profitable for farmers? Evidence on integrated pest management in Bangladesh. *Review of Agricultural Economics* 29 (1): 103–118.

Thapa, G., K. Otsuka, and R. Barker. 1992. Effect of modern rice varieties and irrigation on household income distribution in Nepalese villages. *Agricultural Economics* 7 (3/4): 245–265.

Thirtle, C., L. Lin, and J. Piesse. 2003. The impact of research-led agricultural productivity growth on poverty reduction in Africa, Asia, and Latin America. *World Development* 31 (12): 1959–1975.

Tisdell, C. 1988. Impact of new agricultural technology on the instability of foodgrain production and yield: Data analysis for Bangladesh and its districts. *Journal of Development Economics* 29 (2): 199–227.

Tribe, D. 1994. *Feeding and greening the world: The role of international agricultural research.* Wallingford, U.K.: CABI.

Tripp, R. 2006. *Self-sufficient agriculture: Labour and knowledge in small-scale farming.* London: Earthscan.

Umali, D. 1993. *Irrigation induced salinity: A growing problem for development and the environment.* World Bank Technical Paper 215. Washington, D.C.: World Bank.

Waibel, H. 1999. *An evaluation of the impact of integrated pest management at international agricultural research centres.* Rome: International Association for Engineering Geology Secretariat.

Walker, T. S. 2000. Reasonable expectations on the prospects for determining the impact of agricultural research on poverty in ex-post case studies. *Food Policy* 25: 515–530.

Westley, J. R. 1986. *Agriculture and equitable growth: The case of Punjab-Haryana.* Boulder, Colo., U.S.A.: Westview Press.

World Bank. 2008. *World development report 2007: Agriculture for development.* Washington, D.C.

Chapter 4

Controlling Cassava Mosaic Virus and Cassava Mealybug in Sub-Saharan Africa

Felix I. Nweke

Cassava—commonly known as manioc, yucca, or tapioca—is grown principally for its swollen roots, though in some parts of Africa its leaves are also eaten. It is a perennial shrub that was introduced from Brazil into West Africa in the 16th century and into East Africa in the 18th century (Jones 1959). The roots are 25 to 35 percent starch, but the leaves contain a significant amount of protein and other nutrients. About 95 percent of the cassava produced in Africa is used for human consumption and 5 percent for industrial uses such as starch (Nweke and Haggblade 2009). By contrast, most cassava in Thailand is exported and used as livestock feed or for various industrial purposes. In the northeast region of Thailand, cassava led the export takeoff, with production rising from 1.7 million tons in 1976 to 20.7 million tons in 1996 (World Bank 2009).

The diffusion of cassava can be described as a "self-spreading innovation" in African agriculture. It was initially adopted as a famine-reserve crop because it provided a reliable source of food during droughts, locust attacks, and the "hungry season" (Jones 1959).[1] Currently cultivated in around 40 African countries, cassava covers a wide belt stretching from Madagascar in the southeast to Cape Verde in the northwest.

The cassava mosaic virus disease (hereafter, cassava mosaic disease) was prevalent in East Africa in the 1890s and subsequently spread to most countries in Central and West Africa (Storey and Nichols 1938). On the eve of Africa's independence, Jones (1959) reported that cassava mosaic disease was the only major disease affecting cassava. In the early 1970s, however, two pests—the cassava mealybug and the cassava green mite—were inadvertently introduced to Africa from South America. These pests spread rapidly and threatened the cassava industry in Africa (Yaninek 1994). In the mid-1980s, cassava mosaic disease was brought under control by the breeding and diffusion of the cassava mosaic disease–resistant tropical manioc selection (TMS) varieties. In the late 1970s, biologists launched a search in South America for natural enemies of the mealybug. A small wasp was found to be a parasite that fed on the mealybug; it was multiplied and eventually released in more than 100 locations in Africa. The mealybug problem was thus brought under control in Sub-Saharan Africa through a classical biological control program.

This chapter discusses the cassava mosaic disease and mealybug control programs in Sub-Saharan Africa, drawing mainly on information collected through the Collaborative Study of Cassava in Africa (COSCA). The initial COSCA studies were financed by the Rockefeller Foundation and were carried out from 1989 to 1997 under the aegis of the International Institute of Tropical Agriculture (IITA) in Ibadan, Nigeria. From 1989 to 1992, COSCA researchers collected farm-level data in 281 villages in six countries: the Congo, Côte d'Ivoire, Ghana, Nigeria, Tanzania, and Uganda. From 1993 to 1997, COSCA researchers analyzed the field data and prepared a series of reports on cassava production, processing, and consumption, culminating in publication of *The Cassava Transformation: Africa's Best Kept Secret* (Nweke, Spencer, and Lynam 2002).[2]

Cassava in Africa: An Overview

Annual cassava production in Africa nearly tripled, from 33 million tons in the early 1960s to 90 million tons in the early 2000s (FAO 2009). Most of the dramatic increases in cassava production in Africa were achieved in Nigeria and Ghana. In the early 1960s, Nigeria produced only 8 million tons of cassava per year; it was the fourth-largest producer in the world after Brazil, Indonesia, and the Congo (FAO 2009). In the early 2000s, Nigeria produced 32 million tons per year and became the largest producer worldwide, displacing Brazil, Indonesia, and the Congo. Ghana was the seventh-largest producer in Africa in the early 1960s, with an annual production of only 1.2 million tons. But in the early 2000s, Ghana produced 8 million tons annually and became the third-largest producer in Africa, after Nigeria and the Congo. In Africa, total cassava consumption more than doubled, from 24 million tons per year in the early 1960s to 58 million tons per year in the early 2000s, after accounting for waste (FAO 2009).

Cassava appeals to low-income rural and urban households because it is the least expensive source of food calories and can be used in an increasing array of food products (for example, *gari*), as well as for livestock feed and industrial starch. Compared with grain, cassava roots (fresh or dried) are an inexpensive source of calories. Calories are significantly less expensive from fresh roots of sweet cassava varieties than from maize sold in rural village market centers in Nigeria (Nweke, Spencer, and Lynam 2002).

Cassava Myths

Cassava was a subsistence crop in the era during which 90 to 95 percent of the people in Africa were in farming. Nevertheless, despite the ability of cassava to produce an acceptable yield under low rainfall conditions, several myths have marginalized cassava in research, policy, and donor circles. The view that cassava is still primarily a subsistence crop is a myth. In Ghana, estimates of the income elasticity of demand for cassava, maize, and rice (based on the World Bank Living Standard Survey data) are revealing: the estimate for cassava was significantly greater among urban households (1.46) than among rural households (0.73). Among urban households, the estimate for cassava was about the same as the estimate for rice (1.50) and greater than the estimate for maize (0.83) (Alderman 1990).

Another myth is that cassava depletes soil nutrients. In six African countries, the soils of cassava fields were observed to be as fertile as the soils of other crops (Nweke and Haggblade 2009). The myth that cassava is a "women's crop" is an important half-truth; equally important is the other half-truth, that cassava is also a "men's crop." Both men and women produce cassava. Men are increasingly involved in cassava production, processing, and marketing as cassava increasingly becomes a cash crop (Nweke and Haggblade 2009). The common stigma that some cassava varieties contain lethal cyanogens is also a half-truth.[3] The cases of cyanide poisoning from cassava consumption are rare, and the fear of it should not discourage public or private investment in the cassava food economy. The cyanogens are eliminated during processing and cassava food preparation by using well-known traditional methods (Nweke, Spencer, and Lynam 2002).

Many critics claim that cassava is a nutritionally deficient food because of its low protein and vitamin content. However, one does not define beef as a nutritionally deficient food because it has low carbohydrate content. Without question, the challenge ahead is to increase the productivity of cassava production, harvesting, and processing in order to drive down the cost of cassava to consumers, especially the poor.

Cassava Production, Processing, and Consumption

Cassava is vegetatively propagated with planting sets from stem cuttings. Farmers' most common sources of planting sets are their own previous crops and sometimes purchases in village markets from fellow farmers. COSCA studies revealed that in the early 1990s, fewer than 5 percent of cassava fields were planted with purchased sets, but 15 years later this proportion increased to around 20 percent. In the subhumid savanna zone, planting sets are often in short supply; however, there is an abundance of cassava planting sets in the humid forest zone, where biomass production is high.[4] But in almost all cases, the planting sets used by the farmers are infected or infested with one or more diseases and pests.

Cassava Weeding.   Some scientists report that cassava requires little weeding when planted in optimal plant populations because the cassava canopy suppresses weeds (Onwueme and Sinha 1991). However, both small- and large-scale farmers still consider weeding a major cost because it takes about two to four months for the cassava leaves to close the canopy and suppress weed growth (Dahniya and Jalloh 1998). Commercial producers weed cassava fields twice in the first 12 months and then either harvest the cassava or leave the field to grow into bush along with the cassava.

Harvesting.   Cassava does not have a maturity period. It can be harvested as soon as the root is formed, but if it is not harvested, the root continues to enlarge for up to about three years. When the COSCA studies were designed some 20 years ago, it was assumed that the main labor constraints were at the postharvest stage; however, the high-yielding TMS varieties have shifted labor bottlenecks to the harvesting stage. Cassava harvesting includes chopping off the stem with a machete, pulling up its roots (digging in dry or clay soil conditions), and then cutting the roots off the stump. The farmer moves from one plant to another performing these tasks and later gathers the roots to carry them home, to market, or to a processing center. The labor requirement for most of these harvesting tasks increases in direct proportion to yield, because higher yield means greater bulk and weight. The amount of harvesting labor per hectare is especially high in Nigeria because the cassava yields are higher than in the other COSCA study countries (Nweke and Haggblade 2009). Table 4.1 shows that harvesting cassava is the most labor-intensive field task in Nigeria and Ghana, because TMS varieties have boosted yields by 40 percent.

Common Cassava Foods.   In Africa, there are four common groups of cassava foods: fresh roots, dried roots, pasty products, and granulated products.[5] Dried cassava-root flour is widely prepared and consumed throughout Africa, especially in rural areas (Idowu 1998). There are two broad types of dried cassava roots: fermented and unfermented. Farmers in the savanna zone ferment cassava roots by stacking them, while farmers in the forest zone ferment them by soaking them because of the availability of water. The recent introduction of the grater in processing dried cassava root into flour eliminates fermentation and therefore saves time. The roots are simply peeled, washed, grated, pressed to express effluent and cyanogens, and then sun-dried.

Three forms of pasty cassava products are common in Africa: uncooked, cooked, and steamed. Pasty cassava products are not as bulky as fresh cassava roots and are therefore less expensive to transport.

Table 4.1 COSCA study countries: Cassava production and harvesting labor, 1991 (person-days per hectare)

Task	Congo	Côte d'Ivoire	Ghana	Nigeria	Tanzania	Uganda
Land clearing	66	53	44	40	54	45
Seedbed preparation	21	29	31	41	27	31
Field planting	39	22	28	32	27	28
Weeding	27	28	34	38	28	32
Harvesting	48	44	53	62	46	52
Total days	201	173	191	222	182	187

Source: COSCA (Collaborative Study of Cassava in Africa). Used by permission of the International Institute of Tropical Agriculture.

Gari is a common type of granulated cassava product. *Gari* is a toasted cereal-like cassava food product that is more common in Nigeria than anywhere else in Africa. It is a convenient product because it is stored and marketed in a ready-to-eat form. It can be soaked in hot or cold water depending on the type of meal being prepared. Because *gari* preparation tasks are labor-intensive, a mechanized method of cassava grating is spreading in Nigeria and Ghana.

Labor Bottlenecks at the Peeling Stage.   The processing tasks required for the preparation of the three major cassava food products—dried roots, pasty products, and granulated cassava food products—are time-consuming because they involve peeling for all three products; chipping or grating for dried roots; crushing and sieving for pasty products; grating for granulated products (*gari*); water expressing for pasty products, *gari,* and dried root flour made from fresh roots; sun-drying in the case of dried roots; toasting for *gari;* and finally, milling into flour in the case of dried roots. (Dried cassava-root flour is made from fresh roots that are grated and then sun-dried.) The processing of pasty products ends with water expressing; *gari* is finished with toasting.

The cassava grating machines are made with locally fabricated components at a cost of $100 to $500 per machine. The machines are usually owned by entrepreneurs who provide services to small-scale farmers for a fee based on quantity. In some villages, the graters are located in the market square; in other villages, a grater is mounted on wheels and taken to the fields or homes of farmers who request the service. In many villages, local machine operators in a village processing center provide a comprehensive set of services, including mechanized grating and pressing. In the more comprehensive village processing centers, farmers toast *gari*. Maintenance services for the graters are provided by roadside mechanics and welders at any hour of the day. The replacement of hand-grating with mechanized graters has reduced the cost of making *gari* by 50 percent. The COSCA study found that 51 days of labor were needed to prepare a ton of *gari* by hand, and only 24 days were required with a mechanized grater. Peeling is now the most labor-intensive task, followed by the toasting stage in *gari* preparation (Nweke, Spencer, and Lynam 2002).

Controlling Cassava Mosaic Disease

Development of the Mosaic Resistant Varieties

In the 1920s and 1930s, colonial governments initiated cassava research programs in Africa. Experts on colonial agriculture generally agree that the East African Agriculture and Forestry Research Station at Amani in Tanzania (hereafter called the Amani research station) had the most successful colonial cassava breeding program in Africa. The aim of the British-financed Amani research station was to breed cassava varieties that were resistant to cassava mosaic disease, then spreading rapidly in Africa. The mosaic disease is transmitted by a whitefly, *Bemisia tabaci*, as well as by planting cuttings from infected plants; it reduces cassava yields by 30 to 40 percent (Thresh et al. 1997). In 1935, H. H. Storey, a British researcher, conducted a worldwide search for cassava varieties that were resistant to the mosaic disease and developed various disease-resistant rubber species × cassava hybrids. However, these hybrids had low yields, poor food quality, and poor agronomic characteristics such as lodging. During World War II, from 1939 to 1945, the breeding work at the Amani research station was scaled back (Nichols 1947). In 1951, R. F. W. Nichols was replaced as head of the research station by D. L. Jennings, who developed segregates (for example, 5318/34) from the rubber species × cassava hybrids that showed higher resistance than the hybrids created by Storey.[6]

In 1958, at the Moor Plantation research station in Nigeria, B. D. A. Beck crossed the mosaic disease–resistant Ceara rubber × cassava hybrid, 58308, with high-yielding West Africa selections (Jennings 1976). During Nigeria's independence in 1960, the cassava breeding program at Ibadan was moved to the National Root Crops Research Institute at Umudike in eastern Nigeria, and the breeding work was continued by M. J. Ekandem. Unfortunately, almost all the progeny developed from the Ceara rubber × cassava hybrid, along with the records of the research, were lost during the Nigerian Civil (Biafran) War (1967–70). But the original Ceara rubber × cassava hybrid, 58308, was retained at the Moor Plantation research station (Beck 1980).

Cassava breeding at IITA commenced in 1971, when S. K. Hahn was appointed to head the Institute's root and tuber program. Hahn had access to the rich stock of genetic resources that had been developed at the Amani research station in Tanzania, by Storey and others, from the mid-1930s to the mid-1950s. Hahn drew on Storey's approach of combining the mosaic resistance genes of the Ceara rubber × cassava hybrid, 58308, with genes for high-yield, good root quality, low cyanogens, and resistance to lodging. After only six years of research (1971–77), Hahn achieved the goal of developing high-yielding mosaic-resistant TMS varieties that increased cassava yields on small-scale farms by 40 percent without fertilizer. In 1977 the IITA released the following high-yielding mosaic-resistant varieties: TMS 50395, 63397, 30555, 4(2)1425, and 30572.[7]

Diffusion of the Mosaic-Resistant TMS Varieties in Nigeria

The development of TMS varieties resistant to the mosaic virus in Nigeria is an African success story par excellence. No single factor was responsible for this success. Contributing factors include the pioneering work by Storey in the 1930s and 1940s, as well as Hahn's 23-year leadership in cassava research at the IITA. Other factors include the availability of improved cassava processing and food preparation methods. Further, nongovernmental organizations (NGOs) and the private sector took on the critical role of distributing the cassava plants.

As program director of the root and tuber improvement program at the IITA from 1971 to 1994, Hahn worked to strengthen and expand African agricultural research. He collaborated with National Agricultural Research and Extension programs and invited donors to support human and physical capacity development for nine of these programs.[8] During his tenure he devoted special attention to training African researchers and extension workers (Hahn 2009):

- Forty African scientists and extension workers were trained to the M.Sc. and Ph.D. levels.

- Seven hundred technicians attended short-term training courses at the IITA.

- Several hundred extension workers were trained through in-country training courses.

- IITA scientists were posted to national research programs to help develop national programs.

- Network activities included regular workshops, frequent exchange visits, and publication of workshop proceedings.

- The Institute provided improved genetic materials in tissue culture forms.

In 1977, when the first TMS varieties were released to farmers, improved methods of cassava food preparation and labor-saving mechanical cassava graters were already in place in Nigeria. Farmers' access to mechanical graters reduced the cost of preparing cassava as *gari,* increased the profitability of planting the TMS varieties, and released labor—especially female labor—for planting more cassava (Camara 2000).

The physical presence of the IITA in Nigeria was influential in eliciting help from NGOs and the private sector in the diffusion of the TMS varieties. For example, from 1988 to 1991, multinational oil companies in Nigeria multiplied and supplied TMS planting sets to a large number of farmers, cooperative societies, women's associations, churches, and schools (Hahn 2009).

With the aid of petroleum revenue, the Nigerian government experimented with alternative extension programs and expanded higher education and agricultural research institutions in the 1970s and 1980s. The adoption of the TMS varieties was promoted by Nigeria's national extension program under the National Accelerated Food Production Program and the Agricultural Development Projects (ADPs). In 1986, the Government of Nigeria helped secure a $120 million grant from the International Fund for Agricultural Development (IFAD) and directed the National Seed Service to assist the ADPs in the multiplication of the TMS varieties. The National Seed Service multiplied and distributed free stem cuttings of the TMS varieties to farmers. In 1989 COSCA researchers found that farmers in 60 percent of the surveyed villages in Nigeria had planted the TMS varieties (Nweke and Haggblade 2009). Fifteen years later, the TMS varieties were grown in all of the COSCA-surveyed villages in Nigeria.

Delayed Diffusion of the Mosaic-Resistant TMS
Varieties in Ghana and Uganda

Until the early 1980s, Ghana's food policy favored cereals—maize and rice, the long-time favorites. Widely believed myths about cassava discouraged the government from investing in measures to diffuse the TMS varieties to farmers until its interest in the mosaic-resistant cassava varieties was awakened by a severe drought in 1982 and 1983. Cassava survived the drought and helped people cope with food insecurity (Korang-Amoakoh, Cudjoe, and Adams 1987).

In 1984, Ghana's commissioner for agriculture visited the IITA in Ibadan and met with Hahn. During their discussion, the commissioner described the roles of maize and cassava in food policy in Ghana using the expression, "Monkey de work, Baboon de chop." His meaning was that cassava was feeding the people, but maize was consuming research resources. In 1985, Ghana hosted the Central and Western African Root Crops Network workshop in Accra. The workshop helped government officials grasp the importance of cassava in Ghana (Obimpeh 1994). In 1988, 11 years after the TMS varieties were released in Nigeria, the Government of Ghana imported the TMS varieties from the IITA and turned them over to Ghanaian researchers for field testing. Hahn helped the Government of Ghana obtain IFAD funding for on-farm testing of the TMS varieties from 1988 to 1992.

In 1993, 16 years after the release of the TMS varieties in Nigeria, the Government of Ghana released three TMS varieties to farmers. In February 2001, cassava scientists at the Crops Research Institute in Kumasi reported that the TMS varieties were widely grown by farmers in the eastern, greater Accra, and Volta regions, where farmers prepare *gari* for sale in urban centers. The 16-year delay in Ghana illustrates the need for political leadership in promoting the adoption of new technology from neighboring countries.

In Uganda, government interest in the mosaic-resistant TMS cassava varieties was aroused in 1988 by the appearance of an unknown but severe form of cassava mosaic disease (Ssemakula et al. 1997). This disease has since been designated the Uganda variant (UgV) of the East Africa cassava mosaic disease. Surveys to monitor the progress of the UgV epidemic revealed that the epidemic spread southward along a broad "front" at a rate of approximately 20 kilometers per year. The front was marked by a large number of whiteflies and by a high incidence of recent infection due to whitefly transmission.

To address this problem, Ugandan scientists in the Root Crops Program of the National Agricultural Research Organization (NARO) introduced the mosaic-resistant TMS varieties obtained from the IITA (Otim-Nape and Bua 2000). In 1994, following on-farm tests, three varieties (TMS30572, TMS60142, and TMS30337) were released to farmers (Ssemakula et al. 1997). Multiplication and distribution of the planting sets of the mosaic-resistant TMS varieties were undertaken by NARO, with financial and technical support from several organizations.[9] NARO created the National Network of Cassava Workers (NANEC) to distribute the planting sets to farmers because it believed that existing institutions such as the extension service of the Ministry of Agriculture were inadequate to implement the distribution. NANEC established branches in all cassava-growing districts and brought together cassava stakeholders at the district level—including contact farmers, the district agents of the extension service of the Ministry of Agriculture and NARO, political leaders, and others—and through them distributed the planting sets to farmers.

The University of Greenwich (2000) reported that the area planted with the mosaic-resistant TMS varieties in Uganda increased from 20 percent of the total cassava area in 1993 to 60 percent in 1996 and 80 percent in 1998. Studies of the presence of the UgV of the mosaic disease revealed that the incidence of the disease declined from more than 90 percent on the mosaic-susceptible local varieties to less than 20 percent on the mosaic-resistant TMS varieties. Moreover, the severity of the disease was high on the local varieties but mild on the TMS varieties, with hardly any reduction of root yield (University of Greenwich 2000).

Performance of the TMS Varieties in Nigeria

The TMS breeding efforts of the IITA team were aimed at developing resistance to cassava mosaic disease. But in order to achieve their full-yield potential, the TMS varieties must also be resistant to, or at least tolerant of, other important cassava diseases and pests, notably the cassava bacterial blight, cassava mealybug, and cassava green mite. The TMS varieties must also address other needs of farmers: early harvesting, ability to suppress weeds and suitability for intercropping, ease of harvesting and peeling, low cyanogen content, and suitability for making various food products.

In Nigeria the TMS varieties were more resistant than local varieties not only to the mosaic virus but also to bacterial blight, the mealybug, and the green mite (Table 4.2). The farm-level yield of the TMS varieties, when grown without fertilizer, was 40 percent higher than local varieties (19 tons compared to 13.6 tons per hectare).

The TMS varieties attain their peak yield around 13 to 15 months after planting as compared to 22 to 24 months for local varieties. Nevertheless, Nigerian farmers who produce cassava under increasing demographic and market pressures desire varieties that can be harvested in less than 12 months in order to be able to prepare their fields for continuous cropping.

TMS varieties are mostly branching types with large canopies that are good for weed control. But in spite of the large canopy, COSCA studies have found no significant difference between the TMS varieties and the local varieties in terms of

Table 4.2 Nigeria: Incidences and symptom severity scores (1–4 scale) of cassava problems by local and tropical manioc selection varieties, 2002

Problem	Indicator	Local varieties (N = 93)	TMS varieties (N = 49)	t-ratio[a]
Mealybug	Percentage infested	50	20	—
	Mean severity	2.0	1.2	3.15
Green mite	Percentage infested	26	4	—
	Mean severity	1.5	1.0	3.68
Mosaic disease	Percentage infected	62	73	—
	Mean score	1.9	1.5	2.45
Bacteria blight	Percentage infected	63	71	—
	Mean score	1.9	1.3	4.20

Source: Nweke, Spencer, and Lynam (2002). Used by permission of the International Institute of Tropical Agriculture.
[a]$P < 0.001$ in all cases.

intercropping. For example, 50 percent of the area of the TMS varieties and 55 percent of area planted with the local varieties in Nigeria were intercropped with yams, maize, and other crops.

Nigerian farmers complained that harvesting the high-yielding TMS varieties by hand was laborious. Farmers in southwestern Nigeria, who planted the TMS 30572 to produce *gari* for sale in Lagos, reported that they had to cut back drastically on the area planted with cassava because they lacked enough seasonal labor to harvest and process the crop of the previous season in a timely fashion.

The Nigerian farmers who produced cassava as a cash crop and made *gari* for sale to urban consumers praised the TMS varieties as ideal for *gari* production. However, they complained that peeling the TMS varieties is laborious and results in substantial waste because the roots can be peeled only by slashing the skin and part of the root-flesh with a sharp knife. Mechanized machines have not been developed for cassava peeling because cassava roots vary in size and shape: farm-level yield measurements show that the roots of the TMS varieties ranged in weight from 0.10 kilograms to 1.14 kilograms. There is a need for breeders to develop cassava varieties that produce roots with uniform shapes and sizes and for engineers to develop mechanized peeling machines.

The roots of the TMS varieties are lower in cyanogen content than those of local varieties (an average of 2.20 on a scale of 1.0 to 3.0, compared to 2.35). Sweet cassava occupied roughly 30 percent of the area planted with the TMS varieties and bitter cassava about 70 percent, the same proportion as the local varieties.[10]

The Cassava Mealybug Control Program

The cassava mealybug was accidentally introduced in the Congo in the early 1970s in infested planting materials from South America. The mealybug spread throughout the cassava belt of Africa, sharply reducing cassava yields. In just 10 years, the cassava mealybug threatened to wipe out cassava in Africa (Herren 1981; Norgaard 1988).[11] The pest was spread by the wind as well as through the exchange of infested planting materials. The mealybug feeds on the cassava stem, petiole, and leaf near the growing point of the cassava plant. During feeding, the mealybug injects a toxin that causes leaf curling, slowing of shoot growth, and eventual leaf withering. Yield loss in infested plants is estimated to be up to 60 percent of the roots and 100 percent of the leaves (Herren 1981).[12]

Establishment of the Africa-wide Biological Control Program

The growing concerns of farmers, scientists, agricultural policymakers, and political leaders about the cassava losses from mealybugs were discussed at an international conference in the Democratic Republic of Congo in 1973. One of the recommendations of the conference was that biological control and resistance breeding be undertaken by the IITA and other institutions (Alene et al. 2005).

Researchers and policymakers reviewed the options and decided that the classical biological control solution—the reuniting of predators with their previously dislocated prey—was the best approach to pursue. Few chemicals are used by smallholders in Africa, and the process of resistance breeding was considered too slow to address the emergency problem (Hahn et al. 1981). Following requests from numerous African countries, a regional approach was adopted in 1980. The Africa-wide Biological Control Program (ABCP) for cassava pests was established at the IITA in Nigeria with three objectives: (1) to achieve permanent, ecologically safe, and economically sustainable control of the cassava mealybug and the cassava green mite throughout the African cassava belt; (2) to provide specialized training in biological control techniques; and (3) to initiate national biological control programs.

International Collaboration

In an undertaking of this size, no single institution had the capacity to handle all the essential aspects: foreign exploration, quarantine, rearing, release, field and laboratory studies, monitoring, coordination, training, awareness creation, and impact studies. The IITA therefore organized a network of collaborators in Africa, Europe, and North, Central, and South America as part of the implementation of the ABCP (Wodageneh and Herren 1987). The Inter-African Phytosanitary Council of the African Union provided regulatory and regional liaison services from the beginning of the cassava mealybug project. The IITA coordinated collaboration with the Centro Internacional de Agricultura Tropical in Colombia, Canadian Imperial Bank of Commerce in London, and the Nigerian quarantine service. Agreement was obtained from quarantine facilities all over Africa to import beneficial insects into their respective countries.

Foreign Exploration, Quarantine, and Importation

Because cassava and the cassava mealybug evolved together in South America, the ABCP scientists looked to that continent for a solution to the mealybug epidemic in Africa. Starting in the late 1970s, a systematic search for the cassava mealybug and its potential natural enemies was undertaken in much of Central and South America and from southern California to Paraguay. Although huge areas of South America were scanned, mealybugs were found only in a very restricted area of the continent. ABCP scientists found a natural enemy, a parasitic wasp called *Anagyrus* (*Apoanagyrus, Epidinocarsis*) *lopezi* (hereafter, *A. lopezi*), which uses the mealybug as the site for laying its eggs and whose developing larvae then kill the mealybug (Herren et al. 1987; Herren and Neuenschwander 1991; Neuenschwander 2001).

All the natural enemies of the cassava mealybug destined for introduction in Africa were sent for quarantine to the International Institute of Biological Control at CABI in Silwood Park, England. To ensure that the insects would not become a problem in their new environment, they were reared through one generation and tested for harmlessness to bees and silkworms, absence of pathogens, and relative specificity. This last criterion guarded against the introduction of general natural enemies that could endanger indigenous plants and animals. The ABCP for the cassava mealybug was particularly aimed at the exclusion of hyperparasitoids in order to establish in Africa the natural balance that existed in South America. From quarantine, primary parasitoids and oligophagous predators were sent to the IITA, first in Nigeria and then in Benin, for further study, mass rearing, and finally release (Herren et al. 1987; Herren and Neuenschwander 1991; Neuenschwander 2001).[13]

Rearing and Releasing Biological Control Agents

Mass rearing and distribution techniques were developed for the introduced biological control agents at the IITA in Ibadan, Nigeria. Producing and delivering the biological control agents was a challenge because of the huge size of the project. The timing of operations was also influenced by administrative decisions in various countries, leading to unpredictable requests for the biological control agents. To satisfy the high and shifting demand for the biological control agents, simplified rearing techniques were developed for local scientists (Neuenschwander and Haug 1992). To test the capability of these biological control agents to establish in the new environment, several different agents that had successfully passed quarantine were released at experimental sites: *A. lopezi, H. notata, D. Hennessey, Hyperaspis* sp., *Allotropa* sp., *A. diversicornis, H. jucunda,* and *S. maculipennis.* Releases were made on the ground by pouring the biological control agents onto infested cassava plants. Because ground release was not possible in several locations owing to poor road infrastructure, aerial release techniques were developed at the IITA and adopted for such locations (Herren et al. 1987; Neuenschwander and Haug 1992). From 1981 to 1994, releases were made at 120 sites in about 30 African countries. The releases were all made in collaboration with colleagues from the national research programs. At the release sites, the establishment and spread of the biological control agents were monitored through samplings of mealybugs. *A. lopezi* (the wasp) quickly became the dominant species among all the introduced biological control agents.

Performance of the Biological Control Program

The various biological control agents were monitored to determine their performance in cassava mealybug control and their effect on nontarget species in the fields where they were released. Field experiments demonstrated that the host-finding and aggregation capacity of *A. lopezi* surpassed those of all the other control agents.

The effectiveness of *A. lopezi* in controlling the cassava mealybug populations was evaluated using exclusion experiments, long-term studies of population dynamics, laboratory and field experiments, and large-scale surveys (Neuenschwander 1996). Physical and chemical exclusion experiments demonstrated the effectiveness of *A. lopezi* in southwestern Nigeria. Under rainforest conditions in Ghana, when cassava mealybugs were protected from *A. lopezi,* their populations were much larger. More important, seven years of continuous monitoring in numerous fields in two areas of southwestern Nigeria revealed that the mean cassava mealybug population never reached the height or the duration observed during the first season of release. A survey covering the whole of Nigeria revealed cassava mealybug infestation levels of under 10 mealybugs per cassava tip, with only 3.2 percent of all tips stunted (Neuenschwander and Haug 1992). The mealybug population reduction remained stable, with small peaks at about 10 percent of outbreak levels (Alene et al. 2005).

In a large-scale survey across different ecological zones in Ghana, yield loss due to cassava mealybugs was reduced significantly (Neuenschwander et al. 1989). The presence of *A. lopezi* translated into a reduction in yield loss of 2.5 tons per hectare. The performance assessment was based on surveys using a regular, nonbiased choice of fields and random samples within each field (Schulthess, Baumgartner, and Herren 1989). The field studies revealed that the introduction of *A. lopezi* led to some competitive displacement but not to the extermination of indigenous parasitoids or predators. The introduced organisms were found to fulfill modern safety requirements (Neuenschwander 2001).

Impacts of the Cassava Mosaic and Cassava Mealybug Control Programs

The research culminating in the development and release of the mosaic-resistant high-yielding TMS varieties in Nigeria was achieved with an annual budget ranging from $500,000 to $4.6 million from 1971 to 1977. The annual economic rate of return from that investment was 55 percent throughout a 31-year period (Maredia, Byerlee, and Pee 2000).

Neuenschwander and Haug (1992) reported that, from inception to the end of 1988, the total cost of the cassava mealybug biological control project in Africa was equivalent to $10 per hectare of cassava as a one-time expense to reduce the cassava mealybug for subsequent years. The benefit–cost ratio estimates for this biological control program (estimated by different researchers during different periods using widely different assumptions) range from 94:1 to 800:1. For example, Norgaard's landmark study (1988), using a 24-year time frame, estimated the benefit–cost ratio for the mealybug control program at 149:1.[14] A research team headed by Zeddies et al. (2001), using a 40-year time frame, concluded that the benefit–cost ratio was about 200:1 when cassava was costed at world market prices and ranged from 370:1 to 740:1 at inter-African prices. These findings demonstrate that biological control can play an important role in pest management.

But impact estimates based on internal rates of return and benefit–cost ratios are limited by their assumptions, especially in Africa, where farm-level data are scarce and unreliable (Nweke 2005). For example, the estimate of internal rate of return did not take into account the TMS diffusion costs in Nigeria, including the costs of multiplication and free distribution of planting sets to farmers. Nor did these estimates consider the value of the TMS varieties accruing to other countries, such as Ghana and Uganda, which are now importing, testing, releasing, and diffusing the varieties to farmers. It is not clear, moreover, how the internal rate of return takes account of the possibility that the TMS mosaic-resistant varieties might break down within the estimation period. Likewise, the benefit–cost estimate of the mealybug control program did not account for ecological and health benefits—or for the potential benefits to other cassava-producing regions threatened by mealybugs.

Impact on Production and Consumer Prices

Nigeria is an ideal country for studying the impact of these control programs on cassava production and consumer prices. The TMS cassava varieties were first released to farmers in Nigeria in 1977, followed by a large diffusion program starting in the mid-1980s. During the release of the biological control agents and the rapid diffusion of TMS varieties from the mid-1980s to the early 1990s, Nigeria's per capita cassava production increased (Nweke and Haggblade 2009). The IITA, drawing on data from the COSCA study, calculated that the TMS varieties contributed an extra 1.4 million tons of *gari* per year than would have been available from local varieties—an amount sufficient to feed 29 million people (CGIAR 1996). By the end of the 1980s, cassava prices fell sharply, as reflected in the *gari*–yam price ratio (Nweke and Haggblade 2009). The average inflation-adjusted *gari* price from 1984 to 1992 (18,000 naira per ton) was 40 percent lower than the price prevailing in the prior period, from 1971 to 1983 (29,000 naira per ton)—before the TMS diffusion and before the mealybug was brought under control. This dramatic reduction in *gari* prices represents a significant increase in the real income of the millions of poor rural and urban households that consume cassava as their staple food. The major economic benefit from the control of cassava mosaic disease and cassava mealybugs accrued to consumers (Afolami and Falusi 1999).

In Nigeria, cassava production is a major source of calories and cash income for farm households. The COSCA studies in 1992 revealed that foodcrops contributed 55 percent of the cash income of study households. Cassava, the most important single cash income source, accounted for 12 percent of the total cash income per farm household compared to 8 percent for yams and maize, respectively, and only 6 percent for rice.

These two programs also improved the income position of small producers relative to the large producers (Afolami and Falusi 1999; Johnson, Masters, and Preckel 2006). Cassava sales proved more egalitarian than the alternative staples, such as yams and maize: cassava cash income accrued to more households than did earnings from other major staples. Among rural households, 40 percent earned cash income from selling cassava while 35 percent earned cash from selling maize and 24 percent from selling yams. Although food sales typically remain highly concentrated among an upper stratum of smallholder farmers, cassava sales accrue to a broader spectrum of farm households than do sales of other food staples. The top 10 percent of cash-earning households from the COSCA villages earned 50 percent of all cassava cash income—but they also garnered 60 percent of yam earnings and 70 percent of maize sales.

Sustainability of Cassava Mosaic and Mealybug Control

The Mosaic Disease Control Program

In discussing the sustainability of the mosaic disease control program, two issues arise: sustaining the *quality* of mosaic resistance in the TMS varieties and sustaining the *planting* of TMS varieties by farmers. In crop breeding, disease resistance tends to decline with each generation of seed that is saved and replanted. Therefore, farmers need to collect new seeds either from research institutes or from specialized seed companies each season to continue to realize the benefits of the new varieties. In Nigeria, however, the TMS multiplication and free distribution program assumed that farmers did not need to collect new stem cuttings (planting sets) each season in order to continue to sustain the quality of mosaic resistance in the TMS varieties.

In 2003 the National Root Crops Research Institute (NRCRI) conducted a diagnostic survey of cassava in Nigeria to determine the status of cassava mosaic disease in the country (Ogbe et al. 2004). The results showed that cassava mosaic disease was significantly reduced among the TMS varieties compared to local varieties. In 1989 the COSCA studies conducted by the same NRCRI scientists produced a similar result. Even in the absence of a renewed supply of planting sets (from a research organization or specialized seed company), the TMS varieties sustained their superiority over local varieties in mosaic disease resistance over nearly 15 years, from 1989 to 2003.

The expansion of the planting of the mosaic-resistant varieties by farmers depends on addressing two second-generation problems with the TMS varieties: high costs for harvesting and peeling labor and declining cassava prices as a result of increased production. Both of these problems are due to the high yield achieved with the TMS varieties. Farmers who face increasing costs and declining prices are likely to reduce the size of the area they plant in cassava. Evenson and Gollin (2003) report that, where productivity rose more than prices declined, farm families gained from the Green Revolution in Asia and South America. In the case of the high-yielding mosaic disease-resistant TMS cassava varieties, however, farm wage rates rose faster than cassava prices. For example, in Nigeria over a 10-year period the farm wage rate in real terms nearly tripled, from the equivalent of $1.25 per man-day in 1991 to $3.50 in 2001 (an increase of 180 percent). In comparison, the price of *gari* increased from an equivalent of $185 per ton in 1991 to $255 per ton in 2001—an increase of less than 40 percent. Labor is the main cost of cassava production in Nigeria. Under these conditions, progressive farmers who plant the high-yielding TMS varieties in Nigeria have sometimes suspended planting because they have been unable to hire sufficient labor to harvest previously planted cassava fields (Nweke and Haggblade 2009).

The result is that from the early 1990s to the early 2000s—after the period of rapid diffusion of TMS varieties in Nigeria—cassava production per capita *declined* and cassava prices to consumers *increased* (Nweke and Haggblade 2009). Progressive farmers who were planting the high-yielding mosaic disease-resistant TMS varieties were planting less cassava because they faced serious labor bottlenecks at the harvesting, peeling, and processing stages, with labor requirements that increase in direct proportion to yield.

The high yield obtained with the mosaic disease–resistant TMS varieties has shifted the cassava labor bottleneck from weeding to harvesting and peeling. Farmers who produce cassava as a famine reserve crop or as a rural food staple do not consider cassava harvesting a labor-intensive task because they harvest cassava piecemeal. But farmers who produce cassava as a cash crop for urban markets groan under the burden of high cassava-harvesting labor costs. Harvesting is now proving to be a serious constraint on the planting of mosaic disease–resistant TMS varieties, because labor requirements for cassava harvesting increase in direct proportion to yield.

The Mealybug Control Program

The cassava mealybug control program using biological control agents was a self-spreading innovation that needed only a modest diffusion effort. The biological control agents required no initial investment by farmers, credit programs, or extension services. No manufacturing or distribution system was needed because the biological control agents reproduced and dispersed themselves following their release in the cassava fields (Norgaard 1988). Without question, the biological control of the cassava mealybug with the aid of the biological control agents is one of the important scientific success stories in African history. The speed of dispersal of *A. lopezi* after release was high, proceeding at a rate of between 50 kilometers and 100 kilometers per year. Two years after release, *A. lopezi* was observed in a wide area beyond each original release site (Neuenschwander and Haug 1992). There is no reason to expect that *A. lopezi* will one day disappear from cassava fields unless there are no mealybugs to feed on. Neuenschwander (2001) reported that the much-dreaded "resurgence" of mealybugs (understood here as a permanent increase in host populations following successful biological control) has not been observed, and it is not likely to occur with the cassava mealybug.

Lessons Learned

Five lessons flow from this analysis of two highly successful programs in Sub-Saharan Africa—the programs for control of cassava mosaic disease and control of the cassava mealybug—that provide insights for tackling Africa's food production and poverty problems:

1. *Research is the driving force.* Both of these control programs add evidence that research is the driving force behind cassava production programs in Sub-Saharan Africa. Both success stories highlight the role that research played in expanding food production and thus helping to reduce food prices and rural and urban poverty. This analysis documents how the rapid adoption of cassava varieties with improved resistance to cassava mosaic disease led to dramatic increases in cassava production in the 1980s and 1990s in Ghana, Nigeria, and Uganda. The expansion of cassava has been sparked by demand-side shifts to food products such as *gari*. But Africa still has much to learn about the critical role of research; for example, research in Thailand sparked the export of cassava pellets for livestock feed to the European Union. Thailand's research also includes the development of shorter-season varieties, thus opening the door for double-cropping.

2. *A global approach is important.* Without question, cassava is Africa's most significant "global commodity." It was brought to Africa some 300 years ago from Latin America and is rapidly replacing maize as Africa's most important foodcrop. The research base for controlling cassava mosaic disease stems from colonial research at the Amani research station in Tanzania in the 1930s and 1940s. Thirty years later, the IITA's research on cassava mosaic virus drew on the Amani research findings and developed the high-yielding mosaic-resistant TMS varieties that increased cassava yields by 40 percent without fertilizer. To tackle the mealybug problem, an Africawide biological control center was established at the IITA in Nigeria. The IITA brought together an international group of scientists and donors who crisscrossed Central and South America and eventually found a wasp that fed off the

mealybug. The use of the wasp to control the cassava mealybug reduced yield loss by 2.5 tons per hectare. Both the cassava mosaic disease and the mealybug control programs demonstrate how global partnerships can capture the synergies of local, regional, pan-African, and global cooperation.

3. *Time and continuity of investigation are necessary.* Both the cassava mosaic disease and the mealybug control programs represent a classic case of the incremental benefits of research—in this case, borrowing cassava technology from the global research community. The leader of the IITA's roots and tubers programs carried out a cassava research program for 23 years, providing an extended opportunity to train hundreds of cassava specialists in graduate degree programs and in short-term training programs at the IITA as well as in "tailor-made" in-country programs. The continuity of scientific leadership is also important in pinpointing and addressing second-generation problems, such as the harvesting labor bottleneck that arose from planting high-yielding mosaic-resistant TMS varieties.

4. *Cassava could be transformed into a cash crop for sale in rural and urban markets.* For generations cassava has been conceived and promoted as a famine-reserve crop in Africa. The most unexpected finding of this study is that there was a surge in demand for cassava as a *cash crop,* reflecting the sharp decline in the price of cassava relative to maize and other food staples. Cassava is now an important cash crop in Africa. This transformation is being propelled by the control of the cassava mosaic virus and mealybug problems.

5. *Sustainability matters.* This study shows that both the mosaic disease and the mealybug control programs have been successful and that they reinforce each other. The achievements of both control programs in contributing to high yields of cassava have been sustained for a period of about 25 years. It remains to be seen how much longer the high yields can be sustained. One thing is clear: research on the mosaic and mealybug control programs needs to be continued in order to protect and sustain the substantial achievements that have been attained.

Notes

1. The period before seasonal foodcrops are ready for harvest.

2. Since the original COSCA field studies in the early 1990s, the author has conducted a number of cassava studies in Nigeria and Ghana: a survey of industrial uses of cassava in Nigeria in 2001, financed by the Food and Agricultural Organization of the United Nations (FAO); a survey of government cassava-sector development policy in Nigeria and Ghana in 2002, financed by the International Food Policy Research Institute (IFPRI); a survey of traditional West African cassava snack foods in 2003, financed by the IITA; a 15-year follow-up survey of the original COSCA farm-

ers in Nigeria in 2005, financed by the Rockefeller Foundation; and a study of the market accessibility for cassava products in Nigeria in 2008, financed by the Bill & Melinda Gates Foundation.

3. Cyanogens are poisonous substances.

4. Purchases were made from local farmers in village markets.

5. Cassava leaves are an important vegetable in the Congo, Sierra Leone, and Tanzania.

6. In 1956, one year before the Amani research station program was terminated in 1957, Jennings distributed seeds of these segregates to several African countries (Jennings 1976).

7. The literature on cassava breeding is captured in Evenson (2003a, 2003b) and Evenson and Gollin (2003).

8. The national programs involved were the Zaire National Cassava Program, the Nigeria National Root Crop Research Institute Program, the Cameroon National Root and Tuber Improvement Program, the Ghana National Root and Tuber Improvement Program, the Rwanda National Root and Tuber Improvement Program, the Uganda National Root and Tuber Improvement Program, and the Malawi National Root and Tuber Improvement Program. The donors that supported the national programs were the U.S. Agency for International Development (USAID), the International Development Research Center (IDRC) (Canada), Administration Generale de la Cooperation au Developpement (Belgium), the Gatsby Foundation (U.K.), IFAD, the World Bank, UNICEF, and the United Nations Development Programme.

9. The IITA, the Natural Resources Institute, IDRC, the Gatsby Charitable Foundation, and USAID.

10. Cassava varieties that can be eaten without processing are called sweet; those that must be processed before eating are called bitter.

11. See Neuenschwander (2001) for an excellent review article on the mealybug program.

12. Cassava leaf is consumed as a vegetable in some African countries.

13. Oligophagous predators are insects that feed on a restricted range of food substances, especially a limited number of plants or other insects.

14. Norgaard had access only to West Africa data, and he extrapolated these data for the whole continent.

References

Afolami, C. A., and A. O. Falusi. 1999. Effect of technology change and commercialization on income equity in Nigeria: The case of improved cassava. Paper presented at the International Workshop on Assessing the Impact of Agricultural Research on Poverty Alleviation, September 14–16, in San Jose, Costa Rica.

Alderman, H. 1990. Nutritional status in Ghana and its determinants. In *Social dimensions of adjustment in Sub-Saharan Africa.* Policy Analysis Working Paper 3. Washington, D.C.: World Bank.

Alene, A., D. Arega, P. Neuenschwander, V. M. Manyong, O. Coulibaly, and R. Hanna. 2005. *The impact of IITA-led biological control of major pests in Sub-Saharan African agriculture: A synthesis of milestones and empirical results.* Impact Series. Ibadan, Nigeria: International Institute of Tropical Agriculture.

Beck, B. D. A. 1980. Historical perspectives of cassava breeding in Africa. In *Root crops in Eastern Africa.* Proceedings of a workshop, November 23–27, in Kigali, Rwanda. Ottawa: International Development Research Center.

Camara, Y. 2000. Profitability of cassava production systems in West Africa: A comparative analysis (Côte d'Ivoire, Ghana, and Nigeria). Ph.D. dissertation, Michigan State University, East Lansing, Mich., U.S.A.

CGIAR (Consultative Group on International Agricultural Research). 1996. *Report of the fourth external program and management review of the IITA.* Rome.

Dahniya, M. T., and A. Jalloh. 1998. Relative effectiveness of sweet potato, melon, and pumpkin as live-mulch in cassava. In *Root crops and poverty alleviation,* ed. M. O. Akoroda and I. J. Ekanayake. Proceedings of the Sixth Triennial Symposium of the International Society for Tropical Root Crops—African Branch, October 22–28, 1995, in Lilongwe, Malawi. Ibadan, Nigeria: International Institute of Tropical Agriculture.

Evenson, R. E. 2003a. Modern variety production: A synthesis. In *Crop variety improvement and its effect on productivity: The impact of international agricultural research,* ed. R. E. Evenson and D. Gollin. Wallingford, U.K.: CABI.

———. 2003b. Production impacts of crop genetic improvement. In *Crop variety improvement and its effect on productivity: The impact of international agricultural research,* ed. R. E. Evenson and D. Gollin. Wallingford, U.K.: CABI.

Evenson, R. E., and D. Gollin. 2003. Assessing the impact of the green revolution, 1960 to 2000. *Science* 300: 758–762.

FAO (Food and Agriculture Organization of the United Nations). 2009. FAOSTAT. <http://faostat.fao.org>. Accessed August 20, 2010.

Hahn, S. K. 2009. E-mail message to author, July 2.

Hahn, S. K., E. R. Terry, K. Leuschner, and T. P. Singh. 1981. Strategies of cassava improvement for resistance to major economic diseases for the 1980's. In *Proceedings of 1st Triennial Root Crops Symposium of the International Society for Tropical Root Crops, Africa Branch, September 8–12, 1980,* in Ibadan, Nigeria, ed. E. R. Terry, K. A. Oduro, and F. Caveness. Ibadan, Nigeria: International Institute of Tropical Agriculture.

Herren, H. R. 1981. Biological control of the cassava mealybug. In *Tropical root crops research strategies for the 1980s,* ed. E. R. Terry, K. O. Oduro, and F. Caveness. Proceedings of the First Triennial Symposium of the International Society for Tropical Root Crops, September 8–12, 1980, in Ibadan, Nigeria. Ottawa: International Development Research Center.

Herren, H. R., and P. Neuenschwander. 1991. Biological control of cassava pests in Africa. *Annual Review of Entomology* 36: 257–283.

Herren, H. R., P. Neuenschwander, R. D. Hennessey, and R. D. Hammond. 1987. Introduction and dispersal of *Epidinocarisi lopezi* (Hym. Encytidae), an exotic parasitoid of cassava mealybug, *Phenococcus manihoti* (Hom. Pseudococcidae) in Africa. *Agriculture, Ecosystems, and Environments* 19: 131–144.

Idowu, I. A. 1998. Private sector participation in agricultural research and technology transfer linkages: Lessons from cassava *gari* processing technology in southern Nigeria. In *Post-harvest technology and commodity marketing,* ed. R. S. B. Ferris. Proceedings of a Post-Harvest Conference, November 2–December 1, 1995, in Accra, Ghana. Ibadan, Nigeria: International Institute of Tropical Agriculture.

Jennings, D. L. 1976. Breeding for resistance to African cassava mosaic disease: Progress and prospects. In *African cassava mosaic,* ed. B. L. Nestel. Report of an interdisciplinary workshop, February 19–22, in Muguga, Kenya. Ottawa: International Development Research Center.

Johnson, M. E., W. A. Masters, and P. V. Preckel. 2006. Diffusion and spillover of new technology: A heterogeneous-agent model for cassava in West Africa. *Agricultural Economics* 35 (2): 119–129.

Jones, W. O. 1959. *Manioc in Africa.* Stanford, Calif., U.S.A.: Stanford University Food Research Institute.

Korang-Amoakoh, S., R. A. Cudjoe, and E. Adams. 1987. Biological control of cassava in Ghana: Prospects for the integration of other strategies. In *Integrated pest management for tropical root and tuber crops,* ed. N. S. K. Hahn and F. E. Caveness. Ibadan, Nigeria: International Institute of Tropical Agriculture.

Maredia, M. K., D. Byerlee, and P. Pee. 2000. Impacts of food crop improvement research: Evidence from Sub-Saharan Africa. *Food Policy* 25: 531–559.

Neuenschwander, P. 1996. Evaluating the efficacy of biological control of three exotic Homopteran pests in tropical Africa. *Entomophaga* 41: 405–424.

———. 2001. Biological control of the cassava mealybug in Africa: A review. *Biological Control* 21: 214–229.

Neuenschwander, P., and T. Haug. 1992. New technologies for rearing *Epidinocarsis lopezi* (Hym., Encyrtidae), a biological control agent against the cassava mealybug *Phenacoccus manihoti* (Horn., Pseudococcidae). In *Advances in insect rearing for research and pest management,* ed. T. E. Anderson and N. C. Leppla. Boulder, Colo., U.S.A.: Westview Press.

Neuenschwander, P., W. N. O. Hammond, A. P. Gutierrez, A. R. Cudjoe, J. U. Baumgartner, and U. Regev. 1989. Impact assessment of the biological control of the cassava mealybug, *Phenacoccus manihoti* Matile-Ferrero (Hemiptera: Pseudococcidae) by the introduced parasitoid *Epidinocarisis lopezi* (De Santis) (Hymenoptera: Encyrtidae). *Bulletin of Entomological Research* 79: 579–594.

Nichols, R. F. W. 1947. Breeding cassava for virus resistance. *East African Agricultural Journal* 12: 184–194.

Norgaard, R. B. 1988. The biological control of cassava mealybug in Africa. *American Journal of Agricultural Economics* 70 (2): 366–371.

Nweke, F. I. 2005. *Collection and compilation of Nigeria's official crop statistics.* Ibadan, Nigeria: International Institute of Tropical Agriculture.

Nweke, F., and S. Haggblade. 2009. Africa's cassava surge. In *Successes in African agriculture: Lessons for the future,* ed. S. Haggblade and P. Hazell. Baltimore: Johns Hopkins University Press.

Nweke, F. I., D. S. C. Spencer, and J. K. Lynam. 2002. *The cassava transformation: Africa's best kept secret.* East Lansing, Mich., U.S.A.: Michigan State University Press.

Obimpeh, S. G. 1994. Opening address. In *Tropical root crops in a developing economy,* ed. F. Ofori and S. K. Hahn. Proceedings of the 9th Symposium of the International Society for Tropical Root Crops, October 20–26, 1991, in Accra, Ghana. Ibadan, Nigeria: International Institute of Tropical Agriculture.

Ogbe, F. O., A. G. O. Dixon, J. d'A Hughes, F. Alabi, and R. U. Okechukwu. 2004. *The status of cassava mosaic disease, cassava begomoviruses and whitefly vector population in Nigeria.* Ibadan, Nigeria: International Institute of Tropical Agriculture.

Onwueme, I. C., and T. D. Sinha. 1991. *Field crop production in tropical Africa.* Ede, the Netherlands: Technical Centre for Agricultural and Rural Cooperation.

Otim-Nape, G. W., and A. Bua. 2000. Uganda case study: Cassava multiplication and distribution to needy farmers. Paper presented at the Global Cassava Development Strategy Validation Forum, April 26–28, FAO, Rome.

Schulthess, F., J. U. Baumgartner, and H. R. Herren. 1989. Sampling *Phenacoccus manihoti* in cassava fields in Nigeria. *Tropical Pest Management* 35: 193–200.

Ssemakula, G. N., Y. K. Baguma, G. W. Otim-Nape, A. Bua, and S. Ogwal. 1997. Breeding for resistance to mosaic disease in Uganda. *African Journal of Root and Tuber Crops* 2: 36–42.

Storey, H. H., and R. F. W. Nichols. 1938. Studies of the mosaic diseases of cassava. *Annals of Applied Biology* 25 (4): 790–806.

Thresh, J. M., G. W. Otim-Nape, J. P. Legg, and D. Fargette. 1997. African cassava mosaic virus disease: The magnitude of the problem. *African Journal of Root and Tuber Crops* 2 (12): 13–19.

University of Greenwich. 2004. Uganda: Saving a nation besieged by cassava mosaic disease epidemic. Paper presented at the NEPAD/IGAD regional conference Agricultural Successes in the Greater Horn of Africa, November 22–25, in Nairobi, Kenya. Photocopy.

Wodageneh, A., and H. R. Herren. 1987. International cooperation: Training and initiation of national biological control programs. *Insect Science and Its Application* 8: 915–918.

World Bank. 2009. *Awakening Africa's sleeping giant: Prospects for commercial agriculture in the Guinea savannah zone and beyond.* Washington, D.C.

Yaninek, J. S. 1994. Cassava plant protection in Africa. In *Food crops for food security in Africa,* ed. M. O. Akoroda. Proceedings of the Fifth Triennial Symposium of the International Society for Tropical Root Crops–Africa Branch (ISTRC–AB). Wageningen, the Netherlands / Ibadan, Nigeria: International Society for Tropical Root Crops / Technical Centre for Agricultural and Rural Cooperation / International Institute of Tropical Agriculture.

Zeddies, J., R. P. Schaab, P. Neuenschwander, and H. R. Herren. 2001. Economics of biological control of cassava mealybug in Africa. *Agricultural Economics* 24 (2): 209–219.

Chapter 5

Community Forestry in Nepal: A Policy Innovation for Local Livelihoods

Hemant R. Ojha, Lauren Persha, and Ashwini Chhatre

Community Forestry in Nepal: An Overview

The Community Forestry Program in Nepal encompasses a set of policy and institutional innovations that empower local communities to manage forests for livelihoods while also enhancing conservation benefits. The program was launched in the mid-1970s as part of an effort to curb the widely perceived crisis of Himalayan forest degradation when the Government of Nepal came to the conclusion that active involvement of local people in forest management was essential for forest conservation in the country. Nepal's Community Forestry Program innovations encompass a well-defined legal and regulatory framework, participatory institutions, benefit-sharing mechanisms, community-based forestry enterprises, and biodiversity conservation strategies. The program is considered a global innovation in the field of participatory environmental governance (Kumar 2002), and its history of implementation and program evolution usefully illustrates a path toward meeting the twin goals of conservation and poverty reduction (Pokharel et al. 2007; Kanel and Acharya 2008).

It is worth noting up front that the evidence linking community forestry to improvements in food security is still being accumulated. Despite more than 30 years of innovation in participatory environmental governance through community forestry, the depth and breadth of material on its impact is still somewhat limited, at least in terms of rigorous analysis. There are few comprehensive studies in the academic literature that disentangle the complex causal relationships between community forestry activities and their livelihood impacts and fewer still that account for the confounding influences of unrelated development interventions and other external factors. More rigorous research designs would contribute importantly to filling these knowledge gaps and generating a more precise understanding of the impact pathways through which community forestry affects rural livelihoods in Nepal.

Although there is a paucity of rigorous impact analyses, there are indications from other disciplines and from a range of observers and stakeholders—from academicians to communities themselves—that community forestry in Nepal is a success in terms of improving natural resource management, rural governance, and institutional reform. Community forestry is flourishing in Nepal and is widely believed to play a role in improving the livelihoods of rural households in thousands of communities and nurturing democracy at the grassroots level despite a prolonged insurgency and political upheavals (Ojha and Pokharel 2005; Bk et al. 2009). Three decades of operational innovations, legislative developments, and evolving practice seem to have contributed positively to several aspects of the lives of forest-dependent people and their environment through enhanced access to forest products, improved livelihood opportunities, strengthened local institutional capacity, and improved ecological conditions of forests (Dev et al. 2003; Ojha and Pokharel 2005; Subedi 2006; Pokharel et al. 2007). By April 2009, about 1.6 million households or one-third of the country's population was taking part in the Community Forestry Program, directly managing more than 1 million hectares or more than one-fourth of the country's forest area. The literature indicates, however, that the apparent achievements are not uniform across Nepal's forest communities; there is also evidence of locations with negligible changes in livelihoods and negative impacts on the poor side by side with gains in forest quality (Malla 2000; Malla, Neupane, and Branney 2003; Shrestha and McManus 2008).[1]

In light of the apparent positive effects on welfare and environmental outcomes, community forestry is perceived by many stakeholders and observers as one of the few promising aspects of Nepal's post–World War II history. It has often been used as a face-saving instrument by development actors who have been engaged in, if not responsible for, the five decades of "failed development" in Nepal.[2] In Nepal community forestry has had a positive image not only in the fields of development and natural resource management but also more widely from a governance perspective, with assertions made that the local-level institutions for forest management and their networks provide a model of democratic governance (Ojha and Pokharel 2005).

The wide coverage of the program (Table 5.1) suggests the possible depth of local ownership, action, and empowerment facilitated by the Community Forestry Program within local communities via Community Forest User Groups (CFUGs), local-level forest management entities.

Table 5.1 Summary indicators of community forestry impacts in Nepal, 2009

Indicator	Number	Share
Households directly affected	1,659,775	32% of total population
Number of CFUGs	14,439	
Number of districts with community forestry operations	75	100% of all districts
Area of forest under CFUG management	1,229,669	25% of total forest area

Source: India, DoF (2009).
Note: CFUG means Community Forest User Group.

The Intervention: History and Evolution of Community Forestry

Forests have historically held a central place in local livelihood practices and national politics in Nepal because of their importance in rural livelihoods and state revenues (Ojha 2008). Analysts have usefully delineated three phases of forestry in Nepal: privatization (until 1957), nationalization (1957 to the late 1970s), and decentralization (from the late 1970s onward) (Hobley 1996). Most forests in rural Nepal were controlled and managed by local communities until the late 1950s, when the state took control. The call for citizen participation began in the late 1970s, when the government explicitly admitted that it could not protect the country's forests without the active cooperation of local forest-dependent citizens.

Throughout Nepal's modern history—the past 240 years—the Nepali state has largely been controlled by either the Shaha or the Rana family, except for three brief periods of democracy in the 1950s, the 1990s, and after 2006. Under the control of these families, the state polity retained a strong feudal character, whereby economic surplus flowed from the peasant farmers to the ruling elites through networks of locally based feudal lords (Regmi 1978), although that control apparatus has gradually declined since 1951. Until the Private Forest Nationalization Act was enforced in 1957, all forests were controlled by state-sponsored local functionaries. As the state moved further into the era of planned development after World War II, national bureaucracies assumed political–economic control of resources in ways that served the interests of the ruling elites (Blaikie, Cameron, and Seddon 2002). A number of laws were enacted to enforce national control over forests, which effectively expanded the forest bureaucracy and excluded local people.[3] Although it was implicitly assumed that transferring forests from private groups to the state would enhance people's access to forest resources, in reality the state instituted stringent regulations to exclude people from controlling forest resources and created a strong technobureaucratic field (Malla 2000; Ojha 2008).

Key Policies for Community Forestry in Nepal

Efforts to share power over forests with local people started in 1978 with the institution of panchayat (local government) forest regulations, prompted by the central government's realization that the state forest bureaucracy could not protect forests without engaging the local people.[4] This move was part of the monarchical panchayat system's strategy to thwart growing anti-panchayat resistance by offering people some economic and symbolic spaces in the local panchayat. In the meantime, donors were exerting growing pressure on the government to shift away from centralized practices of development toward more decentralized processes.

During the 1970s, the recognition of Himalayan degradation as a serious environmental crisis (Eckholm 1976) increased the pressure on international development institutions and donor governments to contribute to the conservation of the Himalayas. This led to a shift in the development discourse away from an emphasis on infrastructure and technology transfer toward environmental issues (Cameron 1998). Moreover, Nepal's strategic geopolitical situation (located between China and India) and fragile environmental condition attracted donors (Metz 1995) who viewed forestry and the environment as the key elements of integrated conservation and development projects.[5]

Several international agencies assisted the Nepalese government in formulating the Master Plan for the Forestry Sector (MPFS), which recognized the need for local people's participation in the conservation and management of the country's forest resources. In 1989, as the MPFS was being finalized and formally adopted by the government, an ongoing movement against the panchayat system by the citizenry also culminated in the reinstatement of multiparty democracy in the country. The decisions of subsequent governments further strengthened the regulatory framework of community-based forest management in line with the MPFS.

The most significant regulatory development in support of community forestry was the enactment of the Forest Act in 1993 by the first parliament elected after the 1990 movement for democracy. The Forest Act guaranteed the rights of local people in forest management (MFSC 1995; see Ojha, Persha, and Chhatre 2009 for the details of CFUG rights under the 1993 Forest Act and the 1995 Forest Regulation). Nepal became the world's first country to enact such radical forest legislation, allowing local communities to take full control of government forest patches under a community forestry program (Malla 1997; Kumar 2002). Meanwhile, international agencies continued to support the process of reorienting government forestry officials to work as facilitators of community-based forest management and away from their traditional policing roles (Gronow and Shrestha 1991).

Scaling Up: Policy, Institutional, and Methodological Innovations

The Community Forestry Program in Nepal evolved from a primarily protection-oriented, conservation-focused agenda during its initial years of implementation to a much broader-based strategy for forest use, enterprise development, and livelihood improvement. This occurred through an often conflict-ridden process spread out over more than a decade, during which sustained efforts to engage in policy dialogue with a range of community forestry stakeholders helped to clarify issues and develop a common vision. A well-recognized effort in this regard has been the convening of national workshops every five years since 1987. Along with the evolution of community forestry in Nepal, the government forestry authority (mainly the Ministry of Forest and Soil Conservation and its Department of Forests) has also reinvented itself as a facilitator of community institutions, moving away from traditional policing roles (Niraula 2004a; Kanel and Acharya 2008). Evolution occurred across policies, institutions, and implementation modalities, ultimately leading to a much stronger, more sustainable, and effective Community Forestry system (Pokharel et al. 2008). The experience of community forestry has also been adapted and scaled up in different contexts in Nepal, leading to at least five other institutional regimes of forest governance: leasehold forestry, collaborative forest management, community-based watershed management, integrated conservation and development, and protected-area buffer-zone forestry (Ojha and Timsina 2008).

One of the keys to the establishment and successful outcome of Nepal's community forestry system was the creation of appropriate institutional structures at local, meso, and national levels that included downward accountability and relatively unrestricted decisionmaking at the local level and effective cross-scale interactions among these various institutions. At the local level, this included provisions for subcommittees within CFUGs and the establishment of elected hamlet-level representatives to ensure that the concerns of various constituencies within the CFUGs were expressed (McDougall et al. 2008). Meso-level institutions performed key facilitation, technical, and information exchange functions to meet national priorities in local contexts. Such interactions particularly contributed to better exploitation and wider dissemination of market opportunities by CFUGs (Banjade et al. 2007), which in turn had positive impacts on household livelihoods.

Other institutional factors in the successful evolution of community forestry included efforts to improve the inclusion of all social groups (especially after the mid-1990s, when the Maoist movement also gained momentum through the agenda of inclusion), concomitant democratic processes (Pokharel et al. 2007), and the provision of adequate time and space for frequent discussion, exchange, adaptation, inclusion, and interaction among stakeholders (Banjade et al. 2006).

Policy innovations that enhanced the successful scaling up of the program included progressive legislation (the Forest Act of 1993), which also supported strong, autonomous, self-governed village institutions (CFUGs) and clarification of appropriate property rights arrangements for community members through the provision of an operational plan for community forest management. Deconcentration of authority from the centralized state to the district-level bureaucracy, in which district officials were given the authority to constitute CFUGs, also played an important role (Agrawal and Ostrom 2001).

A variety of methodological innovations also helped to improve community forestry implementation, such as participatory action learning (Malla et al. 2002), adaptive collaborative management (McDougall et al. 2008), participatory and self-monitoring approaches (Banjade, Luintel, and Neupane 2008), and approaches that specifically targeted pro-poor livelihood improvements linking civil society, governance, and democracy with natural resource management (Luintel, Bhattarai, and Ojha 2006). Such innovations addressed relationships among a wide range of participants in the forest governance arena and have triggered pro-poor forestry practices (Kanel and Subedi 2004). The governance practices that were changed encompass a broad range of activities—learning, planning, and decisionmaking; mobilizing marginalized groups to create pressures on elites; developing a clearer vision, indicators, and purpose for the CFUGs and related community forestry organizations; monitoring; promoting transparency; reelecting executive committees; creating ownership in the organizational change processes; improving communication; and promoting public hearings and auditing.

As a result of improvements in governance practices, changes in governance outcomes included more equitable benefit sharing; enhanced transparency, participation, and accountability; and improved pro-poor resource management practices (Luintel, Bhattarai, and Ojha 2006; Shrestha et al. 2009). Many of these innovations came in response to the emergence of second-generation issues in the mid-1990s and were facilitated by national nongovernmental organizations (NGOs), often with technical support from international organizations and bilateral donor projects.

Since 1990, the process of community forestry has been increasingly promoted and scaled up by an expansion of the public sphere, often operating outside of government and donor projects (Ojha, Timsina, and Khanal 2007). There have been increasing instances of proactive engagement of civil groups in forest governance in recent years in Nepal. Of particular importance has been the establishment of a meso-level umbrella institution of CFUGs that represents the interests of local-level actors and serves as an intermediary between national and local processes. This nationwide network of CFUGs, known as the Federation of Community Forestry Users, Nepal (FECOFUN), has emerged as a key player in forest-sector policy debates (Ojha 2002; Ojha and Timsina 2008). These civil society groups have further politicized the practice of forestry and in many respects provided a deliberative bridge between people and the state (Ojha, Timsina, and Khanal 2007). Along with NGO alliances, it has brought civil society perspectives into the policymaking process that was once dominated by technocrats and bureaucrats. The most important policy issue in which FECOFUN has made significant contributions in the past few years concerns the extension of CFUG rights over forest resources in the hills as well as in the Terai. Through FECOFUN, the legal provisions relating to community forestry were spread to areas where there were no prior donor-driven projects or where district forest offices (DFOs) were not as enthusiastic about community forestry implementation (in the Terai, for example). In addition, FECOFUN has played the role of CFUG watchdog in national and international policy arenas. FECOFUN's awareness-raising activities have helped to enhance the political capital of CFUGs beyond the traditional patron–client relationship with the Department of Forests.

Successful scaling up of community forestry also required a nationwide overhaul of local DFOs, with an emphasis on the reorientation of forest officials. This approach enables DFOs and forest officials to reorient their skills toward co-management, extension, and assistance from their previous role as the dominant authorities and decisionmakers in forest management (Acharya 2002).

Finally, scholars have further enumerated a number of specific conditions and factors that played a significant role in the successful evolution of community forestry in Nepal. These include

- the media projection of the crisis of Himalayan degradation and consequent international assistance (Gutman 1991);

- the inaccessibility of Nepal's hill and mountain forests for commercial exploitation;

- the inability of the Forest Department to manage forests effectively, especially in the middle and high hills (Gilmour and Fisher 1991; Subedi 2006);

- the emergence of a multiparty political system in 1990 and consequent expansion of civil society spaces (Ojha 2006);

- the willingness of the elected government to legally empower local communities to manage forests (Ojha 2006);

- the presence of existing forest-based livelihood systems in rural Nepal and incentives for local people to participate in forest management for a range of forest products and livelihood opportunities (Gilmour and Fisher 1991);

- the presence of existing dense social networks and traditional models of collective action around local forest management in Nepal (Fisher 1989; Chhetri and Pandey 1992);

- the continued tradition of piloting new approaches and reflection among Community Forest Program stakeholders, including regular nationwide workshops every five years since the 1980s (Pokharel et al. 2007; Ojha and Timsina 2008);

- the increased research on and scholarly interests in community forestry; and

- the breakdown of traditional power relationships through political movements and emergence of "subaltern" groups taking leadership power at the CFUG level (Bhattarai 2007).

Issues Faced during Implementation

A number of challenging issues were encountered during implementation of Nepal's Community Forestry Program, and some continue today. These include the ongoing evolution of extension skills and technical capacity within government institutions to move beyond a blueprint forest protection model toward a livelihood-oriented system and issues related to equitable distribution of community forestry benefits among local CFUG institutions and households.

Management models, operational plans, and related implementation processes initially adhered to blueprint models provided by the Forest Department and focused on forest protection rather than livelihood improvement (Dougill et al. 2001). Over time, management and operational plans gradually evolved to reflect individual CFUG goals and took on a more livelihood-oriented emphasis. This was also reflected in the design of forestry programs under a livelihoods framework, such as the United Kingdom's Department for International Development's Livelihoods and Forestry Program (LFP), which began in 2000. This new emphasis went hand in hand with the adaptation of appropriate extension skills within the Forest Department and DFOs in order to provide effective technical assistance to CFUGs and to assist them with management decision-making (Dev et al. 2003).

Forest management also became more technically complex beyond the initial simplistic plans. Initially, CFUGs were required to have only one document containing both constitutional aspects and forest management rules (Ojha, Khanal, and Shrestha 1997). Since 1995, two separate documents have been required: a constitution and a forest management operational plan. Additional technical aspects of community forestry included developing readily usable tables to facilitate the estimation of biomass, timber volume, and annual harvesting yields (Acharya 2002). An inventory guideline was enforced in 2000 that was directed more toward technocratic control than to facilitating democratic forest governance (Ojha 2002).

However, stakeholders later came together and developed a common understanding of the format and process of forest inventory, which was incorporated into the revised forest inventory guidelines of 2004 (Paudel and Ojha 2008).

Another set of challenges stemmed from issues related to the distribution of benefits (forest products and income), social exclusion and marginalization of traditionally disadvantaged groups, elite capture of benefits and decisionmaking processes, and lack of transparency in managing CFUG funds (Kanel and Kandel 2004; Chhetri 2006). After these problems were identified, several CFUGs began to include in their operational plans explicit provisions for directing greater benefits to poorer groups, women, lower caste groups, and other marginalized groups (Bhattarai 2007; Banjade, Luintel, and Neupane 2008; Kunwar et al. 2009). Bhattarai (2007) identifies interactions and knowledge networks that influenced the perceptions of local elites about themselves and the poor, thereby triggering pro-poor forest management and use practices in the CFUGs. Pro-poor innovations generally include designating loans, land for cultivation, or areas of the community forest for fodder collection to be explicitly reserved for marginalized groups or the poorest households and setting up female-dominated small business enterprises (Joshi et al. 2006). In several cases, community fundraising through the sale of forest products has also been used to fund rural infrastructure or social development works that address the needs and concerns of the poor and disadvantaged.

Equitable rather than equal distribution of forest products and benefits has been crucial to improving livelihoods, because the poorest households in Nepal are more dependent on forest products and forest-derived sources of livelihoods, having little or no land of their own from which they can obtain such products. An equitable way of distributing forest products, in which the CFUG collectively ranks households on the basis of relative wealth and subsistence needs, should lead to greater livelihood security for the most vulnerable households and should contribute to meeting community forestry's poverty alleviation goals. The LFP in Nepal has supported a number of CFUGs in providing exclusive management rights to groups of poor households for cultivation of income-generating crops and agroforestry. Although currently few in number, some CFUGs do provide community lands to their landless or nearly landless members so they can earn their living through the cultivation of medicinal herbs or other crops. Several CFUGs give preference to poor members or women in locally created jobs, such as processing handmade paper or working as nursery laborers (Subedi 2006). Other examples of pro-poor provisions in CFUG constitutions and operational plans also exist (Bhattarai 2009; Ojha, Persha, and Chhatre 2009). However, a study by Shrestha and McManus (2008) suggests that in certain localities the methods used to promote the equitable distribution of forest products and benefits have had a negligible, and sometimes negative, impact on the poor. This issue of impact is discussed further in the next section.

Impact of Community Forestry

Rural Livelihoods and Welfare

Community forestry appears to have had a net positive effect on livelihoods and a range of other development concerns in Nepal, resulting in direct and indirect positive impacts on rural livelihoods and welfare. However, we caution that rigorous studies demonstrating significant increases in household income as a result of Nepal's Community Forestry Program are sparse in the available literature. Studies that attempt to further disentangle complex relationships among community forestry activities, unrelated development interventions, and economic and other aspects of household livelihoods, particularly through rigorous research designs that control for external factors, would contribute to a clearer understanding of community forestry impacts on household income in Nepal.

Suggestive evidence of direct improvements to livelihoods include an enhanced supply of wild edibles used by the poor, increased availability of forest products to farmers, and more reliable product supply (Acharya 2002; Dev et al. 2003). Indirect but crucially important improvements to livelihoods apparently occurred through increased employment opportunities and more diversified livelihood portfolios. Community forestry has likely enabled households to diversify their livelihood strategies to a greater extent than was possible before through forest-based income-generating activities such as cultivating spices in the forest understory and tapping resin from selected tree species (Dev et al. 2003).

A longitudinal study of 2,700 households from 26 CFUGs in the Koshi Hills may illustrate the large-scale impacts of community forestry on poverty alleviation and livelihood security (Tables 5.2 and 5.3), though issues of research design limit the certainty of the outcomes. In that study, a well-being ranking conducted in the sampled CFUGs during two time periods across five years showed that 46 percent of poor users (very poor and poor) moved into higher well-being categories, facilitated in part through their participation in CFUGs that directly supported livelihood improvement and capacity-building activities (Chapagain and Banjade 2009). A separate study found that the average annual household income of CFUG members increased by 113 percent over a period from 2003 to 2008, from NRs 54,995 in 2003 to NRs 117,075 in 2008 ($710 to $1,512) (Chapagain, Subedi, and Rana 2009). This represented a 61 percent increase after adjusting for inflation and should be seen in the light of Bhattarai and Dhungana's (2008) claim that CFUGs have harnessed less than 40 percent of the revenue potential of their community forests. Chapagain, Subedi, and Rana (2009) found a difference of NRs 16,153 ($208) between household incomes with and without income-generating activities supported by CFUGs, signifying the importance of user-group support. But research design limitations in these studies unfortunately preclude the attribution of

Table 5.2 Changes in well-being of 2,700 households from 26 CFUGs in the Koshi Hills, Nepal, 2002–08 (percent)

Caste	No change			Change (+)			Change (–)		
	VP–VP	P–P	Oth–Oth	VP–P	VP–Oth	P–Oth	P–VP	Oth–P	Oth–VP
Dalit	58	43	100	29	13	51	6	0	0
Ethnic minorities	53	67	100	36	11	32	1	0	0
Advantaged castes	55	59	100	36	7	39	1	0	0
Total	56	61	100	35	9	37	1	0	0

Source: Chapagain and Banjade (2009).
Note: VP means very poor; P means poor; Oth means others.

Table 5.3 Annual employment opportunities in community forestry in the Koshi Hills, Nepal, 2009

Area	Number of CFUGs generating employment	Total population	Employment: Person-days per CFUG per year
Forest management	510	63,888	125.19
Community development	340	9,411	27.66
Office management / office secretaries	161	9,153	56.97
Teachers	172	39,137	226.92
Enterprises	95	15,937	168.64
Total	1,278	137,526	605.38

Source: Adapted from Chapagain and Banjade (2009).
Note: CFUG means Community Forest User Group.

these positive income changes solely to the Community Forestry Program, because control groups were not employed in the longitudinal comparisons of household income, nor were there controls for changes in non-CFUG sources of income over the period of the study.

One recent study that did employ control groups found that economic benefits from Community Forestry did accrue but at the CFUG level rather than the household level (Maharjan et al. 2009). Conducted in eight Community Forestry villages and two control villages across four districts of Nepal, the study did not find evidence of an increase in household-level income as a result of Community Forestry activities. However, it emphasized positive and substantial welfare impacts in terms of improved physical infrastructure, skills development for potential employment, networking and other forms of social capital, and political representation, often made possible through CFUG funding or other Community Forestry activities (Maharjan et al. 2009).

Indirect contributions to household incomes also appear to be realized through revenue-generating activities undertaken by CFUGs, such as the sale of forest products, and are then used for a range of community development initiatives (Dhakal, Bigsby, and Cullen 2005, 2007). One study of 23 CFUGs in the middle hills, in which quantitative surveys about CFUG finances and management were administered to CFUG members and government employees, found that successfully managed community forests (defined by the authors on the basis of comparatively higher revenue returns per hectare of forest managed, among other significant factors included in a hierarchical cluster analysis) generated a mean annual revenue of $18.50 per hectare of forest and that the related CFUGs spent a greater proportion of their revenue on community development (57 percent) and forest management (32 percent) activities than did CFUGs with lower revenue returns (Dongol, Hughey, and Bigsby 2002).

CFUG-based enterprises provide income and employment for community members, and in many cases members of poorer households, who are typically more vulnerable to food shortages because of their socioeconomic status, are given priority for jobs (Dev et al. 2003). For example, 617 CFUGs from five districts in Nepal's midwestern region provided 825,988 person-days of employment over a one-year period (2007–08), and 90 percent of that employment went to people from poor and very poor households (LFP 2008, cited in Bhattarai 2009) (Table 5.4). Employment opportunities may include work as forest watchers or wage laborers during various silvicultural operations, and CFUGs often prioritize jobs as timber workers, nursery technicians, fuelwood sellers, and resin collectors for poorer members (Bhattarai 2009). Nonetheless, the assessment of senior government officials associated with the community forestry program was that "the contribution of community forestry for

Table 5.4 Paid employment generated by CFUGs in midwestern Nepal during the 2064/65 fiscal year (2007/08 on the Gregorian calendar)

Type of work	Number of CFUGs	Person-days of paid employment						
		Dalit	Janajati	Minority	Other	Poor and very poor	Other	Total
Forest management	370	117,921	290,028	11,025	164,242	549,260	33,956	583,216
Nursery management	87	3,375	44,411	261	8,364	53,930	2,471	56,411
Other	160	57,623	60,452	0	68,286	138,892	30,994	186,361
Total	617	178,919	394,891	11,286	240,892	742,082	67,421	825,988

Source: LFP (2008).
Note: CFUG means Community Forest User Group.

poverty alleviation as targeted by the Tenth Plan or PRSP and millennium development goals is limited" (Kanel, Poudyal, and Baral 2005, 81).

Pro-poor mechanisms for the distribution of forest products have also had positive effects on households' ability to meet their livelihood needs. Forest products may be distributed at subsidized rates to poor households and at no cost to female-headed and extremely poor families (Bhattarai 2009). For instance, Mahila CFUG of Kalimati Rampur has moved from a system of equal distribution of forest products among all members to an equitable distribution system that provides forest products at subsidized rates to more vulnerable members of the CFUG. Timber is sold at either 65 or 50 percent of actual price to users from designated poorer households and is freely distributed at no cost to homeless users. Such subsidies provide a more reliable and lucrative source of income for the poorest households than was previously available, because members may buy forest products such as fuelwood at a low rate and then sell them at market for a substantially higher price (Bhattarai 2009). However, as documented by Shrestha and McManus (2008), pro-poor arrangements are not enacted uniformly in all localities.

Community forestry funds have been used for wide-ranging infrastructure and community development projects that have improved market accessibility for remote villages (Table 5.5). Infrastructure projects supported by CFUGs include building or blacktopping of roads and construction of bridges, small-scale irrigation systems, drinking water systems, training centers, and guest houses. Many CFUGs

Table 5.5 Ongoing CFUG activities in the Koshi Hills, Nepal, 2008/09

Activity	Number of CFUGs	Number of households benefiting
Income-generating activities (IGAs)		
Revolving fund	426	8,029
Community forestry land allocation	67	767
Other IGAs	26	319
Health and sanitation activities	23	177
Improved cooking-stove construction and distribution	26	177 (82% poor households)
Drinking water schemes	167	12,480 (68% poor households)
Irrigation canal construction and maintenance	41	7,217 (65% poor households)
Trail or road maintenance to improve market access	297	n.a.
Supporting schoolteacher remuneration	67	n.a.
Electrification projects	10	907 (52% poor households)
Wooden bridge construction	5	n.a.
Office building construction	30	n.a.
Tourism-related activities	5	n.a.

Source: Chapagain and Banjade (2009).
Note: CFUG means Community Forest User Group; n.a. means not available.

in Nepal have used their funds to support the education infrastructure by providing funds for teachers' salaries, school construction, school furniture, scholarships, nutritional enhancement, forest excursions, and cultural programs.

Savings, credit, and microenterprise schemes developed by CFUGs have also made substantial contributions to household livelihoods in some districts in the country, and these are often targeted to benefit the poorest households. Among the most innovative cases are those that provide housing services to the poorest landless households. Although microfinance is currently a relatively uncommon use of CFUG funds, it may be one of the most promising options for future livelihood improvements stemming from community forestry (Dev et al. 2003; McDougall et al. 2008). For example, 312 CFUGs in Parbat District implemented savings and credit activities through which they mobilized more than NRs 5.8 million ($74,935, or $240 per CFUG) and supported more than 2,100 poor, marginalized, Dalit,[6] and conflict-affected households (Luintel, Ojha, and Rana 2009). A total of 14,213 households in the Rapti region benefited from microenterprises developed through community forestry, and 94 percent of those households were from a poor or marginalized group (Luintel, Ojha, and Rana 2009).

In LFP areas, 4,500 CFUGs from 15 of the 75 districts of Nepal raised more than NRs 120 million ($1,521,227, or an average of about $345 per CFUG) from forest and nonforest resources and spent about 70 percent of that income on activities to enhance local livelihoods: 27 percent on forest management, 34 percent on social development activities, 18 percent on pro-poor provisions, and 21 percent on institutional development of the groups (Chapagain, Subedi, and Rana 2009). They estimate that CFUGs spent about NRs 14.4 million on poverty-focused actions during one year in the 15 districts. During 2007–08, these CFUGs generated paid local employment equivalent to NRs 180 million (1.5 million person-days). Of the total employment, 19 percent was for Dalits and 31 percent for women. Overall, 85 percent of employees were from poor households.

Data from the LFP of Nepal shows that Dalit representation on CFUG executive committees was about 6 percent until 2003 but increased to almost 15 percent by the end of 2007, on a par with the overall percentage of Dalits in the 15 program districts of the LFP (Table 5.6) (Chapagain, Subedi, and Rana 2009). Similarly, the representation of disadvantaged ethnic minorities has also increased from 32 percent in 2003 to 44 percent in 2008. The representation of the economically poor on CFUG decisionmaking committees during the period also increased from 31 to 52 percent.

However, it remains to be seen whether increased representation of the poor will translate into greater pro-poor orientation of the spending of community forestry groups. A few years ago a senior official of the government's community forestry program noted that community forestry groups "are now spending 36

Table 5.6 Representation of ethnic people on executive committees of CFUGs (percent)

Caste/ethnicity	Percentage of national population	CFUG representation, 2003	CFUG representation, 2008
Dalits	12.3	6	12
Ethnic minorities	29.9	32	44
Muslims	3.6	0	1
Women	—	21	36
Poor	—	31	52

Sources: Chapagain, Subedi, and Rana (2009); Gurung (1996) for Dalits' percentage of the national population; CBS (2002) for ethnic minorities' percentage of the national population.
Note: CFUGs means Community Forest User Groups; — means data were not available.

percent of their expenditures in community development activities such as school, road, health post and other development activities. . . . The benefits from those activities are minimal to the poor. About three percent is spent on pro-poor programs" (Kanel 2006, 33–34).

Gender-Specific Dimensions

Women hold about 25 percent of executive committee positions within CFUGs (Kanel and Kandel 2004) but are still struggling to rise to decisionmaking positions in the community forestry sphere (Nightingale 2002; Timsina 2002). One strategy toward balancing gender distribution has been to form women-only CFUGs. Luintel and Timsina (2008) report that women-only CFUGs are generally provided with relatively small and marginal land as community forests. These CFUGs have access to forest lands only half the size per household of those available to mixed CFUGs (0.34 hectare for women-only CFUGs, compared with 0.73 hectare for mixed CFUGs) (Rai-Paudyal and Buchy 2004). Another initiative for balancing gender has been to include the names of women on the CFUG member list instead of including only the male household head, as was the earlier practice. Although gender concerns have long been ignored in community forestry (Buchy and Subba 2003), Luintel and Timsina (2008) suggest a rising trend of women's participation in community forestry. An analysis of minutes of assemblies and meetings of 11 CFUGs that they studied shows that representation by women, including Dalits, has increased over time (Table 5.7).

Women's presence at CFUG meetings and assemblies is often constrained by their sociocultural roles in society as well as their limited prior deliberative experience (Nightingale 2002; Luintel and Timsina 2008). However, in recent years greater participation by women—in both quantity and quality—has been

Table 5.7 Increasing trend of women's representation on CFUG executive committees over time

CFUG	Executive committee composition at time of formation		Composition between first and current executive committee		Current composition	
	Women	Men	Women	Men	Women	Men
Dangsera	4	6	11	2	5	6
Tham	0	9	2	9	2	9
Byangdhunga	n.a.	n.a.	1	5	5	6
Mathillo Patle	0	8	1	10	2	9
Pandey	n.a.	n.a.	n.a.	n.a.	2	5
Kagbeni FMSC	0	9	0	9	0	9
Kalo Ban FMSC	n.a.	n.a.	n.a.	n.a.	1	6
Sarbodaya	0	11	5	9	5	6
Nashawa	4	11	4	11	5	10
Kamalpur	5	14	0	15	11	4
Chautari	2	9	0	10	5	8
Total	15	77	24	80	43	78

Source: Luintel and Timsina (2008).
Notes: CFUG means Community Forest User Group; n.a. means not available.

observed in areas where women's ability to make choices is relatively independent at the household level. Many women-only CFUGs are operating successfully, and Community Forestry Program actors clearly prioritize gender-related goals, but still there is a continuing issue of gender discrimination at more subtle levels—cultural, political, and symbolic (Rai-Paudyal and Buchy 2004; Luintel and Timsina 2008). Moreover, when women have greater access to financial assets through various activities, such as income generation and savings and credit, their families benefit from women's participation in the public domain (Luintel and Timsina 2008).

Controversial or Unintended Negative Aspects

Livelihoods seem to have benefited from community forestry in Nepal, although some controversial aspects serve as the basis for ongoing research, critique, and innovation to further strengthen and adapt the LFP. Controversial aspects include insufficient quantitative evidence of an improvement in the income component of livelihoods, particularly for the poorest households and marginalized groups; divergent viewpoints over long-term community forestry management goals; an overly tenuous policy and enabling environment for pro-poor management; land tenure insecurity, especially for the poorest and marginalized groups; and difficulties in implementing community forestry in the timber-rich Terai region of the country.

First, perhaps the most prominent of these aspects is evidence that the poorest households, who are the primary focus of pro-poor development interests, appear to benefit less from community forestry than do wealthier households in a community. Some studies have found that wealthier households not only tend to control forest management decisions but also may make access to forest products disproportionately more difficult for poorer households by making management decisions that serve their own interests. Examples include focusing management efforts on timber production, restricting the amount of nontimber forest products collection, introducing fee-based collection systems, and reducing access to CFUG funds by other members of the group (Pokharel and Nurse 2004; Dhakal, Bigsby, and Cullen 2005).

Other studies show quantitatively that the poorest households in the communities studied, which do not have enough land to support their basic subsistence needs and are thus more reliant on forest products than are other community members, receive disproportionately smaller livelihood benefits from community forestry than do wealthier households (Malla 2000; Adhikari, DiFalco, and Lovett 2004; Dhakal, Bigsby, and Cullen 2007). Certain forest products, such as timber, are too expensive for nonwealthy households to afford, even with subsidies. One suggestion for overcoming this "hidden subsidy" (Iversen et al. 2006) is to shift primary management priorities in community forests away from timber production while increasing the emphasis on fodder production and agroforestry (Dhakal, Bigsby, and Cullen 2005, 2007). Because timber permits and harvesting fees are out of reach for many of the poorer households in a community, another suggestion is to increase the focus on nontimber forest products as sources of revenue (Kanel and Kandel 2004). Other studies of community forestry situations report a decline in the availability of fuelwood and fodder, no evidence of enhanced employment opportunities, little overall increase in household incomes or livestock resources, or a lack of livelihood improvement for the poorest households in a community (Dougill et al. 2001; Timsina 2003; Dhakal, Bigsby, and Cullen 2007; Shrestha and McManus 2008; Maharjan et al. 2009).

Second, some reviews have called for a stronger enabling policy framework to promote pro-poor forest management in Nepal (Acharya, Adhikari, and Khanal 2008; Bhattarai 2009). One strategy for this could be to lease out parts of community forest land to the poorest groups for short-term cash crop cultivation or agroforestry, but community forestry legislation does not allow the planting of annual crops on community forest land.[7] Poor users are encouraged by the local DFO staff to plant forest or wild crops that are not always preferred by the poor or compatible with their requirements. Studies such as those of Bhattarai (2009) and Paudel, Banjade, and Dahal (2009) report a widespread perception among community forest users that the government retains overall forest management authority while simply shedding the responsibility for the day-to-day management of forests to local communities.

Third, as the LFP experience demonstrates (Bhattarai 2009; Kunwar et al. 2009), land tenure security remains a critical issue in relation to providing sustained incentives to the poor to invest in the forest land allocated to them. In the land allocation schemes, tenure is currently defined through an agreement between a poor household (or group of households) and the CFUG committee, but there is no regulatory provision to facilitate and provide legal security for such pro-poor transactions. Given that there is high demand for forest land among households of all wealth categories, Bhattarai (2009) fears that community-level agreements with poor and excluded groups may easily revert under local pressures and politics. Indeed, it may be the case that pro-poor innovation is sustainable over the long run only when there are strong institutional mechanisms built in to favor the poor at village and higher levels. The current political transition is full of inclusion and restructuring agendas, and some of the deeper structural issues of exclusion around forest and natural resource management are being articulated in national politics and the constitutional debate (Himal 2009).

Finally, there has been much controversy over the implementation of community forestry in the Terai region of Nepal, as well as conflict with another competing program called Collaborative Forest Management (Bampton, Banjade, and Ebregt 2008). The commercial value of forests in the Terai greatly exceeds that of forests in the middle hills because Terai forests are dominated by hardwood timber tree species. Terai forests therefore have the potential to generate much greater revenues for CFUGs than do middle hills forests, but they also have a much greater potential as a source of revenue for the state. The presence of products of high commercial value in Terai forests creates greater conflict over forest resource access, benefits distribution, and overall Community Forestry Program implementation. It also prompts concerns that the institutional model for community forestry in the middle hills may need to be substantially altered if it is to succeed in avoiding elite capture or state appropriation of benefits in the Terai (Iversen et al. 2006).

Sustainability of Community Forestry in Nepal

As mentioned earlier, community forestry in Nepal is not an entirely external intervention. It is indeed a negotiated process of forest governance between local communities and the state, with additional developmental inputs from donor-funded programs and advisory and advocacy inputs from NGOs. Thus, when we refer to "community forestry" we do not mean merely a government program but a complex set of socioecological interactions involving local communities and their institutions, government policies and programs, and associated technical, institutional, and political processes at multiple levels that affect forest management choices and actions of local people (Ojha, Agrawal, and Cameron 2009). From this perspective we argue that the question of sustainability should not be focused on external intervention but should concentrate on local processes and then move up the scale to examine the effects of wider contextual drivers. Likewise, sustainability analysis should not be confined to the sustainability of the government program inputs. Sustainability is understood not merely as the net present value of a stream of economic benefits but also as a sense of place and belonging to a community, social identity and power relations, social capital and civic engagement, and local worldviews and knowledge. Seen from these perspectives, community forestry in Nepal largely appears to be moving along a sustainable trajectory.

Community forestry processes have continuously expanded over the past three decades in terms of the number of CFUGs formed, the area of forest handed over to community management, and the number of households or families involved. During this period the level of involvement of donors and government organizations has shrunk while the involvement of NGOs and CFUG networks has expanded. The space for decentralization and community participation in natural resource management has partly been strengthened by the parallel processes of political mobilization and the growing consciousness of people in Nepal of self-governance and democratization. Above all, the immediate livelihood benefits derived by rural households—as an input to agriculture, food security, and cash incomes—are the keys to strong collective action within local communities, allowing them to actively manage their forest resources.

Community forestry is sustained by a legally defined tenurial structure that is well accepted by local communities and wider community forestry program stakeholders. Radical community rights activists do not demand change in the legal system, unlike in many other contexts, but monitor changes in the existing legal framework that may impinge on community rights. Although issues of tenure and power sharing between local communities and the government are legalized and provide secure tenure rights to local communities, there are sometimes tensions between local communities and the government in defining, interpreting, and enacting these formally agreed rights (Shrestha 2001; Ojha 2006). At times this tension overflows in street protests or intense negotiations, cultivating a feeling of instability and confusion over tenurial security even as it strengthens the claims of local communities. The recurrent issue is the extent to which processes of policymaking, program planning, and implementation provide opportunities to local community groups and civil society networks to influence forest governance. The debate is not so much about principles or legal arrangements as about everyday practice, with actors seeking to defend or maximize their self-interests. This is particularly serious when it comes to registering CFUGs, planning forest management, and harvesting and marketing forest products from community forests. See Ojha, Persha, and Chhatre (2009) for a summary of the key risks and opportunities related to the long-term sustainability of the Community Forestry Program in Nepal.

Economic/Financial Sustainability

Nepal's Community Forestry Program is still largely a part of its subsistence livelihood system, with nonmonetary transactions dominating forest management. Due to the absence of any rapid expansion of capitalist production in rural areas of Nepal (Blaikie, Cameron, and Seddon 2002), the opportunity cost of labor is low, which makes it possible to generate the substantial voluntary contributions needed to undertake forest management. Local actors choose to contribute their time and labor largely because forests represent a sociopolitical arena for them to engage in cultural and political exchanges, allowing them to further shape the collective identity of a community. In recent years, local forest-dependent people have become increasingly conscious of civil, political, and economic rights, and marginalized groups such as Dalits and indigenous groups have been seeking proactive involvement in different spheres of forest governance. Clearly, participating in forest management is driven not only by economic benefits but also by a variety of cultural, symbolic, and political benefits that are gained through collective action in the forest governance arena (Ojha, Agrawal, and Cameron 2009).

Here we summarize anecdotal indications of the emerging economic sustainability of the program in the absence of a reliable cost–benefit analysis of the community forestry intervention, noting an increasing expectation of cash benefits from market transactions, such as through the sale of timber and nontimber forest products, to meet the growing livelihood needs of forest-dependent communities (Banjade and Paudel 2009). Depending on the location of a community forest, timber and several high-value nontimber forest products have good local and international markets. Timber in the low-lying Terai and medicinal plants of the higher Himalaya are well known. In recent years, there have been attempts to increase the capacity of CFUGs to promote enterprise-oriented use of forests (Subedi 2006). Despite growing evidence of the positive role of economic incentives in forest conservation through small-scale forestry enterprises, the policy environment is still too restrictive to support and encourage enterprise-oriented management of community forests (Banjade and Paudel 2009; Kunwar, Ansari, and Luintel 2009).

Also leading to reduced pressure on forests is the changing nature of households' dependence on them, and the direction of change varies across different contexts. In areas where out-migration is common, people's dependence on forests has decreased for two reasons: increased access to cash income from distant nonfarm sources and a decline in the supply of active human resources. In other contexts, such as in the fertile Terai, where a large area of de jure government forest is under de facto open access, forest in-migration has continued, adding pressure on forest land. There are instances of squatters organizing as CFUGs and managing forests in a sustainable way (Pokharel 2000), as well as incidents of squatters confronting CFUGs over land for settlements. Such conflicts are highly politicized, and CFUGs and their federations have had to face tremendous pressure from political interests. CFUGs have also organized themselves to protect forests and community forestry.

In the emerging context of climate change, once again, the perceived value of forest land is growing compared with competing land uses. The Government of Nepal and other stakeholders are conducting pilot programs exploring the possibilities of forest carbon marketing from community forestry. Within Nepal's Ministry of Forest and Soil Conservation, a separate government unit has been established to deal with the issues of forest and climate change. Given that the global forest sector contributes one-fifth of global greenhouse emissions through deforestation and degradation (Stern 2006), carbon revenue may provide added incentives for community participation in forestry management (LFP 2009). But there are also fears that the emergence of carbon forestry may trigger a reversal of forest tenure reform, potentially undermining the rights of local communities under community forestry (Dahal and Banskota 2009).

After three decades in place, studies indicate that community forestry may be more effective than government management from a financial point of view. Kanel (2004) argues that, based on a study conducted in 2002, the annual income of the Department of Forests, which controls 75 percent of Nepal's forest area, is about NRs 680 million ($8,684,552), while the income of CFUGs, which control 25 percent of the forest area, is NRs 740 million ($9,450,836) per year. CFUGs are still found to be earning less than they could under a sustainable use approach to forest management (Niraula 2004b). This resonates with the widely held view that community forests are protection oriented and underused (Pokharel et al. 2008).

Despite its apparently greater efficiency than government management, community forestry in its own right has yet to be managed effectively. Because the per capita forest area in the middle hills is relatively low (for example, 0.5 hectare per household in Ramechhap and Dolakha districts), management and use need to be brought to the highest sustainable level (Nurse et al. 2004). The current harvesting level is less than 1 percent of the growing stock (Pokharel et al. 2008). This is again related mainly to the technobureaucratic control of forest management planning and the lack of a service delivery system that is independent of bureaucratic control.

In the initial stage, and to some extent up to the present, community forestry has been largely a donor-funded process, and it is somewhat unclear how financial support for the program will continue as donor funds flatten (Acharya 2002; Pokharel et al. 2008). Donor support for the Community Forestry Program, which covers up to 16 percent of CFUG costs, is currently unsustainable, even after 30 years of implementation. But we argue that sustaining community forestry is a governance problem and not a financial constraint, mainly because the local community forestry processes have sufficient momentum in Nepal, demonstrating the willingness of local communities to make the investments necessary to establish and operate CFUGs. Moreover, the CFUGs can generate a substantial amount of funds even under a protectionist approach to forest management, indicating a potential to meet overhead costs more fully through increased production-based community forestry. If CFUGs believe in the credibility of the Community Forestry Program and own it, and if the government's role is limited to regulation and technical support so that private and nongovernmental service providers are allowed to work directly with CFUGs in other areas of service delivery, the cost of the Community Forestry Program will be low enough to be covered by a levy on community forestry production itself. But to date, CFUGs and their networks have been opposed to paying any extra tax to the government, and this should be seen in the context of the limited credibility and legitimacy of the government.

Another set of evidence of the financial sustainability of community forestry in Nepal is that in several districts community forestry has become functional and perhaps more effective in areas where there has been little or no donor program support. This is the case in Terai districts in general (Dhungana and Bhattarai 2005) and in several hill districts outside of bilateral project areas. CFUGs are also becoming part of subnational, national, and international networks, gaining greater access to information and institutional development services. Since the mid-1990s, civil society groups have taken much of the responsibility for expanding community forestry.

Finally, a key issue for financial sustainability at the CFUG level is equity in sharing forest products and related pricing mechanisms. Recognizing the hidden subsidy (Iversen et al. 2006) accruing to local elites, many CFUGs have begun to adopt differential rates for households with different wealth statuses. Still, wealthier households may make more effective use of benefit quotas on which poor households cannot capitalize, for instance, timber benefits. The issue is whether forest products should gradually be sold at market prices, either within or outside the group, to maximize financial revenue that can then be used to support the livelihoods of the poor in ways that really suit their needs.

Environmental Sustainability

No comprehensive studies are available to assess the environmental outcomes of community forestry, but both case studies and general observations suggest improvement in forest conditions (for example, lower incidence of fire and illegal harvesting of various forest products, better-controlled grazing, higher tree density in formerly degraded forests, increased species diversity, and regeneration of important species (Dougill et al. 2001; Acharya 2002; Dongol, Hughey, and Bigsby 2002; Dev et al. 2003; Yadav et al. 2003).

Amid growing concern about the negative environmental impacts of development activities, the Government of Nepal has also developed environment impact assessment (EIA) and initial environmental examination guidelines. A similar instrument developed by the Ministry of Forest and Soil Conservation outlines EIA procedures for transferring forests to the Community Forestry Program and for undertaking forestry operations. However, this instrument was developed with limited participation or inputs by the CFUG federations and has been opposed by CFUGs and their organizations. Given the dissatisfaction on the part of CFUGs, such instruments of sustainability may ironically have negative effects if imposed without the involvement of the intended targets.

At the level of CFUG management, the issue of forest ecological sustainability is strongly addressed. Forest land is generally put under a protectionist regime immediately after a CFUG is formed, and harvesting is generally done on the basis of block-based management and in combination with an inventory and assessment of mean annual increment. Community forest users also often patrol forests in groups both day and night to protect the forests from external free riders.

Several studies suggest that there have been improvements in forest conditions, forest land uses, and biodiversity following community management. Branney and Yadav (1998) assess the change in condition of community forests between 1994 and 1998 in four districts of the Koshi Hills. They find that the number of stems increased by 51 percent and the basal area increased by 29 percent, whereas grazing intensity declined from 94 to 74 percent compared with public forest. Karna, Gyawali, and Karmacharya (2004) analyze the condition of five community forests at five-year intervals during 1993–2003 and find that several parameters of forest condition, such as tree and sapling density and sapling diameter, increased with subsequent measurements.

Gautam et al. (2003) analyze changes in land use in a watershed covering an area of 153 square kilometers by comparing satellite imagery from 1976, 1989, and 2000. They find that the number of forest patches declined over time (from 395 in 1976 to 323 in 1989 and 175 in 2000), while the average patch area increased over the same periods. This is attributed to the merger of previously isolated small forest patches as previously degraded areas regenerated or came under forest plantation under community forestry. They also find that although 22.5 percent of forest area was converted to other land use during 1976–2000, 37.4 percent of the land under other uses came under forestry during the same period, resulting in a net increase of 14.9 percent in forest area in the watershed. A study of land use change in two central districts using aerial photographs from 1978 and 1992 along with rapid field assessment finds that the area of forest land increased from 7,677 to 9,679 hectares (37.5 percent) over the period assessed (Jackson et al. 1998). The authors attribute the increase primarily to forest plantation establishment as well as some increase in the area of mixed natural forest.

Nagendra et al. (2008) use Landsat imagery from 1989 and 2000 to analyze changes in land cover in three management zones (government control, buffer zone around protected area, and community forestry) using landscape ecology metrics and proportional distribution of land cover categories. The results show significant differences in terms of land cover dynamics and landscape spatial patterns between these land ownership classes and suggest greater improvement of forests managed under community-based institutions. Another study compiles data from 55 forests from the middle hills and Terai plains of Nepal to examine factors associated with forest clearing or regeneration. The results affirm the central importance of tenure regimes and local monitoring, including participation of forest users in the management processes (Nagendra 2007).

At both the subnational and national levels, a continuing issue is a lack of comprehensive monitoring. The Community Forestry Division of the Department of Forests does have a National Community Forestry database, but it contains insufficient biodiversity information and is not updated with sufficient frequency. Data generated by donor projects are specific to project areas, and the collation of similar information across projects is limited because each project collects data that are most relevant to its particular interests. More recently, schemes of forest certification have been introduced by adapting global lessons and methodologies (NFA 2007). But these are not popular among CFUGs, partly because certification is not yet linked to enhanced commercial benefits. Some analysis suggests that biodiversity and community governance are related: if diverse views and preferences regarding forest management are accommodated in the CFUG decisionmaking, there is a greater likelihood of favorable biodiversity outcomes (Banjade 2008).

Social and Political Sustainability

At the broader level, community forestry has been heralded as a success, though issues of inclusion and management effectiveness remain important challenges, even though they have improved since the early years of implementation. The emergence of strong civil society institutions to promote community forestry has politically deepened the community forestry process beyond the technical and largely apolitical approach adopted by government extension agents. CFUGs have organized themselves into strong networks such as FECOFUN and have shown themselves to be politically mobilized actors in the national arena, even participating in street protests defying the king's takeover of power in 2006, in which many FECOFUN activists were detained by the royal regime (Pandey 2009). Today CFUG networks are part of every policy debate that affects local forests and people. Because of this social and political mobilization, political parties strongly support community forestry in particular and decentralization of natural resource management in general. For this reason, community forestry appears to be a viable institution even during periods of conflict (Banjade and Timisina 2005; Pokharel, Ojha, and Paudel 2005). Nevertheless, political interests can overwhelm local dynamics, as when political parties seek to influence elections of CFUG federations (Ojha et al. 2008).

At the level of discourse and knowledge, the mobilization around community forestry has challenged the traditional hegemony of a technobureaucratic ideology. Scholars from multiple disciplines have taken a keen interest in community forestry as it relates to biodiversity, livelihoods, and policy. Such scholars have contributed to deliberative engagement between multiple stakeholders. Noted international scholars in the social and environmental sciences have undertaken case studies of Nepal, thus promoting the discourse of devolution globally. This is considered parallel to the culture of collaborative policymaking through such means as national workshops every five years and regular multistakeholder policy deliberation groups on the part of government officials. Lessons from the successful CFUGs are being scaled up in other sectors.

A critical issue affecting the political sustainability of community forestry is how the relationship between CFUGs and the Forest Department is structured and transformed. Despite some changes in attitude and behavior among the forest officials toward working with local communities, largely as a result of the community forestry movement in the hills, the orthodox image of forest bureaucrats has not changed much (Pokharel, Ojha, and Paudel 2005; Ojha 2006). Power differentials between local people and foresters continue to be large, and ordinary citizens and forest bureaucrats still have problems of mutual mistrust, with limited opportunities for direct deliberative engagement. The majority of foresters still attach great value to what can only be labeled technobureaucratic approaches.

Overall, the strong interest of local communities in forest governance and their adoption of a sustainable approach to forest management are the key foundations for the sustainability of community forestry in Nepal. CFUG networks and civil society actors have challenged the top-down approach of government. Community forestry is respected by political parties despite some strategic influences by politicians, for instance, over CFUG elections. CFUGs have become durable institutions supported by an active and vibrant network of CFUG federations, all contributing to the sociopolitical sustainability of community forestry in Nepal.

Lessons

The Community Forestry Program in Nepal demonstrates an innovation in citizen participation in forest governance from community-level forest management to national-level policymaking, as well as a range of cross-sectoral development institutions. It is characterized by a legally specified tenurial arrangement for community groups to manage and use forests, whereby well-established networks of civil society, social movements, researchers, and government agencies pursue diverse agendas within community forestry, often with conflicting objectives. The emergence of such a complex, multiscale system of governance was triggered in the late 1970s when key forest policymakers and government officials realized that it was not possible for the government to protect forests through the old-fashioned top-down, state-centric approaches without the active support of local people. At the same time, the perceived crisis of Himalayan degradation invited an international response in terms of technical expertise and financing that initially emphasized technical quick fixes such as plantation establishment but later allowed room for experimentation with different strategies, eventually reinforcing community-based forest management. Pilot studies and experimentation were critical aspects of the evolution of community forestry institutions and policy. The devolution of governance through community forestry has yielded both procedural and substantive gains. Procedural gains include democratic deliberation, civic engagement, capacity building, and institutional development. Substantive gains include the apparent creation of livelihood opportunities and contributions to forest ecological regeneration. By reviving the socioecological system of forest and rural landscapes, community forestry in Nepal has significantly improved livelihood systems. It has not only augmented natural capital flows but also nurtured a variety of types of livelihood capitals, such as social, political, financial, and physical capitals, through which local people have been better able to derive food and overall livelihood security.

However, the case also demonstrates that the devolution policy does not guarantee the participation of all. Although the participation of elite members of society has improved forest governance when compared with the state management of forests, the continuing challenge is to understand how marginalized members of society can have equitable access to benefits from community forestry. This is indeed a structural issue of Nepalese society that is characterized by multiple axes of hierarchy—caste, gender, ethnicity, class, and geographic isolation. More recently, community forestry in Nepal has been greatly mediated by swift political transitions in which marginalized groups have been making historic claims of identity and participating in political change. A significant aspect of democratization and tenurial security is related to the interaction between local communities and forest bureaucracy, which has yet to fully transform itself from its colonial legacy of centralized forest management. The changing livelihood context is also affecting community forestry in various directions, including a shift toward commercial use of forest products.

Here we identify various specific lessons learned from program implementation, particularly as they relate to improved livelihoods and welfare. We also outline issues related to the generation of impacts and present lessons for wider replications outside of Nepal.

Key Lessons Learned in Implementing the Intervention

Security of tenure played an important role in the achievements of community forestry in Nepal. CFUGs had clear standing under the 1993 Forest Act, which provided them with legal identity and a high degree of autonomy from arbitrary bureaucratic actions. This act stands in stark contrast to similar initiatives in other countries, such as India, where community forestry groups are not protected by the law (Behera and Engel 2006). This legal protection allowed CFUGs to continue to function through some very tough times, such as during the decade-long Maoist insurgency in some parts of the country, as well as to sustain themselves in situations in which the government was not responsive or even absent. Moreover, CFUGs' independent legal status has enabled them to seek out collaboration with any civil society or private-sector organization of their choice rather than relying solely on the governments Forest Department for diverse services.

During implementation, donors simultaneously invested in creating secondary-level federations of CFUGs and in building the capacity of local and national NGOs. As a career in mainstream politics lost credence in the eyes of midcareer political activists, they entered the emerging civil society domain, where they retained a political outlook and an activist orientation. Because of donor support, as well as an endogenous civic renaissance, community forestry became a platform for all kinds of critical actors—researchers, activists, reformist government officials, and expatriates. This investment paid rich dividends over time, as when CFUGs could intervene directly in national policy debates through FECOFUN, their national network. Besides facilitating horizontal learning and sharing, CFUG networks and NGOs were also instrumental in the expansion of community forestry to districts that were not targeted by donor projects, thus expanding the reach of the program and multiplying the benefits in terms of livelihood improvement.

Key Issues Related to Generating Impacts through the Intervention

Although the legal identity of CFUGs was enshrined in the 1993 law, there was a great deal of flexibility in the choice of institutional structure for each case. This made it possible to tailor the institutional modalities of community forestry in practice to suit the heterogeneity of local contexts across Nepal. At the national level, this allowed for the evolution of differences in the structure of CFUGs across the high hills, the middle hills, and the Terai, for example. When it became clear that the institutional model of CFUGs that worked well in the middle hills was inappropriate for the Terai, there was enough flexibility for groups in the Terai to experiment with alternative institutional forms. Although the current modality for power sharing between government and local communities remains contested, there is now enough experience with diverse modalities—including community, government, and some form of joint management—to inform better formulation of policy for the Terai as well.

Even within local contexts, the law provided a great deal of autonomy for CFUGs to experiment with different institutional arrangements pertaining to a range of issues, including silvicultural practices, access to forest products, benefit sharing, and income generation. Coupled with the horizontal information sharing facilitated by secondary organizations and NGOs, CFUGs were able to learn from a wider base of experience and adapt innovative ideas to their needs, including those that enhanced their income-generating opportunities and enabled greater contributions to household livelihoods.

Through these processes, community forestry in Nepal moved beyond the provision of subsistence—fuelwood and fodder for domestic needs—to incorporate forests as a major source of household incomes. Forests not only contributed directly to rural livelihoods in many forest-dependent communities but also became the sources of investment capital and raw material for new market-oriented livelihoods. This diversification, enabled by CFUG incomes from improved forest management, likely increased households' resilience to external shocks. After the initial period of consolidation, toward the end of the 1990s, community forestry also moved beyond forestry to focus on other second- and third-generation issues (Britt 1998; Ojha et al. 2008). Although forest management remained central to the identity of CFUGs, those groups chose to invest an increasing proportion of their incomes in social and physical infrastructure, thus providing public goods critical to livelihoods in particular and to human development in general. In the context of widespread poverty and poor infrastructure, CFUG investments in education, health, and microcredit, besides roads, buildings, and small-scale irrigation, have created the basis for future improvements in rural livelihoods across large areas of Nepal. However, in past years the benefits of these expenditures accrued disproportionately to the nonpoor, and procedures for more equitable distribution need to be enhanced.

Lessons for Replication

Following are lessons learned from the intervention that can be used to replicate the community forestry experience in Nepal beyond that country.

- *Learning through experience is the key to success.* Community forestry has evolved into a complex institutional network that requires actors to work collectively in a learning mode. Even when there is an absence of political consensus or a well-defined legal framework, collaborative learning has been able to help participants find a way forward.

- *Development of a strong civil society network is a critical part of community forestry success.* Civil society's influence over community forestry has remained critical in the post-1990 political environment (following the advent of multiparty democracy), especially to safeguard community rights and ensure the autonomy of community action from regressive government actions and intrusive private interests. The emergence of community federations at national and subnational levels has nurtured and promoted civic engagement in forest policymaking, defying the traditional top-down approaches. There is also significant empowerment of local citizens to participate in democratic governance.

• *Diverse institutional modalities in practice should be allowed to emerge through flexible regulatory arrangements.* Although community forestry was conceived as a unified program, diverse modalities have emerged in practice. CFUG membership varies from a dozen households to several thousand, and the group structure varies from informal sharing and coordination mechanisms to highly formalized multitiered organizations. This is an adaptive response to the diversity of contexts.

• *A technocratic and "interventionist" approach has given way to a collaborative learning process.* The development of community forestry was in part triggered by the open and responsive attitude of government officials, followed by gradual development and institutionalization of a multistakeholder process of collaboration. Community forestry is no longer a government program alone or a foreign aid–driven activity but is rather a complex governance regime for a forest-dependent socioecological system. Over time, community forestry has grown complex in terms of the range of actors involved, the scale of resources mobilized, the diversity of processes involving conflicts and collaboration, and policy and practical issues encountered. It has covered most of the middle hills of Nepal (with one-third of the country's population) and parts of the low-lying Terai and high hills. And it was resilient to the conflicts that plagued Nepal during 1996–2006.

Nepal's Community Forestry Program is considered an innovation in the areas of natural resource management, participatory local governance, and institutional reform. The evolution of the program demonstrates that conservation and poverty reduction can go hand in hand.

Notes

1. The current forest cover in Nepal is 5.8 million hectares, according to official statistics (Nepal, DFRS 1999). The results of the National Forest Inventory show that Nepal has 4.2 million hectares (29 percent) of forest area and an additional 1.6 million hectares (10.6 percent) of shrub. These forests are distributed across the three geographic regions of the country. The middle mountains represent about 48 percent of the forest area, whereas the forest in the plains (Terai) is nearly 25 percent of the total forest land. The remaining forest area is distributed across the high mountains of the Himalayas.

2. See Shrestha (1998) and Pandey (1999) for an explanation of development failures in Nepal.

3. Two laws are noteworthy here: the Forest Act of 1961 and the Forest Protection Special Act of 1967. The latter even authorized local forest guards to shoot people who used forests illegally.

4. The panchayat system was directly headed by the king. It had three tiers of elected bodies of panchayat politicians—village panchayat, district panchayat, and national panchayat. Despite the election of panchayat members, the real power was derived from the monarchy.

5. The World Bank and the Food and Agriculture Organization of the United Nations initially influenced the national government toward the process of devolution of forest governance, followed by a group of bilateral and international actors.

6. Dalits are considered "untouchables"; they are not allowed to touch food, water, or religious areas in many rural parts of Nepal.

7. Section 49 under Article 11 of the Forest Act clearly states that no person shall attempt to deforest, plough, dig, or cultivate the land in a forest area.

References

Acharya, K. P. 2002. Twenty-four years of community forestry in Nepal. *International Forestry Review* 4: 149–156.

Acharya, K. P., J. Adhikari, and D. R. Khanal. 2008. Forest tenure regimes and their impact on livelihoods in Nepal. *Journal of Forest and Livelihood* 7: 6–18.

Adhikari, B., S. DiFalco, and J. C. Lovett. 2004. Household characteristics and forest dependency: Evidence from common property forest management in Nepal. *Ecological Economics* 48: 245–257.

Agrawal, A., and E. Ostrom. 2001. Collective action, property rights, and decentralization in resource use in India and Nepal. *Politics and Society* 29: 485–514.

Bampton, J. F. R., M. R. Banjade, and A. Ebregt. 2008. Collaborative forest management in Nepal's Terai: Policy, practice, and contestations. In *Communities, forests and governance: Policy and institutional innovations from Nepal,* ed. H. Ojha, N. Timsina, C. Kumar, B. Belcher, and M. Banjade. New Delhi: Adroit.

Banjade, M. R. 2008. Community forestry and biodiversity conservation in Nepal: A critical analysis. In *Biodiversity and livelihoods,* ed. H. Dhungana and J. Adhikari. Chautari, Nepal: Martin Chautari. In Nepali.

Banjade, M. R., and N. S. Paudel. 2009. Economic potential of non-timber forest products in Nepal: Myth or potential. *Journal of Forest and Livelihood* 7 (1): 36–48.

Banjade, M., and N. Timsina. 2005. Impact of armed conflict in community forestry of Nepal. *European Tropical Forest Research Network News* 43–44: 81–83.

Banjade, M. R., H. Luintel, and H. R. Neupane. 2008. Action research experience on democratising knowledge in community forestry in Nepal. In *Knowledge systems and natural resources,* ed. H. Ojha, N. Timsina, R. Chhetri, and K. Paudel. New Delhi: Cambridge University Press / Foundation Books.

Banjade, M. R., N. P. Timsina, H. R. Neupane, K. Bhandari, T. Bhattarai, and S. K. Rana. 2006.

Transforming agency and structure for facilitating pro-poor governance in community forestry. *Journal of Forest and Livelihood* 5: 22–33.

Banjade, M. R., N. S. Paudel, H. Ojha, C. McDougall, and R. Prabhu. 2007. Conceptualising meso-level governance in the management of commons: Lessons from Nepal's community forestry. *Journal of Forest and Livelihood* 6: 48–58.

Behera, B., and S. Engel. 2006. Institutional analysis of evolution of joint forest management in India: A new institutional economics approach. *Forest Policy and Economics* 8: 350–362.

Bhattarai, B. 2007. What makes local elites work for the poor? A case of community forestry user group, Nepal. Paper prepared for the international Conference Poverty Reduction and Forests: Tenure, Market, and Policy Reforms, September 3–7, in Bangkok, Thailand.

Bhattarai, R. C., and S. P. Dhungana. 2008. Economic and market trends in forestry sector in Nepal. Thematic paper, ForestAction Nepal, Kathmandu, Nepal. Photocopy.

Bhattarai, S. 2009. *Towards pro-poor institutions: Exclusive rights to the poor groups in community forest management.* Discussion paper. Kathmandu, Nepal: ForestAction Nepal and Livelihoods and Forestry Program.

Bk, N., R. K. Shrestha, S. Acharya, and A. S. Ansari. 2009. *Maoist conflict, community forestry and livelihoods: Pro-poor innovations in forest management in Nepal.* Kathmandu, Nepal: ForestAction Nepal.

Blaikie, P., J. Cameron, and D. Seddon. 2002. Understanding 20 years of change in west-central Nepal: Continuity and change in lives and ideas. *World Development* 30 (7): 1255–1270.

Branney, P., and K. P. Yadav. 1998. *Changes in community forestry condition and management 1994–1998: Analysis of information from the forest resource assessment study and socio-economic study in the Koshi Hills, Nepal.* Kathmandu, Nepal: Nepal U.K. Community Forestry Project.

Britt, C. 1998. Community forestry comes of age: Forest-user networking and federation-building experiences from Nepal. Paper presented at the biannual conference of the International Association for the Study of Common Property, June 10-14, in Vancouver, Canada.

Buchy, M., and S. Subba. 2003. Why is community forestry a social- and gender-blind technology? The case of Nepal. *Gender, Technology and Development* 3: 313–332.

Cameron, J. 1998. Development thought and discourse analysis: A case study of Nepal. In *New perspectives on India–Nepal relations,* ed. K. Bahadur and M. L. Lama. New Delhi: Har-Anand Publications.

CBS (Central Bureau of Statistics). 2002. *Statistical pocket book Nepal.* Kathmandu, Nepal: Ramshah Path.

Chapagain, B., R. Subedi, and B. Rana. 2009. Beyond elite capture: Community forestry contributes to pro-poor livelihoods in Nepal. Discussion paper. Livelihoods and Forestry Program, Kathmandu, Nepal. Photocopy.

Chapagain, N., and M. R. Banjade. 2009. *Community forestry as an effective institutional platform for local development: Experiences from the Koshi Hills.* Kathmandu, Nepal: ForestAction Nepal and Livelihoods and Forestry Program.

Chhetri, R. B. 2006. From protection to poverty reduction: A review of forestry policies and practices in Nepal. *Journal of Forest and Livelihood* 5: 66–77.

Chhetri, R. B., and T. R. Pandey. 1992. *User group forestry in the far-western region of Nepal: Case studies from Baitadi and Accham districts.* Kathmandu, Nepal: International Centre for Integrated Mountain Development.

Dahal, N., and K. Banskota. 2009. Cultivating REDD in Nepal's community forestry: A discourse for capitalizing potential? *Journal of Forest and Livelihood* 8 (1): 41–50.

Dev, O. P., N. P. Yadav, O. Springate-Baginski, and J. Soussan. 2003. Impacts of community forestry on livelihoods in the middle hills of Nepal. *Journal of Forest and Livelihood* 3: 64–77.

Dhakal, B., H. Bigsby, and R. Cullen. 2005. Impacts of community forestry development on livestock-based livelihood in Nepal. *Journal of Forest and Livelihood* 4: 43–49.

———. 2007. The link between community forestry policies and poverty and unemployment in rural Nepal. *Mountain Research and Development* 27: 32–39.

Dhungana, H., and B. Bhattarai. 2005. *Community forestry in Nepal's Terai: Status of proposed community forests in Terai, Inner Terai and Chure.* Kathmandu, Nepal: ActionAid Nepal and Federation of Community Forest Users, Nepal.

Dongol, C. M., K. F. D. Hughey, and H. R. Bigsby. 2002. Capital formation and sustainable community forestry in Nepal. *Mountain Research and Development* 22: 70–77.

Dougill, A. J., J. G. Soussan, E. Kiff, O. Springate-Baginski, N. P. Yadav, O. P. Dev, and A. P. Hurford. 2001. Impacts of community forestry on farming system sustainability in the Middle Hills of Nepal. *Land Degradation and Development* 12: 261–276.

Eckholm, E. P. 1976. *Losing ground: Environmental stress and world food prospects.* New York: W. W. Norton.

Fisher, R. J. 1989. *Indigenous systems of common property forest management in Nepal.* Honolulu: Environment and Policy Institute.

Gautam, A. P., E. L. Webb, G. P. Shivakoti, and M. A. Zoebisch. 2003. Land use dynamics and landscape change pattern in a mountain watershed in Nepal. *Agriculture, Ecosystems and Environment* 99 (1–3): 83.

Gilmour, D. A., and R. J. Fisher. 1991. *Villagers, forests, and foresters: The philosophy, process, and practice of community forestry in Nepal.* Kathmandu, Nepal: Sahayogi Press.

Gronow, J., and N. K. Shrestha. 1991. *From mistrust to participation—The creation of participatory environment for community forestry in Nepal.* Overseas Development Institute (ODI) Social Forestry Network Paper 12b. London: ODI.

Gurung, H. 1996. Ethnic demography of Nepal. Paper presented at the Nepal Foundation for Advanced Studies Conference, January 10, in Kathmandu, Nepal.

Gutman, J. 1991. Representing crisis: The theory of Himalayan degradation and the project of development in post-Rana Nepal. *Development and Change* (28): 45–89.

Hobley, M. 1996. *Participatory forestry: The process of change in India and Nepal: Rural Development Forestry Network.* London: Overseas Development Institute.

Himal. 2009. [Title unavailable.] *Hima News Magazine,* May 31, 2009.

India, DoF (Department of Forests). 2009. *Community forestry database.* Kathmandu, Nepal.

Iversen, V., B. Chhetry, P. Francis, M. Gurung, G. Kafle, A. Pain, and J. Seeley. 2006. High-value forests, hidden economies, and elite capture: Evidence from forest user groups in Nepal's Terai. *Ecological Economics* 58: 93–107.

Jackson, W. J., R. M. Tamrakar, S. Hunt, and K. R. Shepherd. 1998. Land-use changes in two Middle Hills districts of Nepal. *Mountain Research and Development* 18 (3): 193.

Joshi, M., L. Dhakal, G. Paudel, R. Shrestha, A. Paudel, P. B. Chand, and N. P. Timsina. 2006. The livelihood improvement process: An inclusive and pro-poor approach to community forestry— Experiences from Kabhrepalanchok and Sindhupalchok districts of Nepal. *Journal of Forest and Livelihood* 5: 46–52.

Kanel, B. R., and R. Subedi. 2004. Pro-poor community forestry: Some initiatives from the field. In *Twenty-five years of community forestry,* ed. K. Kanel, P. Mathema, B. R. Kandel, D. R. Niraula, A. Sharma, and M. Gautam. Proceedings of the Fourth National Workshop on Community Forestry, August 4–6, in Kathmandu, Nepal. Kathmandu: Department of Forests.

Kanel, K. R. 2004. Twenty-five years of community forestry: Contributions to millennium development goals. In *Twenty-five years of community forestry,* ed. K. Kanel, P. Mathema, B. R. Kandel, D. R. Niraula, A. Sharma, and M. Gautam. Proceedings of the Fourth National Workshop on Community Forestry, August 4–6, in Kathmandu, Nepal. Kathmandu: Department of Forests.

———. 2006. Current status of community forestry in Nepal. Report submitted to Regional Community Forestry Training, Center for Asia and the Pacific, Bangkok, Thailand. <http://www.recoftc.org/site/fileadmin/docs/Country_profile/NepalCFprofile_3_.doc>. Accessed May 18, 2010.

Kanel, K. R., and D. Acharya. 2008. Re-inventing forestry agencies: Institutional innovation to support community forestry in Nepal. In *Reinventing forestry agencies: Experiences of institutional restructuring in Asia and the Pacific,* ed. P. Durst, C. Brown, J. Broadhead, R. Suzuki, R. Leslie, and A. Inoguchi. Bangkok: Food and Agriculture Organization of the United Nations.

Kanel, K. R., and B. R. Kandel. 2004. Community forestry in Nepal: Achievements and challenges. *Journal of Forest and Livelihood* 4: 55–63.

Kanel, K. R., R. P. Poudyal, and J. P. Baral. 2005. Nepal: Community Forestry 2005. <http://www.recoftc.org/site/fileadmin/docs/publications/The_Grey_Zone/2006/CF_Forum/policy_nepal.pdf>. Accessed May 18, 2010.

Karna, B. K., S. Gyawali, and M. Karmacharya. 2004. Forest condition change: Evidence from five revisited community forests. Paper prepared for the Fourth National Community Forestry Workshop, August 4–6, Kathmandu, Nepal. Kathmandu: Department of Forests.

Kumar, N. 2002. *The challenges of community participation in forest development in Nepal.* Operations Evaluation Department Working Paper 27931. Washington, D.C.: World Bank.

Kunwar, M., P. Neil, B. R. Paudyal, and R. Subedi. 2009. Securing rights to livelihoods through public land management: Opportunities and challenges. *Journal of Forest and Livelihood* 7 (1): 70–86.

Kunwar, S. C., A. S. Ansari, and H. Luintel. 2009. *Non-timber forest products enterprise development: Regulatory challenges experienced in the Koshi Hills of Nepal.* Discussion paper. Kathmandu, Nepal: ForestAction Nepal and Livelihoods and Forestry Program.

LFP (Livelihoods and Forestry Program). 2008. Synthesis of LFP's SFM activities. Kathmandu, Nepal. Photocopy.

———. 2009. *Can forest carbon financing benefit Nepal? A review of potential, challenges, and options.* Kathmandu, Nepal.

Luintel, H., and N. Timsina. 2008. *Is forestry decentralization effective in empowering women's agency?* Kathmandu, Nepal: ForestAction Nepal.

Luintel, H., B. Bhattarai, and H. Ojha. 2006. *Internal group governance in the context of community based forest management: A review of recent innovations and an analytical framework.* Kathmandu, Nepal: ForestAction Nepal and Regional Community Forestry Training Center.

Luintel, H., H. R. Ojha, and B. Rana. 2009. *Community forestry in Nepal: A synthesis of innovative practices.* Kathmandu, Nepal: ForestAction Nepal and Livelihood Forestry Program.

Maharjan, M. R., T. R. Dhakal, S. K. Thapa, K. Schreckenberg, and C. Luttrell. 2009. Improving the benefits to the poor from community forestry in the Churia region of Nepal. *International Forestry Review* 11 (2): 254–267.

Malla, Y. B. 1997. Sustainable use of communal forests in Nepal. *Journal of World Forest Resource Management* 8 (1): 51.

———. 2000. Impact of community forestry policy on rural livelihoods and food security in Nepal. *Unasylva* 51 (202): 37–45.

Malla, Y. B., H. R. Neupane, and P. J. Branney. 2003. Why aren't poor people benefiting more from community forestry? *Journal of Forest and Livelihoods* 3: 78–93.

Malla, Y. B., R. Barnes, K. Paudel, A. Lawrence, H. Ojha, and K. Green. 2002. *Common property forest resource management in Nepal: Developing monitoring systems for use at the local level.* Kathmandu, Nepal, and Reading, U.K.: ForestAction Nepal and University of Reading.

McDougall, C., H. Ojha, M. R. Banjade, B. H. Pandit, T. Bhattarai, M. Maharjan, and S. Rana. 2008. *Forests of learning: Experiences from research on an adaptive collaborative approach to community forestry in Nepal.* Bogor, Indonesia: Center for International Forestry Research.

Metz, J. J. 1995. Development in Nepal: Investment in the status quo. *Geographical Journal* 35 (2): 175–184.

Nagendra, H. 2007. Drivers of reforestation in human-dominated forests. *Proceedings of the National Academy of Sciences* 104 (39): 15218–15223.

Nagendra, H., S. Pareeth, B. Sharma, C. M. Schweik, and K. R. Adhikari. 2008. Forest fragmentation and regrowth in an institutional mosaic of community, government and private ownership in Nepal. *Landscape Ecology* 23 (1): 41–54.

Nepal, DFRS (Department of Forest Resources and Survey). 1999. *The forest resources of Nepal.* Kathmandu, Nepal.

NFA (Nepal Foresters' Association). 2007. *Forest certification in Nepal: Practices and experience.* Kathmandu, Nepal.

Nightingale, A. J. 2002. Participating or just sitting in? The dynamics of gender and caste in community forestry. *Journal of Forest and Livelihood* 2: 17–24.

Niraula, D. 2004a. Current status of recommendations made by the Third National Community Forestry Workshop, 25 Years of Community Forestry. Department of Forests, Kathmandu, Nepal.

———. 2004b. Integrating total economic value for enhancing sustainable management of community forests: A forward looking approach. Proceedings of the Fourth National Workshop on Community Forestry, August 4–6, in Kathmandu, Nepal. Kathmandu: Department of Forests.

Nurse, M., H. Tembe, D. Paudel, and U. Dahal. 2004. From passive management to health and wealth creation from Nepal's community forest. Proceedings of the Fourth National Workshop on Community Forestry, August 4–6, in Kathmandu, Nepal. Kathmandu: Department of Forests.

Ojha, H. 2002. *A critical assessment of scientific and political aspects of the issue of community forest inventory in Nepal.* Kathmandu, Nepal: ForestAction.

———. 2006. Techno-bureaucratic doxa and the challenges of deliberative governance—The case of community forestry policy and practice in Nepal. *Policy and Society* 25 (2): 131–175.

———. 2008. *Reframing governance: Understanding deliberative politics in Nepal's Terai forestry.* New Delhi: Adroit.

Ojha, H., and B. Pokharel. 2005. Democratic innovations in community forestry—What can politicians learn? *Participation* 7 (7): 22–25.

Ojha, H., and N. Timsina. 2008. From grassroots to policy deliberation—The case of Federation of Forest User Groups in Nepal. In *Knowledge systems and natural resources: Management, policy and institutions in Nepal,* ed. H. Ojha, N. Timsina, R. Chhetri, and K. Paudel. New Delhi: Cambridge University Press India and International Development Research Center.

Ojha, H., C. Agrawal, and J. Cameron. 2009. Deliberation or symbolic violence? The governance of community forestry in Nepal. *Forest Policy and Economics* 11 (5–6): 365–374.

Ojha, H., M. Khanal, and B. Shrestha. 1997. *The process of handing over community forestry: The potential role of I/NGOs.* National Workshop on Community Forestry for Rural Development. Kathmandu, Nepal: ActionAid Nepal.

Ojha, H., L. Persha, and A. Chhatre. 2009. *Community forestry in Nepal: A policy innovation for local livelihoods.* IFPRI Discussion Paper 00913. Washington, D.C.: International Food Policy Research Institute.

Ojha, H., N. P. Timsina, and D. Khanal. 2007. How are forest policy decisions made in Nepal? *Journal of Forest and Livelihood* 6 (1): 1–16.

Ojha, H. R., B. Subedi, H. Dhungana, and D. Paudel. 2008. Citizen participation in forest governance: Insights from community forestry in Nepal. Paper presented at the conference Environmental Governance and Democracy, May 10–11, at Yale University, New Haven, Conn., U.S.A.

Pandey, D. R. 1999. *Nepal's failed development: Reflections on the mission and maladies.* Kathmandu, Nepal: South Asia Centre.

Pandey, G. 2009. Interview with author, July 8.

Paudel, K. P., and H. R. Ojha. 2008. Contested knowledge and reconciliation in Nepal's community forestry: A case of forest inventory policy. In *Knowledge systems and natural resources: Management, policy, and institutions in Nepal,* ed. H. R. Ojha, N. P. Timsina, R. B. Chhetri, and K. P. Paudel. New Delhi: Cambridge University Press India and International Development Research Center.

Paudel, N. S., M. R. Banjade, and G. R. Dahal. 2009. Handover of community forestry: A political decision or a technical process? *Journal of Forest and Livelihood* 7 (1): 27–35.

Pokharel, B. K., and M. Nurse. 2004. Forests and people's livelihood: Benefiting the poor from community forestry. *Journal of Forest and Livelihood* 4: 19–29.

Pokharel, B. K., H. R. Ojha, and D. Paudel. 2005. *Assessing development space and outcomes in conflict situation: A reflection from Nepal Swiss Community Forestry Project.* Kathmandu, Nepal: Nepal Swiss Community Forestry Project.

Pokharel, B. K., P. Branney, M. Nurse, and Y. B. Malla. 2007. Community forestry: Conserving forests, sustaining livelihoods and strengthening democracy. *Journal of Forest and Livelihood* 6: 8–19.

———. 2008. Community forestry: Conserving forests, sustaining livelihoods, strengthening democracy. In *Communities, forests, and governance: Policy and institutional innovations from Nepal,* ed. H. Ojha, N. Timsina, C. Kumar, B. Belcher, and M. Banjade. New Delhi: Adroit.

Pokharel, R. 2000. From practice to policy—Squatters as forest protectors in Nepal: An experience from Shrijana Forest User Group. *Forests, Trees, and People Newsletter* 42: 31–35.

Rai-Paudyal, B., and M. Buchy. 2004. Institutional exclusion of women in community forestry: Is women-only strategy a right answer? In *Twenty-five years of community forestry,* ed. K. Kanel, P. Mathema, B. R. Kandel, D. R. Niraula, A. Sharma, and M. Gautam. Proceedings of the Fourth National Workshop on Community Forestry, August 4–6, in Kathmandu, Nepal. Kathmandu: Department of Forests.

Regmi, M. C. 1978. *Land tenure and taxation in Nepal.* New Delhi: Adroit.

Shrestha, K. K., and P. McManus. 2008. The politics of community participation in natural resource management: Lessons from community forestry in Nepal. *Australian Forestry* 71: 135–146.

Shrestha, N. K. 2001. The backlash: Recent policy changes undermine user control of community forests in Nepal. *Forest, Trees, and People Newsletter* 44: 62–65.

Shrestha, N. R. 1998. *In the name of development: A reflection on Nepal.* Kathmandu, Nepal: Educational Enterprises.

Shrestha, R., S. L. Shrestha, S. G. Acharya, and S. Adhikari. 2009. *Improving community level governance: Adaptive learning and action in community forest user groups.* Kathmandu, Nepal: ForestAction Nepal and Livelihoods and Forestry Program.

Stern, N. 2006. *The Stern review: The economics of climate change.* London: Government of the United Kingdom.

Subedi, B. P. 2006. *Linking plant-based enterprises and local communities to biodiversity conservation in Nepal Himalaya.* New Delhi: Adroit.

Timsina, N. 2002. Empowerment or marginalization: A debate in community forestry in Nepal. *Journal of Forest and Livelihood* 2: 27–33.

Timsina, N. P. 2003. Promoting social justice and conserving montane 5 forest environments: A case study of Nepal's community forestry programme. *Geographical Journal* 169: 236–242.

Yadav, N. P., O. P. Dev, O. Springate-Baginski, and J. Soussan. 2003. Forest management and utilization under community forestry. *Journal of Forest and Livelihood* 3: 37–50.

Chapter 6

Agroenvironmental Transformation in the Sahel: Another Kind of "Green Revolution"

Chris Reij, Gray Tappan, and Melinda Smale

This chapter analyzes two agroenvironmental success stories in the West African Sahel (Figure 6.1). The first is the relatively well-documented story of farmer-managed soil and water conservation on the densely populated Central Plateau of Burkina Faso, which achieved rehabilitation of degraded land on a significant scale following devastating droughts in the 1970s and 1980s. Land rehabilitation enabled farmers to develop agroforestry systems to extend their farm area and intensify production. The second story, still not completely documented, is the surprising account of farmer-managed restoration of agroforestry parklands in heavily populated parts of Niger. This process, begun in the mid-1980s, may in fact be one of the largest-scale agroenvironmental transformations in Africa and has been substantiated in recent aerial photography and satellite imagery.

The Sahelian "Green Revolution" began in scattered villages where farmers' practices were rediscovered and enhanced in simple, low-cost ways. Compared with Asia's Green Revolution, this was "barefoot science" (Harrison 1987)—but it similarly required an evolving coalition of local, national, and international organizations to enable large-scale diffusion and maintenance of improved practices. In some instances, outsiders played pivotal roles by facilitating the exchange of knowledge, furnishing start-up capital, or removing technical constraints.

The best evidence of economic viability is that these innovations were sustainably adopted by numerous farming communities in Burkina Faso and Niger. In Burkina Faso the process was documented by field technicians and project staff, albeit before the recent advances in statistical approaches for assessing and attributing impacts.

Figure 6.1 Locations of the two focus areas

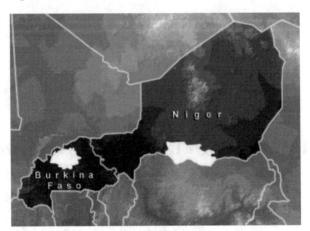

Source: Authors.
Note: The white patches indicate the two focus areas: the Central
Plateau in Burkina Faso and the agricultural plains of southern Niger.

Aerial photographs further attest to its success. In Niger, however, the story has only recently been "discovered" through aerial photography and satellite imagery.[1]

In neither case were all the success indicators measured with scientific rigor, considering the challenge of measuring over a large area across three decades. Some indicators were based on farmers' statements and perceptions; others may not have controlled sufficiently for intervening factors, including the impacts of soil, water, and agroforestry conservation. Clearly, however, the aggregate benefits are great and are likely to increase for years to come—especially in relation to the level of donor and national government funding. Farmers themselves, as well as local and international nongovernmental organizations and bilateral and multilateral donors, bore a substantial share of the total investment costs.

The Sahelian Situation around 1980

The Sahel has long been plagued by droughts. The major droughts of the 20th century occurred during the periods 1910–14, 1942–49, 1968–73, and 1982–84 (Aubréville 1949). This last "signature" drought was followed by persistent dryness through 1993. The decade from 1994 to 2003 remained far drier than the period from 1930 to 1965 (Anyamba and Tucker 2005).

The realities of climate change have been much more severe than indicated by decreased precipitation. The 1968–73 drought caused many deaths, as well as the loss of large numbers of animals and trees (perhaps in part because farmers and herders had "forgotten" how to cope with drought). An acute human and environmental crisis ensued.

Widespread labor migration by men in search of income caused social disruption (Monimart 1989). Between 1975 and 1985, some villages lost up to 25 percent of their families, who migrated to Côte d'Ivoire and to areas of higher rainfall in Burkina Faso. In the early 1980s, groundwater levels on the Central Plateau dropped an estimated 50–100 centimeters per year (Reij 1983). Many wells and boreholes went dry just after the end of the rainy season; in one village, rather than walk 8 kilometers to fetch water, some women left their families (Reij, Tappan, and Belemvire 2005). Average sorghum and millet yields decreased to slightly below 300 kilograms per hectare (Matlon and Spencer 1984; Dugue 1989; Matlon 1990), and a majority of farm households had annual food deficits of 50 percent or more (Broekhuyse 1983).

Meanwhile, the area of barren land on the Central Plateau expanded inexorably. Due to high population densities (50 persons per square kilometer and more), most land was cultivated permanently; agriculture had extended over unsuitable land, and most cultivated soils were lateritic and had low natural fertility. Neither inorganic nor organic fertilizers were used in quantities adequate to maintain soil fertility. Empty, encrusted fields called *zipélé* stretched across significant parts of Yatenga Province. Around 1980 the French geographer Marchal (1985) noted the decreasing productivity of expanded cultivated land, the destruction of vegetation, and the elimination of forests. Useful tree species were lost, with little natural regeneration.

Similarly, in Niger's Maradi region, crop yields were declining while land under cultivation expanded in pace with population growth. The denuded landscape was exposed to severe wind erosion, leaving the agroenvironment increasingly vulnerable to drought (Raynaut 1987, 1997). The crisis in Niger was aggravated by the migration following above-average rainfall from 1950 to 1968, when Hausa farmers moved northward to settle in lands reserved for pastoral communities. Farmers spoke of "fighting the Sahara"—the sand and dust storms that damaged their crops and their health (Larwanou, Abdoulaye, and Reij 2006).

By 1980, agroenvironmental trends in the Sahel had become devastating, threatening the social fabric. Many farmers faced a stark choice: either reclaim their land from the encroaching desert and intensify agricultural production or lay down their hoes and leave.

The Technical Innovations

Soil and Water Conservation on the Central Plateau of Burkina Faso

Before 1980, not much was achieved in soil and water conservation on the Central Plateau of Burkina Faso. Two major projects had been implemented with little short- or long-term impact. The GERES (Groupement Européen de Restauration des Sols) project in the Yatenga region was implemented between 1962 and 1965 to build earthen bunds to reduce erosion over entire catchments, covering 120,000 hectares.[2] Because the project was conceived without their involvement, however, farmers did not maintain the earthen bunds and sometimes deliberately destroyed them (Marchal 1979); the project stopped prematurely in 1965. In 1977 the Rural Development Fund (funded by multiple donors) began once again to construct graded earthen bunds, laid out in small blocks of cultivated village fields (30–60 hectares). However, the conservation technique was designed to fight erosion and not to harvest surface water, and farmers destroyed or breached the earthen bunds because they kept essential water runoff from entering their fields (Reij 1983). An estimated 60,000 hectares of cultivated land was treated, but within three years most of the bunds had disappeared (Sanders, Nagy, and Ramaswamy 1990).

Local farmers, with initial support from nongovernmental organization (NGO) technicians, eventually identified and implemented truly sustainable conservation strategies. During the first half of the 1980s they achieved two major technical advances based on indigenous soil and water conservation practices: improved planting pits and contour stone bunds.[3]

Improved Planting Pits (Zaï).   Around 1980, several farmers in the Yatenga region of Burkina Faso began "innovating out of despair" by experimenting with traditional planting pits (Reij, Tappan, and Belemvire 2005, 648). The innovation was to increase the depth and diameter of the pits and to concentrate nutrients and moisture in them. To reclaim severely degraded farmland impermeable to water, farmers would dig a grid of planting pits across their rock-hard plots and add organic matter at the bottom (Ouedraogo and Sawadogo 2001; Kaboré and Reij 2004).

Planting pits (called *zaï* in Moré, the local language) improve soil fertility in several ways. They capture windblown soil and organic matter that attracts termites, whose channels enhance soil architecture and water retention and make nutrients more readily available to the plant roots (Ouedraogo and Sawadogo 2001). Manure (for nitrogen) and urea may be added, along with mineral fertilizers for phosphorus and potassium.

Rehabilitating degraded land enables farmers to expand the size of their farms. Crop yields on degraded lands are 0 kilograms per hectare without planting pits and 300–400 kilograms per hectare with them, even in a year of low rainfall, and "easily" 1,500 kilograms per hectare in a good year. Water retained in the pits enables plants to survive long dry spells. Because more water is harvested and conserved and organic matter is used in the pits, conditions are improved for using mineral fertilizer to increase yields and biomass production. In the first few years, fields reclaimed with planting pits are hardly infested by *Striga* and other weeds, reducing the amount of labor needed to weed such fields relative to other fields. Because land is prepared during the dry season, farmers do not need to wait until the rains arrive to prepare their land, and some practice dry seeding (Kaboré and Reij 2004).

Because there is no standardized approach to preparing planting pits, quantifying their impacts is difficult. Farmers have adapted pits to meet their own needs, varying the number of pits per hectare and the pit dimensions, as well as the quantity of organic matter used (Hien and Ouedraogo 2001). Planting pits may be used to intensify cereal production, to produce trees, or both. Trees and shrubs start to grow spontaneously from the seeds in the manure and compost, and some farmers even sow the seeds of desired tree species, thus using *zaï* for reforestation.

Contour Stone Bunds. In the early 1980s, farmers helped shift the design of Oxfam's agroforestry project (PAF) in the Yatenga region by prioritizing food production rather than planting trees, setting to work on contour stone bunds.[4] Aside from improving spacing and dimensions, the key innovation was the use of a simple water level made from hose pipe, which cost $6 and could be quickly mastered by farmers, to ensure correct alignment along the contours (Wright 1985).

Following three years of testing, contour stone bunds became the focus of the PAF project from 1982. The traditional technique of creating stone lines was reintroduced along contours, allowing runoff to spread evenly through a field and trickle through the small gaps between the stones, trapping sediments and organic matter and preventing loss of applied manure. Critchley (1991) and Atampugre (1993, xv) provide details of early efforts, referring to the bunds as "Magic Stones."

Farmer-Managed Natural Regeneration in Southern Niger

Unlike the forestry plantations previously promoted in the Sahel, farmer-managed natural regeneration (FMNR) implements ancient methods of woodland management to enable farmers to harvest trees for fuel, shelter, food, and fodder without frequent and expensive replanting (WRI 2008). With the new techniques, farmers were once again managing the regeneration of native trees and shrubs among their crops (Larwanou, Abdoulaye, and Reij 2006).

To produce parklands on cleared land, farmers first identify stumps of useful tree species (Table 6.1). Next they select the tallest stems on each stump, removing unwanted stems and side branches with regular pruning. The original model, developed during the 1970s and 1980s, involved harvesting one-fifth of the original stems each year while selecting replacement stems. Farmers then adapted the technique, even creating woodlands by regrowing many more stems per stump and as many as 200 stumps per hectare (WRI 2008).

Larwanou, Abdoulaye, and Reij (2006) interviewed about 400 farmers in the Zinder region, individually or in groups, and learned that the trees generate multiple benefits. They reduce wind speed and evaporation: farmers no longer have to replant crops covered over by windblown sand. Trees produce fodder for on-farm

Table 6.1 Important tree species in Niger, 2008

Scientific name	Common name
Most commonly regenerated	
Faidherbia albida	Winter thorn or gao
Combretum glutinosum	—
Guiera senegalensis	—
Piliostigma reticulatum	Camel's foot
Bauhinia rufescens	—
Other important tree species	
Adansonia digitata	Baobab
Prosopis africana	Ironwood

Source: Authors.
Note: — means there is no common name.

livestock, as well as firewood, fruit, and medicinal products for home consumption or cash sales. Nitrogen-fixing species such as *Faidherbia albida* (or *gao*) enhance soil fertility as they mature.

The Innovators

Leaders

Charismatic leaders, both farmers and development agents, played key roles in diffusing the innovations. In Burkina Faso, private extension efforts by lead farmers substituted to some extent for the public extension service, which had increasingly concentrated its limited activities in cotton-growing regions (Reij, Tappan, and Smale 2009; Haggblade and Hazell 2010).

Ouedraogo and Sawadogo (2001) describe three models for disseminating planting pits, each spearheaded by an individual farmer:[5]

- Since 1984, Yacouba Sawadogo has organized two market days per year to promote planting pits. First, following the harvest, farmers bring a sample of the crop varieties they cultivated in their *zaï*. Yacouba stores the seeds on his farm. Then, just before planting, farmers select the seeds they want to plant that season. By 2000 Yacouba's market days involved farmers from more than 100 villages.

- In 1992 Ousseni Zoromé began a *zaï* school, training local farmers on a gravelly site next to the road and ultimately attracting the attention of the minister of agriculture. By 2001 his network included over 20 schools and 1,000 members, each school assigned to rehabilitate a piece of degraded land.[6]

- Ali Ouedraogo trained individual farmers in villages around Gourcy, visiting them regularly in their fields. His students, in turn, trained other farmers in improved *zaï* techniques or experimented with their own techniques.

A charismatic leader in FMNR in Niger is Tony Rinaudo of Serving in Mission (SIM, formerly the Society of International Ministries). In 1983 Rinaudo recognized that the stumps and roots in farmers' fields could be regenerated at a fraction of the cost of growing nursery tree stock. During the droughts of 1984 and 1985, he offered food to farmers in return for protecting on-farm natural regeneration. FMNR spread spontaneously as farmers observed the technique in the fields of other farmers.

Local Organization

Fundamental for change was the building of human capital through farmer-to-farmer learning promoted by NGOs and other stakeholders (Critchley 1991; Reij and Smaling 2007; Tappan and McGahuey 2007). Farmers' groups and village associations were also instrumental, especially for larger works such as permeable dams and terraces that affect multiple farms. These projects depend on collective action to organize labor for construction and maintenance and to allocate use rights and responsibilities, strengthening social capital in the process (Ouedraogo 1990; Smale and Ruttan 1994; Tougiani, Guero, and Rinaudo 2009). New governance structures were created to include marginalized social groups and to support community monitoring and management of land and tree regeneration. Farming communities worked to attract donor and public investment through study tours, the sale of tools, subsidized transport of stones, and the provision of food for work for the poorest.

Successful replication resulted from a confluence of efforts by individual farmers, farmers' groups, local NGOs (such as "Groupements Naams"), international NGOs (such as Oxfam), and bilateral and multilateral donors supported by national governments—an illustration of "bridging" capital, enabling local communities to link themselves to national and international institutions.

International Assistance

Since the mid-1980s, all major donors and projects in Burkina Faso have promoted contour stone bunds, *zaï*, or both. At the request of the government, many NGOs intervened in the northern part of the Central Plateau, one of the poorest and most degraded regions of the country (Reij, Tappan, and Belemvire 2005). International assistance has included

- a major Dutch-funded regional development project in the Kaya Region (1982–2000),

- the German-funded PATECORE project in Bam Province (1989–2004),

- a project supported by the International Fund for Agricultural Development (IFAD) in several provinces from 1989 to the present day (Reij and Steeds 2003),

- an IFAD project in Maradi (Niger) promoting FMNR (Larwanou, Abdoulaye, and Reij 2006), and

- other projects funded by the Netherlands and the World Bank.

Government policy and support were also important (Reij and Steeds 2003). The Burkinabé government worked from the mid-1980s to increase awareness of environmental problems and their solutions and instituted sound macroeconomic policies, including a 50 percent devaluation in line with a regionwide exchange rate adjustment of the CFA franc in January 1994 that was designed to stimulate exports. Major road construction linking the capital, Ouagadougou, with two regional capitals reduced transport costs, supporting the marketing of agricultural products (Reij, Tappan, and Belemvire 2005). Trunk roads reduced the costs of traders from Côte d'Ivoire, Ghana, and Nigeria buying produce from the Central Plateau, stimulating the local economy (Reij and Smaling 2007).

Farmers' investments in soil and water conservation (SWC) technologies are seen to respond to two policy-influenced variables: secure land title and intensive livestock management (Kazianga and Masters 2002; Reij, Tappan, and Belemvire 2005). Regarding investment by farmers in trees, the long-standing assumption that trees belong to the government rather than the farmer initially presented an obstacle to grassroots forestry management. Nevertheless, widespread diffusion of FMNR began to occur even before the enactment of property reforms as farmers began to perceive that they owned the trees and tree products in their fields.[7] The "proof of concept" thus originated with farmers and technicians in advance of policy change (McGahuey 2009). National policy also provided incentives for change by involving rural people more in development activities and informing them about the ecological crisis. The combination of farmers' efforts, dedicated work by national and international scientists, and policy dialogue (supported by the Comité Permanent Inter-Etats de Lutte contre la Sécheresse dans le Sahel) eventually led to the official reform of forestry policy (Bretaudeau, McGahuey, and Lewis 2009).

The Scale of Adoption

Stone Bunds and *Zaï*

By 2001, well over 100,000 hectares of strongly degraded land had been rehabilitated by projects and farmers in the northern part of the Central Plateau alone (Reij and

Thiombiano 2003). The total area rehabilitated over the past three decades is estimated at between 200,000 and 300,000 hectares (Botoni and Reij 2009). A recent study shows that in villages with a long history of SWC, 72–94 percent of the cultivated land has been rehabilitated with one or more conservation techniques. In other villages this percentage can range between 9 and 43 percent (Belemviré et al. 2008).

FMNR

Within their project zone in the Maradi region, 88 percent of farmers practiced a form of FMNR, with an estimated 1.25 million trees added each year, according to a 1999 survey by the NGO SIM. Larwanou, Abdoulaye, and Reij (2006) estimated that the affected area was 1 million hectares in three districts of Zinder, with a density of 20–120 trees per hectare; many villages now have 10–20 times more trees than 20 years ago. Farmers have literally "constructed" new agroforestry parklands on a massive scale.

In 2005–06, a team of Nigerian researchers confirmed that FMNR was a locally significant, geographically extensive on-farm phenomenon well correlated with the sandy ferruginous soils of the south central agricultural plain (Adam et al. 2006).[8] More significantly, the highest tree densities were found in areas of high rural population density: since the middle of the 1980s, farmers had been protecting and managing on-farm trees to an unanticipated extent.[9] The high correlation of FMNR to sandy ferruginous soils and areas of intensive cultivation suggests that the FMNR area can be roughly equated to the mapped boundaries of agricultural land use.

Fieldwork and initial analysis of high-resolution satellite imagery indicated that FMNR was present across much of the agricultural plain that dominates the Maradi and Zinder regions (Reij, Tappan, and Smale 2009)—about 6.9 million hectares. However, not all soil types found in the region are conducive to rainfed agriculture and field trees. Estimates by the authors gauged the area of FMNR at 5 million hectares (about the size of Costa Rica); indeed, high-resolution satellite images from 2003 to 2008 show an area of 4,828,500 hectares.[10] Many of the trees are young, the hallmark of a recent and rapidly developing agricultural parkland still increasing in density and cover.

Impacts

Crop Productivity

The immediate economic impacts of planting pits and contour bunds are yield gains. Formerly barren lands can produce a harvest in the first year, with potentially increasing benefits over time. Moreover, fields with planting pits reduce the risk of crop failure in dry years.

The PAF project estimated that contour stone bunds had a positive, statistically significant impact on cereal yields (Wright 1985). One study found an average cereal yield of almost 800 kilograms per hectare at 17 sites rehabilitated with stone bunds—325 kilograms per hectare higher than the average yield on control plots (Matlon 1985).[11] Other studies also found positive impacts of contour stone bunds and *zaï* on cereal yields, varying from 40 percent to more than 100 percent (Kaboré 1995; de Graaff 1996; Kambou 1996; Maatman 1999; Zombré 2003; Zougmoré 2003; Ouédraogo 2005; Sawadogo 2006).[12] Table 6.2 shows that *zaï* alone usually have a greater impact on yields than do contour stone bunds alone and that the combination of either *zaï* or grass strips with contour stone bunds has the greatest impact on cereal yields. The best indicator of positive impacts is simply that farmers continue to dig planting pits and construct stone bunds on their own initiative, without external support.

Farmers who rehabilitate land generally also invest in improved soil fertility management, leading to a stronger integration of agriculture and livestock and even to the emergence of a market for manure (as well as paid transport, by donkey cart). Mossi farmers have begun buying manure from the Fulani, who have little land and much livestock, and they hire Fulani herders to manage their livestock (Sawadogo 2003, 2008).

In Niger, *Faidherbia albida* have a well-recognized impact on soil fertility: increases in nitrogen content were found to range from 15 to 156 percent, with significant increases also in carbon, phosphorus, exchangeable potassium, calcium, and magnesium. Analyzing research data on the impacts of *Faidherbia albida* across Africa, Boffa (1999) mentions increases in millet yields ranging from 49 to 153 percent and increases in sorghum from 36 to 169 percent. This helps explain why farmers in parts of the densely populated southern Zinder region have created a high-density monoculture of *Faidherbia albida*.

Table 6.2 Impact of soil and water conservation techniques on cereal yields in four groups of villages, 2007 (kilograms per hectare)

Village	Control group[a]	Zaï (yield increase)	Contour stone bunds (yield increase)	Stone bunds + zaï (yield increase)	Stone bunds + grass strips (yield increase)
Ziga	434	772 (+ 346)	574 (+ 130)	956 (+ 522)	n.a.
Ranawa	376	804 (+ 428)	531 (+ 155)	922 (+ 546)	n.a.
Noh	486	700 (+ 214)	706 (+ 220)	n.a.	980 (+ 494)
Rissiam	468	716 (+ 248)	649 (+ 181)	n.a.	992 (+ 524)

Source: Sawadogo (2008).
Note: n.a. means not applicable.
[a]Cereal yields in cultivated fields without soil and water conservation techniques.

Food Security

On Burkina's Central Plateau at least 200,000 hectares of land have been rehabilitated, for an additional harvest of 80,000 tons per year.[13] Although some families have become fully food secure, most have merely reduced their structural food deficits—from six months (in the 1980s) to two or three months, an important gain. Farmers here have developed complex livelihood strategies in which migration as well as agriculture plays a major role (Reardon, Matlon, and Delgado 1988).

The southern parts of the Zinder and Maradi regions in Niger are characterized by high population densities of 80–150 persons per square kilometer. About 80 percent of all land is cultivated, and little off-farm natural vegetation is left. Rich families cultivate on average 9 hectares, while poor families cultivate 4 hectares or less. If an average farm household practices FMNR on 4 hectares, about 1.25 million households must be involved in FMNR.

FMNR has both direct and indirect impacts on household food security. Food availability is enhanced through crop yield increases that average an estimated 100 kilograms per hectare—higher in the case of *Faidherbia albida*.[14] At the estimated scale of adoption—1.5 hectare per farm and 8 persons per household—FMNR contributes an additional 500,000 tons of cereals and directly benefits 2.5 million people. (The total population of Niger was about 15 million in 2009.) In 2006, Niger was able to produce 283 kilograms of cereal per capita, almost identical to the 1980 level of 285 kilograms—despite nearly doubling its population over 25 years (Wentling 2008; WRI 2008). This represented a small surplus over the estimated average annual per capita cereal requirement of 200 kilograms.

FMNR has an indirect impact on food security through tree crop products. More fodder and crop residues on-farm allow farmers to keep more livestock closer to their fields, enhancing livestock production (Baoua 2006; Larwanou and Adam 2008). On Burkina's Central Plateau, farmers report having adequate local fodder, stover, and water and reduced deficits of cereal, allowing them to augment their herds. More livestock means that more manure is available to improve soil fertility, especially since firewood from trees on farms has replaced manure as cooking fuel.

Additionally, time spent on fuelwood collection by women can be reallocated to other activities, including food production and preparation and childcare. Tree products may be sold by young men (especially fuelwood and poles) and young women (especially fuelwood, leaves, and fruits), reducing incentives for migration and creating multiplier effects in village economies.[15] Small businesses have emerged related to tree products: medicinal plants, fodder, materials for construction, and sustainable fuelwood (Larwanou, Abdoulaye, and Reij 2006; Tougiani, Guero, and Rinaudo 2009).

Finally, in poor growing seasons and during the hungry period that typically precedes harvest, some tree fruits and leaves assume even greater significance in the local diet, serving as key sources of vitamins and micronutrients during those periods (Savy et al. 2006). In July and August 2005, thousands of women and children filled nutrition centers in the Maradi and Zinder regions; field visits later confirmed that villages practicing FMNR had been much less affected by the food shortages than other villages.[16] Parklands help families survive through consumption and sale of tree products (Larwanou, Abdoulaye, and Reij 2006). Farmers able to stockpile cereals during good years and supplement cereal production with tree products are better insulated against cyclical droughts, which are predicted to increase as a result of climate change (WRI 2008). The village of Dan Saga (Aguié District), with a long history with FMNR, had no drought-related infant mortality in the summer of 2005 because villagers were able to sell tree products to buy expensive cereals.

Equity

Aside from land, labor is the principal factor in crop production in the Sahel. Challenges are met through careful organization of family labor, shared community labor, and hired labor. Farmers are more likely to invest in labor-intensive techniques if they can afford to hire labor or to organize traditional work parties (Hassane, Martin, and Reij 2000). Indeed, those investing in *zaï* had larger households, more means of transport, more livestock, and better housing and equipment than those who did not (Slingerland and Stork 2000)—increasing rather than reducing inequality. Poorer families are more likely to benefit from project-supported interventions that serve multiple households, such as stone bunds or permeable dams (Haggblade and Hazell 2009).

Sahelian women report tremendous benefits from the improved supply of fuelwood and water over the past 20–30 years (Reij and Thiombano 2003; Yamba et al. 2005). Wives of farmer innovators in Burkina Faso stated that because the men concentrated on the *zaï* fields, the sandy soils have been allocated to women for the production of groundnuts and Bambara groundnuts for marketing (Sawadogo et al. 2001). Women also earned income ($210 annually) from the sale of baobab leaves, kapok flowers (*Ceiba pentandra*), shea nuts (*Vitellaria paradoxa*), and locust beans (*Parkia biglobosa*).

Women have free access to deadwood in the fields and to tree products such as gao pods to feed livestock; they may own trees that produce edible products, such as baobab, through inheritance or purchase (Larwanou, Abdoulaye, and Reij 2006, studying villages in Zinder). There is increased livestock investment by both men and women on the Central Plateau (Reij, Tappan, and Belemvire 2005) and in Zinder (Larwanou, Abdoulaye, and Reij 2006). Farmers involved in FMNR report that women have a stronger economic position and better capacity to feed their families a nutritious, diverse diet.

According to the World Resources Institute (WRI 2008), women may be the biggest winners in Niger from FMNR. It requires year-round tending, even though many men still migrate during the dry season. Women save cash by using their own firewood, and they earn income by selling wood and baobab leaves; they then invest in goats and sheep that they feed with pods of *Faidherbia albida* and leaves of *Guiera senegalensis* (Tougiani, Guero, and Rinaudo 2009).

Agroenvironmental Impacts

An analysis by Belemvire (2003) of 68 sites in 12 villages on the northern Central Plateau found evidence of reforestation, summarized in Table 6.3. These data confirm an increase in the number of on-farm trees on rehabilitated land, indicating that farmers have been developing agroforestry systems on formerly barren and crusted land (Reij, Tappan, and Smale 2009).

Reij, Tappan, and Belemvire (2005) compared the evolution of land use in three villages on the northern part of the Central Plateau between 1968 and 2002. Two villages (Ranawa and Rissiam) have a long history of interventions in SWC, unlike the third (Derhogo). In Derhogo, the area covered by cultivated parkland *declined* dramatically, from 32 percent (1984) to 24 percent (2002). During the same period, the two villages with interventions *increased* cultivation (under open parkland and under dense parkland) from around 66 percent to 72 percent. At Ranawa, aerial photographs from 1984, 1996, and 2002 show a clear trend of increasing tree cover, in association with the regular pattern of contour stone bunds (Figure 6.2). The surface under bare soil remains well under 10 percent in Rissiam and Ranawa but

Table 6.3 Analysis of rehabilitation effects on the northern Central Plateau, 2003

Indicator	Rehabilitated land (*n* = 47)[a]	Control plots (*n* = 21)
Number of tree species	33	26
Average tree density (trees per hectare)	126	103
Tree diameter 4 cm or less (%)	56	77.5
Tree diameter 11 cm or more (%)	37	17
Tree species		
Guiera senegalensis (%)	24	17
Leptadenia hastata (%)	12	—
Balanites aegyptiaca (%)	12	—
Piliostigma thonningii (%)	—	30
Combretum glutinosum (%)	—	9.5

Source: Belemvire (2003).

Note: — means either that no information was found or that it represents a negligible percentage.

[a]The initial condition of rehabilitated plots was very poor (almost 0 trees per hectare) compared to the control plots.

Figure 6.2 Time-series aerial photographs over Ranawa, Burkina Faso, 1984, 1996, and 2002 (left to right)

Source: Authors.

Note: This series of aerial views shows the positive impact of contour stone bunds (faint linear features) in increasing on-farm tree density. Trees tend to grow from the soils and seeds trapped by the stone bunds.

has increased in Derhogo from 41.3 to 49.4 percent of the total village territory. Livestock pressure, however, is higher in Derhogo than in the other two villages.

Because the manure and compost used in *zaï* contain seeds of trees, shrubs, and grasses, pitted fields show substantial regeneration of woody and herbaceous species, supporting vegetation diversity. After two years under *zaï*, an initially barren field was growing 23 herbaceous species and 13 species of trees and shrubs (Roose, Kaboré, and Guenat 1995).

A survey in 58 villages in 2002 found that in several regions, including Yatenga and Zondoma provinces, the level of water in wells had improved significantly since the beginning of land rehabilitation (Reij and Thiombiano 2003).[17] Farmers had created small vegetable gardens around several wells. This phenomenon can be attributed mainly to land rehabilitation rather than increased rainfall. Water levels have improved only in wells situated within rehabilitated areas or immediately downslope of these areas. Moreover, rising water levels in some villages began during the dry period, before the mid-1990s. Before rehabilitation, all wells fell dry at the end of the rainy season; after rehabilitation, wells soon had water throughout the year. However, more research is required to establish the scientific relationship between land rehabilitation and groundwater recharge.

Figure 6.3 shows the average tree cover calculated from sampled frames of the Mirriah-Magaria-Matameye (MMM) study area of Niger. The V-shaped trend shows the parkland in 1957, the loss of nearly half the tree cover by 1975, and the restored level in 2005—more than triple that of the 1975 cover. Among the frames sampled, tree cover in 2005 ranged from less than 1 percent to nearly 8 percent, with local concentrations of up to 16 percent tree cover. With a high density of young trees, the cover is expected to increase significantly over the next 10 years, with major implications for accruing still greater quantities of biomass (and therefore carbon).

Figure 6.3 Average tree cover trends in the Mirriah-Magaria-Matameye triangle, Niger, 1957, 1975, and 2005

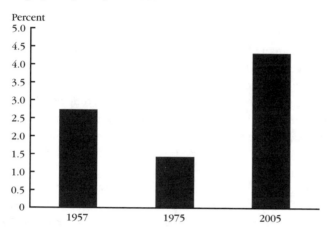

Source: Authors.

Note: The graph shows the tree cover trends for 1957, 1975, and 2005, averaged over the 10- by 52-kilometer study area within the MMM triangle.

Agricultural Intensification, Diversification, and Population Dynamics

It is often assumed that increases in agricultural production in the Sahel reflect an increased area under cultivation. Agricultural statistics for the northern Central Plateau from 1984 to 2001 indicate long-term stability of the area under cereals, with the exception of one area under sorghum in Yatenga Province, which has increased by almost 30,000 hectares since 1984—an extension made possible by careful land rehabilitation (Reij and Thiombiano 2003; Sawadogo 2008).

The complexity of production systems also appears to have rebounded. Although millet and sorghum remain the dominant crops in Burkina Faso, cowpeas and sesame are increasing. Cotton, an important crop in the 1950s and 1960s, disappeared from Yatenga and Zondoma provinces in the 1970s; farmers have begun reintroducing cotton on some rehabilitated fields. More on-farm trees and more livestock, managed more intensively, also contribute to diversity. More vegetable cultivation adds to income and enhances nutrition (Sawadogo 2003).

Intensification and diversification have altered the demographic dynamics in Burkina Faso. From the mid-1970s to the mid-1980s, many farm families abandoned their villages to settle in the valleys in southern Burkina Faso, newly freed from the small fly causing river blindness (McMillan 1988). In addition, men often migrate to urban areas for employment after each harvest. However, census data show stable populations between 1975 and 1985 and a 25 percent population increase from 1985 to 1996.[18] With one exception, the highest growth rates occurred in villages that had invested in land rehabilitation (Reij and Thiombiano 2003). Further study is required to assess the contribution of land rehabilitation to reduced migration.

High-quality time-series aerial photos of landscapes in south-central Niger clearly reveal a pattern of "more people, more trees" (Figure 6.4). The 1957 aerial photographs open a window onto the landscape at the end of the colonial era: during a relatively wet period, the region was, as today, devoted to rainfed agricultural production of cereal grains and peanuts—but with the rural population perhaps a third of the current level. The aerial views show the traditional bush fallow rotation system, with 30–50 percent of the land typically remaining in grassy fallows (medium gray surfaces). Fallow periods were apparently fairly short, perhaps one to three years, judging from the lack of dense bush growth. Farmers maintained cattle corridors of live hedges between village and pasture, and villages were much smaller and fewer than now. Natural depressions forming wetlands were much more pronounced, with aquatic and tall herbaceous vegetation. Trees were unevenly scattered throughout this ecosystem.

The 1975 aerial photographs provide a stark contrast. Niger was just recovering from perhaps the worst drought of the century (1968–73), with the government, international donors, and NGOs gearing up to improve food security. Numerous poorly designed projects attempted to combat desertification through tree plantations, windbreaks, and village woodlots (Rinaudo 2001).[19] The aerial views generally confirm the crisis. Much of the tree cover seen in the 1950s was gone, reflecting human and livestock pressure—including clearing for farming—as well as drought. Fallow land declined significantly, to between 0 and 20 percent of the land area.

Figure 6.4 Landscapes in the Mirriah-Magaria-Matameye study area (south of Zinder, Niger), 1957 and 2005 (left to right)

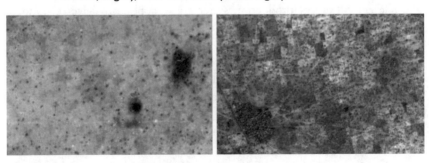

Source: Authors.

Note: Observe the extensive grassy fallow land in 1957 (medium gray surfaces). The dark patch near the center is a wetland, converted to farmland by 2005. Note the large new settlement that appears in the 2005 image.

Rural populations had roughly doubled since 1957, as seen in the sizes of villages. The traditional agricultural parkland was considerably thinner. However, the natural wetlands (dark patches) appear relatively intact, providing sources of standing water and fresh grass during the early dry season.

In 1975 no one could have predicted the extent of the renaissance of the parkland today. By 2005 the sparse tree cover of the 1970s had been replaced by young and fast-growing parkland, with a high density of trees, often in the inner fields around a village. Village sizes have continued to swell, with numerous new settlements in the 2005 image, and the use of fallows has all but disappeared. The natural wetlands have been converted to off-season gardens oriented toward market sales.

Figure 6.5 shows natural wetlands (two dark areas in the 1957 and 1975 photos) that had been converted to off-season farmland by 2005; a large number of small trees had reconstituted the young parkland.

In their definitive study of evolving farming practices in eastern Burkina Faso, Mazzucato and Niemeijer (2000) found evidence of a form of agricultural intensification that allows food production to increase along with the population. Similar findings are reported from other research (Faye et al. 2001; Mortimore et al. 2001; Tappan and McGahuey 2007). These studies demonstrate that in spite of population growth, agriculture has intensified and the environment has improved, with positive consequences for the local economy (Mortimore and Turner 2005).

Limitations

The major disadvantage of planting pits is that they are labor-intensive. Farmers must have access to family or hired labor to dig them, to dig compost pits, and to fill

Figure 6.5 Another landscape in the Mirriah-Magaria-Matameye study area, Niger, 1957, 1975, and 2005 (left to right)

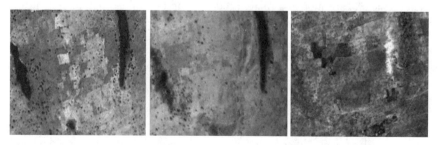

Source: Authors.
Note: Observe the natural wetlands (the two dark areas in the 1957 and 1975 photos) that have been converted to off-season farmland, the decline in grassy fallows (the medium gray areas in the 1957 and 1975 photos), six small settlements in the 2005 image, and the large number of small trees that had reconstituted the young parkland in 2005.

and maintain the pits from year to year. So far, mechanization has not been feasible. The intensive labor investments of *zaï* and contour stone bunds are even higher if both techniques are combined, as in Yatenga and Zondoma provinces. Estimates of the total labor costs for planting pits vary: Roose, Kaboré, and Guenat (1995) estimated 300 hours per hectare, and Maatman (1999) estimated 450 to 650 hours (depending on soil conditions). The high labor requirements mean that relatively rich farmers can more easily rehabilitate degraded land, using hired labor, than can poor farmers, who have to rely on their own labor—a pattern that is likely to contribute to growing inequality. Small farmers can only incrementally rehabilitate degraded land to which they have access.

A systematic study has not yet been undertaken in Burkina Faso to determine the area that can potentially be treated with water-harvesting techniques such as contour stone bunds, *zaï*, and half-moons, but it is not less than hundreds of thousands of hectares. *Zaï* function best in areas with between 300 and 800 millimeters of rainfall (Roose, Kaboré, and Guenat 1993). The soil surface should be barren, flat, and hard to generate sufficient runoff. Because they are labor-intensive, a relatively high population density facilitates diffusion.

There do not appear to be any negative off-site impacts of land rehabilitation, such as decreased water availability downstream. In social and economic terms, there are many winners, including women, who have benefited from local groundwater recharge, more on-farm trees, and the rehabilitation of some of their plots. The Fulani herders are relative losers, with fewer cattle entrusted to them by the Mossi farmers, who prefer to keep their cattle nearby to benefit from the manure. On the other hand, the herders can now sell manure for cash to Mossi farmers.

The only negative impacts of FMNR may include an increase in pests, such as birds that cause damage to crops; competition between trees and crops for nutrients and sunlight; and the negative effect of higher tree densities on local groundwater tables.[20] Larwanou and Adam (2008) found one village that did not engage in FMNR because villagers were afraid of increasing pests. Farmers are aware of competition between trees and crops for nutrients, water, and sunlight and have responded (in one village) by cutting on-farm trees to reduce their densities and generate income from firewood.

Sustainability

Because dryland environments are difficult and market infrastructure is often poor, it is often said that investing in them "does not pay." Estimates show otherwise. An economic analysis of a package tested by the International Crops Research Institute for the Semi-Arid Tropics (ICRISAT), bunding supplemented with a low-dose fertilizer and an improved sorghum cultivar, showed that a yield increment of only 155 kilograms gave a 15 percent return on the labor and cash investment—an increment exceeded by 67 percent of farmers (Matlon 1985). Kaboré and Reij (2004) estimate the gross margins of *zaï* at about $184 per hectare, with returns to labor of $0.19 per hour ($1.14 per six-hour day), about 33 percent higher than dry-season wage levels.

The costs of rehabilitating 1 hectare of degraded land are estimated to be on the order of $200 per hectare. An estimated $40 million or more has thus been invested in land rehabilitation on the Central Plateau. The estimated additional cereal production (at least 80,000 tons) is enough to cover the annual cereal needs of about 400,000 people, valued on the order of $19.2 million per year.[21] An SWC project in Illela District, Niger, cost $1.5 million over seven years (1988–95), with an estimated economic rate of return at completion of 20 percent (Reij and Steeds 2003). The internal rate of return to the PAF project has been estimated at about 40 percent, with a marginal rate of return (for an additional hectare of barren land reclaimed) estimated at 147 percent (Younger and Bonkoungou 1984; Wright 1985).

Abdoulaye and Ibro (2006) calculated an internal rate of return of 31 percent for FMNR based on the value of firewood produced over 20 years and assuming an increase in cereal yields of 5 percent during the first five years only. Even so, this analysis seriously underestimates the benefits of FMNR by excluding its other impacts.

Economic evaluations of *zaï* or contour stone bunds usually compare the yields obtained on rehabilitated fields with the average yields on cultivated fields. But because these fields are otherwise impossible to cultivate, the relevant counterfactual is 0 kilograms per hectare: every kilogram harvested represents a net gain. Similarly, estimated benefits based on single-year yield increases are understated because they do not incorporate long-term advantages, including a decreased risk of crop failure; positive impacts on trees, tree products, other vegetation, and water; off-site effects of increased biomass; groundwater recharge; and stimulation of local markets. There are broader social benefits as well: reducing migration, mitigating climate change, and combating environmental degradation. Project impacts continue after project completion and, given the number of contributing projects, would be more appropriately analyzed at a larger scale.

Reij, Tappan, and Smale (2009) quantify the impacts of FMNR as follows: on a scale of 5 million hectares, if FMNR has increased the number of trees (of all ages) by 40 trees per hectare, it has added about 200 million new trees to Niger's total tree stock.[22] Trees affect the local climate, crop growth and yields, soil fertility, and the availability of fodder, fruit, and other nontimber forest products. Using the authors' estimate that every tree produces an average value of $1.40 per year (in the form of improved soil fertility, fodder, fruit, firewood, and other produce), this would mean an additional value of at least $56 per hectare per year and a total annual production value of $280 million.[23]

Finally, there are numerous examples of how undertaking these environmental efforts—with arduous labor and against tremendous odds—instigated other changes, enabling farmers to create small businesses. In some cases, farmers also provided public goods (training, rehabilitation, demonstration) because they benefited from greater social standing.

The longevity of this innovation process (two to three decades) attests to its social and political sustainability. In fact, the *process* of involving farmers in the development of these technologies may be more easily transferred than the technologies themselves (Haggblade and Hazell 2010).

Despite these advances, land degradation and rural poverty, on the Central Plateau and elsewhere in the Sahel, have not yet been overcome (see Reij, Tappan, and Belemvire 2005). FMNR alone will not enable Niger or other Sahelian countries to keep abreast of the needs of a burgeoning population, though it is one important tool. Current demographic growth rates will lead to a doubling of the population in 20 years or less: Niger now has about 15 million people, and in 2030 it will have about 30 million. No government can be expected to cope with such growth rates. In 2010 Niger will have to feed an additional half-million people, and that number will increase every year.

Lessons for Policy and Practice

Much has been achieved since René Dumont's 1984 report lamenting that Burkina Faso was not developing but disappearing. Much has been invested in land rehabilitation on the Central Plateau of Burkina Faso by farmers, governments, and donor agencies. Ironically, however, although these investments have generated various positive impacts since 1980 and although the battle against land degradation and poverty has not been won, most donor agencies (with the exception of IFAD) have ended or reduced their funding for land rehabilitation.

Land rehabilitation enables sustainable intensification of agriculture and more resilient food production systems. It contributes to maintaining socially and economically viable rural communities, stemming the tide of migration and reducing tensions among groups competing for scarce resources. Continued investments in soil, water, and agroforestry are essential to intensifying agriculture, securing livelihoods in this region, and mitigating global climate change. Much land remains to be rehabilitated, and on other cultivated land, agroforestry systems and soil fertility management can be improved.

Similarly, in 1984 it seemed as if Niger would be blown from the map. Drought and strong winds from the desert (*harmattan*) created a general feeling of despair. No one could have imagined that farmers in densely populated parts of Niger would significantly increase their on-farm tree densities with minimal external support— and on a scale that would not be recognized for many years, precisely because donors did not play a major role. The agroenvironmental transformation was apparently almost invisible to governments and donors because it was derived from the grass-roots, needing little input from outsiders.

FMNR is not a solution to all problems, but it is low in cost and produces multiple long-term benefits. An added advantage is that FMNR is managed and maintained by land users; there are no recurrent costs to governments or donor agencies. Haggblade and Hazell (2010) argue that agricultural development approaches of this type are much less expensive than fertilizer subsidies. Reij, Tappan, and Belemvire (2005) estimate that treating 100,000 hectares with contour stone bunds costs just $200 per hectare for labor, transport, and technical support, comparable to the cost of one major dam to improve the water supply of the capital city of Ouagadougou. In FMNR, too, there is considerable scope for building on existing successes in Burkina Faso, Niger, and other Sahelian countries.

Lessons for Effective Partnerships

These stories carry five important lessons about effective partnerships for agricultural development. First, both "barefoot" science and cutting-edge science are important, particularly in these environments. The most successful innovations are often simple, low-cost improvements on local practices. Second, a single technique or practice, while generally insufficient to achieve meaningful environmental and economic impacts alone, can act as a trigger for other innovations. When farmers undertook multiple innovations simultaneously, they accomplished more rapid environmental change through the synergies of soil, water, and vegetative regeneration, with self-reinforcing feedback. Third, a single "menu" of technical options can be adopted on a large scale, but it must be flexible, adaptable, and testable by farmers under their own social, economic, and environmental conditions. Fourth, in resource conservation, collective action at the level of communities produces more sustainable benefits than adoption on individual farms. Finally, farmers are much more likely to adopt resource conservation innovations if at least one component will provide significant benefits in the first or second year.

Organizational and institutional innovations were needed to attain widespread diffusion of the technical innovations. Successful projects tended to be fairly small in scale at first and involved local farmers closely in the design of technical solutions. Design flexibility, responding to farmers' demands or new opportunities, was a strength. Promoting local leadership has led to positive results. Charismatic leaders, both local and from outside the community, stimulated change through their own actions, serving as role models. Tackling tough conservation problems needs strong village institutions and local leadership.

Lessons for Project Design

No single model of diffusion works in all communities. Innovative practices might be adopted by individual farmers or by groups working across farms or in collective fields. Some approaches will be based on farmer-to-farmer training, while other approaches will use interactive social learning in farmer groups. The poorest in poor communities will require special diffusion mechanisms so they can benefit from innovations. Nevertheless, a few clear lessons emerge:

- Long-term investment in building human and social capital enables rural communities to lead and manage.

- Land users in drylands respond to market opportunities, but intensification and commercialization do not necessarily cause overexploitation.

- Long-term investment in road and information infrastructure, which enhances the ability of farmers to diversify income from sales of tree, crop, and livestock products, can support the investments made by farmers.

- Government support of farmer-led initiatives, public education and awareness about the agroenvironment, and legal frameworks with use rights for farmers have contributed to positive change.

- Macroeconomic policies, such as those on exchange rates and agricultural pricing, will have impacts on farmers' incentives to manage their crops, trees, and livestock more sustainably.

The stories recounted here have global implications as well as examples of poor farmers' successes in enhancing their food security sustainably while adapting to climate change. Investments in tree regeneration and on-farm water harvesting techniques have led to immediate yield increases, helping to reduce rural poverty. The innovation process is sustainable—when led by farmers.

Notes

1. For a complete discussion of why the extent of the transformation in Niger went unrecognized and details about the remote sensing methodology that detected it, see Reij, Tappan, and Smale (2009).

2. Earthen bunds are dikes constructed along the contour lines of slopes that are designed to control runoff and spread water across the fields for better moisture retention. Treatment of watersheds began at the highest points of the slopes, working downward.

3. There were other innovations in the early 1980s. Level, permeable rock dams were developed by French volunteers to control and reclaim gullies, mainly in Bam Province. Half-moons (*demi-lunes*) were also tested by the Oxfam-funded agroforestry project but did not gain immediate popularity. Promoted during the past decade by a project funded by the International Fund for Agricultural Development, they are now increasingly used by farmers.

4. A contour line measures the height of land above sea level.

5. The three models of private extension stopped functioning after a few years due to farmers' age, lack of means, or the fact that most farmers felt they had mastered the technique.

6. This approach is similar to the "Farmer Field School" method that was used successfully to disseminate integrated pest management practices in Asia and is now promoted for a range of development purposes.

7. The underlying reasons for changes in farmers' perceptions are not entirely clear. In 1985 the general perception was that all natural resources, including trees, belonged to the state; however, after the state was weakened by a 15-year economic and political crisis, farmers began to consider the trees in their fields as their own (Larwanou, Abdoulaye, and Reij 2006).

8. The study, supported by Gray Tappan and Chris Reij, was funded by Swiss Development Cooperation, with complementary funding provided by the U.S. Agency for International Development (USAID) for remote sensing (U.S. Geological Survey, Earth Resources Observation and Science Center, South Dakota) and specific research support provided by the International Resource Group.

9. Taylor and Rands (1992) observed that thousands of farmers in the Maradi region protected and managed natural regeneration. Rands (1996) reports that farmer-managed natural regeneration is among the most widely spread natural resource management innovations in Niger. USAID-funded projects that organized farmer visits to the Maradi region helped spread this practice.

10. FMNR is also visible in other regions (it is locally present in Tahoua and Dosso regions and in the northern part of the Niamey region).

11. In the first half of the 1980s, a team from Purdue University researched tied ridges in several villages on the Central Plateau (Sanders and Roth 1985). A main reason farmers never adopted tied ridges was that they require a considerable investment of labor early in the rainy season, when labor is scarce. *Zaï* require even more labor, but this is invested mainly during the dry season, when labor is less scarce.

12. Although annual rainfall and other conditions are quite variable, the findings of these numerous Ph.D. dissertations all point in the same direction: they indicate positive impacts.

13. Using a net average gain in cereal production of 400 kilograms per hectare, a conservative figure.

14. The lower figure is more realistic because *Faidherbia albida* dominates only part of the area under FMNR.

15. The leaves from a single baobab tree are reportedly worth $24–$40, and firewood sales reportedly range from $6 to $120 per year in two surveyed villages (Larwanou and Adam 2008).

16. In 2010 Niger is experiencing a food crisis worse than in 2005, and many more villagers will depend on their trees for survival.

17. The increase in water levels in wells is estimated to be about 5 meters. The reaction of a professor of hydrology at the University of Ouagadougou was "I measured the increase in the well in my village in the Yatenga; it was not 5 meters, but 17 meters."

18. Of 14 study villages in the northern part of the Central Plateau, 9 villages had longer or shorter histories of investment in land rehabilitation and 3 villages lacked such investments.

19. The windbreaks of the Maggia Valley (Niger) are a notable exception (Rochette 1989).

20. Several farmer innovators in Burkina Faso deliberately attract birds during the dry season by providing water. They do this because birds help control crop pests, and their droppings contain seeds of trees and bushes. As for the competition between trees and crops, farmers decide which tree densities they find acceptable on their farms.

21. This calculation is based on an average cereal price of CFA 12,000 per 100 kilograms, or $240 per ton.

22. Average tree densities in villages in the Maradi and Zinder regions, measured by Saadou and Larwanou (2005) and by Larwanou and Adam (2008), were well above 40 trees per hectare.

23. This is most likely an underestimation. The Eden Foundation, which operates in the Tanout, a drier region north of Zinder, calculates that in 2007 farm households harvested fruit, leaves, and berries worth an average of €74 ($103) per household.

References

Abdoulaye, T., and G. Ibro. 2006. *Analyse des impacts socio-économiques des investissements dans la gestion des ressources naturelles: Etude de cas dans les Régions de Maradi, Tahoua et Tillabéry (Niger).* Niamey, Niger: Centre Régional d'Enseignement Spécialisé en Agriculture, Université Abdou Moumouni.

Adam, T., T. Abdoulaye, M. Larwanou, B. Yamba, C. Reij, and G. Tappan. 2006. *Plus de gens, plus d'arbres: La transformation des systèmes de production au Niger et les impacts des investissements dans la gestion des ressources naturelles.* Rapport de Synthèse Etude Sahel Niger. Niamey, Nigeria: Comité Permanent Inter-Etats de Lutte contre la Sécheresse dans le Sahel and Université de Niamey.

Anyamba, A., and C. J. Tucker. 2005. Analysis of Sahelian vegetation dynamics using NOAA-AVHRR NDVI data from 1981–2003. *Journal of Arid Environments* 63: 596–614.

Atampugre, N. 1993. *Behind the lines of stone: The social impact of a soil and water conservation project in the Sahel.* Oxford, U.K.: Oxfam Publications Unit.

Aubréville, A. 1949. *Climats, forêts, et désertification de l'Afrique tropical.* Paris: Société d'Editions Géographiques, Maritimes, et Coloniales.

Baoua, I. 2006. *Analyse des impacts des investissements dans la gestion des ressources naturelles sur le secteur élevage dans les régions de Maradi, Tahoua et Tillabéry au Niger.* Niamey, Niger: Centre Régional d'Enseignement Spécialisé en Agriculture, Université de Niamey.

Belemvire, A. 2003. *Impact de la conservation de l'eau et des sols sur la régénération naturelle assistée. Développement rural et environnement au Burkina Faso: La réhabilitation de la capacité des terroirs sur la partie Nord du Plateau central entre 1980 et 2000.* Rapport de travail 1. Ouagadougou, Burkina Faso: Conseil National pour la Gestion de l'Environnement

Belemviré, A., A. Maïga, H. Sawadogo, M. Savadogo, and S. Ouedraogo. 2008. *Evaluation des impacts biophysiques et socio-économiques des investissements dans les actions de gestion des ressources naturelles au Nord du Plateau Central du Burkina Faso.* Rapport de Synthèse Etude Sahel Burkina Faso. Ouagadougou, Burkina Faso: Comité Inter-Etats pour la Lutte contre la Sécheresse au Sahel.

Boffa, J.-M. 1999. *Agroforestry parklands in Sub-Saharan Africa.* FAO Conservation Guide 34. Rome: Food and Agriculture Organization of the United Nations.

Botoni, E., and C. Reij. 2009. *La transformation silencieuse de l'environnement et des systèmes de production au Sahel: L'impacts des investissements publics et privés dans la gestion des ressources naturelles.* Amsterdam: Comité Permanent Inter-Etats de Lutte Contre la Sécheresse dans le Sahel and Vrije University Amsterdam.

Bretaudeau, A., M. McGahuey, and J. Lewis. 2009. Interview with the author, July.

Broekhuyse, J. T. 1983. *Transformatie van Mossi land.* Amsterdam: Koninklijk Instituut voor de Tropen.

Critchley, W. 1991. *Looking after our land: Soil and water conservation in dryland Africa.* Oxford, U.K.: Oxfam.

de Graaff, J. 1996. *The price of soil erosion: An economic evaluation of soil conservation and watershed development.* Ph.D. dissertation, Wageningen University, Wageningen, the Netherlands.

Dugue, P. 1989. *Possibilités et limites de l'intensification des systèmes de culture vivriers en zone soudano-sahélienne: Le cas du Yatenga (Burkina Faso).* Collection Documents Systèmes Agraires 9. Montpellier, France: Centre de Coopération International en Recherche Agronomique pour le Développement.

Faye, A., A. Fall, M. Tiffen, M. Mortimore, and J. Nelson. 2001. *Région de Diourbel (Sénégal): Synthesis.* Drylands Research Working Paper 23. Crewkerne, U.K.: Drylands Research.

Haggblade, S., and P. Hazell, eds. 2010. *Successes in African agriculture: Lessons for the future.* Baltimore: Johns Hopkins University Press.

Harrison, P. B. 1987. *The greening of Africa: Breaking through in the battle for land and food.* London: Paladin Grafton Books.

Hassane, A., P. Martin, and C. Reij. 2000. *Water harvesting, land rehabilitation, and household food security in Niger: IFAD's soil and water conservation project in Illela District.* Amsterdam: International Fund for Agricultural Development (IFAD)/VU University Amsterdam.

Hien, F., and A. Ouedraogo. 2001. Joint analysis of the sustainability of a local SWC technique in Burkina Faso. In *Farmer innovation in Africa: A source of inspiration for agricultural development,* ed. C. Reij and A. Waters-Bayer. London: Earthscan.

Kaboré, P. D., and C. Reij. 2004. *The emergence and spreading of an improved traditional soil and water conservation practice in Burkina Faso.* Environment and Production Technology Division Discussion Paper 114. Washington, D.C.: International Food Policy Research Institute.

Kaboré, V. 1995. Amélioration de la production végétale des sols dégradés (zipellés) du Burkina Faso par la technique des poquets (zaï). Ph.D. dissertation, Ecole Polytechnique Fédérale de Lausanne, Lausanne, Switzerland.

Kambou, N. F. 1996. Contribution à la restauration et à la réhabilitation des sols ferrugineux superficiellement encroûtés (Zipella) du Plateau Central du Burkina Faso (Cas de Yilou-Province du Bam). Ph.D. dissertation, Université de Cocody, Côte d'Ivoire.

Kazianga, H., and Masters, W. A. 2002. Property rights, production technology and deforestation: Cocoa in West Africa. Paper 19871, presented at the annual meeting of the American Agricultural Economics Association [in 2008 renamed the Agricultural and Applied Economics Association], July 28–31, in Long Beach, Calif., U.S.A.

Larwanou, M., and T. Adam. 2008. Impacts de la régénération naturelle assistée au Niger: Etude de quelques cas dans les Régions de Maradi et Zinder. Synthèse de 11 mémoires d'étudiants de 3ème cycle de l'Université Abdou Moumouni de Niamey, Niger. Photocopy.

Larwanou, M., M. Abdoulaye, and C. Reij. 2006. *Etude de la régénération naturelle assistée dans la Région de Zinder (Niger): Une première exploration d'un phénomène spectaculaire*. Washington, D.C.: International Resources Group for the U.S. Agency for International Development.

Maatman, A. 1999. *Si le fleuve se tord, que le crocodile se torde: Une analyse des systèmes agraires de la région nord-ouest du Burkina Faso à l'aide des modèles de programmation mathématique*. Groningen, the Netherlands: University of Groningen, Centre for Development Studies.

Marchal, J. Y. 1979. L'espace des techniciens et celui des paysans: Histoire d'un périmètre anti-érosif en Haute-Volta. In *Maîtrise de l'espace agraire et développement en Afrique Tropicale: Logique paysanne et rationalité technique*. Mémoires ORSTOM 89. Paris: ORSTOM (Institut de Recherche pour le Développement).

———. 1985. La déroute d'un système vivrier au Burkina: Agriculture extensive et baisse de production. *Etudes Rurales* 99–100: 265–277.

Matlon, P. J. 1985. *Annual report of ICRISAT/Burkina Economics Program*. Ouagadougou, Burkina Faso: International Crops Research Institute for the Semi-arid Tropics.

———. 1990. Improving productivity in sorghum and pearl millet in semi-arid Africa. *Food Research Institute Studies* 22 (1): 1–43.

Matlon, P. J., and D. S. Spencer. 1984. Increasing food production in Sub-Saharan Africa: Environmental problems and inadequate technical solutions. *American Journal of Agricultural Economics* 66 (5): 672–676.

Mazzucato, V., and D. Niemeijer. 2000. *Rethinking soil and water conservation in a changing society: A case study in eastern Burkina Faso*. Tropical Resource Management Papers 32. Wageningen, the Netherlands: Wageningen University.

McGahuey, M. 2009. Interview with author, September 22.

McMillan, D. E. 1988. The social impacts of planned settlement in Burkina Faso. In *Drought and hunger in Africa: Denying famine a future*, ed. M. H. Glantz. Cambridge, U.K.: Cambridge University Press.

Monimart, M. 1989. *Femmes du Sahel: La désertification au quotidien*. Paris: Editions Karthala / Organisation for Economic Cooperation and Development Club du Sahel.

Mortimore, M., and B. Turner. 2005. Does the Sahelian smallholders' management of woodland, farm trees, rangeland support the hypothesis of human-induced desertification? *Journal of Arid Environments* 63: 567–595.

Mortimore, M., M. Tiffen, Y. Boubacar, and J. Nelson. 2001. *Synthesis of long-term change in Maradi Department 1960–2000.* Drylands Research Working Paper 39. Crewkerne, U.K.: Drylands Research.

Ouedraogo, A., and H. Sawadogo. 2001. Three models of extension by farmer innovators in Burkina Faso. In *Farmer innovation in Africa: A source of inspiration for agricultural development,* ed. C. Reij and A. Wayers-Bayer. London: Earthscan.

Ouedraogo, B. L. 1990. *Entraide villageoise et développement: Groupements paysans au Burkina Faso.* Paris: l'Harmattan.

Ouédraogo, S. 2005. Intensification de l'agriculture dans le Plateau Central du Burkina Faso: Une analyse des possibilités à partir des nouvelles technologies. Ph.D. dissertation, Université de Groningen, the Netherlands.

Rands, B. 1996. Natural resources management in Niger: Lessons learned: Agriculture Sector Development Grant, phase II. Report submitted to the U.S. Agency for International Development by the International Resources Group, Washington, D.C.

Raynaut, C. 1987. L'agriculture nigérienne et la crise du Sahel. *Politique africaine* 27: 97–107.

———, ed. 1997. *Sahels: Diversité et dynamiques des relations sociétés-nature.* Paris: Editions Karthala.

Reardon, T., P. Matlon, and C. Delgado. 1988. Coping with household-level food insecurity in drought-affected areas of Burkina Faso. *World Development* 16 (9): 1065–1074.

Reij, C. 1983. *L'évolution de la lutte anti-érosive en Haute Volta: Vers une plus grande participation de la population.* Amsterdam: Institute for Environmental Studies, Vrije University.

Reij, C., and E. M. A. Smaling. 2007. Analyzing successes in agriculture and land management in Sub-Saharan Africa: Is macro-level gloom obscuring positive micro-level change? *Land Use Policy* 25: 410–420.

Reij, C., and D. Steeds. 2003. Success stories in Africa's drylands: Supporting advocates and answering skeptics. Paper commissioned by the Global Mechanism of the Convention to Combat Desertification, Vrije University, and the Centre for International Cooperation, Amsterdam.

Reij, C., and T. Thiombiano. 2003. *Développement rural et environnement au Burkina Faso: La réhabilitation de la capacité productive des terroirs sur la partie nord du Plateau Central entre 1980 et 2001.* Ouagadougou, Burkina Faso: Ambassade des Pays-Bas, German Agency for Technical Cooperation–PATECORE, and U.S. Agency for International Development.

Reij, C., G. Tappan, and A. Belemvire. 2005. Changing land management practices and vegetation in the Central Plateau of Burkina Faso (1968–2002). *Journal of Arid Environments* 63 (3): 642–659.

Reij, C., G. Tappan, and M. Smale. 2009. *Agroenvironmental transformation in the Sahel: Another kind of "Green Revolution."* IFPRI Discussion Paper. Washington, D.C.: International Food Policy Research Institute.

Rinaudo, T. 2001. Utilizing the underground forest: Farmer-managed natural regeneration of trees. In *Combating desertification with plants*, ed. D. Pasternak and A. Schlissel. New York: Kluwer Academic / Plenum Publishers.

Rochette, R. M. 1989. *Le Sahel en lutte contre la désertification: Leçons d'expérience*. Weihersheim, Germany: Josef Margraf.

Roose, E., V. Kaboré, and C. Guenat. 1993. Le zaï: Fonctionnement, limites et amélioration d'une pratique traditionnelle de réhabilitation de la végétation et de la productivité des terres dégradées en région soudano-sahélienne (Burkina Faso). *Cahiers ORSTOM, Série Pédologie* 28 (2): 159–173.

———. 1995. Le zaï, une technique traditionelle africaine de réhabilitation des terres dégradées de la région soudano-sahélienne (Burkina Faso). In *L'homme peut-il refaire ce qu'il a défait?* ed. R. Pontanier, A. M'Hiri, N. Akrimi, J. Aronson, and E. Le Floc'h. Paris: John Libbey Eurotext.

Saadou, M., and M. Larwanou. 2005. *Evaluation de la flore et de la végétation dans certains sites traités et non-traités des Régions de Tahoua, Maradi et Tillabéri*. Niamey, Niger: Centre Régional d'Enseignement Spécialisé en Agriculture, Université de Niamey.

Sanders, J. H., and M. Roth. 1985. Développement et évaluation de nouveaux systèmes de production agricole: quelques résultats de terrain et résultats modèles obtenus au Burkina Faso pour les billons cloisonnés et la fertilisation. In *Technologies appropriées pour les paysans des zones semiarides de l'Afrique de l'Ouest*, ed. H. W. Ohm and J. G. Namy. West Lafayette, Ind., U.S.A.: Purdue University.

Sanders, J. H., J. G. Nagy, and S. Ramaswamy. 1990. Developing new agricultural technologies in Burkina Faso and the Sudan: Implications for future technology design. *Economic Development and Cultural Change* 39 (1): 1–22.

Savy, M., Y. Martin-Prevel, P. Traissac, S. Emyard-Duvernay, and F. Delpeuch. 2006. Dietary diversity scores and nutritional status of women change during the seasonal food shortage in rural Burkina Faso. *Journal of Nutrition* 136: 2625–2632.

Sawadogo, H. 2006. Fertilisation organique et phosphatée en système de culture zaï en milieu soudano-Sahélien du Burkina Faso. Ph.D. dissertation, Faculté Universitaire des Sciences Agronomiques de Gembloux, Gembloux, Belgium.

———. 2008. *Impact des aménagements de conservation des eaux et des sols sur les systèmes de production, les rendements et la fertilité au Nord du Plateau central du Burkina Faso*. Etude Sahel Burkina Faso. Amsterdam, the Netherlands: Comité inter Etats de lutte contre la sécheresse au Sahel and Vrije University.

Sawadogo, H., and M. Ouedraogo. 1996. *Une technologie paysanne pour une agriculture durable: Le zaï*. Institut de l'Environnement et de Recherches Agricoles / Rural Support Project or *Programme Régional Solaire* / Nord Ouest Burkina Faso. Projet Agro-Sylvo-Pastoral / Sécurité Alimentaire Durable en Afrique de l'Ouest Centrale. Groningen, the Netherlands: Université de Groningen.

Sawadogo, H., F. Hien, A. Sohoro, and F. Kambou. 2001. Pits for trees: How farmers in semi-arid Burkina Faso increase and diversify plant biomass. In *Farmer innovation in Africa: A source of inspiration for agricultural development*, ed. C. Reij and A. Waters-Bayer. London: Earthscan.

Slingerland, M. A., and V. E. Stork. 2000. Determinants of zaï and mulching in north Burkina Faso. *Journal of Sustainable Agriculture* 16 (2): 53–76.

Smale, M., and V. Ruttan. 1994. *Cultural endowments, institutional renovation, and technical innovation: The "Groupements Naams" of Yatenga, Burkina Faso.* Bulletin 94-2. St. Paul, Minn., U.S.A.: Economic Development Center, Department of Economics, University of Minnesota.

Tappan, G., and M. McGahuey. 2007. Tracking environmental dynamics and agricultural intensification in southern Mali. *Agricultural Systems* 94: 38–51.

Taylor, G., and B. Rands. 1992. Trees and forests in the management of rural areas in the West African Sahel. *Desertification Control Bulletin* 21: 49–51.

Tougiani, A., C. Guero, and T. Rinaudo. 2009. Community mobilization for improved livelihoods through tree crop management in Niger. *GeoJournal* 74 (5): 377–389.

Wentling, M. 2008. *Niger—Annual food security report and future prospects.* Niamey, Niger: U.S. Agency for International Development.

WRI (World Resources Institute). 2008. Turning back the desert: How farmers have transformed Niger's landscapes and livelihoods. In *Roots of resilience: Growing the wealth of the poor.* Washington, D.C.

Wright, P. 1985. Water and soil conservation by farmers. In *Appropriate technologies for farmers in semi-arid Africa,* ed. H. W. Ohm and J. G. Nagy. Purdue, Ind., U.S.A.: Purdue University, Office of International Programs in Agriculture.

Yamba, B., M. Larwanou, A. Hassane, and C. Reij. 2005. *Niger study: Sahel pilot study report.* Washington, D.C.: U.S. Agency for International Development and International Resources Group.

Younger, S., and E. G. Bonkoungou. 1984. Burkina Faso: The Projet Agro-Forestier—A case study of agricultural research and extension. In *Successful development in Africa,* ed. R. Bheenick, E. G. Bonkoungou, C. B. Hill, E. L. McFarland, Jr., and D. N. Mokgeth. Washington, D.C.: World Bank.

Zombré, N. P. 2003. Les sols très dégradés (zipellé) du Centre Nord du Burkina Faso: Dynamique caractéristiques morpho-bio-pédologiques et impacts des techniques de restauration. Ph.D. dissertation, Université de Ouagadougou, Burkina Faso.

Zougmoré, R. B. 2003. Integrated water and nutriment management for sorghum production in semi-arid Burkina Faso. Tropical Resource Management Papers 45. Ph.D. dissertation, Wageningen University, Wageningen, the Netherlands.

Chapter 7

The Case of Zero-Tillage Technology in Argentina

Eduardo J. Trigo, Eugenio J. Cap, Valeria N. Malach, and Federico Villarreal

Technological innovation can sometimes have a global impact on production processes. Although some technological innovations affect only a few features of production, a far-reaching innovation may affect the whole production process or the organizational and economic logic of an entire productive sector—or even of the economy itself.

Technology is an essential element in both the production processes and the economics of agricultural production (Sábato 1971), and technological change—at any level of the production chain—affects economic interactions among all actors in that chain. The innovation of zero-tillage farming had far-reaching effects on Argentine agriculture and beyond.

Zero-tillage technology allows the farmer to lay seed in the ground at the required depth with a minimal disturbance of the soil structure. Specially designed farm machinery eliminates the need for plowing and minimizes the tillage required for planting. The introduction of this cultivation technique in the Argentine pampean region generated significant changes throughout the agricultural sector on both relevant dimensions, production and productivity, and triggered changes in productive and commercial structures well beyond the farm gate. The adoption of zero-tillage practices had implications for the national economy and, given the substantial role of Argentina's agricultural production in world markets, for global consumers as well. This chapter looks into the process of adaptation and adoption of zero-tillage agriculture in Argentina, addressing the issues raised by these two questions: (1) What factors have contributed to its adoption and diffusion by farmers?

and (2) What are the economic and social impacts of this set of changes at the levels of both the nation and the world? The chapter describes the scenario prevailing prior to the introduction of zero-tillage practices and analyzes the institutional features of the process of innovation, as well as the factors behind its development and con-solidation. It tells the history of zero-tillage technology in Argentina, focuses on its impacts in Argentina and the assessment of its economic benefits at both national and international levels, and presents the conclusions drawn from the study.

The Introduction of Zero-Tillage
Agricultural Practices in Argentina

During the second half of the past century, Argentine agriculture was on a roller coaster. After an extended period of continued growth during the late 1940s and the early 1950s, the country became one of the main players in world agricultural markets. A combination of climatic events and policy changes reversed this trend, however, ushering in a period of production and productivity stagnation. The lack of technological innovation during this time has been explained in terms of the incentive structure, because farmers minimized the use of commercial inputs as a risk management strategy at a time of macroeconomic instability.

Toward the end of the 1960s and throughout the 1970s, a change in overall conditions marked the onset of a new technological cycle involving the mechani-zation of agriculture as well as the use of improved seeds.[1] This development was related to breakthroughs in plant breeding that eventually led to significant produc-tion and productivity increases.

It is in this context that the first soybean varieties adapted for cultivation in the pampean region appeared. This region is home to about 65 percent of the country's population and hosts close to 140,000 farms. It covers an area of about 76 million hectares in the central part of the country (including all of the province of Buenos Aires as well as parts of Córdoba, Entre Ríos, La Pampa, and Santa Fe). The region presents a landscape in which undulating plains predominate, and it is endowed with a temperate climate with 300–1,000 millimeters of rainfall. It is mostly dedi-cated to extensive rainfed production of cereals and oilseeds, as well as livestock (both beef and dairy) and fruit crops. In the "green belts" around the largest cities, however, farming is much more intensive; in these areas, vegetable production is the most important industry.

Although soybeans had been introduced in Argentina in the early decades of the 20th century, it was only in the late 1950s that a basic package of agronomic practices for the crop was developed. At that point, commercial cultivation of soy-beans began to be of some importance.[2] In later years, the continued expansion of international demand for plant protein became the driving force for the rapid and sustained adoption of soybean cultivation by local farmers. Moreover, the introduction of soybeans opened up a process of technological change and overall improvement in the organization of production based on a double-cropping scheme, with soybeans following wheat within a single planting season.

This production scheme led to higher farm incomes, but it also demanded a much tighter management schedule to deal with (among other factors) increased climatic risks, higher demands for weed control strategies, and more efficient use of farm machinery (Senigagliesi and Massoni 2002; Alapin 2008). This new scenario created additional demand for technical assistance at the farm level to manage the relatively unknown crop and the greater complexity of the new cropping systems. Expertise was needed to bring all the parts together in an effective way; access to information and knowledge became a key factor in farmers' success.

This process resulted in a major shift in agricultural output: grain and oilseed production increased by more than 50 percent between the early 1970s and the mid-1990s. However, the new double-cropping system took a considerable toll on soil fertility. Due to the tightness of planting schedules—by planting early in the season, the farmer could minimize the risk of early autumn rains affecting the harvest—the usual practice was to burn the stubble of the preceding crop immediately following the harvest. This minimized the fallow period, but it also had a negative effect on soil fertility (through erosion, loss of organic matter, and so on). The practice began to affect productivity negatively, even in the most resource-endowed areas. A 1995 study estimated that about 36 percent of the total area within the region showed signs of degradation; in two of the most important river basins of the provinces of Buenos Aires and Santa Fe (the Arrecifes and Carcarañá basins), soil degradation was as high as 47 and 60 percent, respectively (SAGyP 1995). This situation recalled the relatively recent episodes of soil erosion and productivity decline in the Kazakhstan plains of the former Soviet Union in the early to mid-1960s. It even brought back memories of the 1930s "Dust Bowl" in the United States (which hit the Great Plains states of Oklahoma, Kansas, Texas, New Mexico, and Colorado), where the rapid but technologically inconsistent expansion of the agricultural frontier made farming unsustainable, triggering social and economic consequences that are still remembered as one of the darkest periods in U.S. agricultural history (Derpsch 1999; Schoijet 2005).

The growing recognition of the effects of these inadequate soil management practices triggered new interest in improved crop management techniques: specifically, it was recognized that a less aggressive approach to soil preparation and planting would provide better protection from soil degradation (and the consequent risk of diminishing productivity). The resulting debate also led to an increasing demand for better information regarding other countries' experiences. Numerous study tours and visits to farm shows were organized for both agricultural scientists and farmers, setting the stage for the development of reduced-tillage technologies (a predecessor of zero tillage) and for the eventual introduction of specialized farm machinery for their effective implementation.

The soil degradation that resulted from the wheat-soybean double-cropping system (and the associated practice of burning stubble) thus prompted a change in crop management practices, along with an increasing reliance on technical assistance to adapt imported technologies (Alapin 2008). Public-sector agricultural researchers, innovative farmers and extension workers, and the associated manufacturing industries (farm machinery, seeds, and agrochemicals) became the core of an innovation network.

This network would play a key role in establishing a new agricultural production strategy based on a completely different approach to soil management and conservation: that is, zero-tillage farming (Ekboir 2002). The various participants, both public and private, shared a common perception of the nature of the problem as well as a strong interest in solving it, facilitating the generation of knowledge and information-sharing within what was at first a rather informal arrangement. This innovation network became a cornerstone of the rapid transformation from stagnation (or even decline) to a rapid *and sustainable* expansion. Local actors— farmers as well as technical assistance providers—played a key role in mobilizing the organizational changes necessary to incorporate the new technologies into existing production systems.[3]

The Institutional Dynamics of the Innovation Process

At the international level, concern for resource conservation dates back several decades. In the United States, the Dust Bowl experience of the 1930s prompted the development of reduced-tillage practices to improve soil coverage and to promote sustainable productivity growth. Zero-tillage technology, however, became possible only when 2-4D (a selective herbicide) became available, because weed control was the main obstacle in switching from existing tillage practices to the new ones. Later, more effective herbicides (such as Paraquat, released in 1961) enabled further development and widespread adoption of zero-tillage production in American agriculture, inducing farm machinery manufacturers to develop new equipment especially designed for zero tillage. The United States Department of Agriculture and the U.S. Universities of Illinois and Kentucky were the leading research institutions that developed the early zero tillage technologies; toward the end of the 1970s, they had produced a comprehensive pipeline of innovations—probably the most advanced in the world at that time. In the United Kingdom as well, a significant amount of work was being done on soil conservation and reduced-tillage practices, alongside an active adoption process (zero tillage reached some 200,000 hectares in

1973). In Latin America, the first efforts were pioneered in Brazil during the 1970s at the Instituto de Pesquisas Agropecuarias Meridional, which later became the Empresa Brasileira de Pesquisa Agropecuaria, emphasizing the adaptation of the new technological concepts to local conditions.[4]

Table 7.1 shows the sequence of events that shaped the development and consolidation of the network behind the massive adoption of zero-tillage practices in Argentina. The main driving forces were in fact individual initiatives: researchers within the National Agricultural Research Systems, as well as universities, farmers, and technical assistance providers, all of them motivated by widespread and increasing concerns with the sustainability of existing production systems.

As early as 1968, the National Institute of Agricultural Technology (Instituto Nacional de Tecnología Agropecuaria, INTA) began to take notice of the soil degradation problems and took steps to develop more environmentally friendly cultivation practices. This effort materialized in an international project, FAO/SEAG/ INTA/ARG/68/526, designed to establish a soil conservation program that would first identify and then discourage the practices that contributed to worsening the problem. This project also played an important role in developing human resources in the field of zero-tillage agriculture, mainly in the United Kingdom and the United States,[5] as well as in supporting the introduction of specialized zero-tillage farm machinery (Senigagliesi and Massoni 2002; Alapin 2008).

These early efforts were followed up in 1986 by a much broader initiative, the Conservationist Agriculture Project (Proyecto de Agricultura Conservacionista [PAC]). This program was aimed at developing a response to the already evident land degradation problem—affecting about 5 million hectares of the best farmland —that had resulted from abandoning the traditional crop-livestock rotation in favor of intensification schemes. This initiative was remarkable in that it represented a policy statement by the largest agricultural research institution in the country on an issue that was already a main concern of both farmers and scientists. It would eventually provide an institutional framework for a whole range of new initiatives. Within PAC, efforts were made to promote crop management techniques aimed at a more sustainable agriculture: a maize-wheat-soybean rotation, reduced and vertical tilling, nutrient replenishment through fertilization (mainly nitrogen and phosphorus), and integrated pest and weed management. It also facilitated the integration and exchange of information among researchers, extension staff, private technical assistance providers, farmers, input suppliers, and related institutions (Senigagliesi and Massoni 2002).

Research was also conducted independent of this institutional effort, both within and outside INTA, by scientists, farmers, technical assistance providers, and farm equipment manufacturers (Ekboir 2002). At INTA, as early as the mid-1960s, scientists at the Pergamino Experiment Station began researching techniques to reduce soil

Table 7.1 Highlights of zero-tillage technologies development and adoption in Argentine agriculture, 1930s–present

Highlight	Year	Events
The "Dust Bowl" devastated the Great Plains of the United States.	1930s	Raising of international awareness of the environmental and social consequences of certain agronomic techniques, prompted by the Dust Bowl
Key actors in the innovation network emerged: farmers, agricultural research institutions, researchers and extension workers, and the agricultural inputs manufacturing industry (farm machinery, seeds, and agrochemicals).	1956	Creation of the Instituto Nacional de Tecnología Agropecuaria (INTA) as part of the international trend to develop national institutions that can apply global knowledge resources to local problems.
	1961	Market introduction of the first systemic herbicide (Paraquat, developed in 1955), leading to the intensification of research on zero-tillage practices
	1968	The first local efforts to find solutions to soil degradation problems, developed by INTA (not yet directed to zero-tillage practices)
	1971	The first zero-tillage experiment in Latin America by the Instituto de Pesquisas Agropecuarias Meridional in Londrina, Estado de Paraná, Brazil, with the cooperation of GTZ (Deutsche Gesellschaft für Technische Zusammenarbeit)
	Mid-1970s	The undertaking of research and development work on different components of zero-tillage technology by individual researchers at INTA, farmers, and technical advisers
	1986	Implementation by INTA of the Conservationist Agriculture Program
The innovation network was institutionalized.	1989	Creation of the Argentine Association of Zero-Tillage Farmers (Asociación Argentina de Productores de Siembra Directa [AAPRESID])
Zero-tillage technology was increasingly recognized as a superior alternative to conventional tilling.	1990–95	Increasing adoption and consolidation of the innovation network
Zero-tillage adoption has rapidly increased to become the "industry standard"; AAPRESID has emerged as a key player in the innovation network.	1996–present	Market introduction of the first genetically modified crop (glyphosate-tolerant soybeans)

Source: Authors, based on literature cited in the text.

degradation induced by agricultural practices. This work was later taken up by fellow INTA researchers at the Marcos Juarez Experiment Station. These initiatives were meant to address what the researchers perceived as critical problems limiting agricultural production in the areas of influence of their respective experiment stations.

In parallel with these activities, farmers, technical assistance providers, and agricultural input and machinery company representatives did experimental work and shared information. Researchers from INTA, for example, worked closely with the farm machinery industry for the purpose of adapting zero-tillage equipment to local conditions, without any formally binding agreements. This informal arrangement played a key role in the development of locally designed zero-tillage machinery (Ekboir 2002).[6] The expansion of zero-tillage practices to other crops was similarly initiated through informal channels. Although most specialists recommended zero tillage only for soybeans (as the second crop in a double-cropping scheme), a group of farmers and extension workers started trials involving other crops, generating the information needed for an even greater adoption of the practice (Ekboir 2002).

At this point, the area under zero-tillage cultivation was negligible; the basic concepts and the institutional structure that would sustain the coming explosive technological expansion were still being established. The informal emerging network identified problems, tried alternative approaches, and shared information about the results. However, the new approach was still being fine-tuned to local conditions, limiting its appeal to farmers. The situation began to change when glyphosate became commercially available in Argentina, greatly facilitating weed control and overcoming the main obstacle to the adoption of zero-tillage technology.[7] Against the backdrop of farmers' experience of soil depletion, the availability of an effective weed control alternative, along with the implementation of the PAC by INTA, had set the stage for the launch of a new zero-tillage innovation cycle.

In this context, the Asociación Argentina de Productores de Siembra Directa (AAPRESID) was created in 1989. Its founding members were mainly medium- and small-scale farmers and technical assistance providers—about 20 people, all of them already involved in the movement to promote zero-tillage agriculture. The new organization focused on the diffusion and exchange of information regarding zero-tillage practices. It was set up as an open institution aiming at integrating representatives of all of the stakeholders (though only farmers were eligible to serve on its board of directors). The new institution grew very rapidly: within two years, the majority of the key players in the agribusiness sector had become members. AAPRESID became the fulcrum around which the development and expansion of zero tillage has continued to evolve until today. Besides working to fulfill its mission of promoting and diffusing the new technology, AAPRESID also acted as a lobbying group for its members, working on such issues as access to more favorable bank loans and tax advantages. For instance, in the 1990s the province of Santa Fe granted tax exemptions for farmers adopting zero-tillage practices (Alapin 2008).

AAPRESID thus became the institutional face of the innovation network supporting zero-tillage development and diffusion in Argentina. Alapin (2008, 117) summarized AAPRESID's institutional evolution by pointing out that it brought together all parties interested in promoting zero-tillage technology and, by ensuring that farmers were in close interaction with others who sought to promote the innovation, assisted farmers in capturing the benefits of zero tillage.

Expansion and Consolidation of Zero-Tillage Practices in Argentine Agriculture

Historical Perspective

Figure 7.1 shows the evolution in planted area (in grains and oilseeds) for the period 1900–2008. Historically, after 35 years of growth, the total area peaked in 1935 at 18.0 million hectares, then began a long phase of decline, hitting bottom in 1952 at 9.5 million hectares. Agriculture patterns stagnated until the mid-1970s; not until 1996 did Argentina get back to 18 million hectares under cultivation, although this time around that figure was achieved with a different basket of crops.

The Policy Shift

Starting at the beginning of the 1990s, the adoption of zero-tillage practices picked up speed. Several factors contributed to the consolidation of zero-tillage farming as the standard for grain and oilseed production in Argentina:

• The new macroeconomic environment of the early 1990s

• The introduction of transgenic soybean crops with herbicide tolerance

• The reduction in the price of herbicides

• The continued research and promotion efforts

Perhaps most significant was the effectiveness of AAPRESID as a consolidated network, bringing together all relevant stakeholders to share technical and economic information and to promote the benefits of the technology.

The change in the macroeconomic environment involved the elimination of export taxes on agricultural commodities and the reduction of import duties on inputs and capital goods. These measures, together with the deregulation of a number of key markets for goods and services, created favorable conditions for the increase of grain and oilseed production, from 26 million tons (1988–89) to over 67 million tons (2000–01). Soybeans became the main cash crop in the Argentine export basket.

Figure 7.1 Historical changes in area of Argentina planted with grains and oilseeds, 1900–2008

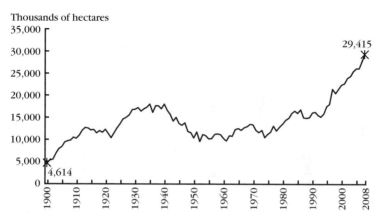

Sources: Ferreres (2005) and SAGPyA (2009).

This expansion took place within a complex international trade environment with erratic agricultural commodity prices and competition from subsidized exports from the OECD countries. It induced both an increase in planted area—at the expense of livestock—and an improvement in crop productivity through techno-logical change. The increase in total planted area may also account for what appears to have been a partial reversal in the decades-long process of rural–urban migration with the creation of some 200,000 new jobs in the agricultural sector between 1993 and 1999 (Trigo et al. 2002).

The second key factor was the introduction of genetically modified (GM) materials. The first GM crop formally approved for commercial use was glyphosate-tolerant (GT) soybeans, in 1996. The herbicide glyphosate, specifically adapted for use with zero-tillage technologies, facilitated the wheat-soybean double-cropping scheme. Its price declined dramatically with the expiration of the patent and the consequent increase in competition from local and foreign manufacturers, helping to consolidate the new production trends.

Since GT soybeans became available to Argentine farmers, their expansion has been one of the major technological events in the country's agricultural history, with an adoption rate higher than in other countries (such as Brazil) and even higher than in their original market in the United States (Galvão Gomes 2008). In Argentina, GM soybeans went from less than 1 percent of total planted area in 1996–97 to virtually 100 percent today (ArgenBio 2007). At the same time, zero-tillage systems expanded from about 300,000 hectares in 1990–91 to more than 22 million hectares in 2007–08 (Figure 7.2).

Figure 7.2 Evolution of the price of glyphosate and the number of glyphosate-based products available in the Argentine market, 1994–2001

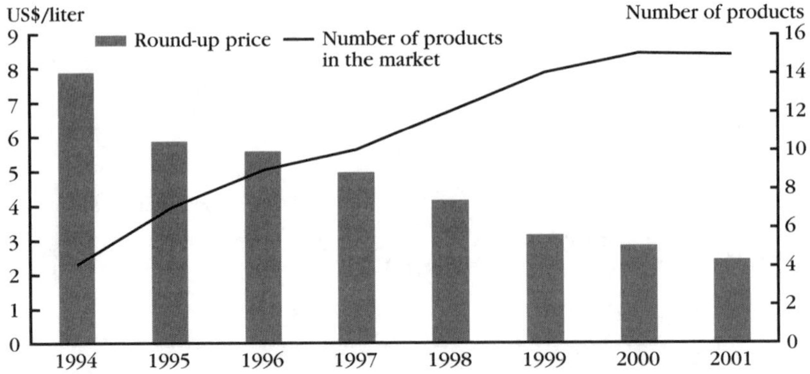

Source: Trigo et al. (2002), reproduced by permission of Libros del Zorzal, as requested by the Inter-American Institute for Cooperation on Agriculture, San José, Costa Rica.

Possibly the most important factor in the expansion of zero-tillage production was a sort of "virtuous intensification" that resulted from having the potential for environmentally friendly increases in productivity through the coupling of zero-tillage planting techniques with herbicide-tolerant soybeans. The new mechanical technologies modified the crop's interaction with the soil, moderating the impact of cultivation, while the new full-range herbicides (with glyphosate in the first place) are environmentally neutral, effectively controlling all kinds of weeds without residual environmental effects. The resulting high factor intensity might be described as a process of "hard" intensification. However, this hard intensification at the same time represents a *virtuous* intensification by reducing the use of atrazine, an herbicide whose residual action has a negative impact on the environment (Figure 7.3). Even with the increased use of agrochemicals, the total use of these products (per hectare of arable land) is still far below that of other countries. The increase in the use of fertilizers during the 1990s was also far below the factor intensity levels recorded in other countries and seems to have stabilized in recent years (Trigo et al. 2002).

Soybeans are a self-pollinated species; consequently, harvested grain can be used as seeds in subsequent plantings, without any significant loss of either genetic characteristics or productivity potential, for at least two to three years after the initial crop. This feature has had an impact on the seed market and thus on the price of GT soybean seeds. Under the terms of the 1978 International Union for the Protection of New Varieties of Plants Convention, farmers are allowed to keep grain for their own use as seeds, a practice that has led to the development of an illegal seed market (the so-called "white bag" soybeans), through which seed multipliers also sell un-

Figure 7.3 Evolution of area of Argentina under zero-tillage production and herbicides used in crop production, 1990–2000

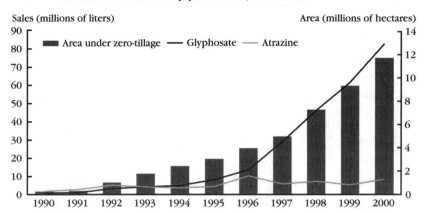

Source: Trigo et al. (2002), reproduced by permission of Libros del Zorzal, as requested by the Inter-American Institute for Cooperation on Agriculture, San José, Costa Rica.

certified seeds of the companies that own the respective intellectual property rights. Some estimates place the market share of this illegal seed at 50 percent or more. This has had a significant effect in driving down the price of GT soybean seeds, further promoting the rapid adoption of GM technology and, indirectly, that of zero-tillage practices (Costamagna 2004). But on the downside, these illegal markets can be a disincentive to companies that conduct research and development on GT soybeans if they are unable to recoup their investments through seed sales in the legal market.

All the factors just discussed help to explain the current dominance of zero-tillage practices throughout the Argentine grain and oilseed production structure. During the past three decades, all the contributing factors have come together: the initial concerns with diminishing soil fertility, new knowledge of how to improve it, the inputs necessary to implement the new technologies (seeds, herbicides, and machinery), the social networks to promote information sharing, and an adequate policy environment. The end result can be defined as a win-win outcome, with positive economic and environmental benefits for both Argentine farmers and world consumers.

The Impacts of Zero-Tillage Technology

The remarkable expansion of total area planted with grains and oilseeds that took place in Argentina during 1971–2008 represented almost a threefold increase, from roughly 11 million hectares to 30 million hectares (Figure 7.4).[8] This period was marked by several milestones.

Figure 7.4 Expansion of area of Argentina planted with grains and oilseeds, 1971–2008

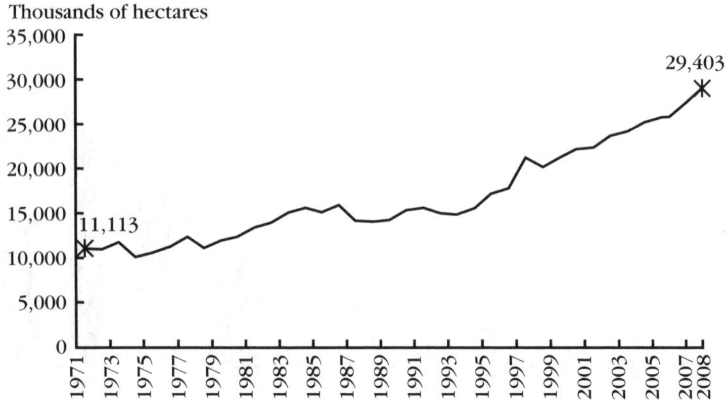

Thousands of hectares

Source: SAGPyA (2009).

Beginning in 1991, this significant trend was closely associated with the high rate of adoption of zero-tillage technology, effectively adapted to local conditions (Figure 7.5). The expansion of one particular crop, soybeans, played a pivotal role in this process (Figure 7.6). The following section assesses the impacts of this expansion, separating out the effects of zero tillage from other technologies (such as GM soybeans).

Some of the effects of the adoption of this technological package are difficult to assess and quantify because they are related to improvements in the quality of the soil, a key natural resource for agriculture. There is ample empirical evidence related to this particular issue: lower rates of depletion of organic matter, greater moisture-holding capacity, and a consequent reduction (or even reversal) of decades-long degradation processes have been reported (Andriulo, Sasal, and Rivero 2001; Sagardoy et al. 2001; Zaccagnini and Calamar 2001; Casas 2003; De Moraes Sá et al. 2004).

The adoption of zero-tillage technologies has had other measurable impacts in terms of changes in income flows of Argentine farmers and improvements in the purchasing power of global consumers. These two kinds of outcomes—the measurable economic impacts and the less measurable physical ones—are probably mutually reinforcing.

There is ongoing debate, however, regarding the nature and magnitude of potentially negative impacts of zero-tillage production on the structure of the soils in more marginal areas. These are areas previously not suited for cultivation that have experienced a major shift in farming systems from (more sustainable) livestock production to relatively intensive (and less sustainable) cropping systems.

Figure 7.5 Area of Argentina under zero-tillage production, 1971–2008

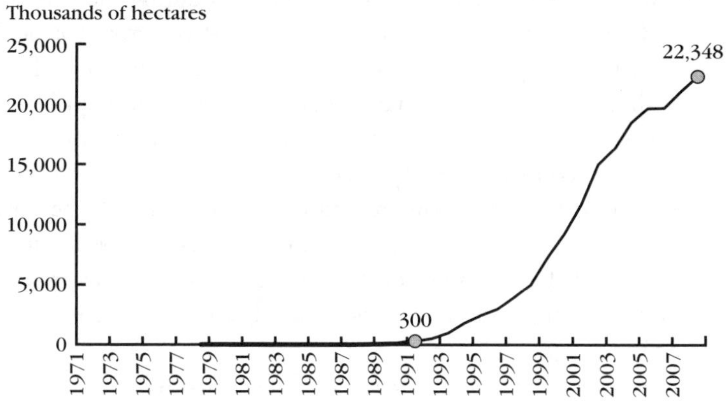

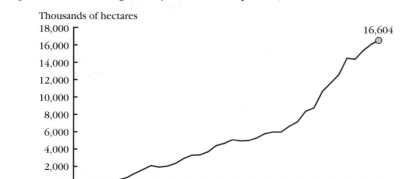

Source: Adapted from AAPRESID (2007).

Figure 7.6 Area of Argentina planted with soybeans, 1971–2008

Source: SAGPyA (2009).

Estimation of Benefits

In this study we will attempt to define the nature and magnitude of the positive impacts of zero-tillage technology in Argentina, taking into account two dimensions: (1) the supply side, that is, benefits to producers in Argentina, and (2) the demand side, in terms of benefits to consumers worldwide. For a detailed discussion of the techniques used to estimate these impacts, see Trigo et al. (2009).

The Supply Side: Benefits to Producers

We assume that the measurable benefits to producers during the period under analysis came from two sources: the supply shock (above-trend expansion of planted area) and the savings in production costs generated by a reduction in factor use intensity. Both are the result of farmers' adoption of zero-tillage practices. The results of these estimations are presented in the following two sections.

Benefits to Farmers from the Supply Shock. Table 7.2 presents the benefits to farmers measured as the increase in gross income associated with the expansion of the area planted in soybeans and maize. Those benefits have been remarkable: about 8.3 percent of the total value of these two crops in 2008 is attributable to zero-tillage-induced supply shock, and the cumulative impact is estimated at $12 billion by 2008.

Benefits to Farmers from Lower Production Costs. As a side effect of the transformation of soybean cultivation, important crops such as sunflowers and wheat (and others) also showed significant rates of adoption of zero-tillage practices

Table 7.2 Zero-tillage-induced supply shock in Argentina: Value of additional production of soybeans and maize, 1991–2008

| Year | Total value of production (TVP) (millions of current US$) | | | Percentage of TVP due to zero-tillage shock |
	Soybeans (S)	Maize (M)	S + M	
1991	31	16	47	1.5
1992	62	36	98	2.8
1993	93	54	148	4.0
1994	121	72	193	4.9
1995	151	107	258	6.1
1996	229	148	377	7.2
1997	227	138	365	7.2
1998	304	190	494	8.0
1999	245	173	418	8.7
2000	278	178	455	8.7
2001	310	197	507	8.5
2002	397	271	668	9.0
2003	552	325	877	8.9
2004	526	355	881	8.8
2005	600	379	979	9.1
2006	637	447	1,083	9.6
2007	1,020	787	1,807	9.7
2008	1,473	898	2,371	9.3
1991–2008	7,255	4,773	12,027	8.3

Source: Authors, based on SAGPyA (2009).

Figure 7.7 Area of Argentina under zero-tillage production planted with grains and oilseeds, by crop, 1991–2008

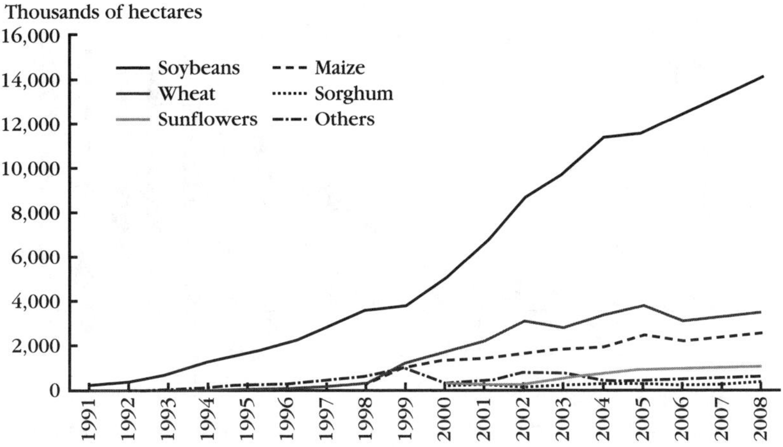

Thousands of hectares

Sources: Adapted from AAPRESID (2007) and SAGPyA (2009).

(Figure 7.7). For these crops, however, there was no concurrent growth in cultivated area; in fact, sunflower cropping evolved in the opposite direction. Nevertheless, the new technology brought about a positive change in the cost structure for all crops; these savings can be counted as an innovation-induced benefit to farmers.

The savings benefit to producers is defined as the cumulative savings from the lower costs of land cultivation and crop production resulting from zero-tillage practices. These savings have been estimated at 1.5 UTA (agricultural tilling unit).[9] With a total of almost 179 million hectares planted under zero tillage since 1991, the cumulative savings amount to $4.7 billion (Figure 7.8 and Table 7.3).

Benefits to Consumers

The benefits to consumers were estimated indirectly from the price effect of the increase in agricultural output. Zero-tillage technology in Argentina is associated with two internationally traded foodgrains, soybeans and maize. Soybean production is almost entirely (95 percent) exported.

We estimate the accumulated savings of consumers worldwide for the period between 1991 (the year of the initial adoption of zero-tillage practices in Argentina) and 2008. (See Trigo et al. 2009 for details of the methodological approach and actual computations for soybeans and maize.) A summary of the results is presented in Table 7.4.

Figure 7.8 Total area of Argentina under zero-tillage production, 1991–2008

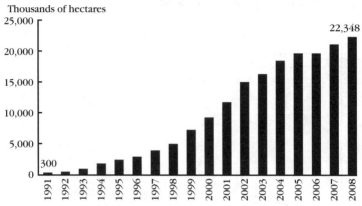

Source: Adapted from AAPRESID (2007).
Note: A total of 178.83 million hectares were under zero-tillage production between 1991 and 2008.

Table 7.3 Cumulative savings from zero-tillage production in Argentina, 1991–2008

Year	Area under zero-tillage (hectares)	UTA annual averages (US$ per hectare)	Total cumulative savings (millions of US$)
1991	300,000	17.37	7.8
1992	500,000	16.87	12.7
1993	970,000	14.86	21.6
1994	1,810,000	14.86	40.4
1995	2,440,000	15.72	57.5
1996	2,970,000	17.55	78.2
1997	3,950,100	19.02	112.7
1998	5,000,000	19.20	144.0
1999	7,269,500	19.20	209.3
2000	9,250,000	19.21	266.5
2001	11,660,000	19.07	333.6
2002	15,000,821	11.21	252.2
2003	16,351,212	15.18	372.4
2004	18,496,446	15.65	434.3
2005	19,683,172	16.09	475.1
2006	19,719,436	15.99	473.0
2007	21,110,471	19.63	621.5
2008	22,348,159	23.74	795.9
1991–2008	178,829,316	17.25	4,708.6

Source: Authors, based on SAGPyA (2009).
Note: UTA means agricultural tilling unit.

Table 7.4 Change in consumer expenditures on soybeans and maize from adoption of zero-tillage production in Argentina, 1991–2008 (millions of US$)

Year	Soybeans (S)	Maize (M)	S + M
1991	−39	−27	−66
1992	−77	−60	−137
1993	−117	−91	−207
1994	−151	−120	−272
1995	−188	−179	−367
1996	−286	−247	−533
1997	−284	−230	−514
1998	−380	−317	−697
1999	−306	−288	−595
2000	−347	−296	−643
2001	−387	−329	−716
2002	−496	−451	−948
2003	−690	−542	−1,232
2004	−658	−591	−1,249
2005	−750	−632	−1,382
2006	−796	−744	−1,540
2007	−1,275	−1,311	−2,586
2008	−1,841	−1,496	−3,337
1991–2008	−9,069	−7,954	−17,023

Source: Authors, based on SAGPyA (2009).

Total Benefits

Table 7.5 presents a summary of total estimated benefits based on the estimates presented in the previous section. The total estimated benefits derived from zero-tillage practices in Argentina amount to $33.76 billion, including impacts on both production and consumption. On the supply side, these benefits (over 1991–2008) include some $12 billion worth of additional gross income and $4.71 billion worth of savings attributed to the reduction in operating costs for Argentina's farmers. On the demand side, $17.02 billion was saved by consumers worldwide as an effect of lower market prices for both agricultural commodities.

Note that the benefits computed for the supply side are likely overestimated. Other technologies involved in the production process (such as improved varieties, new inputs, and more efficient management) have a share in the impact that is difficult to separate out. Moreover, the estimate does not take into account the value of the forfeited production that was displaced (mostly low-productivity livestock production), a value that is difficult to assess owing to the great heterogeneity of the agroecological and microeconomic parameters. Nevertheless, the magnitudes

Table 7.5 Impacts of the adoption of zero-tillage production in Argentina, 1991–2008

Indicator	Measure
Land under zero-tillage cultivation, 2008	22.3 million hectares
Cost reduction for farmers in Argentina, 1991–2008	$4.71 billion
Increased gross income of farmers in Argentina, 1991–2008	$12.03 billion
Reduction in consumer expenditures worldwide, 1991–2008	$17.02 billion
Total benefits	$33.76 billion

Source: Authors, based on SAGPyA (2009).
Note: All dollar amounts are US$.

involved would probably withstand any sensitivity analysis performed on the results. Even if the results on the supply side were overestimated by as much as 50 percent, the total benefit would be reduced by just 23 percent, from $33.76 billion to $26.00 billion—still a very significant figure.

Conclusion

The results presented in this study tell an impressive success story centered on the set of technologies that came to be known as zero-tillage farming. These technological innovations were initially conceived to internalize the negative externalities induced by conventional tillage by reversing the damage to the physical and chemical structure of the soil, thus addressing a threat to the long-term sustainability of agricultural production.

According to the estimates presented in this chapter, a total of $33.76 billion —in savings and additional earnings—can be attributed to this set of innovations, as developed and adapted for Argentina's agroecological conditions and effectively adopted by Argentine farmers. These are measurable, positive impacts of the development, adaptation, diffusion, and adoption of a specific set of agricultural production technologies.

Beyond the quantitative story is the story of the social process that brought about the transformation. A diverse set of stakeholders—farmers, research institutions, technical assistance consultants, and the agricultural inputs and farm machinery industries—came together as an innovation network. Their work evolved from identification of the problem to development of the technical alternatives to solve it, generating not only an information exchange mechanism but also the microeconomic conditions and the policy environment needed to support the process of change throughout its entire cycle.

Notes

1. For an in-depth discussion of the evolution of Argentine agricultural production and productivity during the post–World War II period, see Barsky (1991) and Manciana (2007).

2. In fact, the first shipment of Argentine soybeans, destined for Hamburg, took place in July 1962 (Giorda 1997).

3. Alapin (2008) highlights how farmers participated in mobilizing the process. Farmers in the Rosario area started to adopt zero-till practices by adapting conventional seeding machines. Ricardo Ayerza adapted a zero-till potato seeder in Saliqueló and Río Cuarto (a province of Córdoba). In Córdoba there were attempts in the same line, and toward the end of the 1960s, Santos Alzari, a blacksmith in Ascensión in Junín County, province of Buenos Aires, in response to the demands of farmers in the area, built a seeding machine adapted to the requirements of zero-till agriculture.

4. For a more in-depth discussion of these aspects, see Derpsch (1999, 79–97).

5. In 1974 the United Kingdom was the second country after the United States in total area under zero-till agriculture and was a pioneer on research work in the field (Senigagliesi and Massoni 2002).

6. In highlighting this interaction among private technical assistance providers and local farm machinery firms, Ekboir (2002, 7) states: "Several farmers who had participated in field trials in Marcos Juarez began searching for new sources of information. For example, after a conference by Sherly Philips (University of Kentucky), one of the Rosso brothers traveled to the [United States] to see firsthand what was being done there with regard to zero-till practices, and on this trip they realized that for its development to move forward, they needed sowing machines, so they contacted Schiarre (a local firm) to develop a prototype. At about the same time, Agrometal, another local farm machinery firm, started to work in a collaborative fashion with the Marcos Juarez research group to start the development of a local design. The same type of interaction took place among other firms, farmers, and researchers to improve the design of farm machinery equipment" (Ekboir 2002, translation by the authors).

7. For an extensive review of the characteristics and impact of glyphosate on the environment and on agriculture in general, see Permingeat (2008).

8. Data for a given crop cycle, such as 1990–91, is here presented for the harvest year, in this case 1991.

9. Márgenes Agropecuarios (2009). UTA is an index that compiles the average cost of inputs and labor needed to crop 1 hectare of farmland, from land preparation to harvest.

References

AACREA (Asociación Argentina de Consorcios Regionales de Experimentación Agrícola). 2006. Series de Precios Agropecuarios, Versión 2.0. Buenos Aires. <http://www.crea.org.ar/aacrea/site/Portal Institucional-internet/index.html>. Accessed April 23, 2009.

AAPRESID (Asociación Argentina de Productores en Siembra Directa). 2007. Base Estadística. Rosario, Argentina. <www.aapresid.org.ar>. Accessed April 22, 2009.

Alapin, H. 2008. *Rastrojos y algo más: Historia de la siembra directa en Argentina.* Buenos Aires: Editorial Teseo.

Andriulo, A., C. Sasal, and M. L. Rivero. 2001. Los sistemas de producción conservacionistas como mitigadores de la pérdida de carbono orgánico edáfico. In *Siembra Directa II*. Buenos Aires: Nacional de Tecnología Agropecuaria.

ArgenBio (Consejo Argentino para la Información y el Desarrollo de la Biotecnología). 2007. ArgenBio website. <http://www.argenbio.org/>. Accessed August 20, 2010.

Barsky, O., ed. 1991. *El desarrollo agropecuario pampeano*. Buenos Aires: Grupo Editor Latinoamericano.

Casas, R. 2003. El aumento de la materia orgánica en suelos argentinos: El aporte de la siembre directa. In *XI Congreso Nacional de AAPRESID, Darse Cuenta*. Rosario, Argentina: Asociación Argentina de Productores en Siembra Directa.

Costamagna, O. 2004. Mercado de semillas—Impacto del proyecto de fondo fiduciario (Regalías Globales). Foro de Perspectiva Agroindustrial 2004, Seminario Outlook de la Agroindustria Argentina: El campo como eje de la sociedad argentina. San José, Costa Rica: Inter-American Institute for Cooperation on Agriculture. CD-ROM.

De Moraes Sá, J. C., C. Cerri, M. Piccolo, B. Feigl, L. Seguy, S. Bouzinac, S. Venzke-Filho, and M. Neto. 2004. Acumulo de materia organica no solo em plantio direto como o pasar dos anos. In *XII Congreso de AAPRESID*. Rosario, Argentina: Asociación Argentina de Productores en Siembra Directa.

Derpsch, R. 1999. Expansión mundial de la siembra directa y avances tecnológicos. Proceedings, 7° Congreso Nacional de Siembra Directa de AAPRESID, August 17–20, in Mar del Plata, Argentina. Rosario, Argentina: Asociación Argentina de Productores en Siembra Directa.

Ekboir, J., ed. 2002. *CIMMYT 2000–2001: World wheat overview and outlook—Developing no-till packages for small-scale farmers*. Mexico City: International Maize and Wheat Improvement Center.

Ferreres, O. J. 2005. *Dos siglos de economía argentina (1810–2004): Historia argentina en cifras*. Buenos Aires: El Ateneo.

Galvão Gomes, A. 2008. Benefícios econômicos da biotecnologia no Brasil "O caso da soja RR." Céleres. <http://www.celeres.com.br/SumarioExecutivoBeneficioEconomico_SojaRR29 .01.08.pdf>. Accessed August 1, 2009.

Giorda, L. 1997. La soja en la Argentina.Nacional de Tecnología Agropecuaria, Centro Regional, Córdoba.

Manciana, E., ed. 2007. El campo a fines del siglo XX: Intentos, fracasos y las políticas que vienen. Fortalecimiento de la Organización y Gestión Económica y Social, Buenos Aires.

Márgenes Agropecuarios. 2009. Year 24, no. 285. Buenos Aires.

Permingeat, H. 2008. Los organismos vegetales geneticamente modificados (OVGMs) y el ambiente. *Prospectiva Tecnológica* (número especial). Buenos Aires: Asociación Argentina de Productores en Siembra Directa.

Sábato, J. 1971. *Ciencia, tecnología, desarrollo y dependencia*. Serie Mensajes. San Miguel de Tucumán: Universidad Nacional de Tucumán.

Sagardoy, M. A., H. E. Gómez, F. A. Montero, C. Zoratti, and A. R. Quiroga. 2001. Influencia del sistema de siembra diecta sobre los microorganismos del suelo. In *Siembra Directa II.* Buenos Aires: Nacional de Tecnología Agropecuaria.

SAGyP (Secretaría de Agricultura, Ganadería y Pesca). 1995. *El deterioro de las tierras en la República Argentina: Alerta amarillo.* Buenos Aires.

SAGPyA (Secretaría de Agricultura, Ganadería, Pesca y Alimentos). 2009. Estimaciones agrícolas. <http://www.sagpya.mecon.gov.ar>. Accessed April 23, 2009.

Schoijet, M. 2005. Desertificación y tormentas de arena. *Revista Región y Sociedad* 17 (32): 167–187.

Senigagliesi, C., and S. Massoni. 2002. Transferencia de tecnología en Siembra Directa: Un análisis de lo realizado en el INTA. In *Siembra Directa II.* Buenos Aires: Nacional de Technologia Agropecuaria.

Trigo, E., D. Chudnovsky, E. Cap, and A. López. 2002. *Los transgénicos en la agricultura Argentina: Una historia con final abierto.* Libros del Zorzal. San José, Costa Rica: Inter-American Institute for Cooperation on Agriculture.

Trigo, E., E. Cap, V. Malach, and F. Villarreal. 2009. *The case of zero-tillage technology in Argentina.* IFPRI Discussion Paper 915. Washington, D.C.: International Food Policy Research Institute.

Zaccagnini, M. E., and N. C. Calamar. 2001. Labranzas conservacionistas, siembra directa y biodiversidad. In *Siembra Directa II.* Buenos Aires: Nacional de Technologia Agropecuaria.

Chapter 8

Zero Tillage in the Rice–Wheat Systems of the Indo-Gangetic Plains: A Review of Impacts and Sustainability Implications

Olaf Erenstein

Zero Tillage in Rice–Wheat Systems

The Green Revolution transformed the Indo-Gangetic Plains, which spread from Pakistan through northern India and the *terai* (plains) region of Nepal to Bangladesh, into the cereal basket of South Asia. The technological packaging of improved wheat and rice seed, chemical fertilizer, and irrigation in an overall supportive environment for agricultural transformation led to rapid productivity growth and the advent of rice–wheat systems that now cover an estimated 14 million hectares in the region (Timsina and Connor 2001; Gupta et al. 2003; Gupta and Seth 2007). During the past decade, however, total factor productivity growth has stagnated (Duxbury 2001; Kataki, Hobbs, and Adhikary 2001; Kumar et al. 2002; Ladha et al. 2003; Prasad 2005), leading to concerns over national food security and lagging rural economic growth. These concerns prompted a quest for technologies that conserve resources, reduce production costs, and improve production while sustaining environmental quality (Hobbs and Gupta 2003a; Gupta and Sayre 2007; Gupta and Seth 2007; Erenstein et al. 2008a). One such promising technology is

This chapter synthesizes findings from earlier studies by the same author and associates, as well as secondary sources. It particularly draws from and builds on Erenstein (2009a). The views expressed in this chapter are those of the author and do not necessarily reflect the views of donors or the author's institution. The usual disclaimer applies.

zero tillage, the seeding of a crop into unplowed fields–also known as no-till, direct seeding or drilling, or conservation tillage (Erenstein 2002; Erenstein et al. 2008b). Zero tillage typically saves energy, helps reverse soil and land degradation (such as the decline of soil organic matter, soil structural breakdown, and soil erosion), and leads to more efficient use of water and other inputs (Erenstein and Laxmi 2008).

Zero-tillage planting of wheat after rice has been the most successful resource-conserving technology to date on the Indo-Gangetic Plains, particularly in northwest India and to a lesser extent on the Indus plains in Pakistan (Erenstein et al. 2007c; Erenstein and Laxmi 2008). The interest in zero tillage on the Indo-Gangetic Plains originated from diagnostic studies that highlighted the importance of time conflicts between rice harvesting and wheat planting in both northwest India (Harrington et al. 1993; Fujisaka, Harrington, and Hobbs 1994) and Pakistan (Byerlee et al. 1984). Wheat is grown in the cool, dry winter and is the traditional mainstay of food security on the northwest Indo-Gangetic Plains. Rice is grown during the warm monsoon season, but its introduction and widespread cultivation in the northwest area occurred only in recent decades during the Green Revolution. Zero-tillage wheat has a number of advantages, alleviating a number of constraints in the rice–wheat system: it permits earlier wheat planting, helps control obnoxious weeds such as *Phalaris minor,* reduces costs, and saves water (Erenstein and Laxmi 2008).

Nevertheless, the potential environmental benefits of conservation agriculture have yet to be fully realized; to reap these benefits, the challenges of reducing tillage for rice as well as wheat, retaining crop residues, and diversifying crops must be met. Equity has also posed a challenge so far, and gains need to be extended more rigorously to the less endowed areas and farmers, which calls for a better understanding of livelihood implications and the participation of stakeholders.

The prevailing zero-tillage technology in rice–wheat systems in the area is use of a tractor-drawn seed drill with 6 to 11 inverted-T tines to seed wheat directly into unplowed fields with a single pass of the tractor. This specialized agricultural machinery was originally not available in South Asia. Creating the local manufacturing capacity to supply adequate and affordable zero-tillage drills was a key component of the diffusion of the technology. In Pakistan, adaptive research designed to make zero-tillage methods suitable for local conditions started during the mid-1980s following the importation of a prototype drill with inverted-T openers from Aitcheson Industries, New Zealand, for use by national program scientists from the National Agricultural Research Center and the International Maize and Wheat Improvement Center (CIMMYT). In India, the same inverted-T openers were introduced in 1989 by CIMMYT, and in 1991 a first prototype of the Indian zero-tillage seed drill was developed at G. B. Pant University of Agriculture and Technology in Pantnagar. In both countries, a collaborative program for further development and commercialization of zero-tillage drills by small-scale farm machinery manufacturers was initiated by the national agricultural research system in collaboration with CIMMYT and subsequently with the Rice–Wheat Consortium of the Indo-Gangetic Plains, and its history has been variously documented (Ekboir 2002; Erenstein and Laxmi 2008; Harrington and Hobbs 2009). Surface seeding is one option for employing zero tillage without the use of a tractor or seed drill (Tripathi et al. 2006), but its use is largely confined to low-lying fields with drainage problems on the eastern Gangetic Plains.

The diffusion of the technology has accelerated in the early years of the 21st century, particularly on the northwest Indo-Gangetic Plains of India, where the combined zero- and reduced-tillage wheat area seems to have stabilized at between one-fourth and one-fifth of the wheat area. Several factors make it problematic to reliably measure zero-tillage adoption and impacts on the Indo-Gangetic Plains, however. I estimate that in 2008 the aggregate zero- or reduced-tillage wheat area amounted to 1.76 million hectares and was used by 620,000 farmers. The main driver behind its spread is the cost savings, which makes zero tillage profitable. For instance, because zero-tillage wheat allows for a drastic reduction in tillage intensity, there are significant cost savings as well as potential yield gains through planting the wheat crop at a better time. Wheat farmers who adopt zero tillage are the main beneficiaries, enhancing their farm income by about $100 per hectare. Viable dynamic systems for the diffusion of innovations have been key to the success of zero tillage in India, including a vibrant manufacturing base for zero-tillage drills and drilling service providers.

The promotion of zero tillage in India has been particularly successful, benefiting from the timely congruence of a profitable technological opportunity and several key champions who encouraged adoption (Laxmi, Erenstein, and Gupta 2007a). Several factors proved crucial to its success in India (Seth et al. 2003). A local manufacturing capacity was developed to produce and adapt zero-tillage drills at a competitive cost. The private sector could see substantial market opportunities for its products, whereas the involvement of several manufacturers ensured competitive prices, good quality, easy access to drills by farmers, and a guarantee of repairs and servicing. Close linkages of scientists and farmers with the private manufacturers, including placement of machines in villages for farmer experimentation, allowed rapid feedback and refinement of implements. Strong support from state and local government officials helped with dissemination, including the provision of a subsidy to lower the investment cost and initiate extensive on-farm demonstrations and trials. The Rice–Wheat Consortium played a crucial catalytic role in promoting the public–private partnership, nurturing it through its formative stages and facilitating technology transfer from international and national sources (Erenstein 2009a).

The success of zero tillage in India highlights the importance of institutional support for a technological opportunity to materialize. In Pakistan's Punjab Province, the agroecology and system constraints are relatively similar, and favorable experimental findings led to a pilot production program in the 1990s to promote the use of zero tillage (Aslam et al. 1993). But the spread of zero tillage has been significantly slower there, hampered by institutional controversies within the national system and a lesser presence of the Rice–Wheat Consortium, among other things (Erenstein et al. 2007a).

There is a wealth of information on zero tillage on the Indo-Gangetic Plains, particularly in India (see Erenstein and Laxmi 2008 for a recent review; Gupta et al. 2003; Hobbs and Gupta 2003a; and compendium volumes such as Malik et al. 2002; Abrol, Gupta, and Malik 2005; Malik et al. 2005a, 2005b). Most of the documented evidence relates to the northwestern Indo-Gangetic Plains, where intensive rice–wheat systems prevail and where adoption has been more extensive than in other areas so far. There is ongoing research and development (R&D) work on the eastern Gangetic plains of India (Bihar, Uttar Pradesh, and West Bengal), on the Nepal *terai,* and in Bangladesh. The initial results on the eastern plains are encouraging but not yet well documented or widely published. This chapter synthesizes the documented experiences with zero tillage in the rice–wheat systems across the Indo-Gangetic Plains. It will specifically focus on the impact of zero tillage (particularly diffusion, adoption, and farm-level impacts), its sustainability dimensions (financial and fiscal, environmental, and social and political), and the lessons learned.

The Impact of Zero Tillage in Rice–Wheat Systems

Diffusion and Adoption of Zero Tillage

Adoption of zero tillage for wheat in South Asia started in the second half of the 1990s and accelerated in the early years of the 21st century. Experts estimate that the combined zero- and reduced-tillage wheat area on the Indo-Gangetic Plains amounted to some 2 million hectares in 2004–05, primarily in India. But the actual extent of zero-tillage diffusion on the Indo-Gangetic Plains is not precisely known. Zero-tillage research and the associated data collection and reporting in the area have both increased and improved in the past several years. However, most studies primarily report on the technical aspects of zero tillage at the plot level and often are based on trial data (on-station and on-farm) (Erenstein and Laxmi 2008). Many of the farm surveys are not based on a robust sampling frame that would allow for unbiased diffusion estimates. Instead, most farm surveys contrast a sample of zero-tillage adopters with nonadopters (see, for example, Bakhsh, Hassan, and Maqbool 2005; Jamal, Akhtar, and Aune 2007; Sarwar and Goheer 2007; Singh 2008). Reliable and empirically based zero-tillage diffusion indicators are particularly scarce and problematic. Lacking other estimates, researchers often repeat and cite the expert estimates of zero-tillage adoption. However, in the process, they often omit two important underlying qualifications—that is, they are using expert estimates that are not empirically grounded and that reflect estimated zero-tillage drill use irrespective of tillage intensity.

A problem with estimating zero-tillage adoption is that it is a cultural practice that is sparsely reported in agricultural statistics and studies. It is also ambiguously interpreted, with the term often used interchangeably with the cultural practice and the seeding implement. Yet whereas most zero tillage depends on the use of a zero-tillage seed drill, the practice is not unambiguously embodied in this machinery; the drills are also used on fields where tilling is reduced and even on conventionally tilled fields. Furthermore, zero-tillage use varies over seasons and on farms, with farmers often using it for one part of their farm while adhering to more intensive tillage for other parts, making the reliable measurement of zero-tillage adoption and impacts problematic.

Surveys of manufacturers of zero-tillage drills have revealed that the manufacturing capacity is spatially concentrated in the rice–wheat belt on the northwest Indo-Gangetic Plains, particularly Haryana (India) and the Indian and Pakistani Punjab (Erenstein, Malik, and Singh 2007; Farooq, Sharif, and Erenstein 2007). The first commercial zero-tillage drills therefore originated in the traditional agricultural machinery manufacturing centers in the Green Revolution heartland. The three locations (Haryana, Indian Punjab, and Pakistani Punjab) show a relatively similar increase in the number of manufacturers over time, in both absolute and relative terms. However, aggregate sales figures of the surveyed drill manufacturers in Pakistani Punjab are a fraction of those in Haryana and Indian Punjab. By the end of 2003, a cumulative total of 15,700 zero-tillage drill machines had been sold by 50 surveyed manufacturers in India, with Haryana registering more than half the reported sales in each year—compared with the 2,000 zero-tillage drills sold by 31 manufacturers in Pakistani Punjab. Whereas annual sales figures for zero-tillage drills for surveyed manufacturers in India were still on the increase, the sales figures in Pakistani Punjab were relatively flat, with peak sales in 2002. The wider manufacturing base and significant growth in sales imply healthy competition among manufacturers in India, with favorable implications for price and quality and generally lighter drills. The average sales price of a zero-tillage drill in 2003 amounted to $325 in India compared with $559 in Pakistan (Erenstein et al. 2007a).

Farm household surveys in 2003–04 confirmed significant adoption of zero-tillage wheat in the rice–wheat systems of the northwestern Indo-Gangetic Plains: 34.5 percent of sample farmers in India's Haryana and 19.0 percent in Pakistan's Punjab (Erenstein, Malik, and Singh 2007; Erenstein et al. 2007a; Farooq, Sharif, and Erenstein 2007). The farms were typically only partial adopters of zero tillage—that is, only a share of their wheat area was put under zero tillage, with the remaining area still under conventional tillage. This partial adoption implies that the actual area under zero tillage is typically less than the rate of adoption in terms of number of farmers.

The expert estimates, manufacturer surveys, and farm surveys in India and Pakistan agree that zero tillage has picked up since 2000, but the empirical surveys suggest a much slower uptake and subsequent stagnation in Pakistan (Erenstein et al. 2007a). This reiterates the need for empirical ground-truthing of the technology's uptake and the need for robust and complementary diffusion indicators.

During the winter or *rabi* season of 2007–08 the rice–wheat villages in Haryana surveyed earlier by Erenstein, Malik, and Singh (2007) were revisited (Erenstein 2009a). These visits showed that the zero-tillage wheat area continued to increase, albeit at a slow pace, from an average share of village wheat area of 18 percent in 2004–05 to 24 percent in 2007–08 (Figure 8.1). As reported earlier (Erenstein, Malik, and Singh 2007), the reduced-tillage area in these villages was still marginal, with farmers using either zero tillage or conventional tillage. However, the revisit highlighted the rapid increase in another new tillage system, primarily use of a tractor-drawn rotavator ("other tillage" in Figure 8.1). The rotavator typically implies a single pass of shallow intensive tillage that incorporates crop residues and pulverizes the soil. It may thereby reduce the number of passes compared to conventional tillage, but its tillage intensity goes against the tenets of conservation agriculture (Erenstein 2009a).

In the *rabi* season of 2007–08, an additional wheat tillage monitoring survey was conducted across 120 randomly selected villages in Haryana and Indian Punjab. This study provided a more representative random sample of wheat-cultivating villages and included nonrice–wheat systems. Compared with the Haryana rice–wheat study area (see Figure 8.1), the results show some marked divergences (Figure 8.2). First, the share of zero-tillage area was significantly lower—in both Haryana and Punjab—and showed a small decline during the period 2005–06 to 2007–08. Second, the reduced-tillage area was a multiple of the zero-tillage area and showed a small increase in both states. Extrapolating these estimates would imply a combined zero- and reduced-tillage wheat area of 1.26 million hectares in Haryana and Indian Punjab in 2007–08. With an average area of 3.5 hectares of zero- or reduced-tillage wheat per adopter, this amounts to an extrapolated 360,000 farmers in the Haryana and Punjab alone. The results also suggest that after the initial rapid spread of zero tillage on the northwestern Indo-Gangetic Plains, the zero- or reduced-tillage wheat area stabilized at between one-fifth and one-fourth of the wheat area. The study also shows that, particularly in Haryana (see Figure 8.2), and much as in the other study (see Figure 8.1), there was a marked increase in other tillage systems (primarily rotavator use). The advent of the rotavator merits follow-up research and active engagement with regional stakeholders because it offsets many of the gains implied

Figure 8.1 Evolution of wheat tillage in rice–wheat systems in Haryana, India, 2004–05 to 2007–08 (village survey findings, n = 50)

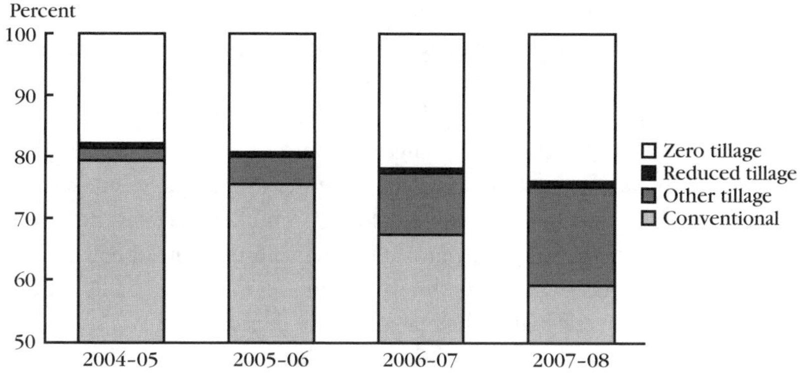

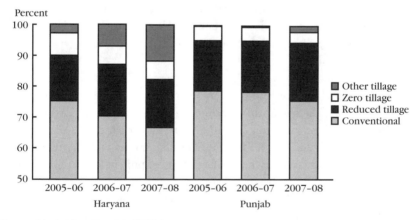

Source: Adapted from Erenstein (2009a).

Figure 8.2 Evolution of wheat tillage systems in Haryana and Punjab, India, 2005–06 to 2007–08 (village survey findings, n = 120)

Source: Adapted from Erenstein (2009a).

by more conservation agriculture–based resource-conserving technologies such as zero or reduced tillage (Erenstein 2009a).

Adoption on the eastern plains is still in its initial stages, with a dearth of empirically based adoption estimates. A random village survey in 2005 found 13 percent of farm households using zero or reduced tillage in the northwestern rice–wheat belt, with still negligible adoption rates elsewhere on the Indian Indo-

Gangetic Plains (Erenstein et al. 2007b, 2007c). In 2006 a village survey of primarily project villages in India reported zero- or reduced-tillage adoption rates of 18 percent of farm households on the northwestern plains and 5 percent on the eastern plains (Teufel, Erenstein, and Samaddar 2007). In 2008 a regional village survey reported zero-tillage use in 14 percent of the wheat area in project villages on the northwestern plains (Pakistan and India), 12 percent on the central plains (India and Nepal), and none on the lower Gangetic plains of Bangladesh (R. Singh et al. 2009). My estimate of the zero- or reduced-tillage area in the Indian Indo-Gangetic states, excluding Punjab and Haryana, would be 0.5 million hectares. Based on an average of 1.9 hectares per adopting farm at the Indian project site of Ballia, Eastern Uttar Pradesh (R. Singh et al. 2009), this would amount to some 260,000 users. For the Indian Indo-Gangetic Plains, the aggregate zero- or reduced-tillage area in 2008 would thus amount to an estimated 1.76 million hectares and 620,000 farmers.

Farm-Level Impacts of Zero Tillage

Erenstein and Laxmi (2008) provide a comprehensive review of the impacts of zero-tillage wheat in India's rice–wheat systems, including effects on land preparation and crop establishment, water use, soils and biotic stresses, yields, and cost savings and profitability. Their review shows that planting zero-tillage wheat after rice generates substantial benefits at the farm level by enhancing farm income from wheat cultivation ($97 per hectare) through the combined result of a yield effect and a cost-saving effect.

The cost-saving effect ($52 per hectare) primarily reflects the drastic reduction in tractor time and fuel for land preparation and wheat establishment (Erenstein and Laxmi 2008). The tractor-drawn zero-tillage drills allow tillage intensity to be drastically reduced for the wheat crop from eight tractor passes to a single tractor pass (Erenstein et al. 2007a, 2008a). This reduction implies a significant, immediate, and recurring cost saving, which makes adoption profitable (corresponding to a 15–16 percent saving on operational costs in Erenstein et al. 2007a).

The review of zero tillage in India found a yield effect amounting to a 5–7 percent yield increase for wheat across studies (including on-station trials, on-farm trials, and surveys in Erenstein and Laxmi 2008). This yield increase provides a further boost to the returns to zero tillage. The yield effect, if any, is closely associated with the enhanced timeliness of wheat establishment after rice. Heat stress at the end of the wheat season implies that the potential wheat yield is reduced by 1.0–1.5 percent per day if planting occurs after mid-November (Ortiz-Monasterio, Dhillon, and Fischer 1994; Hobbs and Gupta 2003a). It is estimated that about one-third of the wheat area on the Indian Indo-Gangetic Plains is sown late—often linked to late-maturing basmati rice on the northwestern Indo-Gangetic Plains (including the Pakistani

Punjab) and generally late rice harvesting on the eastern plains—and zero tillage would potentially alleviate this problem by allowing for timelier establishment.

The literature on zero tillage in Pakistan shows that it is also profitable there, based on a similar combination of yield gains and cost savings (Choudhary et al. 2002; Kahlown, Gill, and Ashraf 2002; Bakhsh, Hassan, and Maqbool 2005; Jamal, Akhtar, and Aune 2007; Sarwar and Goheer 2007) and widespread late planting of wheat (Akhtar 2006; Farooq, Sharif, and Erenstein 2007). In the case of Haryana (India), the large farm surveys confirmed both a significant yield effect and a cost-saving effect (Figure 8.3), thereby making adoption worthwhile and providing a much-needed boost to the returns to wheat cultivation (Erenstein, Malik, and Singh 2007). Similar farm surveys in Pakistani Punjab found zero tillage primarily a cost-saving technology for wheat cultivation, with no significant yield effect (Figure 8.3) (Erenstein et al. 2007a; Farooq, Sharif, and Erenstein 2007; Erenstein et al. 2008a). In fact, the lack of a positive yield response was a major contributor to the slower adoption in Punjab and merits follow-up research.

Whereas zero tillage enhances income for wheat farmers on the Indo-Gangetic Plains, less is documented or known about the associated impacts on household food security and nutrition. The time and additional resources saved are used by the adopting farm households for various productive, social, and leisure purposes (Laxmi, Erenstein, and Gupta 2007a; Erenstein and Laxmi 2008; Erenstein and Farooq 2009). In the Pakistani Punjab, zero tillage reportedly increased families'

Figure 8.3 **Financial advantage of zero tillage over conventional tillage for wheat on farms adopting zero tillage in 2003–04 in Haryana, India, and Punjab, Pakistan (farmer survey findings, *n* = 225)**

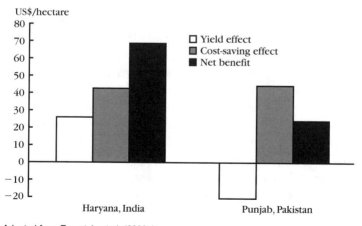

Source: Adapted from Erenstein et al. (2008a).

food consumption, probably through higher disposable income (Farooq, Sharif, and Erenstein 2007; Erenstein and Farooq 2009). In eastern Uttar Pradesh, India, zero tillage generated additional income through sales of the increased production (Jafry 2007). Future research may attempt to more rigorously document such impacts on households' food security and nutrition. Still, such endeavors will present major challenges, starting with the somewhat fuzzy nature of measuring adoption and the immediate crop-level impacts of such crop management practices. The cost-saving effect is relatively robust, but its household impacts will be difficult to trace in view of its fungible nature in relation to outlays forgone and the difficulty of disentangling the various confounding factors. Tracing the impacts of a positive yield effect may be somewhat easier, but such an effect is not always realized by farmers.

Little has been documented about the gender-specific dimensions of zero tillage's impact. To a large extent, this lack of documentation is a reflection of zero tillage's having primarily spread on the northwestern Indo-Gangetic Plains, where households are typically headed by males. In this area, women have limited participation in field-based crop activities and are instead primarily engaged in livestock- and homestead-based activities (Erenstein et al. 2007b; Erenstein and Farooq 2009). The few available studies report that women generally appreciated zero tillage. They acknowledged that after zero-tillage adoption there was less anxiety at the time of wheat field preparation, and this resulted in more peace at home (Laxmi, Erenstein, and Gupta 2007a; Erenstein and Laxmi 2008). One study also reported that more food was available for women and that family diets had improved (Joshi et al. 2007b). Some studies have indicated that women farmers are dependent on their families for knowledge and information about new technologies such as zero tillage (Jafry, Ahmad, and Poswal 2006; Jafry 2007; Joshi et al. 2007b).

Zero tillage was originally perceived as potentially generating higher yields at a lower production cost while being an environmentally friendly practice that saves water and soil (Hobbs, Giri, and Grace 1997; Gupta et al. 2002; Hobbs and Gupta 2003b). The subsequent literature generally confirms significant zero-tillage-induced resource-saving effects on farmers' fields in terms of diesel use, tractor time, and cost savings for wheat cultivation (Erenstein et al. 2008a; Erenstein and Laxmi 2008). Zero tillage saves irrigation water through the combined effect of allowing wheat to be planted without prior irrigation and shortening the duration of the first irrigation, because surface irrigation water flows faster across a nontilled field. However, water savings in farm surveys are less pronounced than expected from the initial on-farm trial data (they are linked to adopters generally reporting a similar number of irrigations and the inability to control for all confounding factors in survey data), but in any case water use is modest compared with that for the subsequent rice crop (Erenstein et al. 2008a). Other technologies, such as laser leveling, can generate more substantial water savings and have recently been taken up by farmers (Jat et al. 2009). Despite zero tillage's ability to reduce turnaround time, farm surveys also did not find that wheat was established earlier, thereby reducing the likelihood of obtaining higher yields with zero tillage (Tahir and Younas 2004; Erenstein et al. 2008a). A better understanding of why farmers did not capitalize on the potential of zero tillage to sow the wheat crop earlier merits follow-up research.

An important issue with zero tillage is that it is a component technology of conservation agriculture. Its current application to only one of the two cropping seasons, without adequate residue management or crop rotation, forgoes many of the environmental benefits associated with conservation agriculture, which is discussed in the next section. In the end, the prime driver of farmers' adoption of zero tillage is monetary gain, not water savings or natural resource conservation (Erenstein et al. 2008a).

Another issue is that the gains currently realized in farmers' fields have long been based on systems built on the premise of intensive tillage. Specifically selecting wheat varieties that grow well under zero-tillage conditions can boost the benefits of zero tillage (Joshi et al. 2007a; Rajaram et al. 2007). Similarly, wheat is adversely affected by the puddling (wet tillage) of the previous rice crop, and implications for the zero-tillage wheat yield are likely to be more favorable when farmers forgo seasonal puddling of the rice crop (Erenstein and Laxmi 2008).

The Sustainability of Zero Tillage in Rice–Wheat Systems

Financial and Fiscal Sustainability

Laxmi, Erenstein, and Gupta (2007a, 2007b) show that investments in the R&D of zero or reduced tillage by the Rice–Wheat Consortium of the Indo-Gangetic Plains and CIMMYT were highly beneficial. They assume that these investments accelerated the adoption of zero or reduced tillage by five years, taking into account only supply-shift gains (excluding social and environmental gains). These investments yielded significant economic benefits: a net present value of $94 million, a benefit–cost ratio of 39, and an internal rate of return of 57 percent. Significant positive spillovers of investments in R&D on zero tillage—both previous investments and those from elsewhere—contributed to the high returns. The zero-tillage case therefore highlights the potential gains from successful technology transfer and adaptation, particularly for a technology that has been widely recommended.

The cost savings, particularly when combined with a yield increase, imply that the returns to the adoption of zero tillage are quite robust, significantly reducing the risk of adoption (Erenstein and Laxmi 2008). Still, in the early diffusion stages, prospective owners of zero-tillage drills used in rice–wheat systems have tended to complain that the machines can be used only during wheat crop establishment (sitting idle for most of the year). This result reflects the prevalence of rice–wheat systems in which rice is typically still being transplanted. However, many of the current zero-tillage drills can effectively seed different crops, including direct-seeded rice. However, even if drills are used only during the limited operational window of wheat establishment, they can be effectively and economically used by tractor service providers on multiple farms. For instance, a survey of zero-tillage drill owners in Haryana showed that each drill had, on average, planted 42 hectares in a single wheat season (Punia et al. 2002). The investment cost of a zero-tillage drill can be rapidly recovered within two seasons through the cost savings alone (Erenstein et al. 2007a).

In India there is a tradition of subsidizing agricultural machinery. Some state governments (such as that of Haryana) therefore provide a subsidy of about 25 percent on the purchase price of a new zero-tillage drill (Ekboir 2002). Such investment subsidies may enhance farmers' access to new implements, but they typically add to the transaction cost, imply opportunity costs, and reinforce paternalistic expectations that reduce farmers' initiative as they wait for subsidies to materialize. Whether conservation agriculture–based technologies such as zero tillage should be entitled to a subsidy remains an open question; the comprehensive social gains need to be weighed against the social costs (Erenstein and Laxmi 2008). In any event, some of the subsidies lack consistency: for instance, subsidies are simultaneously being given for machinery that limits the intensity of soil tillage (such as zero-tillage drills) and machinery that increases the tillage intensity (such as rotavators). In fact, many subsidies in India have been counterproductive to zero tillage and actually have contributed to the sustainability concerns that undermine rice–wheat systems. For instance, subsidies for fertilizer, water, and rural electricity for irrigation undermine the incentives to rationalize the use of such resources and thus the attractiveness of resource-conserving technologies such as zero tillage. The widespread public intervention in produce chains with assured produce prices and marketing channels, particularly in India, also undermines incentives to diversify and rotate crops. Policy reforms to create an enabling environment for sustainable agriculture are likely to be more influential than the machinery subsidies. These reforms imply addressing some of the more thorny policy issues, such as the subsidy and taxation schemes, that currently undermine the sustainability of rice–wheat systems (Erenstein and Laxmi 2008).

The success of zero tillage in India is based on the creation of viable, dynamic innovation systems building on a successful and sustainable business model. Zero tillage is financially attractive to farmers and therefore has created farmer demand. Small-scale machine manufacturers see substantial market opportunities for their products, aided by the wide applicability of this mechanical innovation. They created a wide and competitive manufacturing base and played a key role in meeting the increasing demand. Tractor service providers have further enhanced the spread of the technology by making access to the zero-tillage drill divisible, thus allowing prospective zero-tillage drill owners to defer the investment decision. Adoption surveys reveal that 60 percent of zero-tillage adopters in Haryana did not own a zero-tillage drill (Erenstein, Malik, and Singh 2007). Service providers have the added advantage of having hands-on experience and self-interest in promoting the technology (Erenstein and Laxmi 2008), although the use of a service provider increases farmers' dependence on third parties for the timely establishment of their wheat crops.

So far, the business model has been somewhat less successful elsewhere in South Asia. In Pakistan it has been hampered by more limited demand, a smaller manufacturing base, more expensive drills, and institutional controversy.

In the Nepal *terai,* zero tillage has been hampered by political instability and the lack of a local manufacturing base, thus forcing the use of imported machinery from India, which implies additional transport costs and more problematic after-sale services. Tractor-based zero tillage has largely bypassed Bangladesh, where farmers use primarily two-wheel tractors for their tillage operations. There are ongoing R&D efforts to adapt conservation agriculture–based machinery for two-wheel tractors in Bangladesh, somewhat hampered by the relatively undeveloped local manufacturing base. On the eastern plains there is also some limited use of manual surface seeding, which does not depend on agricultural machinery (Tripathi et al. 2006).

Zero tillage may facilitate or induce other changes in the farming systems in terms of diversification, intensification, or both (Erenstein and Laxmi 2008). Zero tillage also opens the way for a new service industry beyond the farm level, be it for machinery manufacturers or custom hiring services (Dixon et al. 2007). These system changes could generate substantial multipliers, although further studies are needed to substantiate their value (Erenstein and Laxmi 2008).

Environmental Sustainability

Zero tillage has had primarily positive effects on the environment on the Indo-Gangetic Plains (saving fossil fuel and water and reducing emissions of greenhouse gas), although further research is needed to more rigorously substantiate and value these environmental impacts (Akhtar 2006; Sarwar and Goheer 2007; Erenstein and Laxmi 2008; Hobbs and Govaerts 2009; Pathak 2009). The diesel savings are relatively robust—36 liters of diesel per hectare, an 8 percent saving over conventional wheat tillage (Erenstein and Laxmi 2008; Erenstein et al. 2008a). Still, in spite of zero tillage's success on the Indo-Gangetic Plains, the full environmental benefits offered by conservation agriculture, including carbon sequestration, have yet to be fully utilized (Gupta and Sayre 2007; Laxmi, Erenstein, and Gupta 2007a). This is associated with the specific way zero tillage is applied to the rice–wheat systems on the Indo-Gangetic Plains, which distinguish it from related systems elsewhere (see, for example, Ekboir 2002; Erenstein 2002, 2003).

Zero tillage is typically applied only to the wheat crop in the double-crop system on the Indo-Gangetic Plains (Erenstein and Laxmi 2008). The use of zero tillage in wheat has limited spillovers to the productivity and management of the subsequent rice crop (Erenstein et al. 2007a, 2008a). For the rice crop, intensive wet land preparation followed by transplantation still prevails. So the reduction of tillage in rice–wheat systems has thus far been only partially successful, reflecting the increasing acceptance of zero tillage for wheat. Reducing the tillage intensity of the subsequent rice crop still presents a challenge, particularly in terms of water and weed management and available germplasm (Erenstein 2007). Ongoing research (on-station and on-farm) on the Indo-Gangetic Plains is addressing these challenges and attempting to adapt viable "double-no-till" rice–wheat systems (Khan et al. 2009; Saharawat 2009; U. P. Singh et al. 2009).

Zero-tillage wheat does not necessarily imply the retention of crop residues as mulch on the Indo-Gangetic Plains. In fact, the prevailing zero-tillage seed drills are relatively poor in handling trash. However, for better or worse, this limitation has not been a major issue in view of the limited biomass remaining after the rice crop once the prevailing residue management practices have been employed (Erenstein and Laxmi 2008). Leaving more crop residue as mulch implies the need for adaptation of zero-tillage drills. Ongoing R&D has already generated some second-generation zero-tillage seed drills that are able to handle significant residues, but their cost is still relatively high. Leaving more residue also implies the need to address potential trade-offs with existing residue uses (Erenstein et al. 2007b; Erenstein and Laxmi 2008).

In South Asia, crop residues are an integral part of rural livelihoods. Their use provides coherence to the prevailing smallholder crop–livestock systems (Devendra 2007; Erenstein et al. 2007b), serving as important sources of livestock feed for the dominant species in the region (cattle, buffaloes, and goats) and sometimes having other productive uses, such as for fuel, construction material, and mulch. The relative importance of each use varies geographically and by crop and is associated with poverty incidence (Erenstein et al. 2007b; Singh et al. 2007; Thorpe et al. 2007; Varma et al. 2007; Parthasarathy Rao and Birthal 2008). The intensity of residue use as feed has an inverse association with farm size and a positive association with rural poverty on South Asia's Indo-Gangetic Plains (Erenstein 2009a).

The prevailing crop residue management practices on the Indo-Gangetic Plains are largely incompatible with residue retention as mulch, despite significant biomass production (Erenstein et al. 2007b). The ex situ use of crop residues as livestock feed is nearly universal and rigorous, whereas the increasing mechanization of rice and wheat harvesting practices has trade-offs in terms of residue use and management. Wheat is the traditional foodcrop on the northwestern Indo-Gangetic Plains, and wheat residues are the corresponding basal feed for ruminant livestock. This implies significant imbalances in terms of seasonal residue extraction in the area, with surplus rice straw burned in situ during land preparation (Samra, Singh, and Kumar 2003; Gupta et al. 2004; Erenstein et al. 2007c; Bijay-Singh et al. 2008). This is particularly seen after combine harvesting of nonfragrant rice, with 74 percent of rice straw burned according to farmers' estimations compared to 22 percent burned after manual rice harvesting; in the case of wheat, even after combine harvesting, only 10 percent of the residue is burned (Teufel, Erenstein, and Samaddar 2008). Proceeding to the eastern plains, rice is the traditional foodcrop and rice straw the preferred basal feed. On the lower Gangetic plains, where manual harvesting still prevails, only an estimated 4 percent of the rice straw is burned, whereas 62 percent is fed to livestock. In the same area, straw from the less widely cultivated wheat crop is primarily burned (71 percent according to farmers' estimates), either as household fuel or on the field (Teufel, Erenstein, and Samaddar 2008). Its widespread use as feed implies that crop residues have significant value, and residue markets and institutional arrangements have developed accordingly. There are also significant regional variations in the use of nonfeed residues, such as for fuel and construction material. Such potential trade-offs need to be addressed by R&D and will present particular challenges for conservation agriculture–based technologies on the densely populated and poor eastern plains (Erenstein 2009a).

Retention of crop residues as soil cover is imperative in continuous no-tillage systems (Erenstein 2002, 2003). The widespread use of zero-tillage wheat on the Indo-Gangetic Plains without necessarily maintaining some soil cover has so far had limited perceivable negative consequences. However, this is a consequence of the seasonal nature of zero-tillage use, with plots still seasonally tilled for the subsequent rice crop. However, in a year-round—or double-no-till—rice–wheat system, residue retention becomes imperative (Erenstein 2009a). Planting zero-tillage wheat after rice on the Indo-Gangetic Plains does not necessarily entail an increased reliance on herbicides, reflecting that paddy rice fields are relatively weed free at harvest time (Erenstein and Laxmi 2008). In fact, by reducing soil movement, zero tillage serves as an effective measure for control of *Phalaris minor,* a major weed that has reduced wheat yields in the area and in the mid-1990s showed emerging resistance to isoproturon after the continuous widespread use of this herbicide during previous decades (Singh, Kirkwood, and Marshall 1999; Vincent and Quirke 2002; Franke et al. 2003; Yadav and Malik 2005). The ability of zero tillage to control herbicide-resistant *Phalaris* therefore became one of the drivers of adoption of the technology in northwestern India. Used in combination with new herbicides, this technique eventually managed to control the *Phalaris* problem.

Conservation agriculture is a wider concept than zero tillage and involves minimal disturbance of the soil, retention of residue mulch on the soil surface, and a rational use of crop rotations (Harrington and Erenstein 2005; FAO 2007; Hobbs 2007).

These principles of conservation agriculture, along with profitability at the farm level, are increasingly recognized as essential for sustainable agriculture. Alternatively, zero tillage alone is an insufficient condition for conservation agriculture (Erenstein et al. 2008b) and, although attractive in the near term, may be unsustainable in the longer term (Harrington and Erenstein 2005; Erenstein et al. 2008b).

To enhance the environmental sustainability of zero tillage there is a need to use the technology as a stepping-stone to conservation agriculture and start addressing the remaining challenges (Erenstein 2009a). These challenges include the need for more crop rotation, although the combination of secure produce markets and irrigation-ensured yield stability makes rice and wheat production a low-risk venture that has proven difficult to displace so far. Still, there is an increasing scope for diversifying rice–wheat systems, particularly on the northwestern Indo-Gangetic Plains, be it in response to technological developments (Jat et al. 2006), rapidly evolving domestic markets (due to economic growth, urbanization, and emerging marketing chains), or the increasing water scarcity (Erenstein 2009a).

Social and Political Sustainability

Equity still poses a challenge to the impact of zero tillage—both geographically and within rural communities. Zero-tillage wheat has so far primarily benefited the northwestern Indo-Gangetic Plains (Laxmi, Erenstein, and Gupta 2007a), an area that typically has more intensive and productive rice–wheat systems, more favorable institutional support, and markedly less poverty than the eastern Indo-Gangetic Plains (Erenstein et al. 2007b). Rural development indicators in the Indian states of Punjab and Haryana now compare well with those of middle-income countries. Yet large tracts of the Indo-Gangetic Plains remain mired in dire poverty, despite their agricultural potential. The main exponent of this poverty is the pocket of the eastern Indo-Gangetic Plains, an area with 500 million people typically characterized by smallholders (70–90 percent of the farm households have less than 2 hectares) and widespread poverty (30 percent or more are below the official poverty line), where more than two-thirds survive on less than $2 per day (Erenstein 2009a).

Although the early focus of R&D on zero tillage in the rice–wheat system is in part justified in view of the risks inherent in technology development, there is an increasing need to directly target poorer areas and poorer households (Erenstein 2007). In fact, the potential yield gains and cost savings associated with zero tillage are even more pronounced in areas with less intensified agriculture, such as the eastern Indo-Gangetic Plains, thereby potentially reducing poverty and regional inequality (Erenstein and Laxmi 2008). The initial R&D results on the eastern plains are encouraging but not well documented or widely published. Yet most references to zero tillage and the rice–wheat systems on the Indo-Gangetic Plains are still based on or extrapolated from the intensive northwestern situation. This limitation risks ignoring the significant variations across the plains in terms of both biophysical terms (Narang and Virmani 2001) and socioeconomic indicators such as poverty and population density (Erenstein, Hellin, and Chandna 2010).

Resource-saving technologies developed for the capital-abundant northwestern Indo-Gangetic Plains are not necessarily appropriate for less capital-abundant regions of the eastern plains. The popularity of two-wheel tractors in Bangladesh and Nepal is a case in point; these are more appropriate for small farms and plots and are less capital-intensive than four-wheel tractors. Thus there is a need for local adaptation instead of simply transferring "capital-biased" technologies, and the initial results of such adaptations are encouraging (see, for example, Biggs et al. 2004; Hossain et al. 2006; Wohab et al. 2006).

Studies have reported that the benefits of zero tillage are relatively scale neutral, with both large and small landholders adopting the technology (Jamal, Akhtar, and Aune 2007; Erenstein and Laxmi 2008). This zero-tillage adoption by smallholders is facilitated by their ability to contract zero-tillage drill services, just as they do tillage services in general, which makes the tractor-based machinery divisible. Still, zero tillage tends to be adopted first by the better-endowed farmers (Laxmi, Erenstein, and Gupta 2007a; Sarwar and Goheer 2007). Erenstein and Farooq (2009) indicate that zero-tillage adoption in the initial diffusion stage is strongly linked to the wealth of farm households and rice–wheat specialization. The significant wheat area in zero tillage implies larger annual benefits, lower relative learning costs, and earlier payback to a zero-tillage drill investment. Studies in eastern Uttar Pradesh report that all socioeconomic groups of farmers have benefited from using zero tillage, albeit the extent of use has been greater among larger farmers (Jafry 2007; Joshi et al. 2007b). This is in part associated with differential access to information.

The differential adoption of zero tillage calls for a closer consideration of equity implications in future R&D. The structural differences between adoption categories also easily confound the assessment of the impact of zero tillage (Erenstein 2009b; Erenstein and Farooq 2009). Therefore, the access of smallholders to zero-tillage knowledge merits particular attention, and alleviating knowledge blockages can further an equitable access to this promising technology (Erenstein and Farooq 2009). There is an important role here for agricultural service providers, particularly in view of the widespread reliance of zero-tillage adopters on contracted zero-tillage drill services.

Farm surveys have also revealed some discontinuation of zero tillage, and better understanding of the rationale for disadoption is needed (Erenstein et al. 2007a). There was no clear single overarching constraint, but a combination of factors was at play, including technology performance, technology access, and seasonal constraints. The slower diffusion and higher disadoption in Pakistani Punjab are likely associated with the ongoing institutional controversy over zero tillage there.

For instance, Iqbal, Khan, and Anwar (2002, 677) have highlighted that in Punjab "some government agencies . . . have differences of opinion on the usefulness and the benefits of zero tillage." A negative association between the number of extension visits and zero-tillage adoption is also illustrated by Sheikh, Rehman, and Yates (2003, 90), who conclude that "extension workers are not recommending the technology." Provincial agricultural extension providers are indeed not supportive of zero-tillage wheat, and this message is spread through their extension campaigns and by their field staff (Akhtar 2006; Jafry, Ahmad, and Poswal 2006). One of the fears is that by forgoing plowing, zero tillage may encourage overwintering of the stem borer in rice stubble, which may undermine the productivity and competitiveness of basmati rice, a major export crop. However, there is no scientific evidence of such risk (Inayatullah et al. 1989; Srivastava et al. 2005). Filling the institutional vacuum, On Farm Water Management (OFWM, Lahore) has played an important role in promoting this technology. This effort has created institutional rivalry between OFWM and agricultural extension, with unfortunate implications for the farmers and the technology alike in Punjab, particularly in view of conflicting information. In contrast, the initial reluctance of many stakeholders in India vis-à-vis zero tillage has been transformed into significant support for zero tillage at various levels (Erenstein et al. 2007a).

On the Indo-Gangetic Plains, little has been documented about zero tillage's contribution to conflicts between stakeholders. In part, this may reflect a primarily technocentric approach and an inherent diversity among stakeholders that has often resulted in only partial analysis, if any (Erenstein 2007). Still, timely access to a zero-tillage drill is key to ensuring timely wheat establishment and thereby being able to reap the associated yield gains. Differential access to zero-tillage implements and knowledge may therefore have contributed to the differential yield performance of zero-tillage wheat in Pakistani Punjab, where zero tillage resulted in lower yields only on smallholder farms (Erenstein 2009b). Agricultural scientists have increasingly recognized the need to acknowledge differences in the resource base of their target group—although the boundary is often fuzzy between what is considered a large farmer and a smallholder. And, more worrying, the implications of adoption for disadvantaged segments of society in South Asia, such as the landless, are often forgotten (Erenstein 2007).

Labor-saving technologies inherently shift income from laborers to producers. The mechanized land preparation that prevails on the Indo-Gangetic Plains, however, implies that the labor savings associated with zero-tillage wheat are limited. Still, some stakeholders have raised concerns about the perceived labor displacement associated with zero tillage (Jafry 2007; Laxmi, Erenstein, and Gupta 2007a). The concerns raised by migrant and landless laborers about the possible use of zero tillage for rice seem more grounded; these laborers fear a significant loss of earnings were rice no longer transplanted (Laxmi, Erenstein, and Gupta 2007). The gender seg-mentation in the labor market would impose further social costs (Singh et al. 2005). There would also be regional equity dimensions, because the intensive northwestern systems still rely on migrant labor from the eastern plains to alleviate their labor peaks. This issue calls for a better understanding of livelihood implications and a broader stakeholder dialogue in technology development (Erenstein 2007).

Recently international commodity prices for oil, fertilizer, rice, and wheat first increased dramatically, contributing to a global food crisis in 2008, then plunged during the subsequent financial crisis. These developments have only increased the relevance of resource-conserving technologies such as zero tillage. Zero tillage's ability to minimize tilling and fuel use is particularly attractive because it alleviates production costs. Increasing concern about water productivity in agriculture also calls for water-saving technologies (Erenstein 2009c). Technologies based on con-servation agriculture also help regulate the soil ecology (Erenstein 2002; Erenstein 2003; Jat et al. 2009), which is increasingly relevant as an adaptation strategy for climate change. Climate change will indeed exacerbate the heat stress at the end of the wheat season, thereby enhancing the potential payoff to zero tillage in terms of timeliness. At the same time, climate change raises serious questions about the future of rice–wheat systems and agriculture on the Indo-Gangetic Plains in general (Grace, Jain, and Harrington 2002; Aggarwal et al. 2004; Ortiz et al. 2008).

Lessons Learned

Conservation agriculture–based technologies such as zero tillage offer potentially high economic, environmental, and social gains on the Indo-Gangetic Plains, although actually realizing all these gains on the ground has proven challenging (Erenstein and Laxmi 2008; Erenstein 2009a). The vast majority of farmers on South Asia's Indo-Gangetic Plains have adopted zero tillage because it provides immediate, identifiable, and demonstrable economic benefits such as reductions in production costs and timely establishment of crops, resulting in improved crop yields. But in spite of the efficiency gains and the recent diffusion of zero tillage, most farmers, especially small- and medium-scale farmers, have difficulty in follow-ing the wider basic tenets of conservation agriculture, particularly year-round tillage reduction, crop residue retention, and crop rotation. This weak spot implies that the associated environmental benefits are yet to be fully realized. Equity has also posed a challenge so far, and there is a need to extend the gains to the less endowed areas and less endowed farmers.

Thus R&D still faces the challenge of adapting and developing sound, econom-ic conservation agricultural practices that all types of farmers will adopt year round across crops and across regions. But the potential is there to build on the success of zero-tillage wheat and thus to use zero tillage and the associated efficiency gains as a stepping-stone to conservation agriculture and equitable rural development. Still, zero tillage is no panacea, and complementary technologies that are privately and socially attractive are needed. At the same time, technological change can go only so far and needs to be complemented with institutional change to create the incentives necessary to induce change and to align private and social interests.

Despite the wealth of information on zero tillage on the Indo-Gangetic Plains, there are still significant knowledge gaps. Particularly scarce are reliable and empirically based zero-tillage diffusion indicators and documented evidence of zero tillage's socioeconomic, livelihood, and environmental impacts. Addressing these knowledge gaps would significantly enhance our understanding of the sustainability implications and the remaining challenges. A better understanding of livelihood implications and stakeholder dialogue would enhance the ability to keep interventions "pro-poor" and need-based. Addressing the knowledge gaps would also enhance our ability to scale out and replicate the success in a cost-effective, equitable, and sustainable manner.

Keys to Enhancing the Impact of Zero Tillage

Zero tillage's impact on the Indo-Gangetic Plains was enabled by a number of key aspects inherent in the technology, the process, and the context. The technology itself proved attractive from a financial viewpoint to both technology suppliers and users. Without such profitability, the technology's initial public-sector, supply-led nature would not have transformed itself into private-sector supply and demand and to viable delivery pathways and business models in India.

In creating this dynamic system of innovation, the intervention's process combined elements of persistence, flexibility, inclusiveness, and facilitation. Over time, success was built on the concerted efforts of an array of stakeholders that spanned the public and private sectors and national and international research systems and included several persevering zero-tillage champions. The Rice–Wheat Consortium played a pivotal and innovative role as facilitator and information provider, technology clearinghouse, and capacity builder. The Consortium helped provide both resources to get the technology onto farmers' fields and into manufacturers' workshops and an active and dynamic platform for stakeholder interaction. The importance of these policy and institutional aspects in shaping impact pathways and enabling the system of innovation to be built around technologies based on conservation agriculture has been reported in South Asia and elsewhere (Ekboir 2002; Harrington and Erenstein 2005; Erenstein et al. 2008b; Harrington and Hobbs 2009; Sims, Hobbs, and Gupta 2009).

Finally, slowdown in productivity growth in rice–wheat systems and concerns over production costs and sustainability opened the door to zero tillage as a resource-conserving technology. Many farmers became interested in the prospect of enhancing their stagnating bottom line. Selected researchers were interested in and excited by the prospect of enabling change in farmers' fields. Policymakers were interested in technological solutions to enhance the sustainability of South Asia's cereal bowl while avoiding more demanding institutional changes.

The success story of zero tillage in Haryana combines many of these elements. The herbicide tolerance of the weed *Phalaris minor* was so severe that it helped break through the enormous reluctance of farmers to even try zero-tillage technology. Key champions picked up the technology and promoted it despite initial resistance. The private sector made many improvements to the prototypes of the implements based on interaction with and feedback from farmers. The innovation process therefore benefited from many "hinge-of-fate" moments in which coincidence and personalities enabled progress (Harrington and Hobbs 2009).

Scaling Out the Impact of Zero Tillage

Scaling-out strategies will need to build on the local context and stakeholders to establish a dynamic innovation system. On the vast Indo-Gangetic Plains, strategies to expand zero tillage to new regions will thus differ between the more intensive northwestern plains and the eastern plains and between India and neighboring Pakistan, Nepal, and Bangladesh. More concerted efforts and resources are needed to strengthen the R&D in these neighboring countries and on the eastern plains. The experience in Pakistan highlights the importance of bringing on board all relevant stakeholders, including policymakers and extension workers. More research is particularly needed to create year-round zero-tillage (double-no-till) options for rice–wheat systems and to understand and address environmental and social issues.

The zero-tillage experience on the Indo-Gangetic Plains provides a number of useful guidelines for future efforts to replicate it elsewhere. The key to success is having a financially attractive intervention. No matter how attractive an intervention is from the environmental or social point of view, without the private interest to stimulate demand and supply, it will not fly. The development of viable and dynamic innovation systems that can deliver and adapt interventions such as zero-tillage drills will be essential to replicate and extend the success that has been realized in reducing tillage for wheat.

Agricultural scientists and development practitioners therefore need to move away from linear processes to more participatory interventions. Experts often have strong convictions about proposed interventions and what is best for the system. Yet the key to change is creating the demand and enabling it to transform the system. Sowing the seeds for a successful business model is critical. Showing that the technology can deliver its promise in the farmers' villages and on their fields is critical, as is the idea that it can be actively facilitated through on-farm adaptive and participatory R&D work and by encouraging farmer-to-farmer exchanges and extension through traveling seminars. Another critical aspect of success is to link farmers with knowledgeable, accessible, and responsive yet self-interested technology suppliers, including a local manufacturer who can make equipment that will do the job well at a competitive price and can adapt and repair it as needed. Successful business models are also more likely to emerge when the intervention target area and populations are substantial instead of minor niches; the wide applicability of zero-tillage drills across the Indo-Gangetic Plains is a case in point. Moving agricultural experts away from the yield paradigm is another challenge. Indeed, producing the same with less can still be a very attractive proposition for enhancing farmers' bottom line, but it implies a shift in a mindset that has traditionally focused on producing more per unit of area.

A final guideline for scaling out similar agricultural interventions is to start with reaping the easy gains or low-hanging fruits. Such a phased approach generates the successes and momentum that will help address second-generation problems. Such an approach will likely be more successful than tackling environmental or social issues head-on. But this approach does not imply that one can simply ignore the more challenging longer-term and equity issues. In fact, from the onset there should be an emphasis on the more compatible interventions that can serve as a stepping-stone toward sustainable conservation agriculture. This emphasis serves as a salutary reminder that farmers have an outstanding ability to adopt the components of proposed technological packages that best fit their needs and interests. It is also important to emphasize from the start equitable access in terms of including less endowed areas and households and ensuring the appropriateness of interventions. Therefore, a challenge that remains is to minimize the trade-offs with environmental and social goals in an enduring quest for immediate successes.

References

Abrol, I. P., R. K. Gupta, and R. K. Malik. 2005. *Conservation agriculture—Status and prospects.* New Delhi: Centre for Advancement of Sustainable Agriculture.

Aggarwal, P. K., P. K. Joshi, J. S. I. Ingram, and R. K. Gupta. 2004. Adapting food systems of the Indo-Gangetic Plains to global environmental change: Key information needs to improve policy formulation. *Environmental Science and Policy* 7 (6): 487–498.

Akhtar, M. R. 2006. Impact of resource conservation technologies for sustainability of irrigated agriculture in Punjab-Pakistan. *Pakistan Journal of Agricultural Research* 44 (3): 239–255.

Aslam, M., A. Majid, N. I. Hashmi, and P. R. Hobbs. 1993. Improving wheat yield in the rice–wheat cropping system of the Punjab through zero tillage. *Pakistan Journal of Agricultural Research* 14: 8–11.

Bakhsh, K., I. Hassan, and A. Maqbool. 2005. Impact assessment of zero-tillage technology in rice–wheat system: A case study from Pakistani Punjab. *Electronic Journal of Environmental, Agricultural and Food Chemistry* 4 (6): 1132–1137.

Biggs, S., S. Justice, C. Gurung, J. Tripathi, and G. Sah. 2004. The changing power tiller innovation system in Nepal: An actor-oriented analysis. Paper presented at the workshop Agricultural and Rural Mechanization, Bangladesh Agricultural University, Mymensingh, Bangladesh, November 2, 2002. Norwich, U.K.: School of Development Studies, University of East Anglia.

Bijay-Singh, Y. H. S., S. E. Johnson-Beebout, Y. Singh, and R. J. Buresh. 2008. Crop residue management for lowland rice-based cropping systems in Asia. *Advances in Agronomy* 98: 117–199.

Byerlee, D., A. D. Sheikh, M. Aslam, and P. R. Hobbs. 1984. *Wheat in the rice-based farming system of the Punjab: Implications for research and extension.* NARC/CIMMYT Reports Series 4. Islamabad, Pakistan: National Agricultural Research Center / International Maize and Wheat Improvement Center.

Choudhary, M. A., M. A. Gill, M. A. Kahlown, and P. R. Hobbs. 2002. Evaluation of resource conservation technologies in rice–wheat system of Pakistan. In *Proceedings of the international workshop on developing an action program for farm-level impact in rice–wheat systems of the Indo-Gangetic Plains,* September 25–27, 2000, New Delhi, India. New Delhi: Rice–Wheat Consortium.

Devendra, C. 2007. Small farm systems to feed hungry Asia. *Outlook on Agriculture* 36: 7–20.

Dixon, J., J. Hellin, O. Erenstein, and P. Kosina. 2007. U-impact pathway for diagnosis and impact assessment of crop improvement. *Journal of Agricultural Science* 145 (3): 195–206.

Duxbury, J. M. 2001. Long-term yield trends in the rice–wheat cropping system: Results from experiments in Northwest India. *Journal of Crop Production* 3 (2): 27–52.

Ekboir, J. 2002. Developing no-till packages for small-scale farmers. In *CIMMYT 2000–2001 world wheat overview and outlook,* ed. J. Ekboir. Mexico City: International Maize and Wheat Improvement Center.

Erenstein, O. 2002. Crop residue mulching in tropical and semi-tropical countries: An evaluation of residue availability and other technological implications. *Soil and Tillage Research* 67 (2): 115–133.

———. 2003. Smallholder conservation farming in the tropics and sub-tropics: A guide to the development and dissemination of mulching with crop residues and cover crops. *Agriculture, Ecosystems, and Environment* 100 (1): 17–37.

———. 2007. Resource conserving technologies in rice–wheat systems: Issues and challenges. In *Science, technology, and trade for peace and prosperity: Proceedings of the 26th International Rice Research Conference, October 9–12, 2006,* New Delhi, India, ed. P. K. Aggarwal, J. K. Ladha, R. K. Singh, C. Devakumar, and B. Hardy. Los Baños, Philippines, and New Delhi, India: International Rice Research Institute, Indian Council of Agricultural Research, and National Academy of Agricultural Science.

———. 2009a. Adoption and impact of conservation agriculture-based resource conserving technologies in South Asia. In *Proceedings of the 4th World Congress on Conservation Agriculture, February 4–7, 2009,* New Delhi, India. New Delhi: World Congress on Conservation Agriculture.

———. 2009b. Specification effects in zero tillage survey data in South Asia's rice–wheat systems. *Field Crops Research* 111 (1–2): 166–172.

———. 2009c. Comparing water management in rice–wheat production systems in Haryana, India, and Punjab, Pakistan. *Agricultural Water Management* 96: 1799–1806.

Erenstein, O., and U. Farooq. 2009. A survey of factors associated with the adoption of zero tillage wheat in the irrigated plains of South Asia. *Experimental Agriculture* 45 (2): 133–147.

Erenstein, O., and V. Laxmi. 2008. Zero tillage impacts in India's rice–wheat systems: A review. *Soil and Tillage Research* 100 (1–2): 1–14.

Erenstein, O., J. Hellin, and P. Chandna. 2010. Poverty mapping based on livelihood assets: A meso-level application in the Indo-Gangetic Plains, India. *Applied Geography* 30 (1): 112–125.

Erenstein, O., R. K. Malik, and S. Singh. 2007. *Adoption and impacts of zero tillage in the irrigated rice–wheat systems of Haryana, India.* Research report. New Delhi: International Maize and Wheat Improvement Center and Rice–Wheat Consortium.

Erenstein, O., U. Farooq, R. K. Malik, and M. Sharif. 2007a. *Adoption and impacts of zero tillage as a resource conserving technology in the irrigated plains of South Asia: Comprehensive assessment of water management in agriculture.* Research Report 19. Colombo, Sri Lanka: International Water Management Institute.

———. 2008a. On-farm impacts of zero tillage wheat in South Asia's rice–wheat systems. *Field Crops Research* 105 (3): 240–252.

Erenstein, O., K. Sayre, P. Wall, J. Dixon, and J. Hellin. 2008b. Adapting no-tillage agriculture to the conditions of smallholder maize and wheat farmers in the tropics and subtropics. In *No-till farming systems,* ed. T. Goddard, M. Zoebisch, Y. Gan, W. Ellis, A. Watson, and S. Sombatpanit. Bangkok, Thailand: World Association of Soil and Water Conservation.

Erenstein, O., W. Thorpe, J. Singh, and A. Varma. 2007b. *Crop–livestock interactions and livelihoods in the Indo-Gangetic Plains, India: A regional synthesis.* Crop–livestock Interactions Scoping Study Synthesis. New Delhi: International Maize and Wheat Improvement Center, International Livestock Research Institute, and Rice–Wheat Consortium.

———. 2007c. *Crop–livestock interactions and livelihoods in the Trans-Gangetic Plains, India.* Research Report 10. Nairobi: International Livestock Research Institute.

FAO (Food and Agriculture Organization of the United Nations). 2007. Conservation agriculture website. <http://www.fao.org/ag/ca/>. Accessed April 16, 2007.

Farooq, U., M. Sharif, and O. Erenstein. 2007. *Adoption and impacts of zero tillage in the rice–wheat zone of irrigated Punjab, Pakistan.* Research report. New Delhi: International Maize and Wheat Improvement Center and Rice–Wheat Consortium.

Franke, A. C., N. McRoberts, G. Marshall, R. K. Malik, S. Singh, and A. S. Nehra. 2003. A survey of Phalaris minor in the Indian rice–wheat system. *Experimental Agriculture* 39 (3): 253–265.

Fujisaka, S., L. Harrington, and P. R. Hobbs. 1994. Rice–wheat in South Asia: Systems and long-term priorities established through diagnostic research. *Agricultural Systems* 46: 169–187.

Grace, P. R., M. C. Jain, and L. W. Harrington. 2002. Environmental concerns in rice–wheat systems. In *Proceedings of the international workshop on developing an action program for farm-level impact in rice–wheat systems of the Indo-Gangetic Plains,* September 25–27, 2000, New Delhi, India. New Delhi: Rice–Wheat Consortium.

Gupta, P. K., S. Sahai, N. Singh, C. K. Dixit, D. P. Singh, C. Sharma, M. K. Tiwari, R. K. Gupta, and S. C. Garg. 2004. Residue burning in rice–wheat cropping system: Causes and implications. *Current Science* 87 (12): 1713–1717.

Gupta, R., and K. Sayre. 2007. Conservation agriculture in South Asia. *Journal of Agricultural Science* 145 (3): 207–214.

Gupta, R., and A. Seth. 2007. A review of resource-conserving technologies for sustainable management of the rice–wheat cropping systems of the Indo-Gangetic plains (IGP). *Crop Protection* 26 (3): 436–447.

Gupta, R. K., R. K. Naresh, P. R. Hobbs, and J. K. Ladha. 2002. Adopting conservation agriculture in the rice–wheat system of the Indo-Gangetic Plains: New opportunities for saving water. In *Water-wise rice production: Proceedings of the international workshop on water-wise rice production, April 8–11, 2002,* Los Baños, Philippines, ed. B. A. M. Bouman, H. Hengsdijk, B. Hardy, P. S. Bindraban, T. P. Tuong, and J. K. Ladha. Los Baños, Philippines: International Rice Research Institute.

Gupta, R. K., R. K. Naresh, P. R. Hobbs, Z. Jiaguo, and J. K. Ladha. 2003. Sustainability of post–green revolution agriculture: The rice–wheat cropping systems of the Indo-Gangetic Plains and China. In *Improving the productivity and sustainability of rice–wheat systems: Issues and impacts,* ed. J. K. Ladha, J. E. Hill, J. M. Duxbury, R. K. Gupta, and R. J. Buresh. Madison, Wisc., U.S.A.: American Society of Agronomy, Crop Science Society of America, and Soil Science Society of America.

Harrington, L., and O. Erenstein. 2005. Conservation agriculture and resource-conserving technologies: A global perspective. *Agromeridian* 1 (1): 32–43.

Harrington, L. W., and P. H. Hobbs. 2009. The rice–wheat consortium and the Asian Development Bank: A history. In *Integrated crop and resource management technologies for sustainable rice–wheat systems of South Asia,* ed. J. K. Ladha, Yadvinder-Singh, and O. Erenstein. New Delhi: International Rice Research Institute.

Harrington, L. W., S. Fujisaka, P. R. Hobbs, H. C. Sharma, R. P. Singh, M. K. Chaudhary, and S. D. Dhiman. 1993. *Wheat and rice in Karnal and Kurukshetra Districts, Haryana, India: Farmers' practices, problems and an agenda for action.* Mexico City: Haryana Agricultural University, Indian Council of Agricultural Research, International Maize and Wheat Improvement Center, and International Rice Research Institute.

Hobbs, P. R. 2007. Conservation agriculture: What is it and why is it important for future sustainable food production? *Journal of Agricultural Science* 145 (2): 127–137.

Hobbs, P. R., and B. Govaerts. 2009. How conservation agriculture can contribute to buffering climate change. In *Climate change and crop production*, ed. M. Reynolds. Wallingford, U.K.: CAB International.

Hobbs, P. R., and R. K. Gupta. 2003a. Resource-conserving technologies for wheat in the rice–wheat system. In *Improving the productivity and sustainability of rice–wheat systems: Issues and impacts*, ed. J. K. Ladha, J. E. Hill, J. M. Duxbury, R. K. Gupta, and R. J. Buresh. Madison, Wisc., U.S.A.: American Society of Agronomy, Crop Science Society of America, and Soil Science Society of America.

———. 2003b. Rice–wheat cropping systems in the Indo-Gangetic Plains: Issues of water productivity in relation to new resource-conserving technologies. In *Water productivity in agriculture: Limits and opportunities for improvement*, ed. J. W. Kijne, R. Barker, and D. Molden. Wallingford, U.K.: CABI.

Hobbs, P. R., G. S. Giri, and P. Grace. 1997. *Reduced and zero tillage options for the establishment of wheat after rice in South Asia*. Rice–Wheat Consortium Paper Series 2. New Delhi: Rice–Wheat Consortium.

Hossain, M. I., M. A. Sufian, M. E. Haque, S. Justice, and M. Badruzzaman. 2006. Development of power tiller operated zero tillage planter for small land holders. *Bangladesh Journal of Agricultural Research* 31 (3): 471–484.

Inayatullah, C., E. Ul-Haq, A. ul-Mohsin, A. Rehman, and P. R. Hobbs. 1989. *Management of rice stem borers and feasibility of adopting no-tillage in wheat*. Islamabad, Pakistan: Entomological Research Laboratories, Pakistan Agricultural Research Council.

Iqbal, M., M. A. Khan, and M. Z. Anwar. 2002. Zero-tillage technology and farm profits: A case study of wheat growers in the rice zone of Punjab. *Pakistan Development Review* 41 (4): 665–682.

Jafry, T. 2007. *Reaping the benefits: Assessing the impact and facilitating the uptake of resource-conserving technologies in the rice–wheat systems of the Indo-Gangetic Plain*. India's (NDUA&T) final report to DFID. Wallingford, U.K.: CABI.

Jafry, T., B. R. Ahmad, and A. Poswal. 2006. *Reaping the benefits: Assessing the impact and facilitating the uptake of resource conserving technologies in the rice–wheat systems of the Indo-Gangetic Plain*. Pakistan's final report to DFID. Rawalpindi, Pakistan: CABI South Asia.

Jamal, T. N., M. R. Akhtar, and J. B. Aune. 2007. Assessment of conservation agriculture from an agronomic, economic and social perspective in Punjab, Pakistan. In *Institutional and technological interventions for better irrigation management in the new millennium: Proceedings of the INPIM's 9th International Seminar, December 4–8, 2006*, Lahore, Pakistan, ed. I. Hussain, Z. A. Gill, N. Zeeshan, and S. Salman. Islamabad, Pakistan: International Network on Participatory Irrigation Management.

Jat, M. L., R. K. Gupta, O. Erenstein, and R. Ortiz. 2006. Diversifying the intensive cereal cropping systems of the Indo-Ganges through horticulture. *Chronica Horticulturae* 46 (3): 27–31.

Jat, M. L., R. G. Singh, Y. S. Saharawat, M. K. Gathala, V. Kumar, H. S. Sidhu, and R. Gupta. 2009. Innovations through conservation agriculture: Progress and prospects of participatory approach in the Indo-Gangetic Plains. In *Proceedings 4th World Congress on Conservation Agriculture, February 4–7, 2009,* New Delhi, India. New Delhi: World Congress on Conservation Agriculture.

Joshi, A., R. Chand, B. Arun, R. Singh, and R. Ortiz. 2007a. Breeding crops for reduced-tillage management in the intensive, rice–wheat systems of South Asia. *Euphytica* 153 (1): 135–151.

Joshi, A. K., R. Chand, V. K. Chandola, and T. Jafry. 2007b. *Reaping the benefits: Assessing the impact and facilitating the uptake of resource-conserving technologies in the rice–wheat systems of the Indo-Gangetic Plain.* India's (BHU) final report to DFID. Wallingford, U.K.: CABI.

Kahlown, M. A., M. A. Gill, and M. Ashraf. 2002. *Evaluation of resource conservation technologies in rice–wheat system of Pakistan.* Research Report 1. Islamabad, Pakistan: Pakistan Council of Research in Water Resources.

Kataki, P. K., P. Hobbs, and B. Adhikary. 2001. The rice–wheat cropping system of South Asia: Trends, constraints and productivity—A prologue. *Journal of Crop Production* 3 (2): 1–26.

Khan, A. R., S. S. Singh, M. A. Khan, O. Erenstein, R. G. Singh, and R. K. Gupta. 2009. Changing scenario of crop production through resource conservation technologies in Eastern Indo-Gangetic Plains. In *Proceedings of the 4th World Congress on Conservation Agriculture, February 4–7, 2009,* New Delhi, India. New Delhi: World Congress on Conservation Agriculture.

Kumar, P., D. Jha, A. Kumar, M. K. Chaudhary, R. K. Grover, R. K. Singh, R. K. P. Singh, A. Mitra, P. K. Joshi, A. Singh, P. S. Badal, S. Mittal, and J. Ali. 2002. *Economic analysis of total factor productivity of crop sector in Indo-Gangetic Plain of India by district and region.* Agricultural Economics Research Report 2. New Delhi: Indian Agricultural Research Institute.

Ladha, J. K., J. E. Hill, J. M. Duxbury, R. K. Gupta, and R. J. Buresh, eds. 2003. *Improving the productivity and sustainability of rice–wheat systems: Issues and impacts.* Madison, Wisc., U.S.A.: American Society of Agronomy, Crop Science Society of America, Soil Science Society of America.

Laxmi, V., O. Erenstein, and R. K. Gupta. 2007a. *Impact of zero tillage in India's rice–wheat systems. Research report.* New Delhi: International Maize and Wheat Improvement Center and Rice–Wheat Consortium.

———. 2007b. Assessing the impact of natural resource management research: The case of zero tillage in India's rice–wheat systems. In *International research on natural resource management: Advances in impact assessment,* ed. H. Waibel and D. Zilberman. Wallingford, U.K.: Food and Agriculture Organization and CAB International.

Malik, R. K., R. S. Balyan, A. Yadav, and S. K. Pahwa, eds. 2002. Herbicide resistance management and zero tillage in rice–wheat cropping system. In *Proceedings of the international workshop, March 4–6, 2002, Chaudhary Charan Singh Haryana Agricultural University (CCSHAU), Hisar, India.* Hisar, India: CCSHAU.

Malik, R. K., R. K. Gupta, A. Yadav, P. K. Sardana, and C. M. Singh. 2005a. *Zero tillage: The voice of farmers.* Technical Bulletin 9. Hisar, India: Directorate of Extension Education, Chaudhary Charan Singh Haryana Agricultural University.

Malik, R. K., R. K. Gupta, C. M. Singh, A. Yadav, S. S. Brar, T. C. Thakur, S. S. Singh, A. K. Singh, R. Singh, and R. K. Sinha. 2005b. Accelerating the adoption of resource conservation technologies in rice–wheat system of the Indo-Gangetic Plains. In *Proceedings of the project workshop, June 1–2, 2005.* Hisar, India: Directorate of Extension Education, Chaudhary Charan Singh Haryana Agricultural University.

Narang, R. S., and S. M. Virmani. 2001. *Rice–wheat cropping systems of the Indo-Gangetic Plain of India.* Rice–Wheat Consortium Paper Series 11. New Delhi, India: Rice–Wheat Consortium.

Ortiz, R., K. D. Sayre, B. Govaerts, R. Gupta, G. V. Subbarao, T. Ban, D. Hodson, J. M. Dixon, J. Ivan Ortiz-Monasterio, and M. Reynolds. 2008. Climate change: Can wheat beat the heat? *Agriculture, Ecosystems, and Environment* 126 (1–2): 46–58.

Ortiz-Monasterio, J. I., S. S. Dhillon, and R. A. Fischer. 1994. Date of sowing effects on grain yield and yield components of irrigated spring wheat cultivars and relationships with radiation and temperature in Ludhiana, India. *Field Crops Research* 37 (3): 169–184.

Parthasarathy Rao, P., and P. S. Birthal. 2008. *Livestock in mixed farming systems in South Asia.* New Delhi and Patancheru, India: National Centre for Agricultural Economics and Policy Research and International Crops Research Institute for the Semi-arid Tropics.

Pathak, H. 2009. Greenhouse gas mitigation in rice–wheat system with resource conserving technologies. In *Proceedings of the 4th World Congress on Conservation Agriculture, February 4–7, 2009, New Delhi, India.* New Delhi: World Congress on Conservation Agriculture.

Prasad, R. 2005. Rice–wheat cropping systems. *Advances in Agronomy* 86: 255–339.

Punia, S. S., R. K. Malik, A. Yadav, S. Singh, and A. S. Nehra. 2002. Acceleration of zero tillage technology in Haryana. In *Herbicide resistance management and zero tillage in rice–wheat cropping system,* ed. R. K. Malik, R. S. Balyan, A. Yadav, and S. K. Pahwa. Hisar, India: Chaudhary Charan Singh Haryana Agricultural University.

Rajaram, S., K. D. Sayre, J. Diekmann, R. Gupta, and W. Erskine. 2007. Sustainability considerations in wheat improvement and production. *Journal of Crop Improvement* 19 (1–2): 105–123.

Saharawat, Y. S. 2009. Direct seeded rice: Putting double no-till in practice under rice–wheat rotation of Indo-Gangetic Plains. In *Proceedings of the 4th World Congress on Conservation Agriculture, February 4–7, 2009, New Delhi, India.* New Delhi: World Congress on Conservation Agriculture.

Samra, J. S., B. Singh, and K. Kumar. 2003. Managing crop residues in the rice–wheat system of the Indo-Gangetic Plain. In *Improving the productivity and sustainability of rice–wheat systems: Issues and impacts,* ed. J. K. Ladha, J. E. Hill, J. M. Duxbury, R. K. Gupta, and R. J. Buresh. Madison, Wisc., U.S.A.: American Society of Agronomy, Crop Science Society of America, Soil Science Society of America.

Sarwar, M. N., and M. A. Goheer. 2007. Adoption and impact of zero tillage technology for wheat in rice–wheat system—Water and cost saving technology. A case study from Pakistan (Punjab).

Paper presented at the International Forum on Water Environmental Governance in Asia, March 14–15, Bangkok, Thailand.

Seth, A., K. Fischer, J. Anderson, and D. Jha. 2003. *The Rice–Wheat Consortium: An institutional innovation in international agricultural research on the rice–wheat cropping systems of the Indo-Gangetic Plains (IGP).* The Review Panel Report. New Delhi: Rice–Wheat Consortium.

Sheikh, A. D., T. Rehman, and C. M. Yates. 2003. Logit models for identifying the factors that influence the uptake of new "no-tillage" technologies by farmers in the rice–wheat and the cotton–wheat farming systems of Pakistan's Punjab. *Agricultural Systems* 75 (1): 79–95.

Sims, B. G., P. Hobbs, and R. Gupta. 2009. Policies and institutions to promote the development and commercial manufacture of conservation agriculture equipment. In *Proceedings of the 4th World Congress on Conservation Agriculture, February 4–7, 2009,* New Delhi, India. New Delhi: World Congress on Conservation Agriculture.

Singh, J., O. Erenstein, W. Thorpe, and A. Varma. 2007. *Crop–livestock interactions and livelihoods in the Gangetic plains of Uttar Pradesh, India.* Crop–livestock Interactions Scoping Study Report 2. Research Report 11. Nairobi, Kenya: International Livestock Research Institute.

Singh, N. P. 2008. *Adoption and impact of resource conserving technologies on farm economy in India-Gangetic Plains, India.* New Delhi: Indian Agricultural Research Institute.

Singh, R., O. Erenstein, M. K. Gathala, M. M. Alam, A. P. Regmi , U. P. Singh, H. M. U. Rehman, and B. P. Tripathi. 2009. Socio-economics of integrated crop and resource management in the rice–wheat systems of South Asia: Site contrasts, adoption and impacts using village survey findings. In *Integrated crop and resource management technologies for sustainable rice–wheat systems of south Asia,* ed. J. K. Ladha, Yadvinder-Singh, and O. Erenstein. New Delhi: International Rice Research Institute.

Singh, S., R. C. Kirkwood, and G. Marshall. 1999. Biology and control of Phalaris minor Retz. (little-seed canarygrass) in wheat. *Crop Protection* 18 (1): 1–16.

Singh, U. P., Y. Singh, V. Kumar, and J. K. Ladha. 2009. Evaluation and promotion of resource-conserving tillage and crop establishment techniques in rice–wheat system in Eastern India. In *Integrated crop and resource management technologies for sustainable rice–wheat systems of south Asia,* ed. J. K. Ladha, Y. Singh, and O. Erenstein. New Delhi: International Rice Research Institute.

Singh, Y., G. Singh, V. P. Singh, P. Singh, D. E. Johnson, and M. Mortimer. 2005. *Direct seeding of rice and weed management in the irrigated rice–wheat cropping system of the Indo-Gangetic Plains.* Pantnagar, India: Directorate of Experiment Station, G. B. Pant University of Agriculture and Technology.

Srivastava, S. K., R. Biswas, D. K. Garg, B. K. Gyawali, N. M. M. Haque, P. Ijaj, S. Jaipal, N. Q. Kamal, P. Kumar, M. Pathak, P. K. Pathak, C. S. Prasad, M. Ramzan, A. Rehman, M. Rurmzan, M. Salim, A. Singh, U. S. Singh, and S. N. Tiwari. 2005. *Management of stem borers of rice and wheat in rice–wheat systems of Pakistan, Nepal, India and Bangladesh.* Rice–Wheat Consortium Paper Series 17. New Delhi: Rice–Wheat Consortium.

Tahir, M. A., and M. Younas. 2004. *Feasibility of dry sowing technology of wheat in cotton growing districts and impact evaluation of zero tillage technology in rice growing districts.* Publication 364. Lahore, Pakistan: Punjab Economic Research Institute.

Teufel, N., O. Erenstein, and A. Samaddar. 2007. Perceptions and potential of resource conserving technologies in the crop–livestock systems of the Indo-Gangetic Plains. Poster presented at Tropentag, October 9–11, Witzenhausen, Germany.

———. 2008. Impacts of technological change on crop residue management and livestock feeding in the Indo-Gangetic Plains. Paper presented at Tropentag, October 7–9, Hohenheim, Germany.

Thorpe, W., O. Erenstein, J. Singh, and A. Varma. 2007. *Crop–livestock interactions and livelihoods in the Gangetic plains of Bihar, India.* Crop–livestock Interactions Scoping Study Report 3. Research Report 12. Nairobi, Kenya: International Livestock Research Institute.

Timsina, J., and D. J. Connor. 2001. Productivity and management of rice–wheat cropping systems: Issues and challenges. *Field Crops Research* 69 (2): 93–132.

Tripathi, J., C. Adhikari, J. G. Lauren, J. M. Duxbury, and P. R. Hobbs. 2006. *Assessment of farmer adoption of surface seeded wheat in the Nepal terai.* Rice–Wheat Consortium Paper Series 19. New Delhi: Rice–Wheat Consortium.

Varma, A., O. Erenstein, W. Thorpe, and J. Singh. 2007. *Crop–livestock interactions and livelihoods in the Gangetic plains of West Bengal, India.* Crop–livestock Interactions Scoping Study Report 4. Research Report 13. Nairobi, Kenya: International Livestock Research Institute.

Vincent, D., and D. Quirke. 2002. *Controlling Phalaris minor in the Indian rice–wheat belt.* Impact Assessment Series 18. Canberra, Australia: Australian Centre for International Agricultural Research.

Wohab, M. A., K. C. Roy, E. Haque, and M. N. Amin. 2006. Adaptation of minimum tillage seeder as high speed rotary tiller for upland farming. *Bangladesh Journal of Agricultural Research* 31 (4): 525–531.

Yadav, A., and R. K. Malik. 2005. *Herbicide resistant Phalaris minor in wheat: A sustainability issue.* Resource Book. Hisar, India: Department of Agronomy and Directorate of Extension Education, Chaudhary Charan Singh Haryana Agricultural University.

Chapter 9

Shallow Tubewells, *Boro* Rice, and Their Impact on Food Security in Bangladesh

Mahabub Hossain

Since its independence in 1971, Bangladesh has almost tripled its production of cereal grains, despite a continuous decline in arable land. Without this impressive growth in the production of staple grains, poverty and food insecurity would have been much worse than they are today. This growth has contributed to an increase in per capita food availability, kept food grain prices low and stable, and been instrumental in reducing poverty by almost 1 percent a year (Hossain and Sen 1992; Ravallion and Sen 1996; Ahmed, Haggblade, and Chowdhury 2000; Zohir, Shahabuddin, and Hossain 2002; Dorosh 2006; Narayan and Zaman 2009).

Most of the additional cereal production has come from the diffusion of modern rice technology and improved farming practices (Hossain, Bose, and Mustaf 2006; Hossain et al. 2006). The use of modern varieties of seeds has now expanded to three-fourths of the rice-cropped area, which is supported by an expansion of irrigation facilities to two-thirds of the cultivated land. This diffusion of modern rice technology must have surprised many agrarian students in Bangladesh who argued that the small size of farms and exploitative land tenure relationships would seriously impede productivity (Jannuzi and Peach 1980; Van Schundel 1981; Jansen 1986; Boyce 1987).

Progress can largely be attributed to the rapid expansion of groundwater irrigation, which was triggered by a change in government policy in favor of liberalization in the

The author is grateful to three anonymous reviewers for comments on an earlier draft. Research assistance provided by Mahfuzur Rahman and Josephine Narciso is gratefully acknowledged.

procurement and marketing of minor irrigation equipment, such as low-lift power pumps and shallow and deep tubewells (Palmer-Jones 1999; Ahmed 2001; Dorosh 2006). A parastatal, the Bangladesh Agriculture Development Corporation (BADC), had previously controlled access to these resources (Osmani and Quasem 1990; Ahmed 1995; Ahmed, Haggblade, and Chowdhury 2000; Ahmed 2001). The change in policy has radically transformed the production of dry season rice cultivation, known as *boro* rice (Singh et al. 2003). *Boro* rice accounted for only 9 percent of the rice production in 1966–67, when the Green Revolution was initiated. By 2008, however, it had contributed 56 percent to the total rice production in the country. This chapter traces the development of minor irrigation, particularly shallow tubewells, in Bangladesh and its impact on the country's rice production and food security.

Development of Minor Irrigation in Bangladesh

When the government's long-term irrigation policy and water resource development plans were formulated in the 1960s, the general perception among policymakers and in civil society was that private investment-based minor irrigation was inappropriate for Bangladesh because of the dominance of small-scale farmers and scattered holdings. Ghulam Mohammed, a noted Pakistani agricultural economist, conducted a survey on the potential of tubewell irrigation in eastern Pakistan (now Bangladesh) and argued that a total of 26,000 private tubewells could be installed if the size of holding was the main criterion (Mohammed 1966). Today nearly 1.3 million shallow tubewells (STWs) and 31,000 deep tubewells (DTWs) operate in Bangladesh and account for 3.98 million hectares, or about 80 percent of the total irrigated area in the dry season (Bangladesh, Ministry of Agriculture 2008).

Evolution of Policy Changes on Minor Irrigation

Minor irrigation technologies include small-scale devices such as DTWs, STWs, hand tubewells (HTWs), and low-lift pumps (LLPs). Table 9.1 shows the changes in the policy regimes that have influenced the development of minor irrigation.

Low-Lift Pumps.   Modern minor irrigation started in Bangladesh in 1962–63 with the supply of LLPs of 1–2 cusec discharge capacity to lift water from surface sources to adjoining fields (Mandal 1987; Hamid 1988; Palmer-Jones 1989; Ahmed, Haggblade, and Chowdhury 2000). BADC fielded the LLPs in the very low lands (*haor* areas) of Sylhet and Mymensingh regions to reclaim them for dry-season rice farming (locally known as *boro* rice). BADC owned, operated, and maintained these pumping sets, which supplied water to groups of farms on the basis of a flat charge per unit of land per season. The use of the equipment spread quickly in the depressed basins of northeastern and central Bangladesh, where surface water was easily available in the dry season. By the mid-1970s, nearly 35,000

Table 9.1 Time line of changes in the policy regime for the development of minor irrigation in Bangladesh

Year	Public sector	Private sector
1961	Bangladesh Agriculture Development Corporation (BADC) rented low-lift pumps (LLPs) to farmers on an annual basis.	Farmers paid the rental fee and bought fuel at 75 percent subsidy.
1962–66	Water Development Board installed and operated 380 deep tubewells (DTWs) with 4-cusec pumps in the northwestern region.	Farmers got water from the tubewells free of charge.
1972	BADC started renting shallow tubewells (STWs) to farmers' organizations.	Farmers managed operation of the tubewells.
1974–75	BADC started selling STWs to individual farmers with subsidies; Bangladesh Agriculture Development Bank provided credit for the acquisition.	
1979	Private sector was allowed to import and distribute STWs with credit from commercial banks.	Private sector started workshops for repair of irrigation equipment.
1980	Import duty on STW sets was reduced to 15 percent.	Private sector started manufacturing of pumps.
1980	BADC stopped renting LLPs and started selling new and used LLPs to farmers' cooperatives.	
1981–82	BADC started offering rental DTWs for sale at subsidized prices with credit from commercial banks.	Workshops and repair facilities for irrigation equipment grew.
1984–87	Sales of STWs began in the northwestern region, and the Groundwater Management Ordinance was formulated; private sector's import of small diesel engines was banned in response to the draw-down of aquifer during the 1983 drought.	Private sector trade was limited to a few standardized engine brands; sales of STWs dropped due to restricted installation within specified zones and spacing regulations.
1987–89	Private-sector bans on small engine imports were removed, import duties eliminated, standardization requirements for equipment abolished, and tubewell siting restrictions withdrawn.	Private traders started importing less expensive STWs from China; multiple engine brands and sizes entered the market.
1990	BADC started clearing out its stock of irrigation equipment and stopped monitoring the siting of equipment.	Market for engines, pumps, and spare parts spread; repair works mushroomed all over the country.

Sources: Adapted from IWMI/BAU (1996); Ahmed, Haggblade, and Chowdhury (2000).

LLPs had been fielded and the irrigated area reached nearly 0.57 million hectares, or about three-fourths of the total area under modern irrigation at that time. As the operation of a large fleet of LLPs became an unwieldy task, BADC gradually introduced irrigation management groups. BADC later rented the equipment to farmer groups directly and retained only responsibility for maintenance.

Deep Tubewells.   The Bangladesh Water Development Board (BWDB) initiated groundwater irrigation in the early 1960s with the installation of 380 4-cusec-capacity DTWs in Thakurgaon, a northern district. For many years the project had limited success, despite a 100 percent subsidy provided on irrigation. Managing a large number of farmers in irrigation groups was a major problem. BADC began a groundwater irrigation program with smaller-capacity (2-cusec) tubewells. Later the Bangladesh Academy for Rural Development in Comilla also successfully experimented with smaller-capacity tubewells and formed cooperatives of small-scale and marginal farmers. This approach was found to be more appropriate for the Bangladeshi agrarian structure (Howes 1985; Murshid 1985; Sattar and Bhuiyan 1985; Mandal 1989; Palmer-Jones 1992; Hamid 1993). The program was replicated throughout the country by the Bangladesh Rural Development Board (BRDB), which also took on the responsibility of developing appropriate management systems for the efficient distribution of water among the cooperative's members. By 1981–82, 12,000 DTWs were under operation and were irrigating 0.32 million hectares of land.

Shallow Tubewells and Hand Tubewells. The most recent addition to the list of minor irrigation equipment was STWs of 0.25–0.75 cusec capacity and hand tubewells of 0.02 cusec capacity (Hannah 1976; Mandal 1978, 1993; Pitman 1993). Beginning in 1972, BADC imported this equipment, initially renting and later selling it to individual farmers with loans at soft terms from the nationalized commercial banks. The manually operated HTWs initially spread in areas where groundwater was available at upper aquifers. But due to the heavy labor needed for their operation, the hand tubewells were found to be uneconomical—even with the low opportunity cost of labor—and were discontinued (Mandal 1989, 1993). The privately owned STWs proved to be the most appropriate technology for irrigation development in Bangladesh, and their spread continues today.

Until the late 1970s, the procurement, installation and distribution, and management of irrigation equipment were controlled by parastatals such as BADC, BWDB, and BRDB. These entities provided subsidies for their operation. With the expansion of these operations, however, the subsidies—for agricultural inputs such as irrigation, fertilizer, and seeds—began to impose heavy burdens on the government's budget. Their management also put a burden on limited administrative resources and skills and promoted rent-seeking behavior by the government. The farmers' cooperatives that managed the equipment also became inefficient and undisciplined, leading to lower-capacity use of the machines (IWMI/BAU 1996; Ahmed, Haggblade, and Chowdhury 2000).

In 1979 the government decided to change its policy of direct involvement in the input market and to privatize the marketing of irrigation equipment (along with chemical fertilizers). This policy change involved the sell-off of existing and new LLPs and DTWs, initially to farmers' cooperatives and later to individual farmers. The equipment was sold through a number of private dealers with credit from commercial banks and Bangladesh Krishi Bank (BKB), a specialized financial institution set up for the distribution of agricultural credit. BKB started its own program of providing credit to facilitate the purchase of STWs through its appointed private farms.

Although the policy of liberalization was introduced in 1979, the process of its implementation was slow due to (1) the reluctance on the part of civil servants to release control and power and (2) a limited private market for repair and maintenance. Through the early 1980s, there were continued efforts to decrease public-sector involvement in minor irrigation and the momentum grew for private-sector activity (Palmer-Jones 1989, 1992; IWMI/BAU 1996; Ahmed, Haggblade, and Chowdhury 2000). Nonetheless, BADC continued to subsidize spare parts and repairs, which created a disincentive to the development of local repair and maintenance facilities. The process of delivering credit for the acquisition of equipment also remained cumbersome and involved delays and rent-seeking behavior that encouraged defaults on bank loans. In 1983 Bangladesh experienced a severe drought that led to a draw-down of groundwater in the dry season, and there were complaints that STWs were becoming dry in the northern districts (Gill 1983; Palmer-Jones 1999). These incidents led to a setback in the decontrol of the market. The public sector responded with a series of actions in 1984–85 that proved that the policy of market liberalization did not have widespread support within the bureaucracy. These actions included (1) a ban on the sale of STWs in 22 northern subdistricts, (2) an embargo on the importation of small diesel engines used in STWs, (3) the standardization of engine brands, and (4) the formulation of the Groundwater Management Ordinance, which stipulated spacing requirements for all tubewells. In addition, the supply of agricultural credit was reduced in response to reported irregularities in the loan disbursement process and large defaults in loan repayments. As a result of these measures, the expansion of minor irrigation equipment slowed in 1984 and almost stopped from 1985 to 1987.

Policy reform advanced, however, in 1988 when leadership at the agriculture ministry changed. The new secretary of agriculture, Mohammed Abu Sayed, took a direct interest in implementing the liberalization program. He himself visited the diesel engine markets in Dhaka, asked traders about obstacles they faced, directed the formulation of policy changes to remove constraints, and ensured that these changes were approved by the government at the time. A study under the sponsorship of the United Nations Development Programme—the Bangladesh Agriculture Sector Review—supported the policy changes (UNDP 1989). In 1988 the government eliminated duties on diesel engines, withdrew standardization requirements, and allowed imports of agricultural machinery without government permits. The import duties were reimposed in the early 1990s, but the rates were much lower than in the mid-1980s.

Policy Impacts on Irrigation Expansion

Privatizing the procurement and marketing of irrigation equipment in Bangladesh contributed to (1) a mobilization of private savings for irrigation investments;

(2) elimination of the delays in equipment installation, repair, and maintenance that had been caused earlier by bureaucratic procedures and rent-seeking in the public sector; (3) increased competition in the water market, leading to a decline in water charges; and (4) expansion in capacity use of the machines. With the restrictions on private-sector imports removed, farmers realized lower prices for minor irrigation equipment because they could choose lower-cost sources of pumps and motors and could use plastic pipes in place of metal pipes for installation. By early 1989, the cost of installation of a shallow tubewell to irrigate 4–5 hectares of land had fallen to below $600, or about 60 percent of the subsidized price of such equipment through BADC (Gisselquist 1992).

With the reduction in prices, farmers of medium- and small-sized farms could afford to invest in small irrigation equipment such as STWs and LLPs. Because the availability of agricultural credit had not expanded much since the 1980s—due to large-scale defaults and the incapacity of the BKB to recycle loans—the acquisition of irrigation equipment was financed mostly through farmers' savings.

The expansion in the acquisition of minor irrigation equipment can be seen in Table 9.2. The number of STWs under operation increased from 93,000 units in 1982–83 to 189,000 in 1987–88, then expanded exponentially to reach 489,000 units in 1994–95 and 865,000 units in 2000–01. The number has continued to increase until today. The faster growth is observed in years following favorable prices of rice. BADC's latest survey of minor irrigation estimates that in the 2007–08 dry season there were 1,305,000 STWs under operation that were irrigating about 3.2 million hectares of land out of a total irrigated area of 5.0 million hectares (Bangladesh, Ministry of Agriculture 2008). The number of farmers irrigating land with STWs is estimated at 10.2 million out of 14 million farm households in 2007–08.

Private investment in the higher-cost DTWs has remained sluggish. The DTWs acquired earlier by BADC were transferred to the Grameen Bank in the early 1990s, but it could not succeed in operating them efficiently. Recently the government has set up additional units of DTWs under a special project in the Barind Tract area, where the low aquifer cannot be reached by shallow tubewells.[1] The DTWs are supported by subsidies. About 31,000 units are currently in operation, with an estimated irrigated area of 0.8 million hectares (Bangladesh, Ministry of Agriculture 2008).

The LLPs, which are used to lift surface water from creeks and canals in the depressed basins and flood-prone areas in the eastern and central regions of Bangladesh, expanded rapidly under the BADC rental program during the 1960s and 1970s. The equipment was sold to the farmers in the 1980s. Since then, expansion has been slow due to lack of surface water. Only recently has the number of LLPs increased due to the commissioning of the Tista Barrage Irrigation project, a large public-sector irrigation project that took a long time to complete. Gravity canals from the multipurpose surface water irrigation project and the LLPs now irrigate only 1.04 million hectares.

Figure 9.1 shows the trends in the development of modern irrigation facilities. As shown, the expansion of irrigation was accomplished entirely by LLPs until the mid-1970s, but LLP irrigation has not expanded much since then. The area irrigated by tubewells grew slowly until the mid-1980s, but that growth accelerated

Table 9.2 Operation of minor irrigation equipment in Bangladesh, 1976/77–2007/08 (thousands)

Year	Shallow tubewells	Deep tubewells	Low-lift power pumps
1976/77	7	4.5	35
1982/83	93	13.8	38
1987/88	189	20.3	51
1989/90	260	22.6	51
1991/92	309	25.5	50
1994/95	489	26.7	57
1998/99	736	26.7	73
2004/05	1,129	27.2	99
2006/07	1,203	29.2	107
2007/08	1,305	31.3	139

Sources: Bangladesh, Bureau of Statistics (various years) and Bangladesh, Ministry of Agriculture (2008).

Figure 9.1 Expansion of the coverage of modern irrigation and the trend in rice yields in Bangladesh, 1970–2007

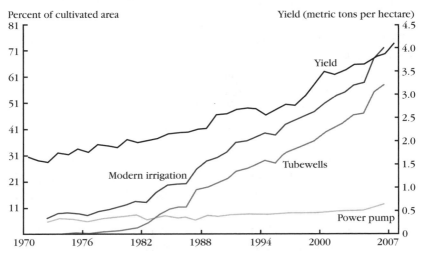

Source: Author, based on data from Bangladesh, Bureau of Statistics (various years).

from the late 1980s as the market for irrigation equipment was liberalized. Since 1990 the expansion of irrigation has been almost exclusively through the exploitation of groundwater by STWs.

The following linear trend lines were fitted on the time-series data on irrigated areas from 1976 to 2008 to assess the degree of acceleration in irrigation that was stimulated by the market privatization of minor irrigation equipment:

Total irrigated area (thousands of hectares) =
540 + 108 Time − 830 Dummy + 59 Time × Dummy
(4.18) (6.68)   (3.55)   (3.40) $R^2 = 0.98$

Tubewell-irrigated area (thousands of hectares) =
−154 + 87 Time − 612 Dummy + 66 Time × Dummy
(−1.52)(5.89)   (−3.73)   (4.16) $R^2 = 0.99$

Dummy is a dummy variable used to represent the period of privatization that takes a value of one for the period 1988–2008, when the irrigation market remained fully privatized, and a value of zero for the earlier period when irrigation development was under the control of BADC. The figures in parentheses are the estimated t-values of the regression coefficients.

The positive and statistically significant coefficient of the interaction term, Time × Dummy, indicates that there has been significant acceleration in the diffusion of modern irrigation since the change in policy in favor of privatization. The values of the coefficients indicate that the irrigated area accelerated from 108,000 hectares per year during the preliberalization period to 167,000 (108,000 + 59,000) per hectare since liberalization in 1988. In the postliberalization period the expansion was almost entirely due to the use of tubewells. The area irrigated by tubewells has increased at a rate of 153,000 (87,000 + 66,000) hectares per year since 1988, which represents a 92 percent expansion of the total irrigated area. In the preliberalization period, tubewells accounted for 80 percent of the expansion of irrigated area.

The question remains whether privatization of the market for irrigation equipment benefited the small-scale and marginal farmers in Bangladesh. In the debates of the early 1980s, the general perception was that privatization would create inequitable access to irrigation water and hence would worsen the distribution of agricultural income (Howes 1985; Quasem 1985; Osmani and Quasem 1990). Using data on ownership of irrigation equipment reported in the Agricultural Censuses of 1977 and 1983–84, Parthasarathy (1988) argued that larger farmers had greater access to irrigation and that smaller farmers had been losing their differential advantage in farm productivity with the progress in tubewell irrigation. A national-level inventory of STWs conducted in 1987 under a Canadian International Development Agency Project reported, however, that the average landholding size of STW operators had declined, indicating that small-scale farmers' access to STW irrigation had been growing (Mandal 1993; IWMI/BAU 1996).

The author of the present study generated household-level data from a national-level sample survey conducted in 62 villages spread all over Bangladesh. The surveys were conducted during 1987–88, 2000–01, and 2007–08 (Hossain et al. 2006). The objectives of the surveys were to assess the impact of the diffusion of modern technologies on income distribution and poverty (1987–88) and the impact of rice research on poverty reduction (2000–01) and to analyze changes in rural livelihoods (Hossain et al. 1994, 2002, 2006). The surveys contained information on owner-ship of different assets, including irrigation equipment.

The information obtained from the survey on the ownership of STWs by different groups of farmers is reported in Table 9.3. As noted, only 4.6 percent of the farms owned STWs in 1988, but this grew to 16.0 percent in 2000, and 22.0 percent in 2007. Thus one out of five farmers now own STWs. The data from the survey confirm that the cost of the tubewells has also declined, from $670 to $220 within the past two decades. This decrease is due to the availability of relatively low-cost machines imported from China and the increasing use of secondhand machines. Before liberalization, BADC imported more durable but higher-cost engines from Japan.[2] The distribution of ownership of the equipment, however, remains unequal. In 2007 almost 90 percent of farmers operating more than 2.0 hectares owned STWs compared to 61 percent of farmers with holdings of 1.0–2.0 hectares, but only 7 percent of marginal farmers operating up to 0.4 hectares owned STWs (see Table 9.3). Marginal farms constitute 52 percent of farm households in Bangladesh.

Although most of the marginal and small-scale farmers in Bangladesh do not own STWs, they have access to irrigation equipment installed by other farmers. Due to fragmented and scattered holdings, land parcels located within the command area

Table 9.3 Distribution of ownership of shallow tubewells (STWs) on the landownership scale of Bangladesh, 1988, 2000, and 2007

Farm size (hectares)	Percentage of households with own STWs			Average replacement cost of STWs (US$ per unit)		
	1988 (*n* = 818)	2000 (*n* = 1,083)	2007 (*n* = 1,131)	1988	2000	2007
Up to 0.4	2.4	2.8	6.8	598	278	194
0.4–1.0	2.1	15.6	22.7	692	263	191
1.0–2.0	3.9	36.5	60.9	560	280	218
Over 2.0	17.2	81.4	89.7	770	343	273
All farms	4.6	16.1	22.1	671	302	223

Source: Author's estimate from primary data collected through a national-level repeat sample survey by the Bangladesh Institute of Development Studies, International Rice Research Institute, and Bangladesh Rural Advancement Committee. The methodology of the survey is described in Hossain et al. (1994, 2006).

of a tubewell are usually owned by a number of farmers besides the tubewell owner. So, the tubewell owner has to sell water to operators of adjoining plots for optimum use of the capacity of the tubewell. As a result, with the diffusion of STWs a market for transacting water has emerged (Palmer-Jones and Mandal 1987; Palmer-Jones 1988, 2002). Palmer-Jones (2002) characterizes this market as "private provision of local public goods within a socially regulated contestable market embedded in local society."

In the water market, several methods for the payment of water charges are practiced. These include sharing one-fourth of the crop with the tubewell owner at the time of harvest; paying a flat charge per season, which is paid in cash in several installments before harvest; or paying an hourly charge for renting the machines, with different rates depending on who provides the fuel for operating the machines.

Mandal (1993, 2000) notes that widespread ownership of STWs has helped break monopolistic control over the supply of irrigation water by the landed rich farmers, who were labeled water lords by an influential civil servant in the early 1980s. In the prevailing social and economic setting, several factors—such as topographical limitations on siting, kinship, and personal relationships—have all contributed to increased competition among prospective water sellers and diminished the scope of unilateral pricing of irrigation water. If in a command area a landowner is not interested in renting water, the owner of the STW may opt for renting the land for tenancy cultivation. The growth of partnerships in the irrigation business formed by small-scale farmers has been another development. Some nongovernmental organizations have organized the landless to invest in STWs with microcredit and to sell water to farmers (Wood 1988).

The expansion of STWs has contributed to the development of rural entrepreneurship, which has led to the growth of other agribusiness services (Mandal 2000). Competition in many areas has reduced the water charge. Over time, arrangements for water sharing in the market have moved from share payments (high-cost) to seasonal cash payments (relatively low-cost) to an hourly rental system. The latter promotes incentives to economize on water use, and overall, there is a trend toward more efficient operation of irrigation services in the market (Hossain et al. 2002; Hossain and Bayes 2009).

Impact on Growth in Rice Production

Diffusion of *Boro* Rice

In Bangladesh rice is grown in three overlapping seasons (Hossain 1989). The monsoon rice *aman* is harvested from late October to December and was the predominant rice crop in the 1960s when the Green Revolution was initiated.

At medium-low to medium-high elevations, the *aman* crop is transplanted; on low-lying lands, it is broadcast-seeded (deepwater *aman*). The early monsoon rice known as *aus* is direct-seeded dry in April when occasional mild rains moisten the soil. The plant survives under mild drought conditions in May, matures with the early monsoon rains in June, and is harvested in July and August. This low-yielding crop is grown on relatively high lands where cultivation of the rainfed transplanted *aman* crop is not possible. The dry-season irrigated rice *boro* used to be grown on extremely low-lying lands in depressed basins. It was not possible to cultivate *aman* rice in these basins because of deep flooding during the monsoon season. The land is usually kept fallow during the period of high floods from June to October. *Boro* is transplanted in knee-deep water from November to December when the floods recede and is harvested from April to May. The yield of *boro* rice is the highest among the three seasonal traditional varieties (TVs) of rice. The development of irrigation infrastructure has contributed to the expansion of *boro* rice in both the traditional *aus* and the deepwater *aman* areas.

The adoption of modern varieties (MVs) of rice was initiated in the *boro* season in 1967 when the Bangladesh Academy for Rural Development in Comilla imported seeds of IR8 from the International Rice Research Institute (IRRI) in the Philippines (Bose 1974). It distributed these seeds to farmers through cooperatives and provided them with support for irrigation. Later IR20 was introduced in the *aman* season. With support from IRRI, Bangladesh was able to develop an active rice research system through the establishment of the Bangladesh Rice Research Institute (BRRI).[3] To date, BRRI has developed 52 MVs to suit the agroecological conditions of all three rice growing seasons (Hossain, Bose, and Mustaf 2006). The development of these MVs is based on parent materials received from IRRI under a program called the International Genetic Evaluation of Rice Germplasm (Hossain et al. 2003).

With the availability of MVs and the spread of minor irrigation, irrigated rice farming during the dry season began to spread rapidly (Hossain 2003). The traditional *boro* has also given way to modern *boro* as improved varieties for the irrigated ecosystem have been adopted. Initially the expansion was limited to low-lying lands in the depressed basins of eastern and central Bangladesh, where surface water could be exploited with LLPs. Later *boro* cultivation was diffused to medium-elevation and high lands with groundwater irrigation through tubewells where surface water was not available. *Boro* has mostly encroached on *aus* land and, to some extent, on deepwater *aman* areas.[4] The area under *aus* cultivation declined from 3.24 million hectares in 1969–71 to 0.96 million hectares in 2006–08. The area under deepwater *aman* declined from 2.08 million to 0.54 million hectares over the same period. In contrast, the area under *boro* rice increased from 0.89 million hectares during 1969–71 to 4.4 million hectares during 2007–08. Because MV *boro* is substantially higher yielding than are traditional *aus* and the deepwater *aman,* the average rice yield has continuously increased with its diffusion (Table 9.4).

Wherever irrigation facilities are available, farmers now grow *boro* with improved varieties in the dry season, except on parcels with dominant light soil that requires more irrigation. On these types of land, farmers grow wheat, potatoes, and recently introduced vegetables and maize, which have comparative advantages over *boro* because of the unusually high cost of irrigating *boro.* Socioeconomic factors such as small and fragmented holdings, tenancy farming, and lack of credit facilities have not constrained the diffusion of modern *boro* varieties, which have been adopted equally by all categories of farmers (Hossain 1989; Hossain et al. 1994; Ahmed 2001; Hossain, Bose, and Mustaf 2006). The sample surveys mentioned earlier estimated that the number of farmers cultivating *boro* rice increased from 34 percent in 1988 to 69 percent in 2000. The latter figure has since remained constant. The preliminary Agricultural Census undertaken in 2008 estimates that there are 14.6 million farm households. Thus *boro* rice is grown by 9.8 million households in Bangladesh (Bangladesh, Ministry of Agriculture 2008).

Bangladesh's rice economy has made respectable progress (Hossain 1989; Abdullah and Shahabuddin 1997; Ahmed 2001; Zohir, Shahabuddin, and Hossain 2002; Dorosh 2006). Bangladesh has an agrarian structure dominated by marginal farmers and share tenants who operate land in scattered, tiny holdings. Many scholars predicted that such a structure would constrain the productive development of agriculture (Jannuzi and Peach 1980; Boyce 1987). But modern agricultural

Table 9.4   Trends in rice area, production, and yield in Bangladesh in different seasons, 1969–2008

	Boro rice (summer)			Aus rice (premonsoon)			Aman rice (monsoon)		
Year	Area (thousands of hectares)	Production (thousands of metric tons)	Yield (metric tons per hectare)	Area (thousands of hectares)	Production (thousands of metric tons)	Yield (metric tons per hectare)	Area (thousands of hectares)	Production (thousands of metric tons)	Yield (metric tons per hectare)
1969–71	894	2,853	3.19	3,242	4,254	1.31	5,860	9,866	1.68
1979–81	1,127	3,648	3.23	3,127	4,692	1.50	5,937	11,348	1.91
1989–91	2,183	9,111	4.17	2,348	3,796	1.62	5,689	13,012	2.29
1999–2001	3,646	16,750	4.59	1,367	2,634	1.93	5,740	15,202	2.65
2006–08	4,315	23,346	5.41	952	2,382	2.50	5,367	15,657	2.92

Source: Author, based on data from Bangladesh, Bureau of Statistics (various years).

technologies have spread rapidly in Bangladesh. Improved varieties are now used on three-fourths of the land under rice cultivation, and their adoption has followed the development of the country's irrigation infrastructure (Hossain and Bayes 2009).

Over the past four decades, the rice-harvested area has increased only marginally, from 10.1 to 10.5 million hectares, but rice production has more than doubled, from 17.1 to 41.4 million tons. The rate of production growth has been about 2.5 percent per year, or much faster than the population growth. The average rice yield reached 4.1 tons per hectare in 2008 from a meager 1.69 tons per hectare during 1969–71, a growth of 2.4 percent per year. The growth in both production and yield was relatively slow during the first two decades after Bangladesh achieved independence but has accelerated since the late 1980s (Figure 9.2). The increase in the average yield for all rice crops was only partly due to the increased yield of individual seasonal varieties arising from the adoption of improved germplasm and better management of crops. The yield increase was mostly due to changes in crop composition that resulted from shifting land from lower-yielding to higher-yielding seasonal varieties (Hossain, Bose, and Mustaf 2006).

Table 9.4 provides information on trends in the production of *boro* rice compared with rice grown in other seasons. The growth in *boro* production and yield was relatively slow in the 1970s and picked up in the 1980s. The increase has been most rapid, however, since 1988, when the area under shallow tubewells grew. As mentioned earlier, the expansion of *boro* areas has been at the expense of the decline in the area under *aus* rice cultivation. The extremely low-lying *aman* areas where deepwater rice used to be grown have also been converted to *boro* rice.[5]

Figure 9.2 Growth in rice production in Bangladesh by season, 1970–90 and 1990–2007

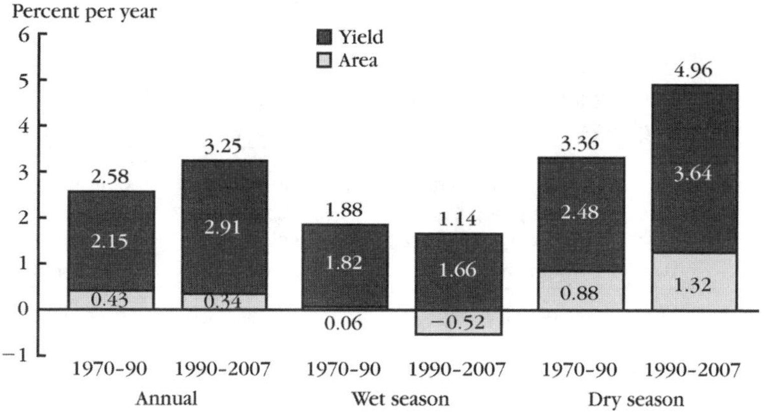

Source: Author's estimates using time-series data from Bangladesh, Bureau of Statistics (various years).

The contribution of *boro* to overall increases in rice production can be assessed from Table 9.5. The share of *boro* in incremental rice production was 34 percent during 1950–70, a period that roughly predated the Green Revolution in Bangladesh. The share increased to 62 percent during 1971–89, the preliberalization period. In the 1990s *boro* accounted for 88 percent of the increase in rice production. In the current decade almost the entire growth in rice production is attributed to *boro* rice.

To assess the impact of the increase in tubewell irrigation in the postliberation period, a linear trend line was fitted on the time-series data on rice production, area, and yield for 1976–2008. The estimated lines are as follows:

Production (thousands of tons) =
 17,769 + 424 Time − 7325 Dummy + 503 Time × Dummy
 (15.74) (2.55)   (−4.01)   (2.83) $R^2 = 0.94$

Area (thousands of hectares) =
 10,777 + 35.7 Time − 114 Dummy − 16.5 Time × Dummy
 (66.6)   (1.60)   (−0.47)   (−0.69) $R^2 = 0.16$

Yield (tons per hectare) =
 1.77 + 0.035 Time − 0.627 Dummy + 0.048 Time × Dummy
 (23.17) (3.07)   (−5.09)   (4.03) $R^2 = 0.97$

The figures in parentheses are estimated *t*-values. The Dummy variable assumes a value of one for the postliberalization period, 1989–2008. The positive coefficients of the interaction term Time × Dummy are statistically significant in the production and yield equations but not in the equation for rice area. The results indicate acceleration in yield and production in the postliberalization period.

Table 9.5 Contribution of *boro* rice to growth in total rice production in Bangladesh, 1950–2008

Period	Increase in total rice production (thousands of metric tons per year)	Increase in *boro* rice production (thousands of metric tons per year)	Contribution of *boro* to incremental rice production (percent)
1950–52 to 1969–71	333	114	34
1969–71 to 1989–91	500	312	62
1989–91 to 1999–2001	867	764	88
1999–2001 to 2006–08	971	942	97

Sources: Bangladesh, Bureau of Statistics (various years).

The estimated value of the coefficient for the yield function indicates that the rice yield increased at a rate of 35 kilograms per year during 1976–88, the pre-liberalization period. The coefficient of the variable Time × Dummy indicates that the yield growth increased by an additional 48 kilograms per year over the period from 1988–89 to 2007–08. The coefficient is statistically significant. We may thus conclude that the policy changes contributed to an additional increase in rice yield of 0.96 tons per hectare over the past two decades, which is about 58 percent of the incremental rice yield during the period.

Rice production increased from 424,000 tons per year during 1976–88 to 927,000 (424,000 + 503,000) tons per year during 1988–2008. In the latter period, the rice-cropped area increased by a mere 19,200 (35,700–165,000) hectares per year. Thus the acceleration in the growth of production was almost entirely due to the acceleration in yield rates. Assuming that 500 grams of milled rice is required to feed a person each day, the 503,000 tons of incremental production per year during 1988–2008 was able to feed an additional 1.84 million people per year. The population of Bangladesh is increasing at a similar rate.[6] From 1988 to 2008, the incremental rice production due to the liberalization policy thus helped to feed nearly 37 million people (1.84 million × 20 years).

For a rigorous estimate of the impact of policy changes, Ahmed (1995) and Ahmed, Haggblade, and Chowdhury (2000) estimated input demand and rice production functions using the seemingly unrelated regression model with time-series data for 1975–76 to 1996–97. This approach helped to establish counterfactuals (i.e., what would have happened without the policy reforms). Ahmed concludes that the gross effect of the liberalization of agricultural equipment market was large, about 2 million tons, while the indirect effect of the expansion of irrigated rice (*boro*) on rainfed rice (*aus*) via the change in crop composition was both large and negative. The net effect of irrigation-related changes, Ahmed notes, was about 1.3 million additional tons of rice (from 1988–89 to 1996–97). This means that annual rice production during the postliberalization period would have been 1.3 million tons less had the market reforms not been undertaken. The net effect was 38 percent of total incremental production (Ahmed 2000). The incremental rice production between 1988 and 2008 was 15.5 million tons. Using Ahmed's parameters, I estimate that 5.9 million tons of the additional production could be attributed to policy changes. With the current level of per capita consumption, this amount of rice would have been able to feed nearly 22 million people.

Socioeconomic Impacts of *Boro* Rice Production

Despite clear food production benefits, how has the diffusion of modern varieties in the *boro* season affected the costs of food production, employment, and farmers' income? To shed some light on these issues, the costs and returns data collected from the repeat surveys mentioned earlier were processed (Hossain et al. 2006). The findings are reported in Table 9.6. Because *boro* rice has expanded mostly at the expense of the *aus* rice, the effect of the diffusion can be assessed by comparing the numbers of these two seasonal varieties. Because paddy prices were abnormally low in 2000, I use the numbers for 1988 and 2007 to assess the effects.

The paid-out costs of production (including the costs of seed, fertilizer, pesticides, irrigation, and machinery rental) for the cultivation of *boro* rice (MVs) were almost three times those for TVs grown under rainfed conditions (*aus* rice) in 1988 and almost double those for TVs in 2007. These higher costs indicate the input-intensive nature of irrigated rice farming. The total cost of production per hectare was about 1.4 times higher in *boro* cultivation in 1987 and 70 percent higher in 2007. But the increase in production from the adoption of MVs was much higher than the increase in cost. Hence the cost per unit of output went down with the adoption of MVs. The unit cost of production was 22 percent lower in the cultivation of *boro* compared to *aus* in 1987–88 and 17 percent lower in 2000–01. The

Table 9.6 Costs and returns in the cultivation of *boro* and *aus* varieties in Bangladesh, 1988–2007 (per hectare of land)

	Boro rice			*Aus* rice		
Elements	1987/88	1999/2000	2006/07	1988	2000	2007
Gross value of production (US$)	747	731	867	274	313	483
Paid-out cost (US$)	380	426	429	129	179	202
Net income (US$)	367	305	438	145	134	281
Total cost of production (US$)	486	445	568	204	282	333
Operating surplus (US$)	261	286	299	70	31	150
Yield (metric tons per hectare)	4.46	4.98	5.36	1.48	2.03	2.63
Unit cost (US$ per ton of paddy)	109	89	106	138	139	127
Harvest price (US$ per ton of paddy)	159	141	152	164	145	171
Labor use (days per hectare)	206	131	120	142	112	98
Labor productivity (US$ per day)	2.20	1.88	2.92	1.37	1.48	1.61
Wage rate (US$ per day)	0.95	1.31	1.81	0.90	1.22	1.56

Source: Author's estimate from primary data collected through a national-level repeat sample survey by the Bangladesh Institute of Development Studies, International Rice Research Institute, and Bangladesh Rural Advancement Committee. The methodology of the survey is described in Hossain et al. (1994, 2006).

Notes: The gross value of production includes the value of paddy straw, which is used as cattle feed. The paid-out cost includes the out-of-pocket expenses for seeds, fertilizers, pesticides, irrigation charges, machinery rental, and wage bills for hired labor. Total cost includes the cost of family-supplied inputs, including family labor. The total costs include the imputed interest charge on working capital but exclude the depreciation of farm equipment and imputed land rent. The land rent is a real cost of rented land for tenant farmers. Nearly 22 percent of the land in 1988 and 38 percent in 1987 was tenant operated. The rent is about 30 percent of the gross produce. Thus, the net returns to the tenant farmer would be much less than the numbers shown in the table.

unit cost of production has declined over time. Faster technological progress has also contributed to a reduction in the unit cost of production, which has helped maintain rice prices at a low level—a major factor behind improving access to food for low-income households.

The MVs are also more labor intensive than the TVs. So the expansion of the area under MVs has contributed to increasing labor demand, which has helped to generate additional employment for labor-selling low-income households. The labor use for MVs compared to TVs was higher by about 64 days per hectare in 1987–88 (see Table 9.6). The *boro* area increased by about 2.2 million hectares over 1988–2008; most of this increase came from the decline in area under TVs. This change would have generated additional employment of 141 million person-days of labor. The actual impact, however, has been much less because farmers have resorted to mechanization of agricultural operations in response to increasing wages and costs of feeding animals.[7] The labor used in *boro* cultivation had consistently declined, from 206 days per hectare in 1987–88 to 131 days per hectare in 2000 to 120 days per hectare in 2007. Despite the spread of labor-saving mechanization, the labor used in the cultivation of *boro* rice was about 28 days more than that in the cultivation of *aus* rice. If we use the recent number, which is a conservative estimate of the effect of the diffusion of *boro* cultivation on employment, it would be 62 million person-days, or about 238,000 full-time-equivalent jobs.

As noted earlier, *boro* rice farming is highly input intensive compared to traditional *aus* farming and entails higher costs associated with irrigation and the use of chemical fertilizers.[8] *Aus* farming, in contrast, is entirely dependent on rainfall. Farmers use fertilizer in low amounts compared to the cultivation of TVs due to the high risk of occasional drought. In 1988, for example, farmers hardly used any fertilizer in the cultivation of *aus* rice, and in 2007 the fertilizer used was one-fourth of that used in the cultivation of *boro* rice. Despite the capital-intensive nature of *boro* farming, the family income from *boro* rice cultivation (gross value of production minus paid-out cost) was substantially higher than that from *aus* cultivation (see Table 9.6). The net income gains from the shift of land from *aus* to *boro* was $222 per hectare in 1988, $171 per hectare in 2000, and $157 per hectare in 2007. The reduction in the cost of production—facilitated by technological progress—has helped keep rice prices within affordable limits for the rural landless and the urban poor.

The trend in the nominal and real prices of rice and its association with the growth in rice production can be seen in Figure 9.3. The real prices showed a consistent downward trend over the entire period. Until the late 1980s, the prices had been fluctuating and the rate of decline was relatively slow. The decline in prices has been more moderate in the current decade. The downward trend in rice prices

Figure 9.3 Trends in the production and price of rice in Bangladesh, 1970–2006

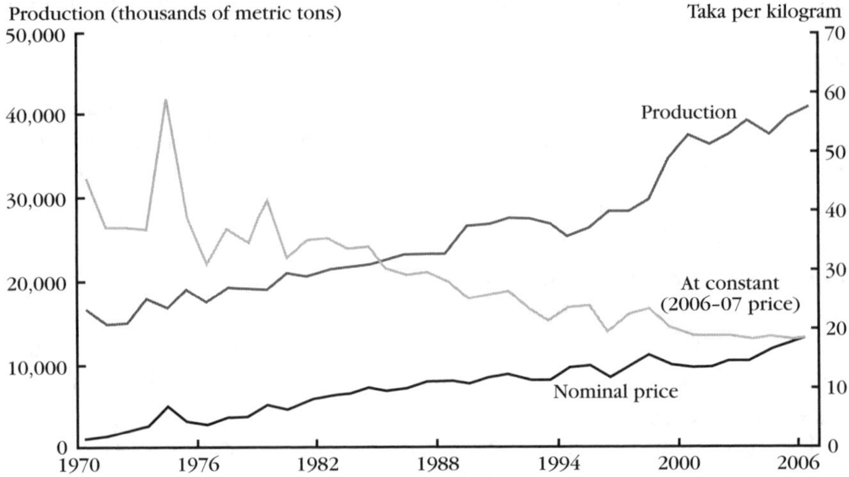

Source: Author, based on data from Bangladesh, Bureau of Statistics (various years).

has been a major factor behind the moderate reduction in poverty that Bangladesh has experienced since the mid-1980s.[9]

The following linear trend in the prices of rice (adjusted for the wholesale price index) is obtained from the time-series data for 1976–2007:

Rice price (taka per kilogram) =
  $37.18 - 0.42$ Time $- 2.62$ Dummy $- 0.13$ Time $\times$ Dummy
  (26.25) (2.02) (-1.14) (-0.58) $R^2 = 0.90$

The values of the parameters indicate that the prices of rice declined by 0.42 taka per kilogram per year during 1976–88 and by 0.55 taka (0.42 + 0.13) per kilogram per year during 1988–2007. However, the decline in prices in the recent period was not statistically significant. The lower price of rice—following higher rice production and lower unit cost of production—kept the vast number of low-income urban consumers and the rural landless and marginal farmers on a stable financial footing. Because the poor spent nearly 50 percent of their budget on rice (30 percent for all consumers), the downward trend in rice prices has been a major factor behind the moderate reduction in poverty that Bangladesh has experienced since the mid-1980s. The incidence of poverty declined by 1 percent per year during the 1980s and 1990s. The progress was more rapid (2 percent per year) during 2000–05 (Ravallion and Sen 1996; Narayan and Zaman 2009).

Environmental Impacts

The expansion of minor irrigation-led *boro* cultivation has not been without its critics, however. In recent years the government has been deemphasizing the cultivation of *boro* in favor of *aman* rice due to some perceived negative environmental effects. The adverse effects mentioned are (1) the pushing out of major noncereal crops—such as pulses and oilseeds—that were important sources of protein and micronutrients for the poor; (2) the decline in soil fertility due to raising two MV rice crops, which are heavy users of soil nutrients; (3) the heavy use of pesticides, which have led to adverse impacts on the quality of surface water and fish habitats; (4) the overexploitation of groundwater resources, leading to adverse impacts on the supply of drinking water; and (5) the arsenic contamination of groundwater that is widely prevalent in Bangladesh (Harvey et al. 2002; Brammer 2009). In this section we briefly touch on these alleged negative effects of the expansion of *boro* rice.

The expansion of irrigation infrastructure and *boro* cultivation gave farmers the opportunity to grow two modern rice varieties on the same parcel of land. But such double-cropping of rice is still not widely prevalent in Bangladesh. The parcel-level data from the 2000–01 sample survey indicate that *boro* rice is single-cropped on about 20 percent of the land, which is mostly in the low-lying areas in the flood-prone ecosystem (Hossain and Bayes 2009). *Boro* is double-cropped with MV *aman* rice on 18 percent of the land and with traditional *aman* on another 9 percent of the area. For parcels that were double-cropped with MV *boro* and MV *aman,* yields were not lower than for single-cropped systems of either *boro* or *aman.* Also, the yields for both crops have increased over time. Because a large proportion of land in Bangladesh is regularly flooded from the overflow of rivers, it benefits from the deposit of silts and recovery from the depletion of organic matter and micronutrients. Thus the validity of the hypothesis that the expansion of *boro* has led to a decline in soil fertility is yet to be confirmed.

The area under pulses and oilseeds has indeed declined over time. But whether this is entirely due to the expansion of *boro* cultivation is again a debatable issue. We have already noted that the major area for the expansion of *boro* has come from the reduction in area under *aus* rice and deepwater *aman.* Only a small fraction of this area has come from other crops and the seasonally fallow land. In fact, it is the expansion of wheat cultivation that has displaced these minor crops more than has *boro* rice. In recent years farmers have been delaying the planting of *boro* rice to benefit from early monsoon rains. This delayed planting also allows them to raise minor crops such as potatoes and oilseeds between the cultivation of *aman* and *boro* crops.

It can be noted in Table 9.7 that the incidence of pesticide use is indeed very high with *boro* cultivation compared to other rice varieties, and the use of pesticides has grown over time. In 2007 more than 80 percent of the farmers used pesticides in the cultivation of modern *boro* compared to only 9 percent for wheat, 16 percent for

Table 9.7 Use of pesticides in the cultivation of different crops in Bangladesh, 1988 and 2007

Crop	Percent of farmers using pesticides			Cost on account of pesticides (US$/ha)		
	1987/88	2000/01	2007	1987/88	2000/01	2007
Rice crop variety						
Aus TV	21	13	33	2.8	0.89	4.41
Aman TV	12	17	16	1.4	1.48	1.92
Aman MV	43	62	50	7.2	7.42	6.40
Boro MV	86	80	82	12.1	3.01	12.7
Other crops						
Wheat	16	12	9	2.09	1.04	0.78
Jute	24	12	37	4.70	1.53	5.70
Potatoes	28	94	96	6.20	48.2	63.4

Source: Data from national-level repeat sample survey conducted by the Bangladesh Institute of Development Studies (BIDS), International Rice Research Institute (IRRI) and Bangladesh Rural Advancement Committee. See Table 9.6.
Note: MV means modern varieties; TV means traditional varieties.

traditional *aman*, 37 percent for jute, and 50 percent for MV *aman*. The rate of the use of pesticides in *boro* cultivation is many times higher than for competing cereal crops but still substantially lower than for other high-value crops such as sugarcane, potatoes, and vegetables. The expansion of *boro* rice has indeed contributed to the increased use of harmful agrochemicals.

It was reported earlier that the expansion of dry-season irrigated rice farming was heavily dependent on the exploitation of groundwater through STWs. Pitman (1993) estimated that by the late 1980s, 0.77 million hectares of land were irrigated with groundwater resources, and an additional 1.2 million hectares could be irrigated with STWs and/or DTWs. The area irrigated by tubewells is two-thirds greater than this number, indicating overexploitation of the groundwater. The National Commission on Agriculture estimated that the available potential recharge of the aquifer that could be extracted by STWs had almost been exploited by 1996 (Table 9.8). Since then, the number of STWs fielded has more than doubled (see Table 9.2), which also suggests that groundwater resources have already been overexploited. Farmers have introduced adjustments in the technology to extend the reach of their STWs, such as digging holes, placing pumps lower in the ground, installing pipes in several places, and moving pumps from one place to another. Recently they have been delaying the planting of *boro* so that they can take advantage of early monsoon rains and reduce their water use. This practice also helps them raise a highly profitable nonrice crop (potatoes or vegetables) during the winter season

Table 9.8 Use of groundwater resources through small-scale irrigation equipment in Bangladesh, 1996 (millions of cubic meters)

Indicator	Region				
	Northwest	Northeast	Southeast	South central	Southwest
Available recharge	12,100	23,100	9,800	3,500	5,600
Maximum extractable through deep tubewells	11,900	14,500	4,700	2,500	4,900
Maximum extractable through shallow tubewells (STWs)	9,900	5,000	1,200	1,000	3,200
Already extracted by tubewells	8,151	4,154	948	524	2,520
Amount required for household and industrial use	185	333	169	79	135
Estimated additional recharge that can be extracted by STWs	1,564	523	83	397	545

Source: Bangladesh, Ministry of Agriculture (1999).

from November to February. Farmers have also started using shorter-maturity MVs of *boro* that require less irrigation.

It has now been established that the arsenic contamination of groundwater is a serious problem for Bangladesh (Kinniberg and Smedley 2001; Harvey et al. 2002; van Geen et al. 2003; Hossain 2005; Brammer 2009). But its link with the exploitation of groundwater for *boro* cultivation is yet to be firmly established. It is suggested that arsenic may have recently been released through sulfide oxidation reactions induced by massive increases in the dry-season pumping of groundwater. A large-scale study conducted by the British Geological Survey and the Bangladesh Department of Public Health Engineering (Kinniberg and Smedley 2001, xvii) concludes, "There is no evidence to support the proposition that the ground water arsenic problem is caused by the recent seasonal drawdown of [the] water table due to [the] recent increase in irrigation abstraction." The study covered samples from 3,534 tubewells from 61 of the 64 districts of Bangladesh and from 433 of the 496 *upazilas* (subdistricts). The greatest concentration of arsenic contamination was found in the southern and southeastern parts of the country, where the coverage of groundwater irrigation is very low. The arsenic contamination was found to be very low in the northwestern part of the country, where groundwater irrigation through STWs has spread most. However, it has recently become apparent that arsenic-contaminated groundwater that is used for irrigation is seeping into soils and rice in several countries of South and Southeast Asia, thus posing a serious threat to sustainable agricultural production and to the health and livelihoods of the affected people in these countries (Brammer 2009). Because groundwater-dependent *boro* rice now accounts for nearly 56 percent of the annual rice production, if arsenic is transported to the grain the risk is indeed very great for Bangladesh.

Summary and Conclusions

The rice economy of Bangladesh has made good progress during the past four decades. The area planted in rice has remained almost stagnant at 10.5 million hectares, but production has increased from 17 to 44 million tons of paddy, almost entirely due to the increase in yield. This yield growth was propelled by the diffusion of *boro* rice, whose cultivation—initially limited to the depressed basins of the country—has gradually expanded from very low-lying lands to medium-low and medium-high elevations. The area under cultivation for MV *boro* rice increased from 0.3 million hectares in 1970–71 to 4.6 million hectares in 2007–08, and the production hectares increased from 1.5 to 26 million tons of paddy rice (17.8 million tons of milled rice). *Boro* now accounts for 44 percent of the harvested rice area and 56 percent of total rice production. The contribution of *boro* to overall increases in rice production grew from one-third in the pre-independence period to more than 90 percent during 1989–2008.

The expansion of *boro* cultivation has proceeded in full force since the late 1980s, when markets were deregulated for minor irrigation equipment. With unrestricted private-sector imports of agricultural equipment and the lowering of prices, private investment in STWs for extraction of groundwater has expanded rapidly, as has the area under *boro* cultivation. The process was helped by private investment in STWs financed by farmers' own savings. The agrarian structure dominated by marginal and small-scale farmers who are mostly tenants did not stand in the way of the dissemination of modern agricultural technology.

Nearly 22 percent of farm households now own 1.3 million STWs that provide irrigation services to 10.2 million out of 15 million farm households. The policy of market liberalization may have contributed to an increase in the rice yield of 0.96 tons per hectare, or 58 percent of the increase in the average rice yield over 1989–2008. It contributed to incremental rice production of 5.9 million tons per year, which would feed about 22 million people. The adoption of improved technology in *boro* rice farming has helped reduce the unit cost of rice production and thereby kept rice prices affordable for the poor. The unit cost of production of *boro* rice was lower by 20 percent in 1988 and 17 percent in 2007 than that of the rainfed *aus* rice that *boro* rice has replaced. Because *boro* cultivation is more labor intensive than that of *aus*, the change in variety composition also generated an additional 52 million person-days of employment, equivalent to 238,000 full-time jobs. The income gains were estimated at $222 per hectares in 1988 and $157 in 2007. The expansion of *boro* farming using STWs has led to some adverse environmental effects, however, such as the increased use of harmful agrochemicals (pesticides) and the overexploitation of groundwater resources.

Bangladesh's progress in the area of modern rice technology can be credited to government policy that encouraged the procurement and marketing of minor irrigation equipment. The diffusion of these new technologies was effective in radically increasing rice production, thus increasing the food security of the country as a whole.

Notes

1. The Barind Tract area is a western, high-elevated, low-rainfall zone.

2. The average price of Japanese engines (4–8 horsepower) for STWs in 1986 was 28,000 taka ($1,000), but these prices would last for more than 12 years. In 1991 a Chinese engine with 4–6 horsepower capacity was imported at 6,300 taka ($160), but it had to be replaced every five years. Still, the cost was within the reach of individual farmers in Bangladesh, and they could acquire the equipment with their own savings. The internal rate of return on investment for Chinese-made STWs was estimated at 37 percent, compared to 13 percent for the Japanese-made machines (IWMI/BAU 1996).

3. BRRI was established in 1972 and was nurtured by an IRRI-managed collaborative project that was funded by a number of donors until 1992. The project financed the development of laboratory facilities and regional stations for adaptive research, degree training of Bangladeshi scientists, and the evaluation of international genetic materials—distributed by IRRI under the International Network for Genetic Evaluation of Rice—for suitability under different agroecological conditions in Bangladesh. It is estimated that almost 70 percent of the improved varieties adopted in Bangladesh are based on genetic materials developed at IRRI (Hossain 1996; Hossain et al. 2003, 2006).

4. Instead of growing the deepwater *aman,* which is often destroyed by floods, the farmer now keeps the low-lying parcels fallow during the monsoon season and grows *boro* rice in the dry season with full irrigation.

5. In earlier years, such land was single-cropped, with *boro* rice grown in the dry season and the land left fallow during the monsoon season. Recently such land has been converted into shallow ponds for aquaculture during the monsoon season and for rice production during the dry season.

6. The population of Bangladesh has increased from 70 million in 1971 to about 145 million at present. The growth rate has declined from 2.5 percent per year in the 1970s to 1.4 percent per year in the 1990s. The population is still growing by 2 million every year. The country needs to produce an additional 0.5 million tons of rice (unmilled) to feed the incremental population.

7. The land preparation is now almost entirely mechanized, and mechanization in threshing operations is progressing quickly.

8. In many areas, STW owners take one-fourth of the harvest for supplying water to the farmers. In 2007, irrigation charges were estimated at 26 percent of the total cost of production and 16 percent of the value of gross produce. The cost of chemical fertilizers was 16 percent of the total cost of production and 7 percent of the gross value of production.

9. The rice price, however, increased rapidly during 2007–08 following the price hike in the international market. The price started declining again in August 2008 following a positive supply response in the 2008 *boro* season from highly favorable farm gate prices. The increase in rice prices in 2007–08 contributed to severe food insecurity in low-income households. It is estimated that the hike in rice prices led to an increase in the poverty ratio from 35 percent in 2007 to 43 percent in 2008.

References

Abdullah, A. A., and Q. Shahabuddin. 1997. Recent developments in Bangladesh agriculture. In *Growth or stagnation? A review of Bangladesh's development,* ed. R. Sobhan. Dhaka, Bangladesh: Dhaka University Press.

Ahmed, R. 1995. Liberalization of agricultural input markets in Bangladesh: Process, impact, and lessons. *Agricultural Economics* 12: 115–128.

Ahmed, R., S. Haggblade, and T. Chowdhury, eds. 2000. *Out of the shadow of famine: Evolving food markets and food policy in Bangladesh.* Baltimore, Md., U.S.A. and London: Johns Hopkins University Press.

Ahmed, R. A. 2001. *Prospects of the rice economy in Bangladesh.* Dhaka, Bangladesh: University Press.

Bangladesh, Bureau of Statistics. 1976. *Agricultural production levels in Bangladesh, 1947–1972.* Dhaka, Bangladesh: Ministry of Planning, Statistics Division.

———. Various years. *Statistical yearbook of Bangladesh, and monthly statistical bulletin.* Dhaka, Bangladesh: Ministry of Planning.

Bangladesh, Ministry of Agriculture. 1999. Report of the National Agriculture Commission. Dhaka, Bangladesh. Mimeo.

———. 2008. *Minor irrigation survey report 2007–08.* Dhaka, Bangladesh: Bangladesh Agriculture Development Corporation, Ministry of Agriculture.

Bose, S. R. 1974. The Comilla cooperative approach and the prospects of a broad-based green revolution in Bangladesh. *World Development* 2 (8): 21–28.

Boyce, J. K. 1987. *Agrarian impasse in Bengal: Institutional constraints to technological change?* Oxford, U.K.: Oxford University Press.

Brammer, H. 2009. Mitigation of arsenic contamination in irrigated paddy soils in south and southeast Asia. *Environment International* 35: 856–863.

Dorosh, P. 2006. *Accelerating income growth in rural Bangladesh.* Washington, D.C.: World Bank.

Gill, G. J. 1983. *The demand for tubewell equipment in relation to groundwater availability in Bangladesh.* Dhaka, Bangladesh: Bangladesh Agriculture Research Council.

Gisselquist, D. 1992. Empowering farmers in Bangladesh: Trade reforms open doors to new technology. Paper presented at the World Bank Annual Conference of the Association for Economic Development Studies in Bangladesh in Washington, D.C.

Hamid, M. A. 1988. Low-lift pump irrigation in Bangladesh. Report prepared for the Agricultural Sector Review, Bangladesh. United Nations Development Programme, New York.

———. 1993. Improving the access of the rural poor to groundwater irrigation in Bangladesh. In *Groundwater irrigation and the rural poor: Options for development in the Ganges basin,* ed. F. Kahnhert and G. Levine. Report of a World Bank Symposium. Washington, D.C.: International Bank for Reconstruction and Development.

Hannah, L. M. 1976. Handpump irrigation in Bangladesh. *Bangladesh Development Studies* 4 (4): 441–454.

Harvey, C. F., C. H. Swartz, A. B. M. Badruzzaman, N. Keon-Blute, W. Yu, M. Ashraf Ali, J. Jay, R. Beckie, V. Niedan, D. J. Brabander, P. M. Oates, K. N. Ashfaque, S. Islam, H. F. Hemond, and M. F. Ahmed. 2002. Arsenic mobility and groundwater extraction in Bangladesh. *Science* 298: 1602–1606.

Hossain, M. 1989. *Nature and impact of the green revolution in Bangladesh.* Research Report 67. Washington, D.C.: International Food Policy Research Institute.

———. 1996. Agricultural policies in Bangladesh: Evolution and impact on crop production. In *State, market and development: Essays in honor of Rehman Sobhan,* ed. A. Abdullah and R. Khan. Dhaka, Bangladesh: University Press.

———. 2003. Development of *boro* rice cultivation in Bangladesh: Trends and policies. In *Boro rice,* ed. R. K. Singh, M. Hossain, and R. Thakur. New Delhi: International Rice Research Institute.

Hossain, M., and A. Bayes. 2009. *Rural economy and livelihoods: Insights from Bangladesh.* Dhaka, Bangladesh: A. H. Development.

Hossain, M., and B. Sen. 1992. Rural poverty in Bangladesh: Trends and determinants. *Asian Development Review* 10 (1): 1–34.

Hossain, M., M. L. Bose, and B. A. A. Mustaf. 2006. Adoption and productivity impact of modern rice varieties in Bangladesh. *Developing Economies* 44 (2): 149–166.

Hossain, M., M. A. Quasem, M. A. Jabbar, and M. Mokaddem. 1994. Production environments, MV adoption, and income distribution in Bangladesh. In *Modern rice technology and income distribution in Asia,* ed. C. C. David and K. Otsuka. Boulder, Colo., U.S.A.: Lynne Rienner.

Hossain, M., M. L. Bose, A. Chowdhury, and R. Meinzen-Dick. 2002. Changes in agrarian relations and livelihoods in rural Bangladesh: Insights from repeat village studies. In *Agrarian studies: Essays on agrarian relations in less developed countries,* ed. V. K. Ramachandran and M. Swaminathan. New Delhi: Tulika Books.

Hossain, M., D. Gollin, V. Cabanilla, E. Cabrera, N. Johnson, G. S. Khush, and G. McLaren. 2003. International research and genetic improvement in rice: Evidence from Asia and Latin America. In *Crop variety improvement and its effect on productivity: The impact of international rice research,* ed. R. Evenson and D. Golin. Wallingford, U.K.: CAB International.

Hossain, M., D. Lewis, M. L. Bose, and A. Chowdhury. 2006. Rice research, technological progress, and poverty. In *Agricultural research, livelihoods, and poverty: Studies of economic and social impact in six countries,* ed. A. Michelle and R. Meinzen-Dick. Baltimore: Johns Hopkins University Press.

Hossain, M. F. 2005. Arsenic contamination in Bangladesh: An overview. *Agriculture, Ecosystems and Environment* 113 (1–4): 1–16.

Howes, M. 1985. *Whose water? An investigation of the consequences of alternative approaches to small scale irrigation in Bangladesh.* Dhaka, Bangladesh: Bangladesh Institute of Development Studies.

IWMI/BAU (International Water Management Institute and Bangladesh Agriculture University–Mymensingh). 1996. *Study on privatization of minor irrigation in Bangladesh.* A report prepared for the Government of Bangladesh and the Asian Development Bank. Dhaka, Bangladesh: Ministry of Agriculture.

Jannuzi, F. T., and J. T. Peach. 1980. *Agrarian structure in Bangladesh: An impediment to development.* Boulder, Colo., U.S.A.: Westview.

Jansen, E. G. 1986. *Rural Bangladesh: Competition for scarce resources.* Oslo, Norway; Oxford, U.K.: Norwegian University Press; Oxford University Press.

Kinniburg, D. G., and P. L. Smedley, eds. 2001. Arsenic contamination of groundwater in Bangladesh. Final Report of the British Geological Survey and Department of Public Health Engineering (WC/00/19). <http://www.bgs.ac.uk/arsenic/bphase2/Reports/Vol2Covers.pdf>. Accessed August 15, 2009.

Mandal, M. A. S. 1978. Hand pump irrigation in Bangladesh: Comment. *Bangladesh Development Studies* 6 (1): 111–114.

———. 1987. Imperfect institutional innovation for irrigation management in Bangladesh. *Irrigation and Drainage Systems* 3: 239–258.

———. 1989. Declining returns from groundwater irrigation in Bangladesh. *Bangladesh Journal of Agricultural Economics* 12 (2): 43–61.

———. 1993. Groundwater irrigation in Bangladesh: Access, competition, and performance. In *Groundwater irrigation and the rural poor: Options for development in the Ganges basin,* ed. F. Kahnhert and G. Levine. Report of a World Bank Symposium. Washington, D.C.: International Bank for Reconstruction and Development.

———. 2000. Dynamics of irrigation water market in Bangladesh. In *Changing rural economy of Bangladesh,* ed. M. A. S. Mandal. Dhaka: Bangladesh Economic Association.

Murshid, K. A. S. 1985. Is there a "structural" constraint to capacity utilization of deep tubewells? *Bangladesh Development Studies* 13 (3–4).

Narayan, A., and H. Zaman. 2009. *Breaking down poverty in Bangladesh.* Dhaka, Bangladesh: University Press.

Osmani, S. R., and M. A. Quasem. 1990. *Pricing and subsidy policies for Bangladesh agriculture.* Research Monograph 10. Dhaka, Bangladesh: Bangladesh Institute of Development Studies.

Palmer-Jones, R. W. 1988. Groundwater irrigation in Bangladesh. Draft report for agricultural sector review. Government of Bangladesh / United Nations Development Programme, Dhaka. Mimeo.

———. 1989. Groundwater management in Bangladesh: Review, issues and implications for the poor. Paper presented at the workshop Efficiency in Groundwater Use and Management, January 30–February 1, in Ananda, India. Ahmedabad, India: Institute of Rural Management.

————. 1992. Sustaining serendipity? Groundwater irrigation, growth of agricultural production and poverty in Bangladesh. *Economic and Political Weekly* 27 (39): A-128–A-140.

————. 1999. Slowdown in agricultural growth in Bangladesh: Neither a good description nor a description good to give. In *Sonar Bangla? Agricultural growth and agrarian change in West Bengal and Bangladesh,* ed. B. Rogali, B. Harriss-White, and S. Bose. New Delhi: Sage.

————. 2002. Irrigation service markets in Bangladesh: Private provision of local public goods and community regulation? Paper presented in a forum on common property resources at the University of Lund, Sweden.

Palmer-Jones, R. W., and M. A. S. Mandal. 1987. *Irrigation groups in Bangladesh.* ODI/IIMI Irrigation Management Network Paper 87/2c, August. London: Overseas Development Institute.

Parthasarathy, G. 1988. Growth and equity issue: Rural poor. Working paper prepared for the *Agriculture Sector Review.* Dhaka, Bangladesh: United Nations Development Programme and Ministry of Agriculture.

Pitman, G. T. K. 1993. National water planning in Bangladesh 1985–2005: The role of groundwater in irrigation development. In *Groundwater irrigation and the rural poor: Options for development in the Ganges basin,* ed. F. Kahnert and G. Levine. Report of a World Bank Symposium. Washington, D.C.: International Bank for Reconstruction and Development.

Quasem, M. A. 1985. Impact of new system of the distribution of irrigation machines in Bangladesh. *Bangladesh Development Studies* 13 (3): 127–140.

Ravallion, M., and B. Sen. 1996. When method matters: Monitoring poverty in Bangladesh. *Economic Development and Cultural Change* 44 (4): 761–792.

Sattar, M. A., and S. I. Bhuiyan. 1985. Constraint to low utilization of deep tubewell water in a selected tubewell project in Bangladesh. *Bangladesh Development Studies* 13 (4).

Singh, R. K., M. Hossain, and R. Thakur. 2003. *Boro rice.* New Delhi: International Rice Research Institute.

UNDP (United Nations Development Programme). 1989. *Bangladesh agriculture sector review: Performance and potential.* New York: United Nations Development Programme.

Van Schundel, W. 1981. *Peasant mobility: The odds of peasant life in Bangladesh.* Asser, the Netherlands: Van Gorcum.

Wood, G. D. 1988. *Social entrepreneurialism in rural Bangladesh: Water selling by the landless.* Dhaka, Bangladesh: Proshika.

Zohir, S., Q. Shahabuddin, and M. Hossain. 2002. Determinants of rice supply and demand in Bangladesh: Recent trends and projections. In *Developments in the Asian rice economy,* ed. M. Sombilla, M. Hossain, and B. Hardy. Los Baños, Philippines: International Rice Research Institute.

Chapter 10

Hybrid Rice Technology Development: Ensuring China's Food Security

Jiming Li, Yeyun Xin, and Longping Yuan

In China, agriculture is the foundation of economic prosperity, social stability, and national independence. Given China's ongoing population pressures and unfavorable population–land ratio, agricultural production is one of the country's top priorities. The amount of arable land per capita has decreased from 0.18 hectare in 1950 to 0.10 hectare today, while the population has doubled over the past 50 years, to 1.3 billion (Riley 2004).

China is the largest rice-producing and -consuming country in the world. Rice accounts for 30 percent of the national foodcrop acreage and 40 percent of production. China has an average of about 30 million hectares under rice cultivation and produces 180 million tons of rice grain. Surpluses and deficits in China's rice production directly affect food prices in China and elsewhere (Qi et al. 2007).

In the 1960s China started to grow semidwarf rice varieties, which resulted in an increase in yields from 2.0 tons per hectare to 3.5 tons per hectare in 1975. The successful commercialization of three-line hybrid rice in the late 1970s brought another revolution in rice production, and rice yields rose to more than 5.0 tons per hectare by 1983. By 1995, with further development of hybrid rice technology, nationwide rice yields averaged more than 6.0 tons per hectare (Figure 10.1).

Hybrid rice has been grown from Liaoning (43° N latitude, cold temperate region) to Hainan (18° N, tropical region) and from Shanghai (125° E longitude)

The authors acknowledge assistance from the following individuals in editing this chapter and preparing a map of China's acreage under hybrid rice: William Lloyd, Kristie Bell, Jennie Shen, and Lang Deng at Pioneer Hi-Bred International and Michael Li at the University of Iowa.

Figure 10.1 Historical changes in rice yield per unit area, 1950–2008

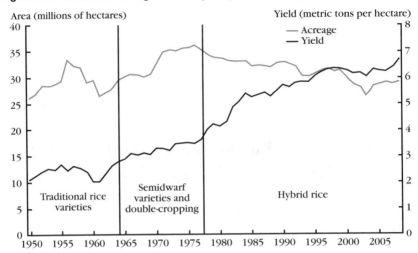

Sources: China, Ministry of Agriculture, various documents, and International Rice Research Institute rice statistics.

to Yunnan Province (95° E) (Yuan and Virmani 1988). There have been dramatic geographic differences in the adoption rates of hybrid rice. In 2003 and 2004 Hunan was the largest hybrid rice-growing province, with 3.0 million hectares (75 percent of the total rice acreage), followed by Jiangxi, with 2.0 million hectares (73 percent of the total rice acreage), and Sichuan, with 1.9 million hectares (91 percent of the total rice acreage).

In 2008 hybrid rice occupied about 63.2 percent of the total rice production area in China, or 18.6 million hectares out of 29.4 million hectares. The yield advantage of hybrid rice over traditional inbred rice ranged from 17.0 to 53.2 percent from 1976 to 2008—equivalent to a 30.8 percent higher average yield (China, Ministry of Agriculture, various documents). Hybrid rice has helped China to save rice land for agricultural diversification while reducing rural poverty and feeding an increasing number of people.

Dramatic geographic and regional differences in the area cultivated in hybrid rice can be attributed to each area's emphasis on agricultural research, adaptive research investments, and the share of rice in total agricultural output (Lin 1990). Regions with more resources dedicated to rice research have developed more rice hybrids and increased the area they devote to growing rice (Lin 1992).

This chapter traces the development and diffusion of hybrid rice technology, the impacts of the higher rice yields and output on Chinese rice farmers and consumers, and the lessons learned from the success of hybrid rice in China.

Development of Hybrid Rice Technology in China

The development of hybrid rice in China begins in the 1960s with exploratory efforts to identify the materials and techniques needed to further the technology. Four decades of research and development have led to widespread cultivation, a vibrant seed industry, and continuous yield gains. See Table 10.1 for a chronology of hybrid rice development in China. The key stages in the development of hybrid rice from 1973 to present are described in detail below.

Initiation and Early Stages (1964–76)

China's hybrid rice development effort began with the basic challenge of producing hybrid vigor or "heterosis" in rice, a phenomenon in which the resulting offspring are superior to their parents in one or more traits. The major obstacle to generating heterosis in rice is that it is a self-pollinated crop: the plant contains tiny florets with both male and female organs in the same floret and thus does not typically

Table 10.1 History of hybrid rice technological development in China, 1965–2006

Phase	Year	Development
Initiation and early stages	1964	Research on three-line hybrid rice was initiated.
	1970	Wild abortive rice was identified on Hainan Island in China.
	1973	Photoperiod- and thermosensitive genic male sterile material was identified.
	1974	First sets of three lines (A, B, and R lines) were developed for three-line system hybrid rice.
	1976	Hybrid rice commercialization started.
Technological improvements and large-scale commercialization	1977	Systematic hybrid rice seed production technique was developed.
	1983	Hybrid rice seed yielded more than 1.2 metric tons per hectare.
Progression to two-line hybrid rice system	1987	Hybrid rice seed yielded more than 2 metric tons per hectare; hybrid rice acreage reached more than 10 million hectares; National Two-Line System Hybrid Rice Program was established.
	1990	Hybrid rice acreage reached more than 15 million hectares.
	1995	Two-line hybrid rice system was developed.
Enhancement of hybrid rice heterosis	1996	National Super Rice Breeding program was initiated.
	1998	Hybrid rice seed yields more than 2.5 metric tons per hectare.
	2000	Super hybrid rice Phase I objective (10.5 metric tons per hectare) is achieved.
	2004	Super hybrid rice Phase II objective (12.0 metric tons per hectare) was achieved.
	2006	Work on super hybrid rice Phase III objective (13.5 metric tons per hectare) was initiated.

Source: Authors.

depend on separate parents to reproduce. Scientists thus needed a sterile line of rice that could be crossed with fertile lines to produce heterosis. Initially they pursued an approach that involved combining three distinct lines, or types, of rice. This approach was known as the three-line, or cytoplasmic male sterility, system. First a male sterile plant (called the A line) would be bred, or crossed, with a genetically identical plant that was not sterile (called the maintainer line, or B line). The resulting plant would be another male sterile plant that could be crossbred with a genetically distinct plant (called the restorer line, or R line). The offspring would be a plant with fertile seeds, commonly referred to as an F1 hybrid (Figure 10.2).

The effort began in 1964, when China initiated research on rice male sterility (Yuan 1966). Breeders observed that strong heterosis in rice occasionally occurred naturally in the field, but it was not until 1970 that a rice researcher working under Longping Yuan, the head of China's hybrid rice breeding program, identified the critical rice germplasm needed for the three-line hybrid rice. This wild abortive (WA) male sterile rice, found on China's Hainan Island, provided a new opportunity for the successful exploitation of rice heterosis (Li 1977).

From the beginning, collaboration and cooperation played major roles in the effort to breed hybrid rice. In 1970 Yuan's team freely distributed the critical WA material to 18 institutes in 13 provinces for collaborative research (Yuan 1973; Yuan 2001). Hunan Province established a collaborative hybrid rice breeding program in 1971, and in 1972 the China National Cooperative Hybrid Rice Research Group was established, led by the Chinese Academy of Agricultural Sciences (CAAS) and

Figure 10.2 The three-line (CMS) system for producing hybrid rice

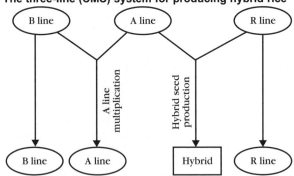

Source: Authors.

Notes: CMS means cytoplasmic male sterility; A line, male sterile plant; B line, maintainer line; R line, restorer line.

Hunan Academy of Agricultural Sciences (HAAS). This collaborative group comprised 14 provinces, municipalities, and autonomous regions.

In 1971 China's Ministry of Agriculture (MOA) selected three-line hybrid rice technology as one of 22 key research projects. This selection facilitated the development of a series of male sterile lines and corresponding maintainer lines from the WA germplasm in 1972. These male sterile lines became the mainstream breeding lines in large-scale commercial production from the mid-1970s to the late 1980s. The year after the establishment of the China National Cooperative Hybrid Rice Research Group in 1972, researchers from different provinces also identified several restorer lines.

In 1974, while working at the HAAS, Yuan developed the first *indica* rice hybrid, Nan-You 2, which initially demonstrated strong hybrid vigor. From 1972 to 1975, HAAS tested 87 hybrids, comparing them with the best inbred varieties. The best hybrids showed a 20–30 percent yield increase over the inbreds in large-scale testing (Lin and Yuan 1980).

In spite of these breakthroughs, however, yields for hybrid rice seed production in the early 1970s were low, sometimes reaching only 83 kilograms per hectare in experimental fields (Li and Xin 2000). Researchers spent two years studying hybrid rice seed genetics, environmental conditions, and water and fertilizer management, significantly increasing yields. In the winter of 1975, the largest group of hybrid rice researchers and technicians in China's history went to Hainan to produce hybrid rice seeds on more than 4,000 hectares of land. This massive seed production campaign enabled China to produce enough hybrid seeds for large-scale commercial production in 1976. This work enabled Chinese breeders to develop a systematic package of hybrid rice seed production techniques, which were further improved by Chinese rice agronomists after the late 1970s. The yield increase of hybrid seed production (Figure 10.3) ensured sufficient quantities of production materials for commercial hybrid rice production, while also lowering costs for seed businesses and farmers (Zhou and Peng 2005).

The MOA formally approved large-scale dissemination of hybrid rice at its 1976 Guangzhou meeting. At this early stage, Shan-You and Wei-You hybrids occupied the greatest acreage under *indica* hybrid rice in China's southern rice-growing region, while Li-You 57 and Zhong-Za 1 were the *japonica* rice hybrids that occupied the greatest acreage in China's northern rice-growing region (CAAS/HAAS 1991).

Technological Improvements and Large-Scale Commercialization of Three-Line Hybrid Rice (1977–85)

In 1982 the MOA established the National Hybrid Rice Advisory Committee, composed of more than 10 hybrid rice experts. After touring hybrid rice-growing regions two to three times annually, these experts provided strategic suggestions to

Figure 10.3 Commercial hybrid rice yield and hybrid rice seed yield in China, 1976–2008

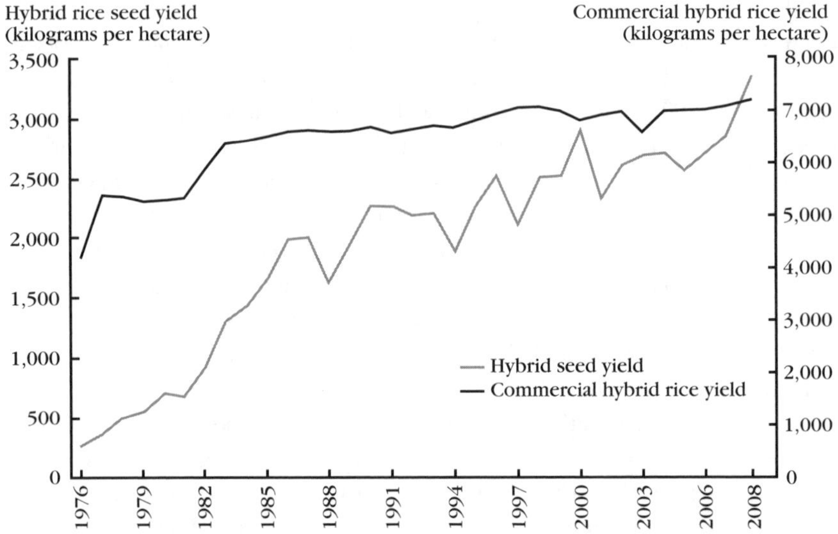

Source: CNHRRDC (2009).

the MOA about the development and use of hybrid rice technology (CAAS/HAAS 1991). In the early 1980s, China's hybrid rice still faced a number of problems that discouraged widespread adoption: they were all based on a narrow genetic base (all of the hybrid rice was based on genetic material from the original WA rice), they were susceptible to disease, they produced only one late crop a year, and they produced relatively few grains when cultivated for seed production. To address the low levels of adoption, hybrid rice breeders developed and released new rice hybrids to replace the first-generation, single-cropping *indica* hybrids. Wei-You 64, in particular, showed high yield potential and resistance to five major rice diseases and insect pests (Yuan and Virmani 1988). Breeders also developed early-cropping hybrids in 1987. The commercialization of new hybrids increased the hybrid rice acreage to 6.7 million hectares in 1983 and 8.4 million hectares in 1985.

In addition to developing improved rice hybrids, hybrid rice breeders developed more genetically diverse male sterile lines in the 1980s (Yuan and Virmani 1986; Cheng, Cao, and Zhan 2005), resulting in rice hybrids that were more resistant to disease and pests. This successful development of diverse parental lines led to the release and commercialization of more and more top-performing rice hybrids. The development of many strongly performing male sterile lines also helped scientists develop rice hybrids with higher-quality grain. In addition, new male sterile lines provided a solid foundation for high-yielding and cost-effective hybrid rice seed production. Breeders used inbred rice varieties from Southeast Asia as the main source of R lines, and advances in understanding the genetic mechanism for male fertility allowed breeders to develop more effective methods for R line breeding. The rapid and steady spread of hybrid rice in China in this period also stemmed from more effective seed production technology and the creation of hybrid rice seed businesses at all levels, from county to state, during the late 1970s.

Hybrid rice technology also revolutionized rice farming practices. Unlike inbred rice, hybrid rice requires different degrees of agronomic management depending on its stage of growth, so researchers needed to develop optimum field management practices. Agronomists thus made recommendations on nursery management, dry seeding practices, fertilizer usage, water management, and integrated pest management (Yan 1988; Lou and Mao 1994; Xu and Shen 2003). Adoption of these practices by Chinese rice farmers played an important role in the rapid growth of hybrid rice (CAAS/HAAS 1991).

Progression to a Two-Line Hybrid Rice System (1987–95)

In the 1980s researchers in China started trying to improve the breeding system for hybrid rice by developing a two-line hybrid rice system. In 1987 China's Ministry of Science and Technology (MOST) initiated the National Two-Line Hybrid Rice Research Program, which established a network of 16 research institutes and universities. The program was renewed in 1991, 1996, and 2001 with substantial funding support.

In a two-line system for hybrid rice (Figure 10.4), researchers cross a male sterile line with the restorer line (R line). They can induce sterility in the male in one of several ways: by changing the plant's environment, length of light exposure, temperature, or light exposure and temperature together. Thus a male sterile line can be an EGMS (environment-conditioned genic male sterile), PGMS (photoperiod-sensitive genic male sterile), TGMS (thermosensitive genic male sterile), or PTGMS (photoperiod- and thermosensitive genic male sterile) line. The R line can be any rice cultivar that restores fertility in the offspring when it is crossed with the male sterile line.

The two-line system for hybrid rice has a number of advantages over the three-line system. First, it is simple and effective because of the removal of the maintainer line from the three-line system. Second, the more diverse sources of male sterile lines increase the probability of developing a commercially sustainable hybrid (Yuan 1998a). Third, EGMS genes are more easily transferred into almost any rice lines. Fourth, using the EGMS line allows for production of more seed on less acreage,

Figure 10.4 The two-line system for producing hybrid rice

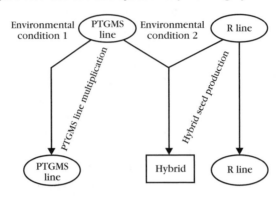

Source: Authors.
Note: PTGMS means photoperiod- and thermosensitive genic male sterile.

reducing hybrid seed costs. And finally, male sterility has no negative effects on the agronomic performance of the EGMS line itself and its resulting hybrids. The dependency of male sterility on temperature and light exposure, however, requires more attention from breeders and seed producers and imposes temporal and geographic limitations on hybrid seed production and EGMS multiplication (Li and Yuan 2000).

Chinese rice scientists found that not only light exposure but also temperature regulates the fertility of the initially dubbed PGMS (Lu et al. 1994). Temperatures above a certain level would induce male sterility, but the critical sterility-inducing temperature (CST) for any EGMS line was relatively high, meaning that pollen would likely remain fertile even in hot seasons. Thus hybrid rice seed production would not be reliable (Yuan 1998a). Consequently, scientists carried out sequential sowing experiments to determine the stable period for inducing sterility in a specific EGMS line at certain locations.

Another risk was that environmental conditions during the sterility-inducing period of seed production would not last long enough to ensure seed purity, because the sterility-inducing period needed to be at least 40 days long. Scientists thus carefully selected seed production locations based on the meteorological data and the CSTs of the specific PTGMS lines to ensure that the sterility-inducing period would be long enough.

In 1995 the two-line hybrid rice technology was successfully commercialized in China (Li and Yuan 2000; Yuan 2004). EGMS lines now offer more freedom in terms of producing hybrids with normal fertility, good rice grain quality, high yield potential, and improved disease resistance. In trials in southern China between 1998 and 2003, a number of two-line hybrids showed remarkable yield increases over the three-line hybrids (Yang, Cheng, and Shen 2004). Dozens of two-line hybrids have been released into commercial production. Between 2002 and 2008, two-line hybrid rice varieties went from occupying 2.8 million hectares to 3.5 million hectares and from 18 to 22 percent of China's hybrid rice acreage. By 2008, 11 percent of China's total rice acreage was being cultivated with commercial two-line hybrids.

As with previous advances, the successful commercialization of two-line hybrid rice was the result of concerted nationwide collaboration, with hundreds of rice scientists from 23 research institutes and universities working together. China's sophisticated three-tier seed system and four-level research extension network was also an important contributor to the success of hybrid rice development. The three-tier seed system included provincial seed companies that specialized in parental line purification, prefectural seed companies for A line multiplication, and county-level seed companies for F1 hybrid seed production. This system ensured the quantity and quality of hybrid rice seed supply for commercial production. The four-level extension network comprised county, commune, brigade, and production teams. This network proved efficient for rapid evaluation, selection, and adoption of hybrid rice, as well as for information diffusion (Lin and Pingali 1994). The government has established stations specializing in seed, agricultural technology, soil and fertilizer, and plant protection to encourage the dissemination of hybrid rice technology at the national, provincial, prefectural, and county levels. Every commune had one or more agricultural technicians to instruct farmers on hybrid rice technologies or new hybrids. This extension network played an important role in the rapid and large-scale commercialization of hybrid rice technology (Xu and Shen 2003).

Enhancement of Hybrid Rice Heterosis (1996–Present)

Research to improve hybrid rice heterosis has proceeded along several paths since the mid-1990s. One research area involves crossing subspecies of rice. Scientists have worked to develop top-performing commercial parental lines by crossing the three subspecies of rice—*indica, japonica,* and *javanica*—in various ways. The discovery of a "wide compatibility gene" by Japanese scientists presented a new opportunity for the use of *indica–japonica* heterosis (Ikehashi and Araki 1986). Currently the most efficient approach to hybrid breeding between subspecies is to cross *javanica* rice or an intermediate type (that is, one with a mixed pedigree of typical *indica* and *japonica*) with typical *indica* or *japonica* rice.

The "Super Hybrid Rice" Program in China. Rice is estimated to have a yield potential of 21.6 tons per hectare under natural conditions (Cao and Wu 1984), implying that large additional yield gains are possible. In 1996 China's MOA endorsed a "super rice" program proposed by Chinese rice scientists (Chen et al. 2007) and established yield targets for the program (Yuan 2003; Yuan 2008) (Table 10.2).

Table 10.2 Yield targets for China's "super hybrid rice" program, 1996–2015

Phase	Hybrid rice yield (metric tons per hectare)			Yield increase (percent)
	Early season	Late season	Single season	
1996	7.50	7.50	8.25	0
Phase I (1996–2000)	9.75	9.75	10.50	> 25
Phase II (2001–05)	11.25	11.25	12.00	> 45
Phase III (2006–15)	n.a.	n.a.	13.50	> 60

Source: Yuan (2008).
Notes: Grain yields must meet these targets in two consecutive years at two locations, each consisting of more than 6.67 hectares. n.a. means not applicable.

The following year the MOA proposed a three-phase "super hybrid rice breeding" strategy as part of the program (1996–2000, 2001–05, and 2006–15) (Yuan 1997). A key component of the strategy was development of an ideal rice plant type, proposed by Yuan, with the following traits: long, erect, narrow, V-shaped uppermost three leaves and large, uniform, and droopy panicles below a taller erect-leaved canopy (Yuan 1998b).

The Chinese government threw its weight behind the effort to breed super hybrid rice. In 1998 Premier Zhu Rongji provided RMB 10 million for the project. The same year, then–Vice Premier Wen Jiabao urged the MOA to enhance research on super rice. As a result, the project received another RMB 10 million in funding in 1999. At the same time, super hybrid rice breeding was included in China's 863 Hi-Tech Plan, a long-term plan for advancing science and technology. The MOST established several programs to support the extension and commercialization of the resulting new super hybrids, as did provincial governments (Quan 2005). In 2005 the Chinese central government included "the extension of super rice" in China's Central Document No. 1, thus further promoting the proliferation of super hybrid rice.

These investments paid off. Chinese rice scientists achieved the Phase I objective (10.5 tons per hectare) in 2000 and the Phase II objective (12 tons per hectare) in 2004, with yield increases of 25 and 45 percent, respectively, over the best hybrids before 1996. For example, the first two-line super rice hybrid, Liang-You-Pei-Jiu, demonstrated high commercial yields across multiple years and locations in large-scale rice production because of the good plant type and the remarkable level of intersubspecific heterosis. This two-line hybrid was the first to reach the Phase I yield goals, and the Chinese Rice Genome Sequence Initiative sequenced the genome of its parental lines (Yu et al. 2002; Quan 2005). The Phase II three-line hybrid Ming-You 8 (Fujian Province) and two-line hybrid P88s/0293 yielded more than 12 tons per hectare in Fujian Province and Hunan Province, respectively, surpassing the Phase II yield target (Yuan, Deng, and Liao 2004). By 2006 the MOA certified 34 rice hybrids as "super rice." Chinese rice breeders are currently working on Phase III super hybrid rice, with a large-scale yield objective of 13.5 tons per hectare.

The Roles of the State and Industry in Hybrid Rice Development. During the development and refinement of hybrid rice technology, the Chinese government provided critical support through both funding and policy (Table 10.3). The MOA and the MOST established national hybrid rice research programs in 1971 and 1972, respectively. Following up on this commitment, the central government leader, Hua Guofeng, provided strong and timely support consisting of special funding, more labor, and more resources for hybrid rice seed production following the successful cross-country demonstration of hybrid rice in 1975.

This support made possible the effort to produce large quantities of hybrid seed on Hainan Island in the winter of 1975. Thanks to this surge in seed production, the planting of hybrid rice increased from 373 hectares in 1975 to 0.14 million hectares in 1976. Also in the mid-1970s, China mounted an information campaign

Table 10.3 Chinese central governmental support for hybrid rice technology, 1971–2008

Year	Government agency	Action in support of hybrid rice technology
1971	MOAFF	Hybrid rice research was listed as 1 of 22 key research programs. A national collaborative research group was established, comprising 14 provinces, municipalities, and autonomous regions and led by the Chinese Academy of Agricultural Sciences and the Hunan Academy of Agricultural Sciences.
1972	MOST	Hybrid rice research was listed as a key national project. A national collaborative hybrid rice program was established with the participation of 19 provinces.
1975	MOAFF	RMB 8 million were invested for 4,000 hectares of hybrid rice seed production on Hainan Island.
1977–87	MOAFF/MOA	The second to sixth national hybrid rice project meetings were held.
1982	MOA	The National Hybrid Rice Advisory Committee was established.
1984	MOF	The Hunan Hybrid Rice Research Center was established.
1987	MOST	Two-line hybrid rice research was listed in the National Hi-Tech or "863" Plan and renewed in 1991, 1996, and 2001.
1994	MOST / Premier Fund	The China National Hybrid Rice R&D Center was established.
1998	Premier Fund	RMB 10 million went to super hybrid rice research.
1998	MOST	The Super Hybrid Rice Breeding Program was established.
2003	Premier Fund	RMB 10 million in special funding went to super hybrid rice research.
2006	Premier Fund	RMB 20 million went to hybrid rice research.
2008	MOST	Super hybrid rice research was listed as National Research Aid plan.

Source: CNHRRDC (2009).
Notes: MOA means Ministry of Agriculture; MOAFF means Ministry of Agriculture, Forestry, and Fishery; MOST means Ministry of Science and Technology.

to encourage the acceptance of hybrid rice technology. High-ranking officials at the central and local levels were assigned to monitor the progress of hybrid rice extension and commercialization.

The Chinese government also played a key role in developing an extensive hybrid seed industry. In 1976 the MOA convened a national meeting in Guangzhou to address problems in hybrid rice seed production. It was during this meeting (and five subsequent meetings held between 1977 and 1987) that the MOA initiated a long-term agricultural shift to hybrid rice breeding, seed production, and extension—a shift that eventually gave rise to the hybrid rice seed industry (CAAS/HAAS 1991).

The evolution of China's hybrid rice seed industry can be divided into three phases. The first phase, from 1978 to 1995, might be called the planned-economy phase. Shortly after the successful development of three-line hybrid rice, the China Seed Corporation was established under the Ministry of Agriculture and Forestry. The total number of hybrid rice seed companies grew quickly during this phase, reaching 1,500 in 1995. Among these companies, 600 county-level seed companies distributed seed supplies to more than 50,000 seed stations at the township level. Annual hybrid rice seed sales reached 0.6 million tons through this seed distribution system.

The next period, from 1996 to 2000, was the early market economy phase. After the 1995 Tianjin National Seed Conference, China's seed production sector underwent several changes. First, seed companies moved from traditional seed production to centralized large-scale production. Second, they shifted from regional to cross-regional seed distribution. Third, previously separate research and business units were integrated into single seed businesses composed of research, production, extension, and sales units.

The consolidation phase began in 2001 following the introduction of China's Plant Variety Protection regulations in 1997 and the publication of China's Seed Law in 2000. These two policies have contributed to the promotion of China's hybrid rice seed industry and led to seed market segmentation and business consolidation. As a result, a number of large hybrid rice seed businesses were established through consolidation in this phase, such as Longping Hi-Tech and Hefei Fengle Seed Company (Yuan, Deng, and Liao 2004).

Future Prospects for Hybrid Rice in China. The future of hybrid rice includes the expansion of hybrid *japonica* cultivation, marker-assisted selection to improve breeding, and the application of biotechnology to introduce new traits into hybrid rice. In terms of the first area of emphasis (*japonica*), there is a potential to further increase the heterosis of two-line and three-line *japonica* hybrid rice. *Japonica* occupies an increasing share of rice-growing acreage in China (Huang et al. 2002), and hybrid varieties have demonstrated strong heterosis and high grain yields. Nonetheless, scientists need to overcome problems of poor grain quality and limited disease resistance, seed production yield, and adaptability.

Molecular marker-assisted selection (MAS) has been shown to be a promising method of shortening the hybrid rice breeding cycle and increasing breeding efficiency for well-characterized traits. With the current low-cost genotyping technology, it will not be long before Chinese hybrid rice breeders routinely use MAS to improve parental lines and hybrids for biotic/abiotic resistance, grain quality, and other traits.

In addition, transgenic hybrid rice will become increasingly important. Chinese rice scientists have already developed and tested transgenic hybrid rice—with herbicide resistance and *Bacillus thuringiensis* (*Bt*) for resistance to rice stem borers—for environmental evaluations. Using a transgenic approach, Chinese rice scientists have developed parental lines that have conferred herbicide resistance, as well as bacterial blight resistance and stem borer resistance on restorer lines using *Bt*. Other genes may be transferred into hybrid rice in the near future, such as genes for drought tolerance, nitrogen use efficiency, and disease resistance.

Social and Economic Impacts of Hybrid Rice

Food Security

The development and diffusion of hybrid rice in China has been an important counterbalance to China's dual pressures of increasing population and decreasing arable land. Over the past 20 years, China's arable land has decreased by an average of 0.2 million hectares annually. From 2004 to 2006, 0.87 million hectares of arable land acreage were removed from agricultural production and returned to forestry (Xue 2007). At the same time, China's population increased from 1.10 billion in 1987 to 1.32 billion in 2007. Because China cannot depend on the relatively small world rice market (about 25 million tons from 2002 to 2003) to meet the consumption needs of its large and growing population, the increase in rice production brought on by hybrid rice has been critical to the country's food security.

Cultivation of hybrid rice enabled China to increase yields by 91.4 percent between 1976 and 2008, from a national average of 3.5 tons per hectare in 1976 to 6.7 tons per hectare in 2008. Total annual rice production rose by 56.7 percent over the same period, from 125.7 million tons to 197 million tons. Put another way, the increase in total annual rice production due to the adoption of hybrid rice in China is close to the total annual rice production from a single high-production province. In cumulative terms, the increases in output attributable to yield gains from hybrid rice totaled 608 million tons between 1976 and 2008—increases that have helped China feed an extra 60 million people every year.

Furthermore, the intensive labor required for hybrid rice seed production has increased both rural employment opportunities and farmers' incomes. Hybrid rice technology has generated more than 100,000 jobs related to hybrid rice research, extension, and seed production and has indirectly generated 10 million jobs in rural areas (Yuan, Deng, and Liao 2004).

Economic Impacts

Hybrid rice has been critical to improving the economic efficiency of China's agriculture. Lin and Pingali (1994) report that hybrid rice had an approximate 15 percent yield advantage over conventional inbreds. According to another study of 209 farms in Jiangsu Province, hybrid rice showed a 37 percent increase in net returns per hectare compared with conventional inbred rice, a 26 percent increase in labor returns, a 12 percent increase in nonlabor returns, and a 30 percent increase in the rate of net returns to total cost.[1]

He and Flinn (1989) confirmed that the higher yields of hybrid rice were largely due to technical innovation instead of differences in management. They found that in Jiangsu Province *indica* rice hybrids were more profitable than conventional rice varieties because of higher returns to both labor and nonlabor inputs. In another small-scale, single-cropping rice-growing area in Hubei Province, experimental results showed that the hybrid Shan-You 2 yielded 2.4 tons per hectare (15 percent) more than a popular inbred, corresponding to an increase of RMB 382 per hectare. The same study found that inbred rice required more water use and labor input. The total input (including labor and nonlabor) was RMB 1,290.3 per hectare for hybrid rice and RMB 1,317.8 per hectare for inbred rice. For farmers, the net return to investing in hybrid rice was RMB 21.9 per hectare compared with RMB 16.2 per hectare for inbred rice—a difference of 35 percent (Tao 1987).

Despite these high rates of return, at least one study contends that the rapid diffusion of hybrid rice in China resulted from pressure by the Chinese government rather than from the economic superiority of hybrid rice. Lin (1991) studied 952 observations in 101 counties of Hunan Province from 1976 to 1986. In this early extensive study, the adoption rate of hybrid rice technology was tested against county-level, time-series data. The study concluded that in the early stage of the collective system, the Chinese government pressured farmers to adopt hybrid rice without consideration of its profitability.

In line with these findings, Lin (1991) reported that some farmers returned to conventional rice varieties in the 1980s. One possible explanation is that the phenomenon was regional in nature and may have occurred only in Hunan Province, where the study was conducted. Another reason may have been the limited availability of hybrid varieties, especially for the early-season rice hybrids and the hybrids with strong biotic and abiotic resistance traits. With the transition from the collective agricultural production system before 1979 to the "household responsibility" system after 1981, it is also possible that farming decisions became less driven by government pressure. Under the household responsibility system, the decision to cultivate hybrid rice or inbred rice shifted from production teams to individual farmers, possibly providing farmers with more opportunity for independent decisionmaking.

As Lin (1991) points out, government pressure may have been a factor in the diffusion of hybrid rice, but given the limitations of Lin's study and the continued diffusion of hybrid rice under the household responsibility system, it is feasible to assume that the economic return to hybrid rice was a more powerful factor in its adoption than was government pressure. Indeed, statistical data from 1976 to 2008 (Figure 10.5) show a constant increase in hybrid rice acreage from 1982 to 1987 after the full adoption of the household responsibility system. The increased acreage in hybrid rice from 1976 to 2008 suggests that the large-scale expansion of hybrid rice was a success not only because of Chinese government intervention but also, and especially, because of the proven high-yielding performance of hybrid rice (see Figures 10.3 and 10.5).

The economic sustainability of China's efforts to develop and disseminate hybrid rice is an important topic for consideration. In the early stages China's advances in hybrid rice depended on public subsidies. For example, after realizing the great potential of hybrid rice from multilocation yield trials, the Chinese government invested RMB 8 million for 4,000 hectares of hybrid rice seed production on Hainan Island in the winter of 1975. This investment resulted in the largest seed production campaign in China's rice-farming history and made possible the rapid expansion of hybrid rice in 1976. Similarly, the government granted tax concessions to seed companies and provided subsidies when seed yield was low during the late

Figure 10.5 Hybrid rice acreage in China, 1976–2008

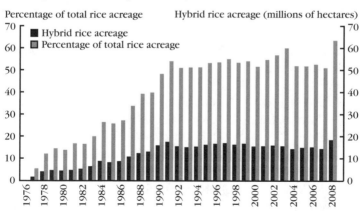

Source: CNHRRDC (2009).

1970s (Yuan 2002). Yet these subsidies and concessions were arguably essential to the long-term success of China's efforts in hybrid rice.

Environmental Sustainability

With pressure to divert an increasing amount of land for nonagricultural use, the threat of food insecurity is becoming severe in China. The country also faces many ecological challenges to agricultural production, including water shortages, soil erosion, land desertification, and environmental pollution. These challenges have irreversibly damaged rice acreage in China. The high-yield potential of hybrid rice enables China to produce more rice on less land and provides opportunities for crop diversification. Specifically, the gains in yield and output associated with hybrid rice allowed China to decrease its rice-growing acreage by 14.5 percent, from about 34.4 million hectares in 1978 to 29.4 million hectares in 2008. Hybrid rice acreage accounted for 63 percent (or 18.6 million hectares) of the total area under rice cultivation in 2008. In addition, the seeding rate of hybrid rice per hectare is much lower than that of conventional inbred rice, and the transplanting density is 40–50 percent less than for inbred rice. As a result, fewer seeds are needed and less land is needed to produce seeds (He et al. 1988).

The rapid extension of super hybrid rice in the past 10 years has greatly boosted the yield per unit of land in China. For example, the average yield of super hybrid rice reached 9.9 tons per hectare on 67 hectares of rice fields in Jin Hua City of Zhejiang Province, 10.5 tons per hectare on 800 hectares of rice fields in Xu Pu County of Hunan Province, and 9.0 tons per hectare on rice fields of Guizhou Province under adverse natural conditions. Yuan proposed to the Hunan provincial government and then to the central government the Planting Three Producing Four program, which aimed to produce the same amount of rice from three units of land as had been previously produced from four units of land (Yuan 2007). By enabling China to produce enough food using less land, this program benefited Hunan, Sichuan, and several other provinces tremendously. So far, hybrid rice technology has helped China to save more than 6 million hectares of land each year for agricultural diversification and nonagricultural use. Also, the decrease in the total rice acreage minimizes the emission of greenhouse gases such as methane and nitrous oxide (Tran and Nguyen 1998). In addition, as a result of heterosis, hybrid rice generally produces more rice straw, which has been used as manure to improve soil texture and fertility.

Hybrid rice has also displayed better adaptability to stress-prone environments than has conventional inbred rice. Because hybrid rice has a vigorous root system, strong culm, thick leaves, and high photosynthetic efficiency, it has a significant yield advantage over conventional rice (CAAS/HAAS 1991). Further, hybrid rice is more adaptable to various climatic (tropical, subtropical, and temperate), topographical (plains, coastal areas, and hilly regions), and ecological (irrigated, drought-prone, and upland) conditions. Hybrid rice has been grown in single-cropping and double-cropping regions. It was reported that hybrid rice was grown with 50 percent less water in irrigated fields but with no significant yield loss (Yuan, Virmani, and Mao 1989), a clear advantage over inbred rice in drought-prone regions. In Liaoning Province, hybrid rice showed a tolerance to soil alkalinity and so was grown in the coastal areas (Yuan, Virmani, and Mao 1989). Selected rice hybrids demonstrated stronger tolerance than inbred rice to wind and flooding in Guangdong Province in the south, which is subject to frequent typhoons and heavy rains (Lin 1989).

Recent developments in hybrid rice have also reduced the degree of vulnerability to rice disease and pests. In the early stage of hybrid rice technological development, WA was the only source of male sterile cytoplasm—a situation that presented potential vulnerabilities from disease or insect epidemics. The diversification of male sterile lines and hybrids helped reduce the risk of epidemics.

Consumer Acceptance

Because China's hybrid rice development efforts before the 1980s were focused strictly on food security (and thus on increasing the quantity of rice production), early-stage hybrid rice generally showed high yields but poor grain quality. It was not until after China's economic reforms and growth took off in the early 1980s that researchers turned their attention to improving the quality of rice. After the Chinese government liberalized the retail rice market in 1993, hybrid rice breeders began developing hybrids with fine grain quality in addition to high yields and multiple resistances to biotic and abiotic stressors. As a result of this breeding effort to produce superior-quality hybrid rice, China developed and released some top-quality male sterile lines and hybrids beginning in the 1990s.

Lessons Learned and Issues Going Forward

Chinese Experiences in the Development of Hybrid Rice Technology

The following lessons learned throughout China's more than 40 years of technological development and improvement of its hybrid rice programs will provide a valuable model for other rice-growing countries to use in developing their own hybrid rice programs.

Supportive Institutions and Policies. The Chinese government's support and commitment were key factors in the success of the country's hybrid rice program. Well-defined policies led to financial support for research, seed production, and extension agencies; guidelines or regulations for hybrid rice seed production; and seed certification standards and distribution. Government policies and standards also made hybrid rice cultivation and seed production attractive, profitable, and sustainable. In addition, the Chinese government subsidized commercial hybrid rice seed production in the early years to ensure a sufficient supply of affordable, high-quality hybrid rice seed to farmers. Farmers also had access to government-subsidized fertilizers and pesticides, which helped ensure their use of appropriate farming inputs in hybrid rice production (Lou and Mao 1994).

A high-ranking scientist (Longping Yuan) with the requisite knowledge, capability, and authority was designated as the national coordinator of China's hybrid rice program. The coordinator worked with the technological steering committee and coordinated and regularly monitored the progress of the research, seed production, and other technology-related programs in the nation. In addition, full-time researchers and extension workers were devoted to the generation and uptake of hybrid rice technology. Hybrid rice breeders played a leading role in the development of China's hybrid rice technology, producing more than 1,000 parental lines and more than 300 large-scale commercialized rice hybrids in the past 40 years. Among these hybrids, 10 three-line rice hybrids have occupied more than 667,000 hectares since 1990 (Qing and Ai 2007).

The Chinese national government and the Hunan provincial government established the CNHRRDC (China National Hybrid Rice Research and Development Center, previously the Hunan Hybrid Rice Research Center), a research institute with expertise in multiple disciplines related to hybrid rice, facilities, and equipment. Close collaboration and strong links among public research institutes, seed production businesses, and extension agencies created an effective network with clearly defined roles and responsibilities. China has established an efficient and coordinated infrastructure for breeding, seed production, certification, and distribution.

Several government agencies and research organizations, along with academia, organized comprehensive training programs for breeders, seed producers, extension workers, and commercial production farmers. Demonstration of the yield improvements generated by hybrid rice has been essential for the successful extension of hybrid rice technology. Once convinced by the performance of hybrid rice, farmers were trained in seed production and high-yielding cultivation. In China's southern rice-growing regions, more than one-fourth of farmers were trained in hybrid rice technology, in addition to more than 400,000 agricultural technicians (CAAS/HAAS 1991). The government also raised awareness about hybrid rice among the general population through workshops, technical briefings, frontline demonstrations, field tours, and mass media campaigns.

The government at the state, provincial, prefectural, and county levels established a reward and recognition mechanism for hybrid rice researchers, seed producers, and extension personnel (Yuan 1993). For example, in 1981 China awarded the *Indica* Three-Line System Hybrid Rice Technology program with its first Extraordinary-class National Invention Prize (Li and Xin 2000). Others who have made significant contributions to hybrid rice technology since the 1980s have also been rewarded at state and provincial levels.

Technological Generation and Uptake. Effective seed production infrastructure has been the foundation for China's success in hybrid rice technology. China's MOA established minimum seed quality standards in 1985, thereby ensuring the long-term genetic purity of parental lines and hybrid seeds in addition to the quantity of hybrid seed supply.

China's experience shows that different ecological rice-growing regions or markets need rice hybrids with different levels of quality and biotic and abiotic resistance. Rice hybrids that perform well in one region may not be useful in another rice-growing region. Therefore, regional research and extension infrastructures were established for hybrid rice breeding and commercialization. Before commercial release, a new hybrid must pass a multilocation and regional trial for two seasons and a production demonstration in a farmer's field.

Pilot regions were identified for each ecogeographic region to demonstrate the new hybrid rice technology. Extensive on-farm demonstrations, large-scale public awareness campaigns, concerted multidisciplinary approaches, and national and regional funding support were organized. For example, in the mid-1970s demonstration plots comparing hybrid rice with inbred rice varieties were set up in target areas to introduce the concept of hybrid rice to traditional farmers in China.

Key Issues for Future Hybrid Rice Production

In spite of the success of hybrid rice in past decades, this technology faces a number of challenges. Increasing urbanization has significantly changed rural social structures: the majority of the educated young and middle-aged labor force has been moving to metropolitan areas with more career and financial opportunities. Moreover, the greatest rural labor flow into urban regions has come from the largest rice-producing provinces in China. The diminished quality and quantity of the remaining rural labor force has made extension and production of hybrid rice a more difficult endeavor. For example, seed producers must have a certain level of education and a good understanding of seed production techniques. China therefore needs to establish a new system for agricultural innovation, technical extension, and social services to accelerate research innovation and advance its hybrid rice technology (Xu and Shen 2003).

Another challenge pertains to environmental stressors. The land and water resources on which farming depends have become more fragile. The irrigation water reservoir system in China is vulnerable due to the poor maintenance over time, and long-term application of chemical fertilizers has contributed to soil erosion.

Also, hybrid rice has gradually lost resistance to diseases and insects. Therefore, Chinese hybrid rice breeders are continuously working to develop hybrids with multiple disease and insect resistances (Yuan, Virmani, and Mao 1989; Zhou, Deng, and Li 2008). With the advance of biotechnology, scientists should immediately seek to develop new traits in hybrid rice to overcome the challenges from increasing biotic or abiotic pressure, increasing fertilizer application, the ever-decreasing water supply, and more severe droughts from global warming. Such an effort will enhance the sustainability of hybrid rice technology and reduce environmental pollution from fertilizers and pesticides.

Note

1. Hybrid rice produced as seed (as opposed to grain) provided farmers with even greater returns —3.8 times higher for net returns per hectare and 2.1 times for labor returns (He et al. 1988).

References

CAAS/HAAS (Chinese Academy of Agricultural Sciences / Hunan Academy of Agricultural Sciences). 1991. *The development of hybrid rice in China.* Beijing: China Agricultural Press.

Chen, L. Y., Y. H. Xiao, W. B. Tang, and D. Y. Lei. 2007. Prospect and practice of three-procedure breeding on super hybrid rice. *Chinese Journal of Rice Science* 21 (1): 90–94.

Cheng, S. H., L. Y. Cao, and X. D. Zhan. 2005. *Techniques for hybrid rice seed production.* Beijing: Jindun Publishing House.

CNHRRDC (China National Hybrid Rice Research and Development Center). Various years. Various official documents. Changsha, China.

He, G. T., and J. C. Flinn. 1989. Economic efficiency of hybrid and conventional rice production in Jiangsu Province, China. In *Progress in irrigated rice research.* Selected papers and abstracts from the International Rice Research Conference, September 21–25, 1987, in Hangzhou, China. Manila, Philippines: International Rice Research Institute.

He, G. T., X. G. Zhu, H. Z. Gu, and J. S. Zhang. 1988. The use of hybrid rice technology: An economic evaluation. In *Hybrid rice.* Proceedings of the International Symposium on Hybrid Rice, October 6–10, 1986, in Changsha, China. Manila, Philippines: International Rice Research Institute.

Huang, J., S. Rozelle, R. Hu, and N. Li. 2002. China's rice economy and policy: Supply, demand, and trade in the 21st century. In *Developments in the Asian rice economy,* ed. M. Sombilla, M. Hossain, and B. Hardy. Proceedings of the International Workshop on Medium- and Long-term Prospects of Rice Supply and Demand in the 21st Century, December 3–5, 2001, in Los Baños, Philippines. Manila, Philippines: International Rice Research Institute.

Ikehashi, H., and H. Araki. 1986. Genetics of F1 sterility in remote crosses of rice. In *Rice genetics*. Proceedings of the International Rice Genetics Symposium, May 27–31, 1985, in Los Baños, Philippines. Manila, Philippines: International Rice Research Institute.

Li, J., and Y. Xin. 2000. Dedication: Longping Yuan—Rice breeder and world hunger fighter. *Plant Breeding Reviews* 17: 1–15.

Li, J., and L. P. Yuan. 2000. Hybrid rice: Genetics, breeding and seed production. *Plant Breeding Reviews* 17: 15–158.

Li, P. H. 1977. How we studied hybrid rice. *Acta Botanica Sinica* 19: 7–10.

Lin, J. Y. 1990. Hybrid rice innovation in China: A study of market demand induced technological innovation in a centrally planned economy. Working Paper 604. Los Angeles: University of California–Los Angeles, Department of Economics.

————. 1991. The household responsibility system reform and the adoption of hybrid rice in China. *Journal of Development Economics* 36: 353–372.

————. 1992. Hybrid rice innovation in China: A study of market-demand induced technological innovation in a centrally planned economy. *Review of Economics and Statistics* 74 (1): 14–20.

Lin, J. Y., and P. L. Pingali. 1994. Economic assessment of the potential for hybrid rice in tropical Asia: Lessons from the Chinese experience. In: *Hybrid rice technology—New developments and future prospects*, ed. S. S. Virmani. Selected papers from the International Rice Research Conference. Manila, Philippines: International Rice Research Institute.

Lin, R. 1989. Increasing rice production through extension of hybrid rice. *Hybrid Rice* 4: 1–2.

Lin, S. C., and L. P. Yuan. 1980. Hybrid rice breeding in China. In *Innovative approaches to rice breeding*. Selected papers from the 1979 International Rice Research Conference. Manila, Philippines: International Rice Research Institute.

Lou, X. Z., and C. X. Mao. 1994. *Hybrid rice in China—A success story*. Bangkok: Asia-Pacific Association of Agricultural Research Institutes.

Lu, X. G., Z. G. Zhang, K. Maruyama, and S. S. Virmani. 1994. Current status of two-line method of hybrid rice breeding. In *Hybrid rice technology—New developments and future prospects*, ed. S. S. Virmani. Selected papers from the International Rice Research Conference. Manila, Philippines: International Rice Research Institute.

Qing, X. G., and Z. Y. Ai. 2007. Consistent independent innovation to achieve a new leap in research and development of hybrid rice. *Hybrid Rice* 22 (1): 1–5.

Quan, Y. M. 2005. An overview of demonstration and extension of pioneer super hybrid rice Liang-You-Pei-Jiu. *Hybrid Rice* 20 (3): 1–5.

Riley, N. 2004. China's population: New trends and challenges. *Population Bulletin* 59 (2): 40.

Tao, S. K. 1987. Evaluation of the economic benefit of single-cropping hybrid rice in comparison with inbred rice. *Hybrid Rice* 1: 10–11.

Tran, D. V., and V. N. Nguyen. 1998. Global hybrid rice: Progress, issues and challenges. *International Rice Commission Newsletter* 47: 16–27.

Xu, K. D., and G. F. Shen. 2003. Promoting Chinese rice production through innovative science and technology. In *Rice sciences: Innovations and impact for livelihood,* ed. T. W. Mew, D. S. Brar, S. Peng, D. Dawe, and B. Hardy. Proceedings of the International Rice Research Conference, September 16–19, 2002. Beijing: International Rice Research Institute.

Xue, L. M. 2007. Yuan Longping and food security. *China Rural Science and Technology* 8: 48–49.

Yan, Z. D. 1988. Agronomic management of rice hybrids compared with conventional varieties. In *Hybrid rice.* Proceedings of the International Symposium on Hybrid Rice, October 6–10, 1986, in Changsha, China. Manila, Philippines: International Rice Research Institute.

Yang, S. H., B. Y. Cheng, and W. F. Shen. 2004. Progress of hybrid rice breeding in southern China. *Hybrid Rice* 19 (5): 1–5.

Yu, J., S. Hu, J. Wang, G. K. Wong, S. Li, B. Liu, Y. Deng, L. Dai, Y. Zhou, X. Zhang, M. Cao, J. Liu, J. Sun, J. Tang, Y. Chen, X. Huang, W. Lin, C. Ye, W. Tong, L. Cong, J. Geng, Y. Han, L. Li, W. Li, G. Hu, X. Huang, W. Li, J. Li, Z. Liu, L. Li, J. Liu, Q. Qi, L. Liu, L. Li, T. Li, X. Wang, H. Lu, T. Wu, M. Zhu, P. Ni, H. Han, W. Dong, X. Ren, X. Feng, P. Cui, X. Li, H. Wang, X. Xu, W. Zhai, Z. Xu, J. Zhang, S. He, J. Zhang, J. Xu, K. Zhang, X. Zheng, J. Dong, W. Zeng, L. Tao, J. Ye, J. Tan, X. Ren, X. Chen, J. He, D. Liu, W. Tian, C. Tian, H. Xia, Q. Bao, G. Li, H. Gao, T. Cao, J. Wang, W. Zhao, P. Li, W. Chen, X. Wang, Y. Zhang, J. Hu, J. Wang, S. Liu, J. Yang, G. Zhang, Y. Xiong, Z. Li, L. Mao, C. Zhou, Z. Zhu, R. Chen, B. Hao, W. Zheng, S. Chen, W. Guo, G. Li, S. Liu, M. Tao, J. Wang, L. Zhu, L. Yuan, and H. Yang. 2002. A draft sequence of the rice genome (*Oryza sativa* L. ssp. *indica*). *Science* 296: 79–92.

Yuan, L. P. 1966. A preliminary report on male sterility in rice. *Science Bulletin* 4: 32–34.

———. 1973. Breeding for cytoplasmic male sterile lines via wild abortive cytoplasm. *Hunan Agricultural Science* 4: 1–4.

———. 1993. China's experience in the development of hybrid rice research programme. In *Hybrid rice: Food security in India,* ed. B. R. Barwale. New Delhi: Macmillan India.

———. 1998a. Hybrid rice breeding in China. In *Advances in hybrid rice technology,* ed. S. S. Virmani, E. A. Siddiq, and K. Muralidharan. Proceedings of the 3rd International Symposium on Hybrid Rice, November 14–16, 1996, in Hyderabad, India. Manila, Philippines: International Rice Research Institute.

———. 1998b. Hybrid rice development and use: Innovative approach and challenges. *International Rice Commission Newsletter* 47: 7–15.

———. 2001. Develop hybrid rice depending on the innovation of science and technology to ensure the food security of our country. *Review of China Agricultural Science and Technology* 3 (2): 54–56.

———. 2002. Recent progress in the development of hybrid rice in China. In *Adoption of hybrid rice in Asia—Policy support.* Proceedings of the workshop on policy support for rapid adoption of hybrid rice on large-scale production in Asia, May 22–23, 2001, in Hanoi, Vietnam. Rome: Food and Agriculture Organization of the United Nations.

———. 2003. Recent progress in breeding super hybrid rice in China. In *Hybrid rice for food security, poverty alleviation, and environmental protection,* ed. S. S. Virmani, C. X. Mao, and B. Hardy. Proceedings of the 4th International Symposium on Hybrid Rice, May 14–17, 2002, in Hanoi, Vietnam. Manila, Philippines: International Rice Research Institute.

———. 2004. Hybrid rice technology for food security in the world—Proceedings of the FAO rice conference: Rice is life. *International Rice Commission Newsletter* 53: 24–25.

———. 2007. Proposal of implementing the "planting-three-produce four" high-yielding project on super hybrid rice. *Hybrid Rice* 22 (4): 1.

———. 2008. Progress of super hybrid rice breeding. *China Rice* 1: 1–3.

Yuan, L. P., and S. S. Virmani. 1986. Status of hybrid rice research and development. In *Hybrid rice,* ed. W. H. Smith, L. R. Bostian, and E. C. Cervantes. Proceedings of the International Symposium on Hybrid Rice held by the International Rice Research Institute, October 6–10, 1986, in Changsha, Hunan, China. Manila, Philippines: International Rice Research Institute.

———. 1988. Status of hybrid rice research and development. In *Hybrid rice.* Proceedings of the International Symposium on Hybrid Rice, October 6–10, 1986, in Changsha, China. Manila, Philippines: International Rice Research Institute.

Yuan, L. P., Q. Y. Deng, and C. M. Liao. 2004. Current status of industrialization of hybrid rice technology. In *Report on China's development of biotech industries.* Beijing: Chemical Industry Publishing.

Yuan, L. P., S. S. Virmani, and C. X. Mao. 1989. Hybrid rice: Achievements and outlook. In *Progress in irrigated rice research.* Selected papers and abstracts from the International Rice Research Conference, September 21–25, 1987, in Hangzhou, China. Manila, Philippines: International Rice Research Institute.

Zhou, C. S., and J. M. Peng. 2005. The development of hybrid rice seed production in China. In *Hybrid rice and world food security,* ed. L. P. Yuan and J. M. Peng. Beijing: China Science and Technology Press.

Zhou, Y. J., Q. M. Deng, and P. Li. 2008. Improvement and resistance analysis of four rice restorers by MAS. *Molecular Plant Breeding* 6 (3): 480–490.

Chapter 11

Pearl Millet and Sorghum Improvement in India

Carl E. Pray and Latha Nagarajan

Sorghum and millets are annual grasses found mainly in arid and semiarid regions of the world that produce small-seeded grains and are often cultivated as cereals.[1] Widely grown in Africa and Asia, sorghum and millets currently constitute an estimated 11.4 percent of the cereal area harvested and 4.1 percent of the total output of world cereals produced (FAO 1995; FAOSTAT 2007). They can be used as either grain or forage. They are resistant to drought, have a short duration (typically three to four months from planting to harvest), and can be grown in a wide range of soil types. Sorghum and millets are often recommended for farmers operating in harsh environments where other crops do poorly, because they can be grown with limited rainfall and often without application of any fertilizers or other inputs. Moreover, these crops constitute the principal source of energy, protein, vitamins, and minerals for millions of the poorest people in these regions.

Public research organizations, international research centers, and the private sector have played an important role in improving sorghum and millet in India. Their work has led to a doubling of average yields of sorghum and pearl millet over the past four decades (albeit with significant geographic variation) and the diffusion of both high-yielding varieties (HYVs), which now account for 80 percent of the sorghum and pearl millet area, and privately bred hybrids (particularly of pearl millet). A succession of more productive and disease-resistant varieties has increased farmers' yields and improved the livelihoods of about 6 million millet-growing households and 3 million sorghum-growing households. This chapter examines the roles of government, private companies, and farmers in making this success possible.

Production and Consumption

Sorghum, or *jowar* (*Sorghum bicolor*), is predominantly grown in the arid and semi-arid regions of India (Maharashtra, Andhra Pradesh, Karnataka, and Tamilnadu), areas with as little as 400 to 500 millimeters of rainfall a year. As many as 100 distinct cultivars of sorghum have been identified in the sorghum-growing regions of India, which is the unique center of origin of the post–rainy season (*rabi*) varieties of sorghum. After sorghum, pearl millet, or *bajra* (*Pennisettum typhoides*), is the next most important millet crop in India in terms of area and production. India is also considered the secondary center of origin of pearl millet, with many distinct cultivars growing throughout the country. Given sufficient rainfall (typically just 500 to 600 millimeters a year), pearl millet tends to be preferred over sorghum and is grown extensively in the dry western and northern regions of the country (Gujarat, Rajasthan, and Haryana). In other parts of India it is grown as a winter crop. Pearl millet is primarily a fodder crop in the western part of Rajasthan and Gujarat—especially during the summer, when green fodder is scarce. In arid regions of India, pearl millet is a major source of food.

Production Trends

India is a major producer of sorghum and other millets. Sorghum and pearl millet account for nearly 5 percent (each) of the total cropped area, but this area is concentrated primarily in the arid and semiarid regions of India, where nearly 60 percent of the rural population lives (ICAR 2006).[2] India ranks second worldwide in both area and production of sorghum and millets (FAOSTAT 2007). Although the area sown with sorghum and millets has steadily declined over the past four decades, yields have increased, especially since the mid-1970s.

Distinct trends in production, area, and yield levels of sorghum and millet define three general periods: postindependence (1947–65), public-supported growth (1966–85), and private-sector-driven growth (1986–present).[3] The postindependence period witnessed a major increase in production of sorghum and pearl millet (Figure 11.1). Official data show that the area and yields rose substantially for both crops (Figures 11.2 and 11.3). Official statistics suggest that yields were a major contributor to increased production of sorghum and millet during the postindependence period; however, this conclusion is based on very low estimated yields in the first few years after independence, which may be unreliable (Evenson and Pray 1991).

During the second period (1966–85), sorghum production increased rapidly while pearl millet production stagnated. Although the total area gradually declined for both crops, yields rose for sorghum but not for pearl millet. By the mid-1960s, new hybrids of sorghum and pearl millet had been developed as part of the All India Coordinated Crop Improvement Projects. It appears that the sorghum hybrids—particularly the CSH (coordinated sorghum hybrids) series from the Indian gov-

Figure 11.1 Production trends for pearl millet and sorghum in India, 1950–2008

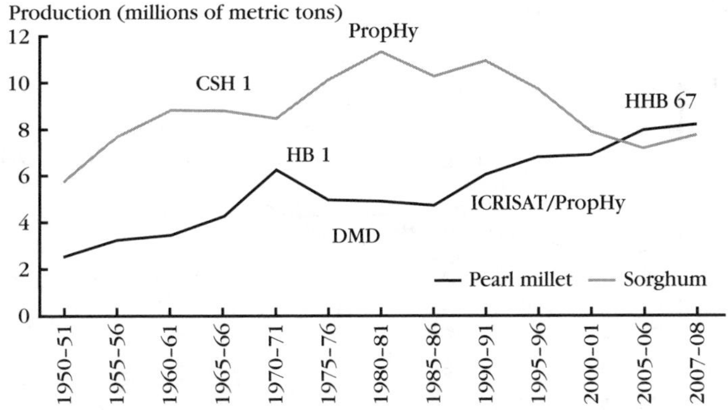

Source: Data from India, Ministry of Agriculture (2008).
Notes: Production amounts are three-year averages. For the year 2007–08, actual data were used
to denote the current trend. CSH 1 means coordinated sorghum hybrid; DMD means downy mildew
(*Sclerospora graminicola*) disease; HB 1 and HHB 67 are the names of hybrid crops; ICRISAT/PropHy
means International Crops Research Institute for the Semi-Arid Tropics PropHy; PropHy is the name
of a hybrid crop.

Figure 11.2 Acreage trends for pearl millet and sorghum in India, 1950–2008

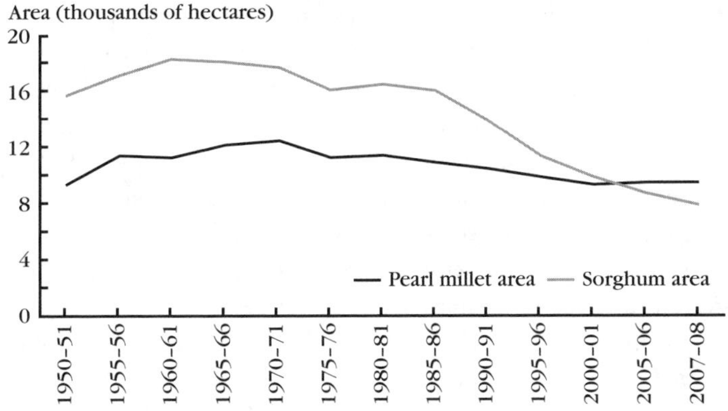

Source: Data from India, Ministry of Agriculture (2008).
Note: Production amounts are three-year averages. For the year 2007–08, actual data were used to
denote the current trend.

Figure 11.3 Yield trends for pearl millet and sorghum in India, 1950–2008

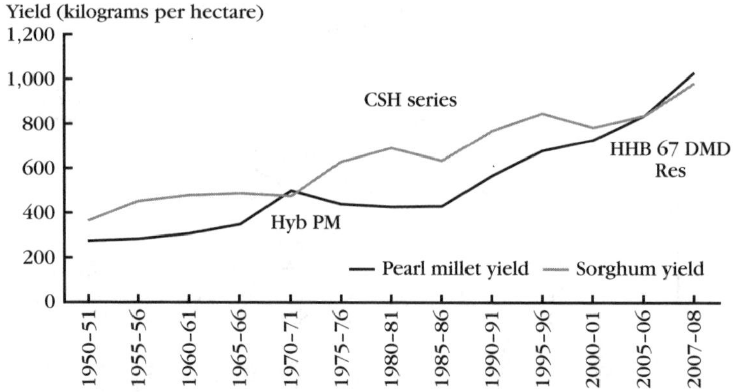

Source: Data from India, Ministry of Agriculture (2008).
Notes: Production amounts are three-year averages. For the year 2007–08, actual data were used to denote the current trend. CSH means coordinated sorghum hybrid; DMD Res means downy mildew disease–resistant; HHB 67 is the name of a hybrid crop; Hyb PM means hybrid pearl millet.

ernment research system, which are short-duration, high-yielding types—were successful in increasing yields. One factor that may have held down pearl millet yields during this period was the recurrence of downy mildew (see Figure 11.3) (Pray et al. 1991; McGaw 2001; Breese et al. 2002).

In the last period, from 1986 to the present, the production trends have been reversed: sorghum production has declined, while pearl millet production has increased. The area declined substantially for both crops but much more for sorghum, and pearl millet yields increased more did than sorghum yields. By 1986 the International Crops Research Institute for the Semi-Arid Tropics (ICRISAT) had made a major contribution to pearl millet research by developing downy mildew–resistant male sterile lines and releasing two hybrids (ICMH 451 and 501). These lines became the basis for numerous hybrids developed through private research, which steadily drove up pearl millet yield and production.

The sorghum story is complicated by a major shift in production from the rainy season (*kharif*) to the post–rainy season (*rabi*).[4] Changing consumption preferences among consumers toward wheat and rice rather than coarse grains reduced the demand for both *rabi* and *kharif* sorghum, creating competition (especially for rainy season sorghum) from modern varieties of food as well as cash crops. *Kharif* sorghum production accordingly declined, despite successful crop improvement efforts by public- and private-sector breeders. *Kharif* sorghum yields, however, are steadily increasing, currently at 900 kilograms per hectare, despite losses to pests and diseases (ergot and mold). Production declines in both seasons are evidently driven mainly by reductions in area (Figure 11.4).

Figure 11.4 Trends in *kharif* and *rabi* sorghum area and production in India, 1962–63 to 2007–08

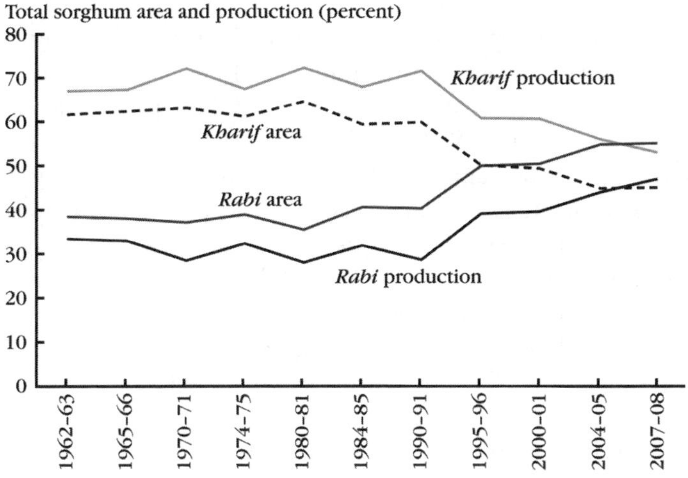

Total sorghum area and production (percent)

Source: Data from India, Ministry of Agriculture (2008).

As Figure 11.3 shows, much like the better-known Green Revolution crops (rice and wheat), dryland crops such as millet and sorghum have also shown increased and stable yields during the past five decades. Millet and sorghum occupy less than 9 percent of the total irrigated area in India—far less than do other cereals (India, Ministry of Agriculture 2006).[5] The technology advancement in millets and sorghum has nevertheless kept production levels stable, despite the decline in area planted.

Consumption Patterns

From 1998 to 2003, the annual per capita consumption of pearl millet in India declined by 57 percent, from an average of 14 kilograms to only 6 kilograms. Per capita consumption of sorghum declined by around 42 percent during the same period; the current level is about 5 kilograms (CWC 2003; Parthasarathy et al. 2006). In the major sorghum-producing regions, however, per capita consumption is still high. In rural Maharashtra, per capita annual consumption of sorghum is around 75 kilograms, accounting for almost half (48 percent) of per capita consumption of all cereals in those districts. Similarly, among the major pearl millet–producing regions, per capita consumption was highest (92 kilograms per year) in rural Rajasthan and the dry areas of Gujarat. In those two regions, pearl millet accounts for more than 50 percent of cereal consumption, contributing about 20 to 40 percent of the total energy and protein intake (Parthasarathy et al. 2006).

The decline in consumption of millets and sorghum reflects rising per capita income levels, along with changing food habits and tastes and the increasing availability of fine cereals at subsidized prices, offered through the government-sponsored Public Distribution System (PDS). The PDS in India distributes primarily rice and wheat, which are less relevant in many areas, especially in the dryland farming areas where millets, sorghum, and pulses were traditionally the staple foods for household consumption (Dayakar Rao, Reddy, and Seetharama 2007).

The demand for sorghum and millet has been boosted, however, by increasing use as feed for poultry (especially laying hens) and livestock. In the past four decades the share of sorghum used as feed has increased from 38 to 50 percent. Dayakar Rao, Reddy, and Seetharama (2007) projected that by 2010 the demand for sorghum for poultry and cattle feed would likely be around 3 million tons. There is also a huge demand for sweet-stalked, high-energy sorghum as a major bioenergy crop for the production of industrial alcohol, gasohol, and electricity.

Emerging Trends in Cultivation and Production

Recent statistics reveal a considerable decline in sorghum area, production, and consumption in the primary growing regions of India. Sorghum production slowly increased from 6 million metric tons during the 1950s to a peak of 11 million tons in the early 1980s. It started declining thereafter and currently stands at around 9.2 million tons (see Figure 11.1). During the rainy season in particular, sorghum acreage has declined because of competition from other high-value crops, such as maize, cotton, and soybeans.

The story is different, however, for *rabi* sorghum, especially in Maharashtra and northern Karnataka, where the cultivated area is expected to stabilize at between 4.5 and 5.0 million hectares. Public and private seed suppliers recognize this crop as an investment opportunity. For example, Pioneer Hi-Bred Seeds in India developed two sorghum hybrids (Pi-8703 and Pi-8704) exclusively for post–rainy season growers. These two hybrids yield 30 percent more than the existing traditional cultivar Maldandi; with one or two supplemental irrigations, the yield increase would exceed 50 percent. Other seed companies (JK Agri Genetics and Proagro Seeds) have also engaged in post–rainy season sorghum research.

Sorghum and Millet Research in India

Indian public-sector agricultural research agencies have been breeding improved millet varieties since the early 20th century. The development of hybrid sorghum in India started in the early 1960s with the establishment of hybrid breeding programs at a number of agricultural research centers. The spread of modern varieties and hybrids of pearl millet and sorghum has had an important impact on small farmers' welfare in India. The success and sustainability of these varieties resulted from three types of interventions (corresponding to the three historical periods defined earlier): (1) public-sector research on sorghum and millet plant breeding (especially the development of increased resistance to diseases and pests) and crop management, (2) government support for seed production by both the public and the private sectors, and (3) government policies that allowed the private seed industry to grow.

Government-Supported Research Efforts

Indian public research agencies have been breeding improved sorghum and millet varieties since the 1960s with the establishment of research programs at a number of sites, including the Indian Agricultural Research Institute; the State Agricultural Universities in Andhra Pradesh, Haryana, and Karnataka; the National Dry Land Research Center in Hyderabad; the Small Millets Research Program at the University of Agricultural Sciences in Bangalore; the Directorate of Sorghum and Millet Research in Hyderabad; and the All India Coordinated Pearl Millet Improvement Project and the All India Coordinated Sorghum Improvement Project (AICSIP), established in 1967 and 1969, respectively. In 1972 ICRISAT was established, further spurring sorghum and millet improvement research.[6]

These agencies and programs organized collaborative research efforts, drew on related science conducted around the world, collected and conserved germplasm from the areas of origin as well as areas of cultivation, and conducted multilocation testing for improved characteristics of hybrids and varieties. Table 11.1 shows that public-sector research in millet and sorghum has resulted in the release of many improved varieties.

Table 11.1 Number of pearl millet and sorghum crop varieties released in India, 1961–2005

Release period	ICAR		ICRISAT		Other notified varieties	
	Sorghum	Pearl millet	Sorghum	Pearl millet	Sorghum	Pearl millet
1961–70	n.r.	n.r.	n.r.	n.r.	9	5
1971–80	1	3	2	n.r.	39	10
1981–90	n.r.	3	8	14	53	23
1991–2000	32	79	13	28	58	7
2001–05	4	7	2	3	22	54
Total	37	92	25	45	262	207

Source: Authors, compiled from India, Ministry of Agriculture (2002) and ICRISAT annual reports.
Notes: The number of varieties includes both varieties and hybrids. Notified varieties include releases from state agricultural universities. ICAR means Indian Council of Agricultural Research; ICRISAT means International Crops Research Institute for the Semi-Arid Tropics; n.r. means no releases.

Sorghum Research. Public sorghum research led to several early successes, including the release of the first sorghum hybrid bred in India, CSH 1, in 1964. These successes were followed by the release of more popular hybrids, such as CSH 5 and CSH 6 in the mid-1970s and CSH 9 in the early 1980s, augmenting the spread of sorghum HYVs and open-pollinated varieties and boosting productivity. Hybrids CSH 1 through CSH 23 not only enhanced yields but also diversified parental lines and advanced breeding resistance to major pests and diseases. The hybrids played a major role in pushing up productivity and production, particularly in the case of *kharif* sorghum. CSH 1, CSH 5, CSH 6, CSH 9, CSH 14, and CSH 16 showed dramatic increases in productivity. From CSH 5 and CSH 6, with a yield potential of 3.4 tons per hectare, yield potential was raised to 4.0 tons per hectare in CSH 9 and to more than 4.1 tons per hectare in CSH 16 and CSH 23 (NRCS 2007).

The worth of these improved cultivars is demonstrated by their successful adoption by farmers. The use of improved varieties of sorghum decreased the unit cost of production during the 1980s and 1990s compared with the early 1970s, because the productivity gains from improved cultivars more than compensated for the cost of the additional inputs used for their cultivation (Bantilan and Deb 2002; Reddy et al. 2007). In the 1990s the cost of production per ton fell by 40 percent and 37 percent in Maharashtra and Rajasthan, respectively, compared with the early 1970s. The cost–benefit ratio of producing improved cultivars in India is 1:1.4 (Bantilan and Deb 2002).

Public research institutions such as the University of Agricultural Sciences (Dharwad, Karnataka) and the National Research Center for Sorghum (NRCS) in Hyderabad are also involved in developing specific cultivars for the *rabi* season. *Rabi* sorghum is highly valued as food because of its excellent grain quality and because it is produced during the post–rainy season. It commands higher prices in the market than *kharif* sorghum, often on a par with or higher than those of the local durum wheat. *Rabi* sorghum is also highly valued as fodder during lean months and is grown without irrigation. In addition, the *rabi* sorghum stover is of high quality and is much more important than *kharif* sorghum stover because its harvest precedes the lean summer months. Five *rabi* hybrids and five open-pollinated varieties have been released so far by public sector research institutions. The yield potential of the newly bred cultivars, however, is only marginally higher than that of M 35-1, the widely grown local cultivar. More recently released hybrids (CSH 15R and CSH 19R) are more productive, but their acceptability is limited because farmers are reluctant to invest in hybrid seeds during *rabi* without irrigation (NRCS 2007).

The economic contribution of fodder to the total income from *rabi* sorghum is estimated at 45 to 57 percent in open-pollinated varieties and 39 to 47 percent in hybrids in Maharashtra and Andhra Pradesh (NRCS 2007). Thus, even at a low productivity level, *rabi* sorghum is far more profitable for producers than is *kharif* sorghum. Both the grain and the stover enjoy strong demand, which may further expand.

Millet Research.   India, which produces more than half the world's pearl millet, has been the center of millet research efforts since the 1960s, when the availability of cytoplasmic-genetic male sterile lines made available a succession of hybrids (FAO 2009). Collaborative research efforts led to the release of the first pearl millet hybrid (HB 1) in 1965. Today hybrids cover more than 50 percent of the total national pearl millet area of 24.7 million acres (Thakur et al. 2003).

Epidemics of downy mildew (*Sclerospora graminicola*) disease (DMD) constitute the most significant risk to the cultivation of pearl millet hybrids, a risk that can be reduced by effective crop improvement research and proper agronomic practices. Losses, which can approach 100 percent on individual fields, are estimated to average 14 percent across India. When one hybrid is overcome by rapidly evolving pathogen populations, other hybrids with a genetically identical parental line soon follow—and pearl millet hybrids in India are in fact based on a narrow range of closely related parental lines. This pattern results in rapid cultivar turnover, driven mostly by disease pressure rather than yield or quality improvements, to the detriment of pearl millet consumers, producers, and all those involved in the seed trade (Breese et al. 2002).

For example, in 1974–75 a heavy attack of DMD reduced pearl millet production dramatically to 3.3 million tons—a decrease of 57 percent from previous years. A second series of millet hybrids was released during 1974–75 (PBH 10 and 14) and proved more tolerant than previous cultivars, but they also broke down quickly. Two hybrids introduced in the mid-1980s—BJ 104 and BK 560, known for their short duration and drought resistance—slowly brought production levels back up to 6 million tons. These cultivars also suffered from DMD, however, and the Government of India withdrew both of them from commercial use in 1986 (Pray and Ribeiro 1990).

In economic terms, a DMD epidemic can be disastrous. For HHB 67—a publicly bred hybrid developed and released by Haryana Agricultural University in 1995–96 and widely adopted by farmers in Haryana and Rajasthan, covering nearly 400,000 hectares—a DMD epidemic would amount to an estimated loss of at least $7.7 million in just its first year.[7] But the introduction of a hybrid resistant to DMD (HHB 67 Improved) promises an additional estimated return of $2.6 million, reflecting an improved yield advantage of 10 percent over the existing cultivar (Hash et al. 2007).

Public and Private Roles in Seed Industry Growth

A major contributor to the diffusion of improved sorghum and pearl millet cultivars, particularly hybrids, has been the emergence of a vibrant seed industry in India.

Until the late 1980s, public agencies played the most significant role in varietal development, multiplication of seeds, and distribution through seed outlets (by state departments of agriculture, national and state seed corporations, and farmer cooperatives). India's seed sector was deregulated in 1971, with relaxed restrictions on seed imports and entry of private firms. Private companies began to breed their own millet varieties in the 1970s, but it took a decade to produce the first commercially successful improved cultivars. In 1988 a new seed policy spurred enormous growth in private-sector seed supplies in India.

Strong public-sector research on sorghum and millet was a major driver of the private-sector seed industry. After the establishment of ICRISAT in 1972, public research on pearl millet and sorghum grew more rapidly than that on other hybrids. International agricultural research centers such as ICRISAT exchanged breeding material with both public and private research institutions. National agricultural research centers such as ICAR and the agricultural universities provided breeder seed to the national and state seed corporations as well as the private seed companies, to be multiplied as foundation seed and distributed through company outlets, farmer cooperatives, and private dealers. The first private pearl millet hybrids were based on local lines developed in the public sector and exotic lines brought in by ICRISAT.

Developing inbred lines or restorers takes a long time—up to nine seasons. The association of private firms with ICRISAT or ICAR and state universities was thus invaluable because the public institutions developed inbred lines and provided them gratis. In the absence of public-sector research, the private seed industry would have started much later and developed much more slowly (Pray and Ribeiro 1990). By developing hybrid sorghum and millet in the 1960s, public-sector research generated demand that the private commercial seed industry would build on. Moreover, in the late 1960s and 1970s government programs provided subsidies and technical advice to small and medium-sized firms to produce and multiply seeds. The 1970s were a period of experimentation for all companies: private firms invested in research and breeding activities to produce various public hybrids using various seed-growing locations and production strategies. The only companies that produced and sold their own hybrids were Mahyco, Nimbkar, and Pioneer.

Another key seed policy instrument enhanced the participation of private firms: firms could multiply and sell seeds to farmers without going through the regular certification process by selling their hybrids or varieties as "truthfully labeled" seed. Seed certification procedures for most crops also favored private firms' participation in seed markets. Farmers could have some assurance of a minimum quality of seed even if they did not know the company that produced it. But because certification was (and is) voluntary, it did not slow the development of private hybrids of millet and sorghum. Companies always had an option of selling their seed as truthfully labeled rather than certified.[8]

Over the past two decades, research and development (R&D) in pearl millet have become increasingly privatized, reflecting a general shift in India's agricultural research system from publicly dominated to privately driven seed development and distribution (Pal and Byerlee 2003). Interviews with personnel of private firms found that in 1970 just 4 companies had their own sorghum and pearl millet breeding programs; by 1985 the number had grown to 10 companies (Pray et al. 1991).

Pray et al. (1991) estimated that in the late 1980s private investments in pearl millet improvement were equal to public investments, and their share has increased considerably since then. This finding might be surprising because pearl millet is grown largely by subsistence farmers in India. Nonetheless, the large size of the market, together with the fact that farmers were already used to regular seed replacements, provided a sufficient business incentive. Moreover, because all pearl millet hybrids periodically succumb to diseases such as DMD, there is ongoing demand for new and better products.

The expansion of the seed industry in the late 1980s and early 1990s brought about a significant increase in investment research, along with growth in the supply of and demand for improved seeds in Maharashtra and other pearl millet–growing states. The cultivars that had dominated during the 1980s were mostly replaced by new varieties and hybrids. The benefit–cost ratio of shifting from public hybrids (of sorghum and millet) to private hybrids was much higher for small farmers than for large farmers. In the 1990s the seed market was dominated by ICRISAT-based hybrids. While the adoption of privately released hybrids of pearl millet and sorghum increased (developed by private firms such as Mahindra Hybrid Seed Company, Pioneer, and ProAgro), most of these hybrids contained parent materials from ICRISAT and other public research agencies (Bantilan and Deb 2002). Other public research hybrids, such as those in the CSH series, also remained popular with farmers.

ICRISAT now has a consortium approach to providing private companies with parental lines for breeding hybrid sorghum and pearl millet. ICRISAT's Hybrid Parents Research Consortia brings together small and medium-sized domestic firms for the purpose of commercializing sorghum, millet, and pigeonpea hybrids, thus contributing to the commercial viability of both domestic seed firms and the wider seed market in India. Consortia members include both international corporations and a large number of domestic seed companies, and all members may obtain breeding materials from ICRISAT through the Consortia on a nonexclusive basis.

Today the Indian market for agricultural seed is considered one of the largest in the world, with annual sales of around $1,080 million. The domestic market accounts for $975 million in sales, and international trade (mainly with developing countries) accounts for the remaining $20 million. The Indian seed industry has now evolved from public-sector domination into a multifaceted industry, with broad involvement by private firms and increasing emphasis on research and development. A recent Government of India report on the status of Indian agriculture claimed that nearly 80 percent of the commercial seed sales of pearl millet and sorghum are made by private seed companies (Reserve Bank of India 2005).

Matuschke and Qaim (2008) estimated the determinants of pearl millet adoption and the impact of increasing privatization on technology diffusion, based on a comprehensive survey of 266 pearl millet farmers in the state of Maharashtra. The Government of Maharashtra (2005) reports that distribution of hybrid pearl millet seed by public and private sources tripled between 1990 and 2000. The study identified three factors that contributed to the adoption of pearl millet hybrids over recent decades: farmers' education level, distance to the main source of information, and good market infrastructure. In addition, the increasing role of private companies in seed development and distribution had a positive effect on innovation rates. The study refutes the notion that privatization of seed markets would hamper technological progress in the small farm sector and suggests that even in the production of typical subsistence crops, such as pearl millet, the private sector can play an important role.

Economic Impacts

Maize, sorghum, and pearl millet are the three most widely planted cereals in India after rice and wheat. Pearl millet and sorghum together constituted nearly 25 percent of the total cropped area and about 12 percent of the total value of all crop seeds sold commercially in 1999–2000 (India, Ministry of Agriculture 2000).

For all cereal crops except sorghum, saved seed was formerly the dominant source of seed but has dramatically declined. Similarly, for sorghum the proportion of purchased seed to saved seed increased during the 1990s (Gadwal 2003). Sales of proprietary hybrids for pearl millet increased nearly eightfold in the 1990s, but sales of proprietary hybrids for sorghum rose less dramatically, around 20 percent. In contrast to proprietary hybrids marketed by private companies, sales of publicly bred sorghum and pearl millet hybrids declined considerably. There was also a significant reduction in the sale of open-pollinated varieties (OPVs) of pearl millet from 1990 to 1999, but an increase in OPVs of sorghum during the same period (Nagarajan, Smale, and Glewwe 2007).[9]

During 1992–94, about 55 percent of the area under sorghum and pearl millet cultivation in India was planted with HYVs, nearly doubling the productivity of both crops compared with the pre-HYV era. The area under HYV cultivation continues to increase, and so does productivity, with no yield plateau in sight. In addition, cultivar diversity has increased substantially, giving farmers more appropriate choices of cultivars and hence more stable yields. But these positive changes in adoption scales and cultivar diversity have occurred primarily in relatively favorable environments and in states with well-developed seed production infrastructure (Rai et al. 1999).

Six million hectares (more than 60 percent of the total pearl millet area) is planted with more than 70 hybrids, of which at least 80 percent are hybrids from the private sector (Dar et al. 2006). More than 60 of these hybrids are based on ICRISAT-bred hybrid parents (mostly seed parents) or on the proprietary hybrid parents developed from ICRISAT-bred improved germplasm. It has been conservatively estimated that the annual return to India's farmers from pearl millet varieties developed by ICRISAT totals $50 million—more than 12 times the cost of its investment in pearl millet research (CGIAR 1996). Unlike research on pearl millet, sorghum research in India is implemented mostly by public-sector institutions such as ICAR and state agricultural universities rather than by the private sector, and the estimated returns on the research investment are also higher. The annual return to India's farmers from government investment in sorghum crop improvement and development of HYVs (by NRCS and AICSIP) for 1981–99 is estimated at Rs. 11,450 million (or about $275 million)—nearly 30 times the cost of the investment (NRCS 2007).

Hybrid technology has also contributed to employment generation and to farmers' income at the seed production stage. Each year the production of pearl millet hybrid seeds in India occurs primarily during the summer season in one district of Andhra Pradesh. In that district it generates an additional annual income of $1 million for the seed-producing farmers. According to the Seedsmen Association of Andhra Pradesh, nearly 90 percent of all sorghum hybrid seeds and 65 percent of pearl millet seeds were produced by contract seed growers from the state in 2004–05. Seed production directly or indirectly employs more than 200,000 of the state's farmers.

Maharashtra state, with a large number of private seed companies and an aggressive state seed corporation, had about 18 improved pearl millet cultivars at various scales of cultivation during the mid-1990s, compared with 3 during the mid-1980s. Similar changes in pearl millet cultivar diversity occurred in Gujarat. These two leading states now have 85–90 percent of the total pearl millet area under HYV cultivation. In Gujarat these HPVs are mostly hybrids; in Maharashtra a substantial proportion are still OPVs (ICTP 8203).

Pearl millet constitutes an important staple crop, especially for marginalized households, for which coarse cereals account for a larger share of daily diets than do wheat and rice (Ramaswami 2002). Pearl millet hybrids are widespread and have been increasingly adopted over the past decades (Bantilan, Deb, and Singh 2000; Thakur et al. 2003). Pray, Ramaswami, and Kelley (2001), in a study in Andhra Pradesh, Karnataka, and Maharashtra, found that the share of coarse cereals (millets, maize, and sorghum) in total cereal expenditure was highest for the poorest 30 percent of the population. As a result, any yield improvement in coarse cereals would have a direct impact on the poorest households. Especially in the states of Karnataka and Maharashtra, where coarse cereals are more important in the diets of poor households than are rice and wheat, productivity increases in coarse grains are more important in increasing the welfare of the poor than are productivity increases in rice and wheat.

In sum, the contribution of private hybrids to agricultural productivity is significant in both production and distribution of seed. These results are especially striking because they pertain to semiarid tropical regions, where the Green Revolution based on HYVs of wheat and rice has had limited impact. Given that the semiarid tropics tend to be poorer than the more favorably endowed growing regions (the Punjab and the Indo-Gangetic Plains) and given that private hybrids have had the most impact in subsistence crops, it is likely that poor farmers in semiarid areas have gained from the spread of private hybrids.

Western Rajasthan, a dryland zone with sandy soils, is one of the major pearl millet–growing regions in India. In the early 1990s, ICRISAT, in collaboration with the Rajasthan Agricultural University, a local nongovernmental organization, and farmers in selected villages in western Rajasthan, started a program of farmers' participatory breeding of improved pearl millet cultivars that continued for about 10 years. Major benefits perceived by households in villages of western Rajasthan included an improved choice of varieties to suit the weather. These more suitable varieties helped them manage the risk of rainfall failure and further stabilized their long-term yields. Improved technology, including improved varieties and crop management practices, allowed greater soil augmentation, increasing the yields of pearl millet; more stable yields further enabled farmers to shift a portion of their farmed area from millets to cash and other crops (Bantilan, Parthasarathy, and Padmaja 2003). Researchers showed that the increased adoption of new and improved varieties also resulted in increased asset generation by individual households in western Rajasthan, such as the building of *pucca*, or concrete houses, with the surplus cropping income. The participatory rural appraisals conducted by ICRISAT researchers in these villages also found that rates of school enrollment increased up to 20–22 percent within four years of adoption, and enrollment rose especially among girls (Parthasarathy and Chopde 2000).

Studies on the economic impacts of seed industry reforms found that farmers gained most from the resulting increase in private research (Pray et al. 1991; Pray and Ramaswami 2001). They found that in 1986–87 yields of private pearl millet and sorghum hybrids were higher than public hybrids and OPVs in all-India coordinated yield trials conducted by the ICAR in farmers' fields. For instance, Mahindra's pearl millet hybrid MBH 110 outyielded the publicly bred check hybrid

BJ 104 by an average of 23 percent. Researchers examined returns on several crops using the increase in net income of seed firms and farmers from the sale and use of private rather than public hybrids as an estimate of the total benefits from private varietal improvement research. They found that the seed companies captured no more than 18.5 percent of the benefits from using improved sorghum varieties. Similarly, for hybrid pearl millet seed firms captured only about 6 percent of benefits, with more than 90 percent of benefits accruing to farmers. Another study on the maize seed industry in India found similar results regarding the benefit shares to farmers versus seed supply companies (Singh, Morris, and Pal 1997).

The social internal rates of returns to private pearl millet and sorghum research were at least 50 percent (Pray and Ribeiro 1990). The annual returns to Indian pearl millet farmers from cultivating varieties from ICRISAT and private firms are estimated at $54 million (ICRISAT 1998). The impact of private-sector research became much more evident during the late 1990s, with increased area under private hybrids of cotton, pearl millet, sorghum, maize, and fodder crops.

Although crop improvements have been less pronounced for millets and sorghum than for rice and wheat in India, progress in these crops has nevertheless been significant (Evenson and Gollin 2003). A study of the impacts of ICRISAT's research showed that privately released hybrid millet and sorghum varieties relied heavily on ICRISAT-developed male sterile lines and restorers (Bantilan and Deb 2002).

Nagarajan, Pardey, and Smale (2007) examined the relationship between biological (varietal) diversity of pearl millet in the farm communities of semiarid regions of Andhra Pradesh and Karnataka. They found that communities with high combined farm and off-farm income levels maintained greater richness of millet varieties across their farms, perhaps because of greater access to improved materials and greater capacity to grow them. The educational levels of farmers was also higher in these communities and had a positive effect on crop diversity at the community level. Higher seed-to-grain price ratios also enhanced millet profits among village communities, reflecting the use of modern varieties. Formal seed transactions through dealers also correlated with improved millet diversity among the village communities surveyed.

A key finding is that the presence of active local formal and informal seed markets enhances the millet profits of farming communities. These findings suggest that, through judicious introduction of improved varieties that complement local varieties by providing a needed trait, it may be feasible to enhance farmers' income while supporting millet crop diversity to promote the resilience of farming communities in these marginal environments (Nagarajan, Pardey, and Smale 2007). The long-term influence of proprietary hybrids and varieties is apparent not only in favorable environments but also in drylands.

Sustainability of the Interventions

In the more favored growing environments of India (such as the states of Haryana, Maharastra, and Punjab), where farmers have access to irrigation and rising incomes are changing food consumption patterns, the area sown to sorghum and other millet crops is gradually giving way to rice, wheat, maize, and other specialty crops (Seetharam, Riley, and Harinarayana 1989). In the arid and semiarid regions (including the states of Karnataka, Andhra Pradesh, Rajasthan, and Gujarat), however, farmers' demand for a range of millet crops and millet varieties is unlikely to diminish in the near future because there are currently few substitute crops for these harsh growing environments.

The Food and Agriculture Organization of the United Nations has estimated that 55 percent of the world's semiarid lands with rainfed farming potential are located in Sub-Saharan Africa and South Asia (including India), and these areas are characterized by the lowest per capita nutrition levels and the highest population growth rates (FAO 2004). These semiarid regions are likely to be home to an additional 400 million people by 2025.[10] Soil salinity and drought remain major abiotic stressors that pose a threat to agricultural production in this part of the world. Water is becoming an even more scarce resource, and significant expansion of irrigation does not seem feasible in many of these semiarid countries. Furthermore, public irrigation systems need substantial investments in rehabilitation, modernization, operation, and maintenance. Desertification may be aggravated over time either by overexploitation by native populations or by regional climatic changes. These factors underscore the need for concerted efforts to develop crops that are more tolerant of stressful environments.

Millet and sorghum are reasonably tolerant of extreme soil and weather conditions. They also have other desirable attributes: higher nutritive value (including in micronutrients such as iron, calcium, and zinc) than most major cereals, higher fodder value, and higher tolerance of pests and diseases. For these reasons, a case can be made for conserving as well as promoting cultivar diversity for these two major dryland cereals to help meet future food and feed needs, especially those of subsistence producers in these less favored economic and physical environments.

The emerging trends in the use of sorghum for alternative purposes (such as for biofuels and animal and poultry feeds) provide some evidence of increased demand for these crops in India. Although sorghum is used largely as a feedgrain throughout the world, in India the cost of production and quality limitations make it less attractive as a feedgrain than maize. Current feed production grew from 2.7 million tons in 2004–05 to 5.2 million tons in 2010 (India, Ministry of Agriculture 2009). India's huge livestock population and the increasing demand for milk and animal products also create pressure for the production of green and dry fodder and forages. Under semiarid conditions, sorghum and millet are the major suppliers of green and dry fodder and forage, especially during the lean season; 20–60 percent of the dry fodder supply in the semiarid regions of India is provided by sorghum crops alone. Moreover, the diversification of rainy season sorghum as a bioenergy crop has vast potential for helping to meet the growing demand for fuel.

Part of the reason for the stagnating production of sorghum and pearl millet is the growing competition in dry regions from other major cereals (including maize) and cash crops, which benefit from government price support programs. The per capita consumption of sorghum in rural India declined from 1.59 kilograms per month in 1973 to 0.45 kilograms per month in 2003–04 (NRCS 2007). Some of this decline was due to governmental policies that excluded sorghum from public procurement at the minimum support price (MSP) and from supply through the PDS (NRCS 2006). Government policies encouraged increased consumption of wheat or rice in the regions where sorghum was traditionally valued as the preferred cereal. Over time, the calculation of the MSP shifted against sorghum and coarse cereals. In 1980–81 the MSPs for rice and sorghum were equal, but by 1995–96 the MSP for sorghum was 9 percent lower than that for rice. Moreover, government policies on pricing sorghum and millet compared with pulses, oilseeds, and other dryland crops were similarly unfavorable and further accelerated the diversion of the *kharif* sorghum or millet area toward other commercial alternatives such as cotton and maize. The implication for sustainable sorghum and pearl millet production is clear: if the government decides that it cannot afford to continue subsidizing wheat, rice, and maize production, demand for sorghum and pearl millet is likely to increase.

The public-sector research system continues to provide new technological opportunities for the public and private seed industries to develop profitable products. More than 50 private companies are marketing about 75 hybrids of pearl millet, and 11 companies are marketing 20 hybrids of sorghum based on seed and pollen parents from ICRISAT. ICRISAT's public–private pearl millet and sorghum Consortia has helped increase cultivar adoption while helping mobilize resources for public-sector research. As of December 2005, ICRISAT had generated more than $2 million in investment since the Consortia program was initiated in 2000. The funds generated augment ICRISAT's core funds to support crop improvement research for developing elite sorghum, pearl millet, and pigeonpea hybrid parents to serve both the public and the private sectors. This resource mobilization is particularly significant in light of diminishing core funding for crop improvement research at ICRISAT.

ICAR and state agricultural university breeding programs are the major source of germplasm as well as of finished inbred lines to private breeding programs. Abandoning public breeding programs could therefore lead to less technological diversity and higher seed prices, with negative implications for agricultural development in general and smallholder farmers in particular. Decisions on appropriate public- and private-sector roles must be country- and crop-specific to achieve desirable welfare and distribution effects (Matuschke and Qaim 2008).

Strong public–private partnerships such as the ICRISAT Consortia model, as well as government-sponsored science parks, represent a strategic approach to providing the necessary infrastructure for research as well as the skilled human resources needed for technology exchanges.

Conclusion

Although India's combined sorghum and millet production has been stagnant since 1965, new technology has had an important impact on improving small farmers' productivity in some of the poorest areas of India. Yields have doubled since 1966, largely because of improved genetics and crop management, initially spearheaded by public research (1966–85) and then by the private sector (1986 to the present). Unlike in the case of the major Green Revolution crops, little of the increase in yields can be attributed to irrigation because at least 90 percent of these crops are grown under unirrigated, rainfed conditions. The doubling of yields allowed farmers to grow the same amount of food on half the land, and they often shifted the rest of the land to valuable cash crops and increased their incomes. The improved sorghum and millet crops also contributed to food security because they are considerably more resistant to drought than the other major foodgrains. Furthermore, it is clear that these improved varieties benefit primarily poor consumers, because the wealthy tend to eat rice or wheat.

Improved hybrid cultivars in these crops can be very valuable to small farmers who grow crops in dryland conditions. Pray et al. (1991) showed that 80–90 percent of the benefits from the adoption of hybrids of these crops went to farmers rather than to the seed companies. The takeover of pearl millet and sorghum seed markets by proprietary hybrids nevertheless shows that private firms capture sufficient benefits to induce them to invest in research to develop cultivars for small farmers in unirrigated regions.

The lessons from the Indian experience with improving sorghum and pearl millet hybrids highlight three important interventions:

1. *Investments in public-sector plant breeding and crop management research by the national government, state governments, and international agricultural research centers.* In the early days of the development of sorghum and millet hybrids, all three contributed OPVs, hybrid cultivars, and inbred lines that benefited farmers directly while providing the basis for private researchers to develop new cultivars.

2. *Government investments in seed production through government and private institutions.* The Indian national and state governments, with the help of donors, made major investments in government seed corporations for the production of

pearl millet and sorghum, as well as Green Revolution cultivars of wheat, rice, and maize. At the same time, small private seed companies (apart from large and medium-sized firms) were permitted to enter the seed business and make profits. The government provided training for seed business personnel in both public and private institutions.

3. *Liberalization of the seed market beginning in the mid-1980s.* Instead of allowing state seed companies to become regional monopolies, the government opened the doors to investment by large Indian firms, which had been excluded from this sector until 1986. It also allowed foreign direct investment in the sector at about the same time.

Liberalization has been coupled with ongoing indirect government support for private operations: continuing public investment in hybrid breeding, public–private partnerships, provision of inbred lines and germplasm for development of proprietary hybrids, and a seed law that allowed truthfully labeled seed instead of mandatory registration and government testing of new cultivars. This approach has led to a vibrant and sustainable supply of new cultivars and seeds that are resistant to important diseases and pests and tolerant of drought.

Notes

1. Millet generally refers to a group of small, seeded cereal crops that includes pearl millet, finger millet, little millet, fox tail millet, and other minor crops. This chapter discusses only pearl millet and its cultivars. Pearl millet is also known as bulrush millet or spiked millet; sorghum is referred to as great Indian millet.

2. Arid and semiarid regions constitute more than 50 percent of the total area of India.

3. Three-year averages were used in analyzing the data to compensate for weather variability.

4. This research focused mainly on *kharif* (rainy season) sorghum.

5. For wheat, 90 percent of total planted area is under irrigation; for rice, 56 percent; and for maize, 21 percent.

6. ICRISAT was established at Hyderabad, India, to focus research on the arid and semiarid regions of the world. Its mandate crops are sorghum, pearl millet, finger millet, groundnuts, chickpeas, and pigeonpeas.

7. This figure represents 30 percent of the harvest of 550,000 hectares yielding 0.7 tons per hectare, valued at Rs. 3,000.

8. Well-established enterprises with reputations to protect may sell seed that has no official seed certification. Such seed is often described as "truthfully labeled" and bears a label describing minimum seed quality standards. It is self-certified rather than certified by an official agency (Tripp 2001).

9. The adoption of pearl millet hybrids increased both because of its yield advantage (compared with OPVs) and because of active promotion by the private companies. A survey by Nagarajan (2004) found that private companies foresee further area expansion under pearl millet in new areas, especially in Gujarat and some parts of Maharashtra.

10. The semiarid lands are characterized by unpredictable weather, long dry seasons, inconsistent rainfall, and soils that are poor in nutrients. They include parts of 48 countries in the developing world, including most of India, locations in Southeast Asia, much of southern and East Africa, and a few locations in Latin America.

References

Bantilan, C., D. Parthasarathy, and R. Padmaja. 2003. Enhancing research–poverty alleviation linkages: Experience in the semiarid tropics. In *Agricultural research and poverty reduction: Some issues and evidence,* ed. Shantanu Mathur and Douglas Pachico. CIAT Publication 335, Economics and Impact Series 2. Cali, Colombia: International Center for Tropical Agriculture.

Bantilan, M. C. S., and U. K. Deb. 2002. Grey to green revolution in India: Role of public–private–international partnership in research and development. Paper presented at the BAEA–IAAE conference Public–Private Sector Partnership for Promoting Rural Development, October 2–4, in Dhaka, India.

Bantilan, M. C. S., U. K. Deb, and S. D. Singh. 2000. Farm-level genetic diversity in pearl millet in India. Poster paper presented at the 3rd International Crop Science Congress, August 18–22, in Hamburg, Germany.

Breese, W. A., C. T. Hash, A. Sharma, and J. R. Witcombe. 2002. *Defeating downy mildew: Improving pearl millet, the staple cereal crop of some of the world's poorest people, whilst keeping one step ahead of downy mildew.* Poster presented at the conference Plant Pathology and Global Food Security, July 8–10, at Imperial College, London.

CGIAR (Consultative Group on International Agricultural Research). 1996. A new generation of pearl millet on the horizon. *CGIAR News* 3 (3). <http://www.worldbank.org/html/cgiar/newsletter/Oct96/6millet.html>. Accessed August 20, 2010.

CWC (Central Water Commission). 2003. *Statistical report.* New Delhi: Department of Statistics, Ministry of Agriculture, Government of India.

Dar, W. D., B. V. S. Reddy, C. L. L. Gowda, and S. Ramesh. 2006. Genetic resources enhancement of ICRISAT mandate crops. *Current Science* 91 (7): 880–884.

Dayakar Rao, B., S. Reddy, and N. Seetharama. 2007. Reorientation of investment in R&D of millets for food security: The case of sorghum in India. *Journal of Agricultural Situation in India* 64 (7): 303–305.

Evenson, R. E., and D. Gollin. 2003. Review: Assessing the impact of the green revolution, 1960 to 2000. *Science* 300: 758–762.

Evenson, R. E., and C. E. Pray, eds. 1991. *Research and productivity in Asian agriculture.* Ithaca, N.Y., U.S.A.: Cornell University Press.

FAO (Food and Agriculture Organization of the United Nations). 1995. Sorghum and millets in human nutrition. FAO Food and Nutrition Series 27. Rome.

———. 2004. FAOSTAT–agriculture. Rome.

————. 2007. FAOSTAT–agriculture. Rome.

————. 2009. FAOSTAT. <http://faostat.fao.org>. Accessed August 20, 2010.

Gadwal, V. R. 2003. The Indian seed industry: Its history, current status and future. *Current Science* 84 (3): 399–406.

Government of Maharashtra. 2005. *Economic survey of Maharashtra 2004–2005.* Mumbai, India: Government Press.

Hash, C. T., R. S. Yadav, A. Sharma, R. Bidinger, K. M. Devos, M. D. Gale, C. J. Howarth, S. Chandra, G. P. Cavan, R. Serraj, P. S. Kumar, W. A. Breese, and J. R. Witcombe. 2007. *Release of pearl millet hybrid HHB 67-2: An improved downy mildew resistant version of HHB 667 produced by marker assisted selection: ICRISAT highlights.* Patancheru, Andhra Pradesh, India: International Crops Research Institute for the Semi-Arid Tropics.

ICAR (Indian Council of Agricultural Research). 2006. *Annual report.* New Delhi, India: Krishi Bhavan.

ICRISAT (International Crops Research Institute for Semi-Arid Tropics). 1998. *Annual report.* Patancheru, Andhra Pradesh, India.

India, Ministry of Agriculture. 2000. *Agricultural statistics in India.* New Delhi.

————. 2002. *Agricultural statistics at a glance.* New Delhi.

————. 2006. *Agriculture statistics at a glance.* New Delhi: Directorate of Economics and Statistics, Government of India.

————. 2008. *Agricultural statistics in India.* New Delhi.

————. 2009. *Agricultural statistics in India.* New Delhi.

Matuschke, I., and M. Qaim. 2008. Seed market privatisation and farmers' access to crop technologies: The case of hybrid pearl millet adoption in India. *Journal of Agricultural Economics* 59 (3): 498–515.

McGaw, E. M. 2001. *Fine tuning the progeny: More miracles from the millet molecules.* London: Department for International Development.

Nagarajan, L. 2004. Managing millet diversity: Farmers' choices, seed systems and genetic resource policy in India. Ph.D. dissertation, Department of Applied and Agricultural Economics, University of Minnesota, St. Paul, Minn., U.S.A.

Nagarajan, L., P. Pardey, and M. Smale. 2007. Seed systems and millet crops in marginal environments of India: Industry and policy perspectives. *Quarterly Journal of International Agriculture* 46 (3): 263–288.

Nagarajan, L., M. Smale, and P. Glewwe. 2007. Determinants of millet diversity at the household–farm and village–community levels in the drylands of India: The role of local seed systems. *Agricultural Economics* 36 (2): 157–167.

NRCS (National Research Centre for Sorghum). 2006. *Annual report.* Rajendra Nagar, Hyderabad, India.

————. 2007. *Perspective plan: Vision 2025.* Rajendra Nagar, Hyderabad, India.

Parthasarathy, D., and V. K. Chopde. 2000. Building social capital: Collective action, adoption of agricultural innovations, and poverty reduction in the Indian semi-arid tropics. Paper prepared for the Global Development Network. International Crops Research Institute for the Semi-Arid Tropics, India.

Parthasarathy Rao, P., P. S. Birthal, B. V. S. Reddy, K. N. Rai, and S. Ramesh. 2006. Diagnostics of sorghum and pearl millet grain-based nutrition. *International Sorghum and Pearl Millet Newsletter* 47: 93–96.

Pray, C. E., and B. Ramaswami. 2001. Liberalization's impact on the Indian seed industry: Competition, research, and impact on farmers. *International Food and Agribusiness Management Review* 2 (3): 407–420.

Pray, C. E., and S. Ribeiro. 1990. *Government seed policy, the development of private seed industry, and the impact of private R&D in India: The final report of the Indian seed industry project.* New Brunswick, N.J., U.S.A.: Department of Agricultural Economics, Rutgers University.

Pray, C. E., B. Ramaswami, and T. Kelley, 2001. The impact of economic reforms on R&D by the Indian seed industry. *Food Policy* 26 (6): 587–598.

Pray, C. E., S. Ribeiro, R. A. E. Mueller, and P. Parthasarathy Rao. 1991. Private research and public benefit: The private seed industry for sorghum and pearl millet in India. *Research Policy* 20 (4): 315–324.

Rai, K. N., D. S. Murty, D. J. Andrews, and P. J. Bramel-Cox. 1999. Genetic enhancement of pearl millet and sorghum for the semi-arid tropics of Asia and Africa. *Genome* 42: 617–628.

Ramaswami, B. 2002. Understanding the seed industry: Contemporary trends and analytical issues. *Indian Journal of Agricultural Economics* 57 (3): 417–429.

RBI (Reserve Bank of India). 2005. *Economic survey of India.* Mumbai, India.

———. 2006. *Annual report 2005–06.* Mumbai, India.

Reddy, B. V. S., S. Ramesh, S. T. Borikar, and K. H. Sahib. 2007. ICRISAT–Indian NARS partnership sorghum improvement research: Strategies and impacts. *Current Science* 92 (7): 909–915.

Seetharam, A., K. W. Riley, and G. Harinarayana. 1989. *Small millets in global agriculture.* New Delhi: Oxford and IBH Publishing.

Singh, R. P., M. L. Morris, and S. Pal. 1997. Efficiency and equity considerations in the maize seed marketing system: Role of public and private sectors in India. *Indian Journal of Agricultural Marketing* 52 (4): 28–34.

Thakur, R. P, V. P. Rao, K. N. Amruthesh, H. S. Shetty, and V. V. Datar. 2003. Field surveys of pearl millet downy mildew: Effects of hybrids, fungicide, and cropping sequence. *Journal of Mycology and Plant Pathology* 33: 387–394.

Tripp, R. 2001. *Seed provision and agricultural development.* London: Overseas Development Institute.

Chapter 12

Institutional Reform in the Burkinabè Cotton Sector and Its Impacts on Incomes and Food Security, 1996–2007

Jonathan Kaminski, Derek Headey, and Tanguy Bernard

Cotton and Economic Development in West Africa

The history of cotton in West Africa is both an economic and a political story. On the economic side, cotton has always been a principal cash crop used to barter or purchase other tradable goods (Schwartz 1996). In the colonial era it gained further importance as an internationally traded good, although usually with limited results (Bassett 2001; World Bank 2004).[1] Starting in the 1950s, however, cotton cultivation expanded sharply.[2] From 1960 to 2000 production in the African Financial Community (CFA) zone grew at a compound rate of 9 percent per year, so rapidly that by 2000 Francophone Africa accounted for 70 percent of all cotton lint produced in Sub-Saharan Africa, 4.4 percent of total world production, and 13 percent of international cotton fiber exports (a sixfold increase from 1960). As

We acknowledge the support of David Spielman from IFPRI, the comments of two anonymous referees, and several students' reviews of the first draft of this chapter. We also thank a number of people: Jose Tissier (AFD), François Traoré (UNPCB), Wilfried Yaméogo from the Ministry of Trade (Burkina Faso), as well as senior officials in CNCA, local executives of the AFD and the World Bank, and a number of people in GPCs and local government. Kaminski thanks ODI (London) and David Booth for the Africa, Power, and Politics Project (a 2009 survey) and Jean-Paul Azam and ARQADE Toulouse for funding his 2006 survey. Bernard acknowledges the support of the World Bank and the Norwegian Trust for his 2002 survey.

of today, 16 to 20 million people in West Africa depend directly or indirectly on cotton cultivation in a cotton belt stretching across at least 11 countries.

On the political side, cotton is a vitally important sector in a number of these economies, making it the subject of intense competition between different groups. For most of the postindependence era, central governments heavily taxed the cotton sector, which has been the most important source of foreign exchange earnings in countries such as Mali and Burkina Faso (Baffes 2007). However, cotton producers have gradually gained a political voice and learned to organize their action to influence policymaking (Roberts 1977; Schwartz 1996; Bassett 2001). Another factor of politicization is the "catalytic mobilization effect" that arose from the early organization of cotton markets at the global and the local scale, which enabled farmers to interact with officials, bilateral donors, and large cotton traders at a scale that is rare in African agriculture (Bingen 1994, 1996, 1998).

The political importance of cotton has meant that West African governments have always taken a keen interest in managing the sector and benefiting from its success. For most of their history, West African cotton economies have used monopsonistic state companies for extension services, input and credit provision, and the marketing of output, with prices for seed cotton set panterritorially and panseasonally. This state-led strategy ensured a high degree of government control but also addressed several of the most important market failures that smallholders face, such as poor access to credit, inputs, and information and highly variable output prices. But despite early successes, the state-led strategy came to be increasingly criticized by the late 1980s. Foremost among the cited problems were (1) growing financial insolvency due to bad management and corruption in cotton parastatals; (2) stagnating yields (related to bad management); (3) low profitability for farmers due to substandard margins resulting from both explicit and implicit taxation, as well as declines of the world price of cotton fiber; and (4) the overall expansion of the cotton sector, which rendered national economies vulnerable to external shocks, such as volatility in cotton prices.[3]

As a result, many African countries (and their donors) considered reforming their cotton sectors in the 1990s with the aim of addressing these problems, although the direction and outcomes of reforms varied tremendously (Goreux and Macrae 2003; Poulton et al. 2004; Tschirley, Poulton, and Labaste 2009). In particular, a number of countries opted for a radical liberalization of the sector, which often met with disappointing results largely attributed to coordination failures. For example, in spite of better-managed ginning firms and better price incentives for farmers, the stability of tying input credit to output procurement broke down due to the emergence of multiple buyers, which in turn provided the opportunity for farmers to break their input-for-output contracts and engage in side-selling, which further triggered a cascade of detrimental effects along the supply chain (see Lele, Van de Walle, and Gbetiobouo 1989; Jayne et al. 1997; Poulton et al. 2004; or Tschirley, Poulton, and Labaste 2009, which provided early warnings of the dangers of wholesale liberalization).

Burkina Faso notably went down a very different route by trying to address the government failures plaguing the system (corruption, mismanagement, and principal–agent problems) without unleashing the market failures lurking beneath the surface (coordination problems, public goods, economies of scale, information asymmetries). After considerable debate about the direction of reform among multiple stakeholders, the Burkinabè cotton sector underwent a series of sequenced institutional reforms.

This chapter assesses the effectiveness of these reforms. We argue that new institutional arrangements led to noticeable improvements for households in the agricultural sector, despite an almost unprecedented flow of return migrants from war-torn Côte d'Ivoire (most of whom were absorbed into the agricultural sector) and declining cotton prices along with the declining value of U.S. dollars with respect to euros. Of course the challenge of all nonexperimental empirical research is proving causal linkages between reforms and these highly favorable outcomes. We attempt to do so in two ways. First, we use microeconomic survey data to innovatively establish counterfactual scenarios in which reforms did not take place. Second, we use the admittedly imperfect natural experiment provided by comparisons between Burkina Faso and neighboring Mali, a country that did not engage in substantial reform but otherwise has similar agroecological conditions, common world prices and a common currency, a highly comparable economic structure, and an almost identical institutional history in the cotton sector. Pertinent comparisons with other West African cotton-producing countries are also provided.

The basic conclusion from our research is that the reforms were well designed and fundamentally sustainable and that they had large positive impacts on development outcomes in Burkina Faso. Of course we should not claim that the reform process is unfinished or that the reforms solved all of the problems facing the Burkinabè cotton sector or West African cotton in general. Difficulties experienced since 2006 justifiably raise concerns about the financial sustainability of the reforms. In our penultimate section we review these and other problems, including political sustainability, declining world cotton prices, unfair competition from Organization of Economic Cooperation and Development cotton subsidies, and governance issues. These problems remind us that successful reform is ultimately an ongoing rather than a one-time process. That said, the new institutional arrangements that have emerged in the sector offer a deliberative forum for debating key policy issues before implementing additional reforms.

The Reforms

Early Successes and the Pressure for Reform

According to oral traditions, the cultivation of cotton has always had a specific place in Burkina Faso for several ethnic and social groups (Schwartz 1996; see Kaminski, Headey, and Bernard 2009 for further details). This early specialization shaped the later interaction of farmers with officials, with the specialized ethnic groups the first to expand cultivation and to organize collective action. Cotton then became a coercive tool of the French Upper Volta colony when cultivation became compulsory in the 1920s, but farmers circumvented the "forced corvée" through out-migration and by selling their cotton on the local parallel market or exporting it to Ghana (named Côte d'Ivoire). So in spite of several adjustments, the failure of the colonial policy in the Upper Volta led to its being dismantled in 1932.

Starting in late 1950s, the partnership between the French Company for the Development of Textile Fibers (CFDT) and the Exotic Textiles Research Institute (IRCT) allowed substantial improvements in cotton varieties, and marketing became profitable through the progressive recognition of CFDT quality standards on the world market. Due to substantial increases in yields and to growing interest among farmers, the cotton area increased rapidly after the independence of the Upper Volta in 1960. The CFDT remained a key player, associated through a partnership with the young state and with bilateral donors who funded several development projects for cotton in the 1970s. This association was then replaced in 1979 by the Société des Fibres Textiles du Burkina Faso (West) (SOFITEX), the new parastatal firm also in charge of rural development projects. Meanwhile, the rural communities were progressively organized under a cooperative mode through village groups, called the GVs, that enabled farmers to self-manage their cotton marketing to SOFITEX and to access input credit through village-level joint-liability schemes. The introduction of new production techniques such as the ox-plow, mineral fertilization, and pesticides, along with high-yielding seeds, ensured a twofold increase in agricultural yields over the 1980s, increased cereal production, and improved food security among smallholders. With ample cotton profits, both SOFITEX and the state invested in rural infrastructures such as roads, education, and health, thus further improving farmers' living standards.

Despite early successes, however, state-led contract farming and farmers' groups gradually became less and less successful. Falling cotton prices in the late 1980s revealed serious structural problems that were overlooked under better circumstances. These revealed excessive costs arising from waste, overcharging, duplication of responsibilities, inadequate financial management, and a lack of incentives to control costs (Tefft 2008). Accusations of corruption in cotton parastatals were also common, as well as the opportunistic behavior of farmers. In several countries these reviews of the sector coincided with political change, farmer protests, and the adoption of broader structural adjustment plans at the behest of the World Bank, the International Monetary Fund (IMF), and other donors. The World Bank and the IMF also noted the large gap observed between the producer price and the international price in the aftermath of the CFA devaluation, which suggested that cotton farmers did not benefit from the devaluation (ICAC 1998; Bourdet 2004).

Key Players in the Reform Process

Along with the aforementioned problems due mainly to political interference, the price paid to producers declined from 1988 to 1992, GVs accumulated large debts, and production started to collapse at the beginning of the 1990s.

Setting up the reform agenda, however, required overcoming a number of disagreements among donors, farmers, and the government. Policymakers in Burkina Faso and the Agence Française de Développement (AFD) advocated partial privatization of the sector through local monopsonies. The World Bank did not support the idea, arguing that it would fail to produce the "competitive pressures" that are the crux of the system (ICAC 1998). More details of the market reorganization debate are provided by Kaminski, Headey, and Bernard (2009).

Another potential sticking point was the role that farmers would assume in policy decisions in the country. The government obviously recognized the emergence of politicized farmers, particularly through the Fédération Nationale des Organisations de Producteurs du Burkina Faso (FENOP), although they arguably viewed the incorporation of farmers as a necessary step to bring the World Bank on board. However, a fairly close alignment of interests between the Burkinabè government and AFD suggests that they sought to minimize the political risk of reform through the establishment of a more government-friendly farmers' union. Hence they promoted the formation of a new union, the UNPCB (Union Nationale des Producteurs de Coton du Burkina Faso) in lieu of supporting the more unionist FENOP. That said, the leadership of the UNPCB was drawn partly from FENOP—including the head of the UNPCB, François Traoré—and FENOP representatives were included in some initial fact-finding missions to neighboring countries (an example of cross-country learning in this context). The establishment of cotton cooperatives and the UNPCB was then supported by AFD capacity-building programs.

Because of these negotiations and institutional changes, it took five years for the state to find an acceptable compromise and persuade SOFITEX and producer groups that reforms were in their best interests.

Details of the Reforms and Their Economic Rationales

Table 12.1 shows the chronology and basic content of the reforms. The principal conclusion we draw from Table 12.1 is that Burkina Faso's reform process was

Table 12.1 The chronology of cotton reforms in Burkina Faso, 1992–2006

Phase	Period	Reform
Institution building from the ground up	1992–93	Formal commitment by the Société des Fibres Textiles du Burkina Faso (West) (SOFITEX) to engage producers' representatives in the reform debate through the "Contrat-Plan Etat SOFITEX."
	1994	Amendment of the laws pertaining to the establishment of farmer groups.
	1996–99	Introduction of free adhesion–based mechanisms for local groups of cotton farmers, replacing former village groups with market-oriented organizations, *groupements de producteurs de coton*, with the implementation of new local governance rules.
	1996–2001	Progressive establishment of the national cotton union (UNPCB) with the support of French aid, the government, and SOFITEX based on the membership of local groups and their integration into regional unions.
	1998	Establishment of the Accord Inter-professionnel (Inter-professional Agreement) among SOFITEX, the state, the UNPCB, donors, and the financial consortium (Caisse Nationale du Crédit Agricole, Banque Internationale pour le Commerce, l'Industrie, et l'Agriculture–Banque Internationale du Burkina Faso), replacing the former Contrat-Plan and defining the reallocation of responsibilities.
Progressive reallocation of responsibilities toward new institutions	1999	Partial withdrawal of the state and partial privatization of SOFITEX, with the government allocating half of its SOFITEX shares to the UNPCB.
	2000–06	Progressive delegation of economic activities from SOFITEX and the government to the UNPCB: provision of cereal input credit, management assistance for cotton groups, and participation in quality grading, financial management, and price bargaining; the state's downsizing of its involvement in public goods investment (research and extension services).
	2002–06	Progressive introduction of new players: private input providers, regional private cotton monopsonies (Société Cotonnière du Gourma, East, and Société Cotonnière du Faso, Centre), and private transport companies.
	2004–06	Establishment of an inter-professional association, the Association Interprofessionnelle du Coton du Burkina Faso (AICB), with cooperation among well-represented stakeholders: cotton farmers, banks, private stakeholders, government, and research institutes.[a] Establishment of an association of cotton firms (the Association Professionnelle des Sociétés Cotonnières du Burkina Faso) interacting with the UNPCB within the AICB.
	2006	Change in the price-setting mechanism with more correspondence relative to world price levels and the creation of a new smoothing fund, operational in 2008 and managed by an independent organization.[b]

Source: Authors.

[a] Each stakeholder has representatives in the two committees of the inter-professional body: the management and the decisionmaking committees. A growing number of decisions have been decided and enforced by these two committees since 2004. The model is that of a hybrid public–private regulatory agency.

[b] This smoothing mechanism is a new generation of smoothing funds that Burkina Faso was the first to adopt as a strategy against the current cotton crisis in the region. The fund is linked to the price-setting mechanism and is used to compensate for discrepancies between pre-harvest set prices and observed world prices at the time of commercialization. The French cooperation originally put money into the fund in 2008, and the mechanism involves the smoothing of the set price of cotton by cotton companies according to world price variations, compensation when the world price declines substantially, and a contribution from farmers and companies when the trend is positive. The Banque d'Afrique de l'Ouest is in charge of the management of the smoothing fund to ensure a balanced budget in the long run.

distinguished by a mix of *gradualism, sequencing,* and *institution building*. The gradualism is self-evident from the fact that reforms were carried out over a 14-year period (1992 to 2006) in a step-by-step fashion (Bourdet 2004). The interrelated importance of sequencing and strengthening institutions requires a more careful appraisal, including an appraisal of alternative reform paths. A particularly tempting avenue of reform for donors could have been wholesale liberalization, and with varying degrees of structure and speed, this was a path pursued by many other cotton-producing countries in Africa (Tschirley, Poulton, and Labaste 2009). As we noted in the introduction, however, the problem with this path is the trade-off between competition and a series of market failures—such as information asymmetries and principal–agent problems in credit markets—that could quickly have led to side-selling, credit rationing, a self-defeating loss of competition, and the underprovision of various public goods (research and development, quality control, marketing, and price stabilization). On the other hand, maintaining the status quo was also undesirable. But if some degree of public or private monopsony needed to be maintained, strengthening smallholder participation would also be vital to avoid exploitation of farmers—a position supported by the French cooperation.

Hence the first stage of reform was about building up farmers' organizations rather than prematurely introducing competition into the sector. In 1996 the government legally replaced involuntary GVs with market-oriented free adhesion groups for cotton farmers called groupements de producteurs de coton (GPCs). The significance of changing the rules for grassroots group formation should not be underestimated. Although collective decisionmaking and collective responsibility are widely thought to be highly ingrained features of rural African economies, farmers' groups in Africa have a checkered history from an economic viewpoint (Bernard et al. 2008). In particular, the former joint-liability system of GVs grouped cotton and noncotton growers from the same villages for their input needs, but the input cost was deducted from the value of cotton sales rather than the sales of all products, meaning that farmers had weak incentives to produce cotton. Moreover, GVs were formed at the village level, despite a high degree of heterogeneity within each village and limited capacity for members to monitor each other. This eventually led to very high default rates across the GVs, as well as financial diversion from the GV activities to locally rural development projects rather than cotton projects specifically. Worse still, the joint-liability system for credit repayment left the door wide open to all kinds of abuses, such as the misappropriation of fertilizer and farm equipment to either resale or other crops. As of September 30, 1995, farmers' debt to the National Agricultural Bank reached CFAF 2.1 billion, not including internal outstanding debts (ICAC 1998).

Once the reform of farmers' groups was under way, reform efforts focused on the parastatals. The approach was pragmatic and piecemeal, without brushing aside the problems that had been identified with the cotton parastatal. The proposed solution was to grant regional monopolies or monopsonies to private firms. This "zoning" was essentially a compromise reached by the Burkinabè government between the strong privatization leanings of the World Bank and the reluctance of farmers and SOFITEX to let go of the integrated commodity chain model. Consensus-building was made relatively easy for the government because of its oligarchic structure and the consensus-building process based on field missions to Benin, Côte d'Ivoire, and Mali.[4]

The economic challenge of zoning is the low induced level of competition, which can result in weak incentives both for producers and for local monopolies. But in Burkina Faso it enabled an effective cooperative framework to ensure market coordination while producers gained more power. New initiatives have also been established to expand cotton production areas and improve yields, with farmers consistently involved in the deliberation process (Kaminski, Headey, and Bernard 2009).

Burkina Faso's Intervention in Comparative Perspective

Burkina Faso is only one of a dozen or so significant cotton producers in Sub-Saharan Africa. Moreover, like Burkina Faso, most other African cotton-producing countries have made some attempt at reforming their cotton sectors, and several have engaged in quite drastic reforms. To make comparisons more meaningful, we now more narrowly focus on the case of Francophone countries, which share a number of institutional and historical similarities.

According to Bingen (1998) and Fok (2008), the Francophone model addressed the interests of farmers, states, and the CFDT through its institutional specificity. This model is characterized by four key features. First, the joint CFDT–state parastatals were meant to attract foreign public capital (from the World Bank notably). Second, it benefited from 50 years of French investment in cotton research (by the Institut de Recherche pour le Développement and the Centre de Cooperation Internationale en Recherche Agronomique pour le Développement, with central laboratories in France). Third, national parastatals were linked to the Compagnie Parisienne de Coton (COPACO),[5] the marketing unit of the CFDT, which ensured a market identity for Burkinabè cotton on the world market and recognition of quality. Fourth, monopsonistic power over national production allowed parastatals to purchase agricultural supplies at concessional terms or through public subsidies.

Clearly this viewpoint identifies a strong economic rationale for the Francophone model, especially insofar as it addressed farmers' liquidity constraints, risk aversion, and high transaction costs and the shortage of public goods. These features explain why Francophone West African countries questioned the World Bank's critique of the sector and instead adopted a more cautious approach to reform. Farmers' groups and regional and national unions were also set up in Benin in 1993 (three years before Burkina Faso) and in Cameroon in 2000, but the process is still under way in Mali. In Mali, social conservatism has hampered the emergence of more flexible farmers' groups that could help increase repayment rates on credit (as have the GPCs in Burkina Faso), and a more contestable democracy has slowed down the reform process. Mali and Cameroon are also trying to follow the Burkinabè model by allocating a significant share of the capital of privatized cotton companies to the farmers' union, and all other West African countries are also trying to transfer more responsibilities for input supply and/or extension services to farmers' groups or the farmers' unions.

However, apart from Burkina Faso, only Benin (in 1995) and Côte d'Ivoire (in 1998) have so far permitted private control of cotton companies, although both countries have precluded competition for the supply of seed cotton, whereas the rights to purchase seed cotton in Burkina Faso are allocated to three regional monopolies. The approach in Benin was to give responsibility for activities at different stages of the supply chain to separate private firms, then to establish coordinating institutions to ensure that the different stages of the supply chain function together. Thus, input supply was privatized first, starting in 1992, followed by ginning in 1995, then transportation. Multiple actors, mostly local firms, have now been allowed to enter these different stages, but the prices of different goods and services are still centrally determined, as is the distribution of seed cotton across ginneries (Baffes 2007). Observers also point to the continued favoring of the poorly managed parastatal (Bourdet 2004), estimated to amount to an extra cost of CFAC 65 per kilogram sold (Salé, Tosbé, and Waddell 2001). Bourdet (2004) also mentions the possibility of excess entry into the sector, which reduced profits. Another factor in Benin may be the more limited capacity to expand cotton areas, although we note that in other regards Benin has advantages over Mali and Burkina Faso (such as lower transport costs).

In Côte d'Ivoire the zoning system collapsed because of a lack of an established regulatory scheme, as well as the disastrous effects of the political crisis. In addition, the selection of the new operators in Côte d'Ivoire was not based on long-term development objectives; besides, there was a growing conflict of leadership among producers and ginneries, and the inevitable disputes were difficult to resolve in the absence of any interprofessional committee. Add to that the political crisis in Côte d'Ivoire, and it is little surprise that cotton firms experienced financial difficulties and producers received lower prices. One consequence was that a significant amount of Ivorian cotton was smuggled into Burkina Faso and Mali.

The Intervention's Impact

In this section we assess the impact of the cotton reforms on several important indicators of growth and development. According to stakeholders, the purpose of the reforms was to reorganize the cotton sector, recover management and financial performance, and make outgrower schemes viable. Although addressing food security concerns was not an explicit objective, any increase in income should obviously impact food consumption, so we also closely examine available food security statistics.

Growth in Agricultural Production and Exports

An indirect indicator of the success of the reforms is that Burkina Faso has overtaken Egypt and Mali to become the African leader (in 2006 and 2007) in cotton production and exports of lint cotton, based on a threefold increase in production since the early 1990s. As for its contribution to growth, cotton production accounted for 3.3 percent of national agricultural production before the reform and reached over 8.0 percent in 2006 (FAO 2009).

In neighboring countries, production has followed very different patterns. Benin and Mali experienced some production growth, but at much slower rates than in Burkina Faso. After a short-term positive effect following early reforms, production collapsed in Benin after 2003 due to coordination failures in the newly liberalized sector, including difficulties in recovering input loans and lower investment in critical functions such as research and extension services. In Mali production has stagnated over the past decade after strong growth in the early 1990s. As discussed earlier, the sector has been characterized by uncertainty and gridlock over reform, particularly in the form of political tensions between the parastatal and farmers.

As to the sources of growth, an unusual feature of Burkina Faso's success story is that production growth has been largely based on increased land use for cotton and the entry of many new producers. This extensive growth has been driven mostly by the intervention itself, because it significantly improved incentives for cotton production through better contractual relationships within farmers' groups and between farmers' groups and cotton firms. Kaminski and Thomas (2010) show that the direct effects of the reform involved earlier payments of cotton seed to farmers, easier access to inputs, and a guarantee of selling. Thus the lower risk profile of the cotton crop, together with better use of inputs, has been instrumental to Burkina Faso's success.

Another potential source of growth lies in the pricing issue. As Baffes (2007) showed, West and Central African Francophone countries typically went from taxing producers to a more or less neutral stance (except in the late 1980s). The shift from taxation to supportive governmental policies is often explained by the world prices levels and the willingness of states to give farmers sufficient incentives to keep growing cotton. Nevertheless, the change in the price determination mechanism over the past decade eliminated a major part of governments' intervention in pricing. From the World Bank perspective, this was one of the goals of reform, so the reform was successful in this respect. However, because several countries engaged in these reforms, it is notable that Burkina Faso achieved much higher production growth than its peers (Figure 12.1 and Table 12.2). This implies that simply fixing the prices was not sufficient to achieve production growth and that institutional reforms were key to Burkina Faso's success (Kaminski and Thomas 2010).

We note that the indirect contributions of cotton sector growth to GDP growth might also be significant, especially in terms of foreign exchange (Kaminski, Headey, and Bernard 2009).

Finally, one significant concern in interpreting Burkina Faso's export performance is the side-selling problem associated with Côte d'Ivoire's political crisis over 2002–06. Although this was substantial (see more detailed evidence in Kaminski, Headey, and Bernard 2009), it actually made very little difference to the growth rate of cotton production in Burkina Faso. Indeed, because there appears to have been no diversion in 2007, the average annual growth rate of production over 1995–2007 is unchanged.

Contributions to Employment Creation

As noted earlier, growth in Burkina Faso has involved an extensive process, including greater use of labor. Indeed, the share of cotton farmers (and related household

Figure 12.1 Five-year average of seed cotton production in six African countries, 1980–84 to 2005–07

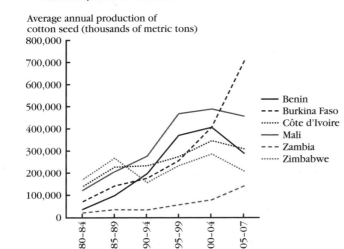

Source: FAO (2009).

Table 12.2 Cotton production performance in Benin, Burkina Faso, and Mali, 1995 and 2007

Performance (measure)	Benin	Burkina Faso	Mali
Seed cotton production			
1995 (metric tons)	328,227	150,451	405,907
2007 (metric tons)	313,500	690,000	414,965
Average annual growth rate (percent)	−0.38	13.53	0.18
Cotton areas			
1995 (hectares)	246,789	145,362	335,463
2007 (hectares)	285,000	699,797	479,734
Average annual growth rate (percent)	1.21	13.99	3.03
Yields			
1995 (kilograms per hectare)	13,297	10,346	12,072
2007 (kilograms per hectare)	11,000	9,857	8,645
Average annual growth rate (percent)	−1.57	−0.40	−2.74
Production per worker			
1995 (kilograms)	90.0	16.6	47.9
2007 (kilograms)	73.9	53.9	37.7
Average annual growth rate (percent)	−1.63	10.31	−1.98
Export earnings			
1995 (millions of U.S. dollars)	169	68	201
2006 (millions of U.S. dollars)	110	235	256
Average annual growth rate (percent)	−3.83	11.93	2.22

Source: FAO (2009).

members) has almost doubled between 1994 and 2003, from 11.3 to 19.9 percent of agricultural employment.

In terms of raw numbers, the acceleration in cotton production has absorbed more than 200,000 new farmers (the number of households of cotton farmers has almost doubled in these 10 years), some of whom were already cropping land and others of whom were migrants.[6] This absorption of new labor was remarkable not only because previous influxes of return migrants have been associated with unemployment and economic turmoil (for example, that in Ghana in the early 1980s) but also because the cotton reforms allowed returning migrants from Côte d'Ivoire to quickly access inputs and form their own farmers' groups (Kaminski 2006), which have been integrated into the UNPCB and regional unions. According to Savadogo and Sakurai (2007), the impact of the political crisis in Côte d'Ivoire on Burkinabè agriculture was the influx of 0.7 to 0.8 additional active workers per rural household. Overall, it is estimated that cotton growth in Burkina Faso has created around 235,000 full-time jobs, so cotton farmers represent almost one-sixth of all rural households in Burkina Faso, the largest employment group in the country (Table 12.3).

Table 12.3 Estimates of the numbers benefiting from the intervention, 1996 and 2006

Measures of those benefiting	1996	2006
Number of households cultivating cotton	98,520	176,570
Number of people living with cotton earnings	837,250	1,845,300
Number of full-time "cotton jobs" in the agricultural sector	345,000	580,000
Average land cultivated by a household that produced cotton (hectares)	6.06	6.92
Average land share dedicated to cotton by cotton farmers (percent)	0.34	0.56
Number of *groupements de producteurs de coton*	6,600[a]	9,100
Number of ginneries	5	18

Sources: The table displays estimates of farmers' characteristics obtained by a survey of representative households in cotton areas of Burkina Faso (Kaminski 2006) and from official data (DGPSA 2008).
Notes: The land shares are gross estimates but overestimate land use because other crops are often associated with cotton on the same plots (intercropping and mixed farming). On average, the net land share dedicated to cotton has shifted from roughly 0.3 to 0.5 percent (figures given by a local World Bank officer in 2006), but it is difficult to precisely estimate the net land share.
[a]This figure is for 1999 rather than 1996.

Repayments of Loans and the Financial Performance of Ginning Firms

An important goal of the reforms was to ensure that farmers repaid loans. In this regard, the results have been very impressive. Repayment rates have risen from around 40 percent before the reforms to around 95 percent (under standard climatic conditions) thanks to a better management of outgrower schemes and more cohesive farmer groups.[7] Input use was already quite high among cotton growers before the reform, so input use per hectare has not risen markedly (Kaminski, Headey, and Bernard 2009). However, many new farmers and new areas of cultivation have benefited from greater access to inputs, which, in turn, have supported the expansion of cotton cultivation.

The expansion of modern inputs has also contributed to the increased availability of inputs for cereal production (on credit and linked to cotton production), which are now cautiously managed by the UNPCB.[8] GPCs are central at a time of rapid commercialization of cotton production because they reduce the costs of collection by centralizing members' products (eventually gathering products from other local GPCs as well), check weights and grades with purchasers, and even engage in legal action in case of disagreements. Overall, more than 80 percent of GPCs are engaged in input provision, nearly 100 percent in output commercialization services for their members' cotton. In many cases, such services are also provided for crops other than cotton.

This positive outcome should be attributed to the establishment of GPCs, although there are caveats. As shown by Gray (2008), some groups have incurred higher internal debts (cleared by the most productive members of the groups), which has been associated with some disenchantment with cotton production since 2005. As of 2009, internal debts have grown substantially over the past three years, which has raised concern about the cohesiveness of GPCs (Kaminski 2009b).[9]

The financial sustainability of the cotton sector also requires competitive ginning costs, efficient transportation and marketing, research and extension investment, improved management of cotton firms, and realistic but fair price determination mechanisms. According to Bourdet (2004), production costs at SOFITEX increased only slightly in nominal terms and most likely decreased in real terms, which explains the greater profitability in Burkina Faso as compared to Benin and Mali. Other factors include the greater competition among trading companies in charge of the sale of cotton from Burkina Faso and a more successful policy of forward sales in the early 2000s (Goreux and Macrae 2003). The increasing production costs of the Compagnie Malienne du Développement des Textiles in Mali are also well documented by several audits (that reported in Gergely 2004, as an example) and are often attributed to the perennial mentality that the state or donors will always bail the sector out if it gets into trouble (Bourdet 2004).

Income and Poverty

The ultimate test of the broader success of this intervention is its implied reduction of poverty. However, poverty outcomes remain difficult to assess with descriptive data alone. In fact, although the overall trend of rural incomes in Burkina Faso has been positive, the poverty impacts of the intervention have clearly been distorted by the negative effects of the political crisis in Côte d'Ivoire,[10] decreasing cotton world prices, and increasing prices of inputs (especially fertilizers), as well as by the positive effect of the unusually large price margins received by farmers. Hence, assessing counterfactual scenarios, as we do in a subsequent section, is also important.

In recent times cotton has developed a poor reputation with respect to poverty reduction because of the so-called Sikasso Paradox (Kaminski, Headey, and Bernard 2009). Grimm and Gunther (2004) revised national poverty estimates to correct for various problems and found that although poverty increased from 1994 to 1998, the poverty headcount over 1998–2003 fell from 62 to 48 percent (68.7 to 53.3 percent in rural areas).[11]

Clearly these broad national poverty trends bear only an indirect relation to the cotton intervention. However, Grimm and Gunther (2004) found that the poverty headcount among cotton farmers was reduced by one-fourth over 1995–2003, from 62.1 percent of cotton producers in 1994 to 46.8 percent in 2003. In contrast, the incomes of other people in occupations increased minimally or even declined, depending on the deflator used (Table 12.4). Data from Kaminski (2006) show that the poverty rate (headcount ratio) among cotton households was around 47 percent in 2006.[12] Kaminski argues that the conventional poverty statistics arguably under-

Table 12.4 Percentage change in household expenditure by occupation, 1995–2003

Occupation	Deflated with consumer price index	Deflated with household price index
Public	13	5
Private	4	-3
Informal	3	-7
Subsistence	33	10
Cotton	43	19

Source: Grimm and Gunther (2004).

estimate poverty reduction, because the cotton producers in 2006 included new entrants with higher poverty rates than those of more experienced cotton farmers. His data also show that the distribution of income among cotton producers was reasonably equal (with a Gini coefficient of 0.41), with a relatively small subset of large land-holders occupying the highest quintiles of the distribution.

Subjective measures of well-being also confirm the positive impacts on cotton farmers. Kaminski (2009a) observed that Burkinabè cotton farmers expressed significantly higher levels of satisfaction with their levels of wealth after the reforms (see Kaminski, Headey, and Bernard 2009 for further discussion).

Unlike food production, cotton growth does not impact poverty by lowering the price of food, which is one of the most direct determinants of poverty reduction. We estimate that the spill-overs of cotton growth onto the local economy are also relatively modest. From data collected by Kaminski (2006) and informed guesses about which expenditure items are produced in the region and which are not, we estimate that there has been roughly a 7 percent increase in the demand for food in the region and a 4 percent increase in the demand for nonfood, or a 6 percent increase in demand overall (Table 12.5). These spill-over effects are reasonable but are probably not large enough to have a major poverty reduction impact on the noncotton population.

Food Production and Food Security
The rapid growth in cotton production could have impacted food security and nutrition through two principal channels. First, the cotton intervention could have induced or spilled over into more rapid growth in food production due to land extension and productivity increases. Second, the higher incomes of cotton producers would have increased their own household food security.

Regarding the first of these channels, food production has obviously not increased as rapidly as cotton production. However, several products, which are

Table 12.5 Back-of-the-envelope estimates of the increased demand for noncotton goods and services in cotton-growing areas, 1996–2006 (percent)

Cotton households' characteristics	Food	Nonfood	Total
(1) Share of purchases that are locally produced	70	40	57.1
(2) Share in expenditure items in total expenditure	57	43	100
(3) Increase in cotton expenditures	30	30	30
(4) Share of productive population in cotton production	50	50	50
Total increase in demand within region = (1) × (3) × (4)	10.5	6.0	8.6

Sources: (1) and (2) are numbers based on household survey data from Kaminski (2006) and best guesses based on local knowledge. (3) and (4) are based on averaging the two estimates from Grimm and Gunther (2004), also presented in Table 12.4, and on information from other sources on the amount of income growth among cotton producers.

grown mostly in cotton–cereal systems, have experienced a strong increase. In the case of maize, the average annual growth rate in the value of production has been around 10 percent over the cotton reform period. Significant production growth has also been achieved for sorghum. The pertinent question here is whether the cotton intervention resulted in the production growth of these products.

We have some tentative evidence that this is indeed the case. From Table 12.6 we see that growth in overall demand for food in cotton regions in Burkina Faso could have been 7 percent or more and that a significant portion of this could have been in locally produced food (often perishable). But a second channel of impact could have been increased grain production from cotton farmers themselves through the rotation of cotton production with grain production or through the extension of cultivated land (via mechanization and additional labor). Before the reforms, cereal and cotton input credit were managed jointly by SOFITEX but with significant credit rationing. However, the formation of cotton-specific farmers' groups did not lead to a neglect of cereal production or to the farmers' demand for diversifying production between cash crops and food staples, precisely because the groups themselves make autonomous decisions and also have influence on the management of the sector via the cotton unions. As a result of these institutions, fertilizer application to maize increased by 20 kilograms per hectare among cotton producers (Direction Générale des Prévisions Statistiques Agricoles [DGPSA] 2008). However, these positive effects on food production may have been counterbalanced by the reallocation of larger shares of land from food production to cotton production (Table 12.7). Thus the overall impact on food production is ambiguous.

Kaminski and Thomas (2010) estimated the evolution of land use (cotton versus foodcrops) by cotton households. This enabled them to estimate the total evolution

Table 12.6   Evolution of nutrition-relevant food consumption for representative producers, 1996–2006

| Foods consumed | Changes in consumption by households (percent) | | | | |
	Large increase	Slight increase	Constant	Slight decrease	Large decrease
Fat nutrients	5	48	31	10	6
Dairy products	5	21	43	16	15
Animal proteins	17	47	14	14	7
Fruits	6	34	36	18	6
Vegetables	10	44	28	15	3
Tubers	5	33	37	18	7
Cereals	19	53	17	7	4

Source: Authors' estimates.

Table 12.7   Evolution of land use for a representative sample of cotton households, 1996 and 2006

Measure of land use	1996	2006
Total cultivated area (hectares)	5.51	6.92
Land share dedicated to cotton (percent)	20	56
Average cultivated area under foodcrops (hectares)	4.39	3.07
Average cultivated area under cotton (hectares)	1.12	3.85

Source: Authors, based on estimates computed from Kaminski and Thomas (2010).
Note: These means account for new cotton producers whose land share was 0 percent and land under cotton was 0 hectares before the reform and are weighted according to land cultivated by households to estimate the average increase in foodcrop-cultivated areas.

of foodcrop areas for the average household cultivating cotton in 2006 (see Table 12.7). Moreover, a significant share (10 percent of total cultivated cotton areas in 2006, according to Kaminski 2006, and 20 percent among new producers) of new cotton fields have been intercropped with foodcrops, meaning that the increase in land shares dedicated to cotton should be treated with caution because it is only a gross measure. This fact, together with the increase in food production yields, explains why the expansion of cotton production (mostly extensive) appears to have had a fairly neutral impact on food production among cotton growers.

The impact on food security also depends on how increased cotton incomes contribute to food consumption. Hence, both cotton (cash income) and food production (minus the input costs) determine food security. Cotton incomes may be particularly useful in enabling households to face familial and social expenses without having to sell their foodcrops or their livestock. Using data from a retrospective survey from Kaminski (2006) that tracks consumption patterns in 1996 and 2006, we define a threshold of household food security and then estimate the evolution of food security across households over the reform period.[13] We find that a large share of cotton households appear to have improved their food security. In 2006 around 69 percent of households that cultivated cotton could be classified as food secure. Of these around 49 percent have increased their consumption in cereals, while 3 percent of the remaining 31 percent who are food insecure have decreased their cereal consumption. Overall, an improvement in food security is likely to characterize up to 46 percent of households in the case of cereals.

Focusing on the households located close to the food security curve, we also found that 12 percent of the population crossed the line upward during the reform, while 2 percent fell below (based on cereal consumption).[14] These can be interpreted as lower bound estimates. On average, it is likely that food security has improved for 30 percent of the cotton population and deteriorated for 4–5 percent. In 1996, around 40–45 percent of the population that cultivated cotton in 2006 was food secure, while 70 percent could be so declared in 2006. Hence, the rural population dependent on cotton cultivation has become much less vulnerable in terms of food needs. However, we have no evidence that nutrition has improved in the broader population. In fact, Demographic Health Survey data for Burkina Faso indicate that malnutrition rates for children under 5 years old rose in the 1990s and then declined somewhat in only a few regions, including the cotton belt in the west.

Estimating the Effects of the Reforms Based on a Counterfactual Scenario

As we noted in previous sections, the cotton sector in Burkina Faso has been subject to a number of adverse exogenous shocks that have distorted the overall impact of the cotton intervention itself. The principal shocks over the period in question were the falling level of world cotton prices, the decrease in value of the U.S. dollar with respect to the euro and producers' prices, and the effects of a large influx of return migrants from Côte d'Ivoire. In this section we attempt to plausibly simulate what would have happened to cotton production without the intervention and without the Ivorian crisis.

Methodology

Our strategy for simulating these different counterfactual scenarios is described in Figure 12.2 and as follows.

Our first step was to consider the most important and most tangible benefits of the intervention, as well as the impact the Ivorian crisis would have had on the

Figure 12.2 A methodology for estimating counterfactual scenarios

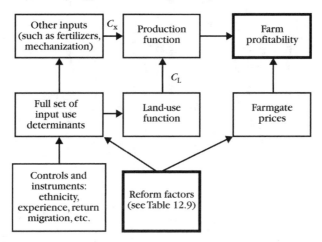

Source: Authors.
Notes: C_x is the elasticity of output with respect to other inputs and C_L is the elasticity of output with respect to land. In the simulations these elasticities transmit the effect of the various assumptions that we make in Table 12.9 about the benefits of reform.

various determinants of farm production and profitability. The characteristics of the reform and counterfactual scenarios are listed in Table 12.8. The channels of impact include access to and use of inputs, prices received by farmers, and land use patterns, including access to land (which is important for return migration scenarios and ethnicity factors). Clearly the differences between the counterfactual and reform scenarios are a matter of judgment, but we believe the previous two sections and existing research (Brambilla and Porto 2005; Savadogo and Sakurai 2007; Kaminski and Thomas 2010) have provided a sound basis for the assumptions listed in Table 12.8. All counterfactual results are displayed in Table 12.9.

Counterfactuals on Land Use

Using the bivariate ordered probit model of Kaminski and Thomas (2010) we estimate changes in land use patterns and in total cultivated land by recalibrating the values of the model's explanatory variables according to the various scenarios assumed in Table 12.8.

Our first finding is that much more new land was cultivated because of the intervention and the Ivorian crisis. With the growth of the active labor force, migration, and mechanization, we estimate that the new land cultivated was 215,000 hectares greater than it would have been without the reforms (this amount is calcu-

Table 12.8 Main hypotheses for the counterfactual and reform scenarios

Channels of impact	Counterfactual scenario (C)	Reform scenario (R)	Comments
Purchase price of cotton	Decrease of 5% in real value	Increase of 5% in real value	The difference is due to the lack of farmers' influence in C.
Local price of maize	Increase of 20% in real value	Increase of 10% in real value	The difference is due to less production and more demand in C.
Input access and use	Input credit decreasing by farm for experienced cotton growers; marginal input access for cereals and new cotton growers	Opportunity to borrow inputs for new land under cotton and also for cereals (mostly maize)	Input access is severely rationed in C due to low credit repayment rates under village groups compared to *groupements de producteurs de coton*.
Contractual relationships in the cotton sector	Uncertainties about the date of payment, recurrent late delivery of inputs, disputes about quality grading and weighing	Earlier payment for cotton seed, earlier delivery of inputs, better quality-grading process, and transparent weighing	The conditions in C were the ones prevailing before the reform.
Technical assistance	Farmers encouraged to moderate land allocation to cotton and to improve farming systems	Farmers encouraged to allocation moderate land to cotton and improve farming systems	These statements are drawn from estimations by Kaminski and Thomas (2008) and interviews by Kaminski (2006).
Mechanization	Only experienced farmers who pursued cotton farming seen adopting animal farming at a 50% lower rate than in the actual scenario	Mechanization correlated to experience in cotton growing, livestock assets, and technical assistance	The slower rate of mechanization is due to less capital accumulation and lower farmers' incomes.
Number of cotton households	With Ivorian crisis: 110,000 in 2006; without Ivorian crisis: 99,000 in 2006 (10,000 exiting and 10,000 new cotton households)	98,500 in 1996; 176,600 in 2006 (without Ivorian crisis: 164,800 in 2006)	According to the estimates of Brambilla and Porto (2005) for entry to and exit from cotton production, Savadogo and Sakurai (2007) for the impact of the Ivorian crisis.
Number of active workers per household	With Ivorian crisis: 6.3 in 2006; without Ivorian crisis: 6.05 in 2006	5.8 in 1996; 6.6 in 2006 (without Ivorian crisis: 6.35 in 2006)	This is due to fewer incentives for out-migration in R and to the influx of returnees.
Ethnicity effects	Significant for input access but less so for land access (less demographic pressure)	Dependent on access to land; more difficult for nonresident ethnic groups and migrants	Nonresident ethnic groups have limited access to land in R and to input access in C.

Source: Authors.

Table 12.9 Comparing actual events to counterfactual outcomes, 1996 and 2006

Outcome	Actual events (reform + crisis)			Counterfactual (no reform + crisis)	Counterfactual (reform + no crisis)	Counterfactual (no reform + no crisis)
	1996	2006	Change	Change	Change	Change
Household cultivated land (hectares)	5.51	6.92	+1.4	+0.33	+0.94	–0.03
Total cultivated land (thousands of hectares)	973.1	1,222.1	+249	+36.3	+154.7	–5.1
Household cotton land share (percent)	20	56	+36	–5	+28	–15
Cotton area (thousands of hectares)	194.6	684.4	+489.8	–25.1	+336.2	–88.2
Food area (thousands of hectares)	778.5	537.7	–240.8	+51.4	–181.5	+83.1
Cotton production (thousands of metric tons)	208.1	766.0	+557.9	–45.6	+390.7	–111.5
Maize production (thousands of metric tons)	459.7	380.7	–79.0	–17.7	–54.9	–6.3
Millet production (thousands of metric tons)	159.4	218.4	+59.0	+96.7	+71.0	+91.6
Groundnut production (metric tons)	89.7	65.0	–24.7	–3.1	–18.5	+2.0
Sorghum production (metric tons)	286.4	243.9	–42.5	+9.7	–24.3	+20.8
Household one-year cotton income (2006 US$)	418.3	1,627.6	+1,209.3	–171.8	+916.8	–418.9
Household one-year food income (2006 US$)	975.8	983.6	+7.8	+420.7	+197.4	+485.0
Household credit cost for one year (2006 US$)	156.4	668.2	+511.8	–32.4	+389.5	–140.4
Household agricultural one-year income (2006 US$)	1,237.7	1,943.0	+705.3	+281.3	+724.7	+206.5
Agricultural one-year income per active worker (2006 US$)	213.4	294.4	+81.0	+24.5	+96.8	+23.1
Average net one-year cotton income per active worker, 1996–2006	n.a.	n.a.	80.2	58.7	72.8	50.8
Percent food-secure households	45	70	+25	+5	+25	+3
Number of food-secure people (thousands)	675.5	1291.8	+616.3	+146.6	+520.7	+79.7

Source: Authors.

Notes: These estimates are computed for the average household cultivating cotton in 2006, including those who were outside cotton production in 1996, to derive total net effects. The 1996 levels are different across scenarios because they include cotton households that entered cotton production during the reform and exclude those that exited. Intrafood crop allocation on food areas is assumed to be constant over time and across scenarios. n.a. means not applicable because the row is for averages.

lated by summing the reform values in Table 12.9 and subtracting the "no reform" values). Of these 215,000 hectares, the Ivorian crisis drove the cultivation of around 95,000 hectares, or slightly over two-fifths.

In the case of cotton, around 500,000 hectares were cultivated because of the combined effects of the Ivorian crisis and the intervention. However, of these 500,000 hectares, the Ivorian crisis accounted for about one-third of the increase in cotton areas, so the increase was largely driven by the intervention. Without the intervention, the land allocated to cotton would have stagnated. This is actually consistent with what happened in other West African countries where reforms were unsuccessful or postponed. After the 1994 CFA devaluation, the cultivation of cotton land increased in all Francophone countries. However, in the other three countries (Benin, Côte d'Ivoire, and Mali) the amount of land allocated to cotton stagnated after around 1998, while in Burkina Faso it increased steadily up to 2007, by which time it was 3.5 times larger than in 1994.

Perhaps the only caveat to the success of the intervention's impact on land use is that without the intervention the area allocated to foodcrops in the cotton-growing areas of Burkina Faso would have been around 290,000 hectares more. However, as we noted in the previous section, the total land area cultivated with foodcrops in all of Burkina Faso still increased over the reform period. Moreover, our counterfactual estimates suggest that without the intervention there would have been no increase in total land use in cotton areas. This is because, with no reform, the slower rate of mechanization and the lower quality of technical assistance would have counterbalanced the impact of return migrants and other migrants on land extension. And because Burkina Faso is a land-abundant country, making greater use of land (in a sustainable fashion) constitutes an efficient development strategy.[15]

Counterfactuals on Yields of Cotton and Food Crops

To estimate counterfactuals on crop yields, we use DGPSA (2008) data from 1996 to 2004 on production, plot characteristics, and input use to estimate a Cobb-Douglas production function, in the spirit of Fan (1991). The production function is estimated for the main crops cultivated by cotton households in 2004, which were cotton, maize, groundnuts, white sorghum, and millet. The full production function methods and results are presented in Kaminski, Headey and Bernard (2009).

Figure 12.3 shows the estimated annual growth in crop yields under reform and no-reform (with crisis) scenarios. Here again, the results show that the reforms were the decisive factor in the significant increase in farm yields, whatever crop is considered. As noted earlier, stagnant yields of cotton should be understood as an increase in yields for the most experienced farmers counterbalanced by the entry of marginal land and less experienced farmers. The counterfactuals are especially use-

Figure 12.3 Counterfactual estimates of annual average growth in crop yields, 1996–2006

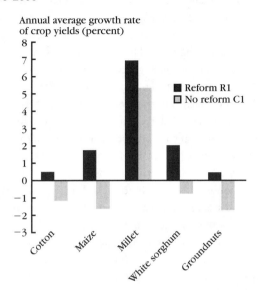

Source: Authors' estimates based on counterfactual simulations.

Notes: Computations are done to account for all cotton households in 2006, including those that did not cultivate cotton in 1996, to derive the total net gain in crop yields over the reform period. Obviously this does not apply to cotton. Growth rates were derived by comparing yields, as measured in kilograms per hectare.

ful in this regard because they show that without the reforms, cotton yields would actually have decreased. Another interesting finding is that cereal yields increased much more rapidly because of the cotton reforms, notably for maize and sorghum, the crops that benefit more from rotation with cotton. Note that in this context the effect of the Ivorian crisis is rather marginal. We attribute this to the fact that many return migrants joined already-established households, so the more labor-intensive use of the land would have increased yields, thus counterbalancing any negative effect on yields from the operation of new farms by return migrants.

Turning to the production results, the reform scenarios are characterized by a slight decrease in food production, which was more than compensated for by a substantial increase in cotton production (the extent of this net benefit is assessed later when we look at farm incomes). Also of interest is the effect of the Ivorian crisis, which accounts for about one-third of the growth rate in cotton production, with the intervention accounting for two-thirds.

Counterfactuals on Farm Incomes and Food Security

Based on the previous estimates, we are able to measure changes in overall agricultural incomes for each scenario for the average household, according to price data collected at the village level in 2006 by Kaminski (2006) and the cost of input credit.

The results indicate that the effect of the intervention alone accounts for a significant change in the agricultural incomes generated by cotton households. The average worker increased his or her agricultural income by around $56. From a poverty reduction perspective, the rural poverty line was set at CFAF 100,000 per consumption unit in 2006, so around 25 percent of all basic needs were derived from the net increase in agricultural incomes that resulted from the reforms in addition to the base income and other agricultural and off-farm activities, such as livestock.

As for food security outcomes, in the previous section we estimated that 70 percent of the cotton households were food secure in 2006 but that only 40–45 percent had been food secure before the reforms. Using the counterfactual estimations on agricultural incomes, we compute the counterfactual rates of food security among cotton households. The results indicate that the cotton reforms played an extremely positive role in the reduction of food insecurity among cotton-producing households. They benefited 460,000 additional people if we account for the effects of the Ivorian crisis, 365,000 additional people if the Ivorian crisis is not considered. Overall, that means that at least 5 percent of the food-insecure population in Burkina Faso became food secure because of the reforms.

The Intervention's Sustainability

Sustainability relates to the economic or financial viability of an intervention, its political sustainability, and also its environmental sustainability. In this section we try to briefly consider the financial and political facets.[16]

Political Sustainability

The political sustainability of reform is all too often overlooked in economic appraisals of "good" policies. Nevertheless, political sustainability is a necessary condition for economic success, if only because the survival of a reform depends on it (Rodrik 1996). Indeed, many periods of growth have not been sustained precisely because institutional arrangements capable of resolving conflicts did not emerge (Rodrik 1999). The West African cotton sector itself provides a long list of examples of political unrest, especially tensions between cotton farmers and governments (for example, that in Mali in 1973, the early 1990s, and 2000 but also that in Burkina Faso in 1992). Resolving these conflicts is therefore a major challenge in the cotton sector, especially in the context of an imperfectly competitive market in which monopsonies have considerable potential to exploit farmers in a laissez-faire setting.

Perhaps the first dimension of sustainability simply relates to completing the reform process. This is especially important in a gradualist process in which reforms proceed very incrementally. Although the UNPCB has emerged as an important player at a national level, the local farmers' groups (the GPCs) still face a number of constraints in their economic development. Most important, they are limited in the scope of the services they can offer due to lack of resources. This is apparent from their low level of capital stock: in 2002 just 7 percent had warehouse facilities and fewer than 2 percent provided their members with occasional access to a tractor.[17] This lack of resources is itself enhanced by the GPCs' environment and their partners. First, they can be constrained by their own village environment: as shown by Bernard et al. (2008), egalitarian norms sometimes impede the development and the effective functioning of market-oriented groups. Second, NGOs and other external partners sometimes perceive GPCs as community organizations more than as professional ones. As a result, support is conditioned on the GPCs' engaging in "social" actions that sometimes impair their capacity to pursue their economic strategies. Finally, during times of limited government resources in rural areas, local administrations often rely on GPCs to help finance investments and services (schools, police office, health posts, etc.) via a tax on their earned "ristournes," although noncotton farmers, traders, and civil servants are less often expected to do so.

There are also doubts about the role of the UNPCB and how effectively it still represents the interests of cotton farmers. Certainly farmers have tangibly benefited from the GPC–UNPCB relationship through participating in quality grading with SOFITEX executives, discussing financial issues with extension agents through local and regional credit committees, representing farmers in the claim instigated against U.S. cotton subsidies in Cancún in 2004, and working for a larger say in determining price outcomes (Gray 2008). But lack of accountability is an increasingly common charge leveled against the UNPCB (see Kaminski, Headey, and Bernard 2009 for more details).

The governance problems of SOFITEX (although production and management costs were better in 2004) are also linked to a lack of transparency and recently began to affect farmers harshly, while late payments have caused the trust relationships between GPCs and other cotton firms to deteriorate. In 2009, for example, farmers ordered inputs without having information about the price they would be charged (Kaminski 2009b).

So although the progressive establishment of UNPCB and GPCs delivered many positives for farmers, it seems probable that further reforms will be needed in order to increase accountability and transparency and improve the functioning of the UNPCB, as well as SOFITEX. Furthermore, because Burkina Faso's political system is highly oligarchic, farmers have few means to influence policy outcomes if the UNPCB fails to voice their concerns (or indeed, if the UNPCB is the principal problem).

Financial Sustainability

If Burkina Faso's cotton boom is built on an unsustainable financial model, its success will certainly be found to be illusory. Of particular concern is that the growth of participation in the cotton sector was driven by unsustainably high prices being paid to farmers (Tschirley, Poulton, and Labaste 2009). In that sense the empowerment of farmers and the increased role of the UNPCB in price setting might have resulted in an overcorrection to the previously low prices paid to farmers in the 1970s and 1980s. In all West African countries, the price mechanism was also linked to a stabilization fund designed to support producers' prices when the world market was low and to be replenished when the world market was high (by paying farmers lower prices than could otherwise be paid). The rationale for these funds was to avoid dramatic drops in producers' prices and to limit market risks for cotton companies. According to Tschirley, Poulton, and Labaste (2009), these support funds functioned well in Burkina Faso and Cameroon until 2004, when the world price of cotton fell precipitously. Since then, however, they have fallen victim to the unsustainably high prices agreed to among cotton companies and the producer union. In Burkina Faso, the fund was depleted and could not cover the deficits during the 2005 and 2006 seasons, and recently a three-year cumulative deficit of more than €100 million was revealed.

The reasons for the depletion of funds partly related to overly generous prices for farmers, but there were also allegations of corruption because the replenishment of funds from high-price years was less than expected.[18] Also, the price determination mechanism was not sufficiently flexible and was based on world prices of the past crop season. As a result, the government and donors were required to bail out the fund. Because of this and the large regular expenditures of the cotton sector, it is no surprise that the Burkinabè cotton sector's net budgetary contribution is actually negative. However, unlike in Mali, the value added per capita in Burkina Faso is the highest in Africa ($10.52). This means that for every dollar of value added generated, the Burkinabè government had to pay roughly 8 cents, whereas the Malian government had to pay six times as much (41 cents).

Unsustainable pricing induced overly large income benefits for farmers in recent years (Kaminski, Headey, and Bernard 2009), but most of the benefits were sustainable in the long run. Further counterfactual exercises show that the cotton boom was 25 percent artificial when considering both the Ivorian crisis and the unsustainable price in the peak production period of 2006–07.

Lessons from the Burkinabè Cotton Reforms

A healthy degree of caution should always temper our enthusiasm about the lessons to be learned from a success story. Of particular concern in this case is that successful outcomes were not driven by the intervention itself (the so-called functionalist fallacy); that the reform was unsustainable environmentally, financially, or politically; and that the reform cannot be generalized to other contexts because its design was highly context specific (the external validity issue). In this chapter we have tried to show—through survey evidence, quasi-experimental comparisons with the situations in neighboring countries, and counterfactual simulations—that the remarkably successful outcomes observed in Burkina Faso's cotton sector over the past 10 years were indeed primarily driven by the reform process but were only partially sustainable.

This leaves us with the important question of generalizability: To what extent can policymakers in other contexts learn from the Burkinabè experience? In this regard we distinguish between general principles and specific practices that could potentially be adopted by other cotton producers.

With regard to principles, Burkina Faso's success story confirms much of the existing evidence on the process of reform. Specifically, gradualism facilitates institution building (prices can be changed overnight but institutions cannot) and allows policymakers to observe and improve upon the outcomes of reform. In contrast, wholesale liberalization can be dangerous in the presence of uncertainty because liberalization is often difficult or costly to reverse. Gradualist or partial liberalization outcomes therefore offer policymakers some opportunity to gauge the impact of liberalization while keeping deeper liberalization on the slate for future reform.[19]

Sequencing also appears to have been important. In most of African agriculture, farming is beset by several fundamental market failures relating to information asymmetries, coordination failures, and insufficient provision of public goods. Reforming prices before reforming farmers' incentives could therefore have had little impact on production if government-induced price distortions were not the binding constraint (ICAC 1998). In other contexts, liberalization in rural Africa has hurt farmers because of information asymmetries between farmers and agricultural traders or marketers (Kherallah et al. 2002). These empirical accounts are clear reminders of the theory of second best (Lipsey and Lancaster 1956), which warns us that removing one distortion (such as a monopoly) can lead to adverse outcomes if other distortions (such as coordination problems, information asymmetries, and public goods) are also present.

Finally, as in a number of Asian countries that used "deliberation councils" to bring together policymakers and private-sector (industrial) representatives (World Bank 1993), Burkina Faso's Inter-Professional Committee provided a useful platform for swapping information, resolving conflicts, and applying "soft" policies of persuasion rather than coercion. Even prior to the formation of the inter-professional committee in the sector, reformers encouraged SOFITEX personnel and farmers' representatives to tour neighboring countries and learn from their reform experiences. Thus, consensus building was a major feature of the reform process right from the start.

But given that Burkina Faso's reforms were indeed context specific, it is important to raise the question of how much other developing countries can learn from Burkina Faso. Undoubtedly Burkina Faso's experience is mostly relevant to cash crop sectors in other African countries, especially cotton sectors. Even there, however, a high degree of caution is certainly warranted because of the very different historical contexts of cotton sectors across the continent (Tschirley, Poulton, and Labaste 2009). In West Africa the common historical and institutional background of Francophone cotton producers and their joint membership in the CFA union and other regional bodies hint at the considerable scope to adopt the Burkinabè cotton model, albeit with appropriate modifications. Indeed, because of their similarity to producers in Burkina Faso, it is no surprise that neighboring producers such as those in Mali are interested in following in Burkina Faso's footsteps. Reforming farmers' groups, adopting inter-professional committees, granting regional monopolies, and outsourcing the management of cotton stabilization funds are all plans that are either already in existence in Mali or on the way to being adopted.

Perhaps the main stumbling block to successfully adopting the Burkinabè model is political. There is thus room to study more narrowly the specific key political realities of each country and their impact on the effectiveness of reform and the functioning of institutions.

Notes

1. According to Moseley and Gray (2008, 5), the cotton market was one of the first commodity markets to be organized at a global scale when the textile sector was industrialized in Great Britain in the late 18th century. Sub-Saharan Africa was "increasingly seen as a potential supplier for European mills" after the Civil War in the United States.

2. The expansion of cotton in Sub-Saharan Africa is an outcome of state-led efforts and state–cotton trader partnerships (more evident in Francophone Africa) and "export promotion policies pushed by the World Bank and IMF" (Moseley and Gray 2008, 6).

3. Some authors believe that it is the very combination of cotton's economic and political successes that may have contributed to the politicization of its management and the emergence of rent-seeking behavior (Tschirley, Poulton, and Labaste 2009).

4. It is worth noting that producers were consulted on the choice of the two private actors. In fact, they even rejected one of the initially chosen firms for its poor reputation in neighboring countries.

5. Note that the zoning approach of Burkina Faso relied on already established large traders on the export market of cotton fiber, COPACO and Reinhart. As noted by Bassett (2008), these traders act as an oligopoly and can benefit from more stable prices on the world market, securing some of their commercial margins.

6. This figure relates to full-time employment in the agricultural sector, where active men have a coefficient of 1.0 and women and children from 6 to 18 years of age have a coefficient of 0.5, as do elderly people within households.

7. It is now possible to track input, land use, profit, and debt for individual cotton farmers

through the data of GPCs. Managers of GPCs are offered specific training programs, and the local credit committees have substantially improved the performance of outgrower schemes for cotton farmers according to their input needs for both cotton and cereals. Farmers themselves express more confidence in their local leaders and trust in the new governance systems.

8. This program may still be scaled up once financial institutions have been convinced by the encouraging results induced by the cotton reform.

9. The growth in debt was driven by later payments of seed cotton, pushing the most cash-constrained farmers to sell some of their agricultural inputs on the black market at the beginning of the next crop season, and also by lower purchase prices and higher costs of inputs.

10. Grimm and Gunther (2004) provide a good overview of the impact of the Ivorian crisis and the many controversies over the impact. The less controversial numbers indicate that around 1.2 million persons of Burkinabè origin were thought to have lived and worked in Côte d'Ivoire before the crisis and that the share of households receiving remittances from Côte d'Ivoire decreased from 20.7 percent in 1998 to 12.7 percent in 2003 (rural, 24.6 to 14.6 percent; urban, 7.0 to 5.3 percent). The figures on loss of remittances are more contested because some government sources suggest that returning migrants brought significant amounts of savings with them upon their return. Another source shows that transfers decreased by approximately 68 percent in real terms between 1998 and 2003, from approximately 4.0 to 1.3 percent of GDP. Grimm and Gunther (2004) estimate that without any reduction of remittances from Côte d'Ivoire the poverty headcount in 2003 would have decreased by an extra 2 percentage points.

11. The extensive work of Grimm and Gunther (2004) is in line with that of Lachaud (2005), whose contribution improved tremendously the quality of available data on poverty and incomes, which can be used as consistent data for our study.

12. We evaluated the rural poverty line at FCFA 100,000 per capita in cotton regions in 2006 after performing a comparison with the standards of Burkina Faso, INSD (2003) and an evaluation of a basket of basic goods at their local market price (Kaminski 2006).

13. This threshold corresponds to a food security curve accounting for basic cash needs and the value of food needs per capita, according to local market prices (Kaminski 2006). We then compare the per capita income measure of the households' agricultural production to this value, calibrated at CFAF 60,000 per capita. This value is estimated from a set of values and expenses of basic goods, as reported by Kaminski (2006).

14. If we base the estimates on consumption of animal proteins, we obtain an upward move of 10 percent of the population and a downward move of 3 percent.

15. Note that Gray and Kevane (2001) reported that cotton regions are subject to demographic pressure on land due to the extensive agricultural growth and migration. Several conflicts among ethnic groups are likely to emerge.

16. Kaminski, Headey, and Bernard (2009) discussed environmental sustainability in depth. Specifically, their study found contradictory evidence on the matter, although it concluded that the reforms were mostly sustainable on this front. Burkina Faso still appears to have large quantities of arable land suitable for cotton production, such that the sector's extensive growth made use of the country's comparative advantage. The reforms also encouraged more fertilizer use, which helps maintain the fertility of the land. Of greater concern is the still relatively low use of other land-care practices, particularly given that climate change already seems to be affecting Burkina Faso.

17. Rainfall can negatively damage cotton quality once harvested and therefore the price paid by collectors.

18. Note that profit-sharing has continuously increased in favor of producers over the reform period along the cotton value chain (see Baffes 2007 and Table 12.4) and that this might have

threatened the profits of cotton firms under a more rigid price determination mechanism. See Bassett (2008) for more information on the pricing issue.

19. Gradualism, in turn, is facilitated by political stability, which has characterized Burkina Faso since the early 1990s, but much less than its neighbors.

References

Baffes, J. 2007. Distortions to cotton sector incentives. In *Distortions to cotton incentives in Benin, Burkina Faso, Chad, Mali, and Togo.* Washington, D.C.: World Bank.

Bassett, T. 2001. *The peasant cotton revolution in West Africa, Côte d'Ivoire, 1880–1995.* Cambridge, U.K.: Cambridge University Press.

———. 2008. Producing poverty: Power relations and price formation in the cotton commodity chains of West Africa. In *Hanging by a thread: Cotton, globalization and poverty in Africa,* ed. W. G. Moseley and L. Gray. Athens, Ohio, U.S.A.: Ohio University Press.

Bernard, T., M.-H. Collion, A. de Janvry, P. Rondot, and E. Sadoulet. 2008. Do village organizations make a difference in African rural development? A study for Senegal and Burkina Faso. *World Development* 36 (11): 2188–2204.

Bingen, J. R. 1994. Agricultural development policy and grassroot democracy in Mali—The emergence of Mali's farmer movement. *African Rural and Urban Studies* 1 (1): 57–72.

———. 1996. Leaders, leadership, and democratisation in West Africa: Observation from cotton farmers movement in Mali. *Agricultural and Human Value* 13 (2): 24–32.

———. 1998. Cotton, democracy, and development in Mali. *Journal of Modern African Studies* 36 (2): 265–285.

Bourdet, Y. 2004. *A tale of three countries–Structure, reform and performance of the cotton sector in Mali, Burkina Faso, and Benin.* Country Economic Report 2. Stockholm: Swedish International Development Cooperative Agency.

Brambilla, I., and G. G. Porto. 2005. Farm productivity and market structure: Evidence from cotton reforms in Zambia. Working Paper 5. New Haven, Conn., U.S.A.: Economic Growth Center, Yale University.

Burkina Faso, INSD (Institut National des Statistiques et de la Démographie). 2003. *Les comptes economiques de la nation.* Ouagadougou, Burkina Faso: Ministry of Economics and Development.

DGPSA (Direction Générale des Prévisions Statistiques Agricoles). 2008. *Permanent agricultural survey, 1994–2004.* Ouagadougou, Burkina Faso: Ministry of Agriculture.

Fan, S. 1991. Effects of technological change and institutional reform on production growth in Chinese agriculture. *American Journal of Agricultural Economics* 73 (2): 266–275.

FAO (United Nations Food and Agriculture Organization). 2009. *AGROSTAT.* Rome.

Fok, M. 2008. Cotton policy in Sub-Saharan Africa: A matter of institutional arrangements related to farmers' constraints. Paper presented at the conference Rationales and Evolutions of

Cotton Policies, May 13–17, at the Centre International de Recherche Agronomique pour le Développement in Montpellier, France.

Gergely, N. 2004. *Étude comparative sur les coûts de production des sociétés cotonnières au Mali, au Burkina Faso et au Cameroun.* Paris: Agence Française de Développement.

Goreux, L., and J. Macrae. 2003. *Reforming the cotton sector in Sub-Saharan Africa.* Africa Region Working Paper Series 47. Washington, D.C.: World Bank.

Gray, L. C. 2008. Cotton production in Burkina Faso: International rhetoric versus local realities. In *Hanging by a thread: Cotton, globalization and poverty in Africa,* ed. W. G. Moseley and L. Gray. Athens, Ohio, U.S.A.: Ohio University Press.

Gray, L. C., and M. Kevane. 2001. Evolving tenure rights and agricultural intensification in southwestern Burkina Faso. *World Development* 29 (4): 573–587.

Grimm, M., and I. Gunther. 2004. *A country case study on Burkina Faso.* Research paper for the Operationalising Pro-Poor Growth project. A joint initiative of AFD, BMZ (GTZ, KfW Development Bank), DFID, and the World Bank. Gottingen, Germany: University of Gottingen.

ICAC (International Cotton Advisory Committee). 1998. Statements of France and the World Bank. Statements of the ICAC at its 57th plenary meeting, October 12–16, in Santa Cruz, Bolivia.

INSD (Institut National de la Statistique et de la Démographie). 2006. National demographics census and living standards survey data. Ouagadougou, Burkina Faso.

Jayne, T., J. D. Shaffer, J. M. Staatz, and T. Reardon. 1997. Improving the impact of market reform on agricultural productivity in Africa: How institutional design makes a difference. MSU International Development Working Paper 66. East Lansing, Mich., U.S.A.: Michigan State University.

Kaminski, J., 2006. *Retrospective survey on ten years of changes in the cotton sector of Burkina Faso: Interviews of stakeholders, field survey of representative producers and GPCs, and price information on local markets.* Toulouse, France: Toulouse School of Economics.

————. 2009a. Subjective wealth and rural development: Evidence from the cotton reform in Burkina Faso. CAER Working Paper. Jerusalem: Hebrew University of Jerusalem.

————. 2009b. *Survey of individual cotton farms, GPCs, leaders of 5 cotton villages in Burkina Faso, and exchanges with Malian officials in Bamako for the "Power and Politics in Africa" Project.* London: Overseas Development Institute.

Kaminski, J., and A. Thomas. 2010. Land use, production growth, and the institutional environment of smallholders: Evidence from Burkinabè cotton farmers. *Land Economics,* forthcoming.

Kaminski, J., D. Headey, and T. Bernard. 2009. *Institutional reform in the Burkinabè cotton sector and its impacts on incomes and food security: 1996–2007.* IFPRI Discussion Paper 920. Washington, D.C.: International Food Policy Research Institute.

Kherallah, M., C. Delgado, E. Gabre-Madhin, N. Minot, and M. Johnson. 2002. *Reforming agricultural markets in Africa.* Baltimore: Johns Hopkins University Press.

Lachaud, J.-P. 2005. A la recherche de l'insaisissable dynamique de pauvreté au Burkina Faso. Une nouvelle évidence empirique. Documents de travail 117. Bordeaux, France: Centre d'Economie du Développement de l'Université Montesquieu Bordeaux IV.

Lele, U., N. Van de Walle, and M. Gbetiobouo. 1989. *Cotton in Africa: An analysis of differences in performance*. MADIA Discussion Paper 7. Washington, D.C.: World Bank.

Lipsey, R. G., and K. Lancaster. 1956. The general theory of second best. *Review of Economic Studies* 24 (1): 11–32.

Moseley, W. G., and L. Gray, eds. 2008. *Hanging by a thread: Cotton, globalization and poverty in Africa*. Athens, Ohio, U.S.A.: Ohio University Press.

Poulton, C., P. Gibbon, B. Hanyani-Mlambo, J. Kydd, W. Maro, M. N. Larsen, A. Osorio, D. Tschirley, and B. Zulu. 2004. Competition and coordination in liberalized African cotton market systems. *World Development* 32 (3): 519–536.

Roberts, R. L. 1997. *Two worlds of cotton: Colonialism and the regional economy in French Sudan, 1800–1946*. Cambridge, U.K.: Cambridge University Press.

Rodrik, D. 1996. Understanding economic policy reform. *Journal of Economic Literature* 34 (1): 9–41.

———. 1999. Where did all the growth go? External shocks, social conflict, and growth collapses. *Journal of Economic Growth* 4: 385–412.

Salé, M., R. Togbé, and A. Waddell. 2001. *Étude sur la crise financière cotonnière*. Cotonou, Benin: Association Interprofessionnelle du Coton.

Savadogo, K., and T. Sakurai. 2007. War induced covariate shocks and natural resource degradation in Burkina Faso: Evidence from the Ivorian crisis. Working paper. Ouagadougou, Burkina Faso: University of Ouagadougou.

Schwartz, A. 1996. Attitudes to cotton growing in Burkina Faso: Different farmers, different behaviours. In *Economics of agricultural policies in developing countries*, ed. M. Benoit-Cattin, M. Griffon, and P. Guillaumont. Paris: Revue Française d'Economie.

Tefft, J. 2008. Mali's White Revolution: Smallholder cotton from 1960 to 2006. In *Successes in African agriculture: Lessons for the future*, ed. S. Haggblade and P. Hazell. Washington, D.C.: International Food Policy Research Institute.

Tschirley, D., C. Poulton, and P. Labaste. 2009. *Organization and performance of cotton sectors in Africa: Learning from reform experience*. Washington, D.C.: World Bank.

World Bank. 1993. *The East Asian miracle: Economic growth and public policy*. New York: Oxford University Press.

———. 2004. Cotton cultivation in Burkina Faso, a 30-year success story. Paper presented at the conference Scaling Up Poverty Reduction, a Global Learning Process, May 25–27, in Shanghai, China.

Chapter 13

Private-Sector Responses to Public Investments and Policy Reforms: The Case of Fertilizer and Maize Market Development in Kenya

Joshua Ariga and T. S. Jayne

Overview of Interventions in Kenya's Fertilizer and Maize Marketing Systems

In Kenya, as in most of Africa in the early 1990s, the key agricultural policy challenges revolved around the classic "food price dilemma": how to keep food prices at tolerable levels for consumers while maintaining adequate incentives for producers to feed the nation and raise farm incomes. Kenyan policymakers had for many years attempted to strike a balance between these two competing objectives through controlling maize and maize meal prices. The National Cereals and Produce Board (NCPB), a state-run maize marketing agency, generally offered maize prices higher than those prevailing in parallel markets and sold to millers below prevailing market prices in urban areas. The marketing margin for the NCPB became insufficient to cover its costs, and it consequently incurred massive deficits during the 1980s (Jayne and Jones 1997). A similar storyline applied to

This study builds on and is made possible by over a decade of sustained data collection, capacity-building, and policy analysis conducted under the Tegemeo Agricultural Monitoring and Policy Analysis project, a joint collaboration between Egerton University and Michigan State University, funded by the U.S. Agency for International Development–Kenya. The authors also acknowledge the Bill & Melinda Gates Foundation for its support of research and capacity-building activities related to agricultural markets in Africa through an agreement between Egerton University and Michigan State University.

state-influenced agencies supplying fertilizer inputs that accumulated debts resulting from inefficient management and corruption.

The unsustainable fiscal deficits caused by state agencies, a global pull away from state-led development models toward liberalization advocated by development partners (Bates 1981; World Bank 1981; Williamson 1997; de Soto 2000), and the growing perception that the government was misusing public funds for patronage purposes (Bates 1981, 1989) set the stage for reforms in input and output markets in the late 1980s and early 1990s. Suppression of free speech engendered a growing antigovernment sentiment among the populace, which pushed for free political association and the formation of competing political parties. These forces culminated in a range of policy reforms geared toward increased private-sector participation and fewer government restrictions on trade, which, combined with complementary public investments, induced a substantial response by the private sector in the input and maize markets. This resulted in measurable improvements in smallholder maize productivity and rural farm incomes between 1997 and 2007 in Kenya.

Though a number of sectors were targets of reforms, none had the elevated attention that has always been bestowed on maize, a staple grown by more than 90 percent of rural households and accounting for more than 40 percent of fertilizer use (Ariga et al. 2008). In the period from the early 1990s to 2007, national fertilizer use nearly doubled (Figure 13.1), from a mean of roughly 250,000 tons per year to over 400,000 tons in 2006 and 2007. According to nationwide farm survey data, smallholders' use of fertilizer per cultivated hectare of maize has grown by 33 percent in the past 10 years.

This study documents the factors driving the growth in fertilizer use and maize productivity in Kenya. The following section describes the reform process in fertilizer and maize markets. Kenya's case, in which the liberalization of input and maize markets has contributed to smallholder farm productivity growth and national food security, stands in contrast to experiences elsewhere in Africa. The concluding section of this chapter explains why Kenya's efforts have proved relatively successful and the degree to which their success could be replicated more broadly elsewhere in Africa.

Data

The analysis in this study is based largely on nationwide surveys of 1,260 smallholder farm households in 24 districts surveyed in 1997, 2000, 2004, and 2007. The panel household survey was designed and implemented under the Tegemeo Agricultural Monitoring and Policy Analysis Project implemented by Egerton University's Tegemeo Institute, Kenya. The sample frame was developed by Tegemeo Institute

Figure 13.1 Trends in fertilizer use, commercial imports, and donor imports in Kenya, 1990–2007, with projections for 2008

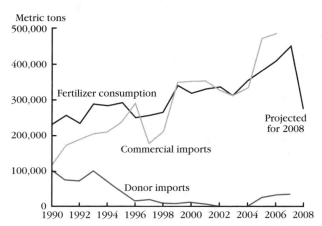

Source: Ariga et al. (2008); 2008 projections from interviews with fertilizer importers.

in consultation with the government's Central Bureau of Statistics, now the Kenya National Bureau of Statistics.[1] The survey sample has been classified into zones for analytical convenience, based on agroecological characteristics, districts, and agricultural production potential (Table 13.1).

Other data have been obtained from the Ministry of Agriculture (MoA), such as monthly maize price levels, annual fertilizer consumption, and fertilizer prices at the port of Mombasa and Nakuru. The NCPB provided data on its annual maize purchases, sales, and price levels. We also compare our results with those from farm surveys implemented jointly by the International Maize and Wheat Improvement Center and the Kenyan Agricultural Research Institute (KARI), the 2005 Rockefeller-funded household survey in western Kenya, and the government's nationally representative Kenya Integrated Household and Budget Survey, undertaken in 2005.

Description of the Interventions

Figure 13.2 describes the synergies between the liberalization of input and maize markets and public investments in support of smallholder agriculture, leading to substantial private-sector investment in fertilizer retailing and maize marketing, which resulted in an impressive increase in smallholder fertilizer use and maize yields on small-holder farms over 1997–2007. These reforms are described in the following sections.

Table 13.1 Sampled districts in agroecological zones of Kenya

Agroecological zone	District	Categorization	Number of households
Coastal Lowlands	Kilifi, Kwale	Low potential	70
Eastern Lowlands	Machakos, Mwingi, Makueni, Kitui, Taita-Taveta	Low potential	143
Western Lowlands	Kisumu, Siaya	Low potential	149
Western Transitional	Bungoma (lower elevation), Kakamega (lower elevation)	Medium potential	148
Western Highlands	Vihiga, Kisii	High potential	128
Central Highlands	Nyeri, Muranga, Meru	High potential	240
High-Potential Maize Zone	Kakamega (upper elevation), Bungoma (upper elevation), Trans Nzoia, Uasin Gishu, Bomet, Nakuru, Narok	High potential	345
Marginal Rain Shadow	Laikipia	Low/medium potential	37
Overall sample			**1,260**

Source: Tegemeo Institute of Agricultural Policy and Development / Michigan State University (1997, 2000, 2004, 2007).

Figure 13.2 Synergies among public goods investments, policies, and private-sector responses in promoting fertilizer use and maize yield improvements by Kenyan smallholder farmers

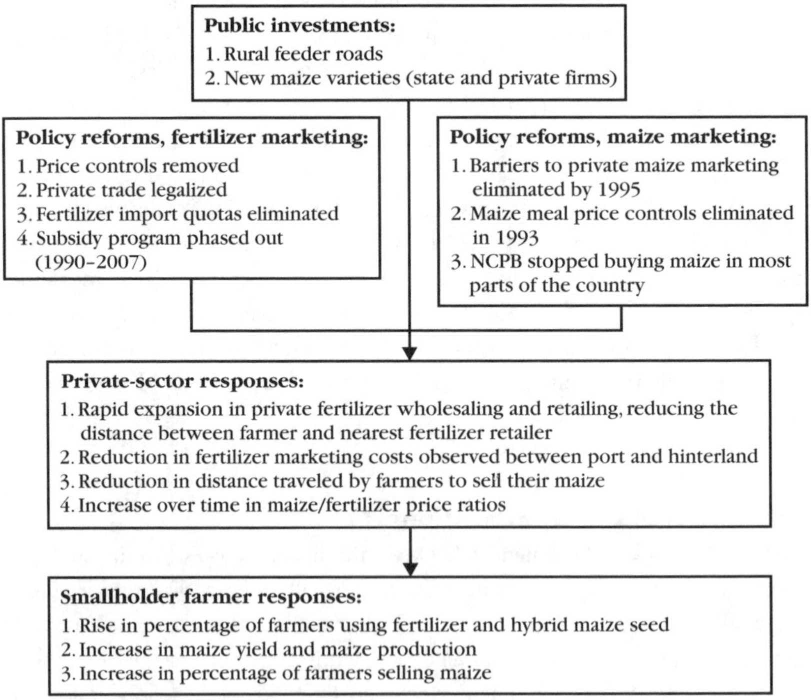

Public investments:
1. Rural feeder roads
2. New maize varieties (state and private firms)

Policy reforms, fertilizer marketing:
1. Price controls removed
2. Private trade legalized
3. Fertilizer import quotas eliminated
4. Subsidy program phased out (1990–2007)

Policy reforms, maize marketing:
1. Barriers to private maize marketing eliminated by 1995
2. Maize meal price controls eliminated in 1993
3. NCPB stopped buying maize in most parts of the country

Private-sector responses:
1. Rapid expansion in private fertilizer wholesaling and retailing, reducing the distance between farmer and nearest fertilizer retailer
2. Reduction in fertilizer marketing costs observed between port and hinterland
3. Reduction in distance traveled by farmers to sell their maize
4. Increase over time in maize/fertilizer price ratios

Smallholder farmer responses:
1. Rise in percentage of farmers using fertilizer and hybrid maize seed
2. Increase in maize yield and maize production
3. Increase in percentage of farmers selling maize

Source: Ariga and Jayne (forthcoming).
Note: NCPB means National Cereals and Produce Board.

Fertilizer and Maize Market Reforms

Prior to the onset of liberalization in the late 1980s and early 1990s, the fertilizer and maize markets were monopolized by state agencies with little competition from private trade.[2] The Kenya Farmers Association, consisting of large-scale producers, and the Kenya National Trading Corporation, a state-run corporation, were the main participants in the fertilizer input market. On the output side, the NCPB controlled maize prices at all levels of the market chain (Nyoro, Kiiru, and Jayne 1999). By setting fixed panterritorial prices for all market participants, these entities stifled private trade by removing opportunities for arbitrage. Private traders were required to apply for movement permits to let them transport grain across district boundaries. However, despite government restrictions, private maize trade was being conducted in Kenya even before the liberalization process began.

The licensing requirements, setting of fixed prices, and allocation of foreign exchange reserves to importers by the government created rent-seeking opportunities for public-sector officials, the costs of which had to be absorbed by firms forced to operate within the narrow margins resulting from price controls (Kimuyu 1994). The fiscal deficits resulting from the operations of these bureaucracies (Kodhek 2004) eventually led the state to abolish licensing requirements for fertilizer imports and eliminate subsidies (in 1992, 1993), remove fertilizer duties and taxes (in 1994), and allow private trade in maize (1993) alongside that of the NCPB, which concentrated on buying from large-scale producers. According to the Tegemeo / Michigan State University household surveys, fewer than 4 percent of smallholder households sold maize directly to the NCPB within four years after the liberalization of maize markets. However, in some good production years the NCPB has indirectly protected smallholder farmers against downside price risks by buying surplus production from large-scale farmers at prices above market prices, which has helped to stabilize maize prices (Jayne, Myers, and Nyoro 2008).

Coupled with the freeing of the foreign exchange regime in 1992, these changes in the policy environment led to substantial new entry of private-sector firms into the importing, wholesaling, distribution, and retailing of fertilizer and maize. The International Fertilizer Development Center (IFDC) (2001) estimates that by the late 1990s there were more than 500 fertilizer wholesalers and 7,000 fertilizer retailers operating in the country. Although major risks and uncertainties persist due to continued discretionary government policies and the reintroduction of fertilizer subsidies starting in 2008, there is concrete evidence of private-sector response and smallholder satisfaction with the main elements of the liberalization process in Kenya, as we describe in a subsequent section. Table 13.2 gives a tabular chronology of policy events in the maize market. Since 2005 Kenya has complied with regional initiatives under the Common Market for Eastern and Southern Africa and the East

Table 13.2 Evolution of maize marketing and pricing policy reforms in Kenya, starting in 1988

Year	State marketing agency	Year	Market regulation and pricing policy
1988	National Cereals and Produce Board (NCPB) financially restructured. Phased closure of NCPB depots. NCPB debts written off; crop purchase fund established but not replenished.	1988	Cereal Sector Reform Program envisaged widening of NCPB price margin. In fact, margin narrowed. Proportion of grain that millers were obliged to buy from NCPB declined. Limited unlicensed maize trade allowed.
		1991	Further relaxation of interdistrict trade.
		1992	Restrictions on maize trade across districts reimposed. NCPB unable to defend ceiling prices.
		1993	Maize meal prices deregulated. Import tariff abolished.
1995	NCPB restricted to limited role as buyer and seller of last resort. NCPB market share declined to 10–20 percent of marketed maize trade. NCPB operations confined mainly to high-potential areas of western Kenya.	1995	Full liberalization of internal maize and maize meal trade; maize import tariff of up to 30 percent reimposed.
		1996	Export ban imposed after poor harvest.
		1997	Import tariff imposed after poor harvest.
		1997 onward	External trade and tariff rate levels changed frequently and became difficult to predict. NCPB producer prices normally set above import parity levels.
		2000 onward	NCPB provided with funds to purchase a greater volume of maize. NCPB's share of total maize trade rose to 25–35 percent of total marketed maize.
		2005 onward	The government withdrew the maize import tariff from maize entering Kenya from East African Community member countries. An official 2.75 percent duty was still assessed. Variable import duty still assessed on maize entering through Mombasa port.

Source: Adapted from Ariga and Jayne (2008).

African Community (EAC) to eliminate cross-border tariffs within the region and harmonize regional and international trade policies.

Investment in Seed Technology and Market Development

The release and dissemination of improved maize seed varieties has been another important dimension of Kenya's success in input intensification. Suri (2007) documents the release of at least 10 new maize hybrids or open-pollinating varieties released since 1995 by KARI and other firms. Hassan, Mekuria, and Mwangi (2001) show a fivefold increase in the number of private seed companies between 1992 and

1996. However, recent hybrid releases offer less dramatic yield advantages over previous varieties (Karanja 1996). Research shows significant yield increases from using fertilizer in areas of at least medium potential in Kenya (de Groote et al. 2005; Ariga, Jayne, and Nyoro 2006; Suri 2007; Ariga et al. 2008; Marenya and Barrett 2009).[3] Strong research evidence reviewed by Suri (2007) also indicates that the marginal product of fertilizer is heightened by the use of improved modern seed varieties.

The Tegemeo farm household panel surveys collected information from respondents in each wave of interventions on access to markets, infrastructure, and services. This information enables us to track changes over time in these "access" variables. Some of these indicators show changes in private-sector investment, such as the distance from farms to the nearest fertilizer retailers. Other indicators are measures of public goods investments, such as distance from the farm to the nearest motorable road, tarmac road, clean water supply, health facility, and electricity grid.

Figure 13.3 shows changes over 1997–2007 in mean household distance from farms to these public- and private-sector services. The results are plotted as a percentage of the initial 1997 survey values. For virtually every indicator, there was a clear reduction in the distance traveled by households to markets and services. There was an especially large decline in mean household distance from farms to public- and private-sector services based on initial 1997 survey values. This trend reflects the rapid expansion in the number of fertilizer retailers noted by the IFDC earlier and represents the expanded incentives for investment in fertilizer retailing in smallholder areas and the response by private input-retailing firms.

The greatest improvements in many of these infrastructure and service variables occurred between 2004 and 2007, such as the reduction in distance to a motorable

Figure 13.3 Changes in the distance from farm households to selected services and markets in Kenya, 1997–2007 (kilometers)

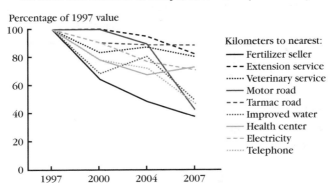

Source: Tegemeo Institute of Agricultural Policy and Development / Michigan State University (1997, 2000, 2004, 2007).

road, water supply, and veterinary services. The mean distance from farms to motor-able roads fell by one-half between 2004 and 2007. The Constituency Development Fund (CDF), under which local authorities were given increased control of budget resources for local development, was established in 2003/04. All of the 210 constitu-encies in Kenya are allocated 2.5 percent of the total government revenue for CDF funding. The sharp reduction in the distance to motorable roads and clean water between the 2004 and 2007 surveys is associated with this administrative reform, although causality cannot be inferred.

Outcomes of the Interventions

This section provides evidence of the synergistic effects of public and private invest-ments and policy reforms on a number of outcomes measured by farm- and market-level indicators including (1) increased private-sector investment, (2) a reduction over time in the distance to the nearest fertilizer retailer or maize buyer, (3) a decline in fertilizer marketing margins, and (4) improved maize–fertilizer price ratios facing farmers. Other outcomes detected in household panel survey data point to (1) a rise in the percentage of farmers using fertilizer, (2) increasing maize yields over time, and (3) an increase in the percentage of farmers relating to markets as sellers.

Increased Fertilizer Investment and Reduction
in the Distance Traveled to Buy Inputs

As explained in the previous section, changes in the policy environment led to a significant new entry of private-sector firms into the importing, wholesaling, distribution, and retailing of fertilizer (Kimuyu 1994; Allgood and Kilungo 1996; IFDC 2001; Wanzala, Jayne, and Staatz 2002). Evidence of private-sector response in input and maize markets is also revealed in the farm panel survey data reported in subsequent sections.[4] The mean distance of small farmers to the nearest fertilizer retailer declined from 8.1 kilometers to 3.4 kilometers between 1997 and 2007 (Table 13.3). This is likely to be an important factor in increased fertilizer use by smallholders. A similar trend is observed in the distance to hybrid maize seed, which declined from 5.6 kilometers in 2000 to 3.4 kilometers in 2007.

The Decline in Fertilizer Marketing Margins
between Mombasa and Upland Markets

Figure 13.4 plots trends in real cost, insurance, and freight (c.i.f.) prices of diammo-nium phosphate (DAP) fertilizer at Mombasa port and corresponding real whole-sale price at Nakuru in the interior. The difference between the Nakuru and Mombasa prices reflects domestic fertilizer marketing costs. Although world pric-es remained roughly constant over 1990–2007, prices at Nakuru have declined

Table 13.3 Mean distance to fertilizer and hybrid maize seed retailers in Kenya, 1997–2007 (kilometers)

Zone	Distance to fertilizer seller				Distance to hybrid maize seed retailer		
	1997	2000	2004	2007	2000	2004	2007
Coastal Lowlands	30.6	24.3	18.4	11.3	21.8	18.7	9.5
Eastern Lowlands	9.8	5.4	4.2	2.7	6.4	3.7	3.0
Western Lowlands	16.0	11.6	7.5	3.8	9.1	5.4	3.8
Western Transitional	6.3	4.6	2.8	3.6	4.2	2.7	3.7
High-Potential Maize Zone	5.0	4.0	3.0	3.6	4.5	3.0	3.7
Western Highlands	3.3	2.2	1.4	2.4	2.6	1.6	2.4
Central Highlands	2.7	1.5	1.4	1.3	1.9	1.5	1.5
Marginal Rain Shadow	26.2	5.8	5.4	2.3	5.2	4.3	2.3
National sample	8.1	5.7	4.1	3.4	5.6	3.9	3.4

Source: Tegemeo Institute of Agricultural Policy and Development / Michigan State University (1997, 2000, 2004, 2007).

Figure 13.4 Price of diammonium phosphate in Mombasa and Nakuru, Kenya, 1990–2008

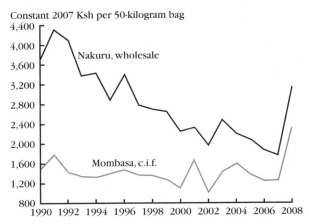

Source: Kenya Ministry of Agriculture. FMB weekly fertilizer reports for c.i.f. Mombasa.
Note: Nakuru is a maize-producing area in the Rift Valley of Kenya, 400 miles (645 km) by road from the port of Mombasa.

substantially, from roughly Ksh 3,800 to Ksh 2,000 in constant 2007 shillings. Although both import prices and upcountry prices shot up in 2008 in relation to the general price index, DAP prices in 2008 were about equal in real terms to where they stood in the mid-1990s, about the time that the substantial decline in marketing costs began.

Recent interviews of key informants in Kenya's fertilizer sector undertaken for this study identify four factors responsible for the declining fertilizer marketing costs just described: (1) taking advantage of cheaper backhaul transportation when trucks return from Mombasa to Rwanda or Congo, (2) private importers' accessing relatively less expensive international sources of credit, (3) mergers between local and international firms in which knowledge and economies of scope enabled cost savings, and (4) increased competition among local importers and wholesalers, which has caused inefficient firms to exit. However, because of high world fertilizer prices in 2008, maize–fertilizer price ratios plunged, contributing to a drop in fertilizer use by Kenyan farmers in 2008.

The Rise in the Percentage of Farmers Using Fertilizer and Hybrid Maize Seed

The proportion of sampled smallholder farmers using fertilizer on maize during the main season grew from 56 percent in 1996 to 70 percent in 2007 (Table 13.4). These rates vary considerably throughout the country, ranging from fewer than 10 percent of households surveyed in the drier lowland areas to up to 95 percent of small farmers in Central Province and the maize surplus areas of western Kenya.[5]

Overall, the fertilizer dose rates for maize have not increased appreciably. The mean dose rate was 56 kilograms per acre in 1997, rising to only 59 kilograms in 2007 (Table 13.5). Dose rates even appear to be declining somewhat in the lowland zones, though they are increasing in the moderate- and high-potential areas. Matsumoto and Yamano (2009) conclude that fertilizer application rates in Kenya are at near-optimal levels in light of the maize–fertilizer response rates and price ratios.

Table 13.4 Percentage of farm households in Kenya using fertilizer on maize, 1996–2007

Agroregional zone	1996	1997	2000	2004	2007
Coastal Lowlands	0	0	3	4	14
Eastern Lowlands	21	27	25	47	43
Western Lowlands	2	1	5	5	13
Western Transitional	39	41	70	71	81
High-Potential Maize Zone	85	84	90	87	91
Western Highlands	81	75	91	91	95
Central Highlands	88	90	90	91	93
Marginal Rain Shadow	6	6	12	11	16
Total sample	56	58	64	66	70

Source: Tegemeo Institute of Agricultural Policy and Development / Michigan State University (1997, 2000, 2004, 2007).

Table 13.5 Fertilizer dose rates in Kenya, 1997–2007
(kilograms per acre on maize fields receiving
fertilizer, main season)

Agroregional zone	1997	2000	2004	2007
Coastal Lowlands	11	5	3	7
Eastern Lowlands	10	18	15	16
Western Lowlands	24	14	10	12
Western Transitional	54	48	62	71
High-Potential Maize Zone	65	67	74	75
Western Highlands	31	36	46	47
Central Highlands	68	64	64	58
Marginal Rain Shadow	12	15	43	43
National sample	56	55	60	59

Source: Tegemeo Institute of Agricultural Policy and Development / Michigan State University
(1997, 2000, 2004, 2007).

The mean dose rates in the six districts sampled in the High-Potential Maize Zone in 2007 were 75 kilograms per acre (187 kilograms per hectare), comparable to or higher than the post–Green Revolution dose rates on rainfed grain crops in the relatively productive areas of South and East Asia. Overall, Kenya's agricultural extension system recommends a uniform 100 kilograms of fertilizer per acre of maize, which may not conform to agricultural conditions, rainfall, and prices. These rates based on the Tegemeo sample are comparable to those from the Rockefeller-funded study, Marenya and Barrett's (2008) study of Vihiga and South Nandi districts, and the Obare, Omamo, and Williams (2003) study of Nakuru District.[6]

The proportion of households planting hybrid maize seed over 1997–2007 rose from 70 percent in 1997 to 74 percent in 2007. Analysis by zone reveals that the greatest increase was in the lowland and midaltitude zones, where particular progress has been made in the release of improved varieties. National estimates show that sales of improved maize seed rose from 45,000 tons in 1996/97 to 51,000 tons in 2006/07, a 13 percent increase (Kenya, Ministry of Agriculture 2008), despite a nearly constant annual maize area cultivated over this period.

Trends in Fertilizer Application Rates for Monocropped and Intercropped Maize Fields

Tables 13.6 and 13.7 present fertilizer use rates and doses per acre for different kinds of maize fields depending on the intensity of intercropping. Roughly, over time, an increasingly higher proportion of maize area has been seen in maize fields intercropped with four or more other crops (see Table 13.6). The proportion of maize area under fertilization has risen dramatically.

Table 13.6 Proportion of smallholder maize area fertilized in Kenya, 1996/97–2006/07 (percent)

Category of maize field	Percentage of maize area receiving fertilizer (total acres in sample)			
	1996/97	1999/2000	2003/04	2006/07
Maize fields planted with maize alone	74	73	76	80
	(518)	(429)	(332)	(473)
Maize fields intercropped with fewer than four other crops	63	71	70	85
	(1,432)	(1,012)	(1,057)	(790)
Maize fields intercropped with four or more other crops	21	53	49	55
	(310)	(1,118)	(894)	(1,049)
All maize fields in sample	60	63	63	70
	(2,260)	(2,560)	(2,283)	(2,312)
Percentage of total maize area planted with maize alone	22.9	16.8	14.5	20.4

Source: Tegemeo Institute of Agricultural Policy and Development / Michigan State University (1997, 2000, 2004, 2007).

Table 13.7 Fertilizer use rates in Kenya on maize fields cultivated by smallholder farmers and dose rates on fertilized maize fields, 1996/97–2006/07 (kilograms per acre)

Category of maize field	Mean fertilizer use rates on all maize fields, both fertilized and unfertilized (mean dose rates, fertilized maize fields only)			
	1996/07	1999/2000	2003/04	2006/07
Maize pure-stand fields planted with maize alone	37.9	36.4	49.3	53.7
	(72.6)	(64.2)	(71.0)	(74.1)
Maize fields intercropped with fewer than four other crops	36.1	37.5	46.7	59.4
	(60.9)	(61.9)	(66.4)	(74.2)
Maize fields intercropped with four or more other crops	13.5	30.7	32.2	33.3
	(42.1)	(60.7)	(58.0)	(56.1)
All maize fields in sample	33.6	34.2	41.1	44.7
	(61.3)	(61.6)	(64.1)	(63.5)

Source: Tegemeo Institute of Agricultural Policy and Development / Michigan State University (1997, 2000, 2004, 2007).

Table 13.7 presents trends in the intensity of fertilizer application on different categories of maize fields, with a significant increase for intercropped fields. When counting all fields, both fertilized and unfertilized fields in the category of maize field, mean application rates rose from 36.1 kilograms per acre in 1997 to 59.4 kilograms per acre in 2007 (see Table 13.7, second row), a 65 percent increase.

The Relationship between Household Farm Size and Fertilizer Use Rates

A common worry is that even if fertilizer use rates are increasing in Kenya, it may not have much of an impact on poverty if the poor cannot afford to purchase fertilizer. Landholding size is one of the most important indicators of wealth in Kenya. Across the 1997, 2000, and 2004 surveys, the majority of all households had 75–100 percent of the value of their total assets in land (Burke et al. 2007).

Using locally weighted smoothing techniques (Cleveland 1979) to depict the relationship between area and fertilizer application, Ariga and Jayne (2009) show that Zone 3, the most productive region, has higher mean fertilizer use than Zone 1, the semiarid lowland region. Many small farms use fertilizer more intensively (more kilograms per acre) than others. Household characteristics that are associated with fertilizer use are discussed in subsequent sections.

We now examine the profitability of fertilizer and benefits accruing from increased fertilizer intensity. Based on an earlier study that estimated a translog production function for maize using the same data (Ariga 2007), we generated marginal product for fertilizer and other factors including seeds and labor and controlling for semifixed factors (see Ariga and Jayne 2009 for details). Value cost ratio (VCR) results are summarized in Table 13.8. If the response function were known with certainty, the incentive would be to apply nitrogen to the point at which the VCR was 1.0. However, there is clearly substantial weather uncertainty about the outcome of applying fertilizer. For these reasons, researchers have suggested that a VCR of 2.0 or greater is generally required for farmers to use fertilizer in appreciable amounts (Kelly 2006). Our chapter adopts this convention and considers a VCR of at least 2.0 as an indicator that fertilizer use is likely to be profitable.

In the High-Potential Maize Zone (Zone 3), for the first tercile of fertilizer users, an investment of Ksh 10 in fertilizer contributes Ksh 65 worth of maize, while the same investment contributes Ksh 30 and Ksh 19, respectively, for the second and third terciles. An additional kilogram of fertilizer generates less additional output for households using nearly optimal amounts compared with those that are currently using relatively less fertilizer per acre, following the theory of diminishing marginal returns. The VCR estimates in Table 13.8 appear to provide an explanation of the observed variations in application rates across zones; the recommendation of a blanket 100 kilograms of fertilizer per acre would appear to be inefficient for many farmers but profitable for many others due to differences in agroecological conditions, management practices, and market conditions. In the drier areas, such as along the coast, even very low fertilizer application rates would not provide a VCR greater than 2.0; hence the very low fertilizer use in such areas would not necessarily reflect suboptimal use.

Therefore, combining the findings that the proportion of maize area being fertilized by smallholders in Kenya is rising over time, that application rates are also

Table 13.8 Value cost ratios (VCRs) for fertilizer in Kenya, 1997–2004 (by terciles of fertilizer use)

Zones		1st Tercile				2nd Tercile				3rd Tercile			
		1997	2000	2004	Total	1997	2000	2004	Total	1997	2000	2004	Total
1	N	14	12	33	59	3	1	3	7	—	—	3	3
	VCR	4.25	8.79	8.58	7.60	1.59	1.77	3.83	2.58	—	—	0.74	0.74
2	N	67	77	57	201	36	47	65	148	29	23	75	127
	VCR	5.31	14.65	4.72	8.72	1.75	3.92	2.48	2.76	1.22	1.91	1.78	1.67
3	N	51	39	34	124	110	88	107	305	68	79	120	267
	VCR	6.21	4.29	9.49	6.51	2.48	2.77	3.85	3.05	1.60	2.02	1.98	1.90
4	N	45	32	42	119	22	16	17	55	40	25	39	104
	VCR	8.86	8.68	9.45	9.02	2.22	4.29	2.44	2.89	1.52	1.95	1.90	1.89
Total	Total	177	160	166	503	171	152	192	515	137	128	236	501
		6.39	10.49	7.66	8.11	2.28	3.28	3.26	2.94	1.50	2.10	1.91	1.85

Source: Tegemeo Institute of Agricultural Policy and Development / Michigan State University (1997, 2000, 2004, 2007).
Notes: N, sample size. Shaded cells show VCR estimates over 2.0, indicating the likely profitability of fertilizer use. — means no data were available.

rising, and that VCR estimates are generally well over 2.0 for at least the first and second terciles of fertilizer users in most zones, we can infer that increased use of fertilizer in the postliberalization period has benefited households in most zones, particularly the medium- to high-potential zones.

Increasing Maize Yields over Time

Particular attention has been focused on the widespread perception that maize yields have declined since the partial adoption of reforms in the 1990s. This is based on the fact that NCPB operations, which were primarily designed to support maize price levels in maize-surplus areas of the country, have been scaled down since the mid-1990s. The perception that real maize prices have declined slightly over the past 15 years is indeed correct. However, we feel that the evidence of declining maize yields is not very strong, and the nationwide household panel survey data available actually indicate the reverse.

National maize yield trends based on Food and Agriculture Organization statistics are presented in Figure 13.5. These estimates are based on the "best guesstimates" of local agricultural extension agents and have been aggregated to the district level and then to the province and national levels by MoA staff. The official national estimates show that after rising between 1965 and 1980, maize yields have largely stagnated over the past two decades.

There are two reasons that MoA national yield estimates are likely to underestimate actual maize yield growth. First, the MoA figures show that the fraction of maize area in marginal areas has increased over time. Second, and more subtle, is the fact that the proportion of maize area on intercropped land has increased dramatically since the early 1990s, as shown earlier and in Ariga et al. (2008). MoA

Figure 13.5 National maize yield, Kenya, 1965–2007

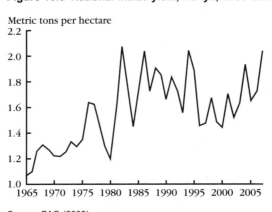

Source: FAO (2009).

yield estimates do not capture the composite output of all crops on the mixed-crop plots.[7] Maize yields on intercropped fields are almost certainly lower than those on monocropped fields (although the total value of crop output per unit of land may be either higher or lower). These survey data findings indicate that there is a general trend among Kenyan farmers to cultivate maize in more complex intercrop patterns over time.[8] In fact, nationwide field-level panel data indicate that both monocrop maize yields and intercrop yields generally rose between 1997 and 2007 (Kibaara et al. 2008; Ariga and Jayne 2009).

To further break down how maize yields are evolving in Kenya by technology set, we examine yields from farm survey data in four groups: (1) fields using both hybrid seed maize and conventional fertilizer ("combination"), (2) fields using hybrid seed but no fertilizer, (3) fields using open-pollinated varieties with fertilizer, and (4) fields using traditional seed and no fertilizer ("neither").

We use two estimates of yield. We first convert all crops harvested from mixed-crop plots into maize equivalents using relative price ratios as weights (Figure 13.6a). This provides a more complete picture of output per unit of land in areas devoted to maize. In the second method we ignore the production of other crops and count only the yield from maize (Figure 13.6b).

Maize yields generally appear to have been increasing across the years from 1997 to 2007 for each of these four categories. But the year 2000 stands out as recording the highest yields for each of these classes of technology use. Moreover, and most important, maize yields were consistently lowest in the neither category and highest in the combination category. The stark difference between the neither group and the other groups for every year shows the effect of hybrid and fertilizer use on maize yields. The group using both fertilizer and hybrid seed maize had the highest average yield, 15 bags per acre (see Figure 13.6a). A multivariate analysis of the contribution of fertilizer to maize yield, holding geographic and other factors constant, is found in Kibaara et al. (2008).

The Decline in the Distance Traveled by Farmers to the Point of Maize Sale

The liberalization of maize trade in Kenya has been associated with increased penetration by private maize assemblers into rural areas (Table 13.9), with over 90 percent of maize sales going to private traders. In the Eastern Lowlands the mean distance between farms and private buyers declined from 6.55 kilometers in 1997 to 1.62 kilometers in 2007, and in the High-Potential Maize Zone it was from 1.80 kilometers in 1997 to 0.40 kilometers in 2007. Because per-kilometer marketing costs tend to be highest at this stage of the marketing chain, where road quality is poorest, the improved penetration of maize assemblers into rural smallholder areas has most likely brought tangible benefits to smallholder farmers.

Figure 13.6a Maize yields in Kenya, converting other crops on intercropped maize fields to maize equivalents, by seed and fertilizer technology category, 1997–2007

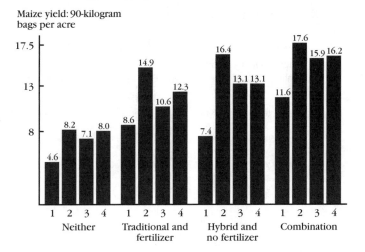

Source: Tegemeo Institute of Agricultural Policy and Development / Michigan State University (1997, 2000, 2004, 2007).

Notes: Key to bar numbers: 1 = 1997 season; 2 = 2000 season ; 3 = 2004 season; 4 = 2007 season.

Figure 13.6b Maize yields in Kenya, not converting production of other crops into maize equivalents, by seed and fertilizer technology category, 1997–2007

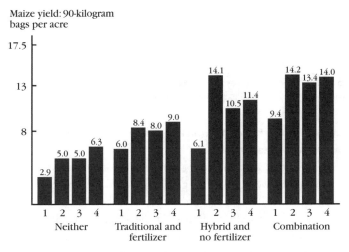

Source: Tegemeo Institute of Agricultural Policy and Development / Michigan State University (1997, 2000, 2004, 2007).

Notes: Key to bar numbers: 1 = 1997 season; 2 = 2000 season ; 3 = 2004 season; 4 = 2007 season.

Table 13.9 Mean distance from farm to maize buyer in Kenya, 1997–2007

Zone	Number of households selling maize	Kilometers from farm to point of maize sale, by buyer type				
		Private trader	NCPB	Millers/ processors	Cooperative	Consumers
Eastern Lowlands						
1997	58	6.55	—	—	—	1.27
2004	94	3.15	—	—	—	1.46
2007	88	1.62	0.00	—	—	1.28
Western Lowlands						
1997	21	1.83	—	—	—	1.00
2004	48	2.48	—	—	—	2.50
2007	50	1.04	—	—	—	0.26
Western Transitional						
1997	41	0.71	—	—	—	0.00
2004	108	0.25	—	—	4.67	0.34
2007	90	0.07	—	—	—	1.55
High-Potential Maize Zone						
1997	230	1.80	12.77	29.88	2.00	0.59
2004	313	1.13	18.57	9.48	32.00	2.88
2007	312	0.40	13.50	9.75	—	2.69
Western Highlands						
1997	40	3.15	—	—	—	2.70
2004	116	2.62	—	—	—	2.24
2007	105	1.81	—	—	—	0.96
Central Highlands						
1997	82	0.94	0.00	—	—	2.07
2004	85	1.32	—	19.33	—	0.26
2007	125	0.42	0.25	24.00	—	0.50
Marginal Rain Shadow						
1997	1	—	—	0.00	—	—
2004	15	0.71	—	—	—	0.00
2007	24	0.00	—	—	—	0.20

Source: Tegemeo Institute of Agricultural Policy and Development / Michigan State University (1997, 2000, 2004, 2007).
Notes: NCPB means National Cereals and Produce Board; — means no data were available.

The Reduction in Marketing Margins between
Maize Grain and Maize Meal Prices

Figure 13.7 shows that since the inception of the market reforms in the 1990s, the marketing margin between wholesale grain prices and retail maize meal prices has declined substantially. Retail maize meal prices have been declining at a trend rate of $0.57 per year (statistically significant at the 1 percent level), while wholesale prices in Nairobi have been declining at only $0.11 per year (a trend not statistically different from zero). Increased competition at the milling and retailing stages of the maize value chain has greatly benefited low-income consumers in Kenya (Jayne and Argwings-Kodhek 1997; Ariga and Jayne 2008).

Figure 13.7 Trends in maize grain and maize meal prices, Nairobi, 1994–2006

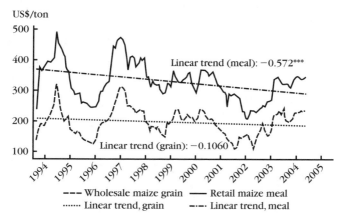

Source: Jayne and Chapoto (2006).
Notes: Prices were recorded each year in May. *** means 1 percent level of significance.

The Improved Affordability of Maize Meal for Consumers

Mason et al. (2009) examined trends in wage rates relative to the retail prices of maize grain, maize meal, and wheat bread between 1993 and 2009 for urban consumers in Kenya. They found a high correlation among wage rate series for various government and private-sector categories. For all categories of wage earners, formal-sector wages rose at a faster rate than the prices of maize grain, retail maize meal, and bread from 1994 to mid-2007, as evidenced by the upward trajectory in the quantity of these commodities affordable per daily wage (Figure 13.8). Although the recent food price crisis partially reversed this trend, the quantities of staple foods affordable per daily wage in urban Kenya during the 2008/09 marketing season were still roughly double their levels of the mid-1990s for formal workers.

Perceptions of Farmers

Respondents in the Tegemeo household panel survey were asked the following questions regarding their access to markets for maize: (1) "The government has liberalized the maize market since 1992. Compared to 5–10 years ago, is it now more convenient or more difficult to sell your maize?" and (2) "Overall, would you prefer to go back to the controlled grain marketing system as it existed in the 1980s or do you prefer the current liberalized marketing system?" The first of these questions was asked only in 1997, while the second question was asked both in 1997 and in 2000 (Table 13.10).

The overwhelming majority of households in all regions (88 percent) stated that it was more convenient to sell grain after liberalization. This is because most

Figure 13.8 Kilograms of maize meal and maize grain affordable per daily wage in Nairobi and loaves of bread affordable per daily wage in Kenya, 1994–2009

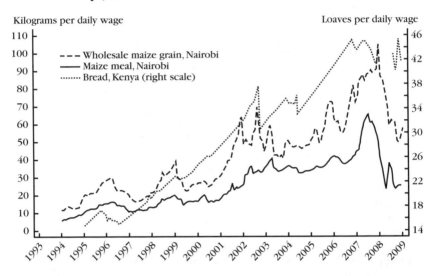

Source: Mason et al. (2009).

Note: Numbers were computed each year in January as mean wage rates (Ksh per day) divided by price of commodity (Ksh per kilogram).

farmers are being paid cash on the spot, unlike when they sold their grain under the NCPB, which often took months to pay, and because 70 percent of farmers can now sell their produce at the farm gate instead of moving produce to NCPB depots (Kọdhek et al. 1998). The median distance traveled by farmers selling their maize was zero in 2004 and 2007, revealing the extent to which assembly traders have penetrated deep into rural areas to buy maize from smallholders. The preference for the current liberalized system was especially clear in grain-deficit areas such as the Central Highlands because grain is easier to purchase now than during the control period. Interestingly, the majority of households nationwide also stated a preference for the current system over the former control system, by a margin of almost 2 to 1. This was true even in the High-Potential Maize Zone, the region where most farmers are maize sellers who formerly enjoyed the maize price supports of the NCPB.

The survey data also indicate that the proportion of households that either sell only maize or are net maize sellers rose progressively over 1997–2007, consistent with the general picture of rising maize yields over this same period. Although 32 percent of the panel sample were either sellers or net sellers of maize in 1997, this figure rose to 43 percent in 2007.

Table 13.10 Household perceptions of the performance of the current marketing system compared with the controlled marketing system in Kenya, 1997 and 2000 (percentage of households)

Zone	Year	Convenience of selling grain at time of survey compared with control period			Preference for current marketing system versus system during control period		
		Better now	Better during control period	No change	Prefer current system (h)	Prefer control system (i)	No change
Coastal Lowlands	1997	50	10	40	67	23	10
	2000	—	—	—	69	19	13
Western Lowlands	1997	81	14	5	52	44	4
	2000	—	—	—	59	40	2
Eastern Lowlands	1997	87	3	10	75	17	8
	2000	—	—	—	68	31	2
High-Potential Maize Zone	1997	93	5	2	61	36	3
	2000	—	—	—	62	37	1
Western Highlands	1997	84	11	5	53	44	3
	2000	—	—	—	67	31	2
Western Transitional	1997	99	1	0	37	61	2
	2000	—	—	—	63	36	1
Marginal Rain Shadow	1997	90	5	5	71	27	2
	2000	—	—	—	74	24	3
Central Highlands	1997	82	8	10	76	16	8
	2000	—	—	—	75	22	3
National average	1997	88	7	5	61	34	5
	2000	—	—	—	66	32	2

Source: Tegemeo Institute of Agricultural Policy and Development / Michigan State University (1997, 2000).
Notes: The districts included in each zone grouping were as follows: Coastal Lowlands (Kalifi, Kwale); Western Lowlands (Kisumu, Siaya); Eastern Lowlands (Mwingi, Makueni, Machakos, Kitui, Taita Taveta); High-Potential Maize Zone (Nakuru, Trans Nzoia, Uasin Gishu, Bungoma [Kimilili and Tongaren divisions]); Kakamega (Lugari division); Western Highlands (Vihiga, Kisii); Western Transitional (Bungoma [Kanduyi division]); Kakamega (Kabras and Mumias divisions); Marginal Rain Shadow (Laikipia); and Central Highlands (Muranga, Nyeri, Meru). — means no data were available.

Lessons Learned, Sustainability, and Potential for Replicability

This study documents the factors driving growth in fertilizer use and maize productivity in Kenya from the early 1990s to 2007. The basic story is one of synergies between the liberalization of input and maize markets and public investments in support of smallholder agriculture, leading to tangible private-sector investment in fertilizer retailing and maize marketing, which, in turn, encouraged an impressive increase in fertilizer use and maize yields on smallholder farms over 1997–2007.

As shown in Figure 13.2, the Government of Kenya implemented a number of policy reforms affecting the incentives for investment by private fertilizer distribution firms. The government also legalized the domestic and regional maize trade, although other actions adopted by the government during the 1990s partially eroded the potential response by the private sector. In spite of the rather mixed government stance toward maize market liberalization during the 1990s and early 2000s, evidence of increased private-sector investment is tangible. The IFDC estimates that more than 500 wholesalers and 7,000 retailers are operating in the country. This has led to a denser network of rural retailers and a major reduction in the distance between farms and fertilizer sellers, which contributed to the impressive growth in fertilizer use by Kenyan smallholders from the early 1990s to 2007. Analysis of wholesale maize grain prices and retail maize meal prices indicate that the miller–retail marketing margin has declined significantly over time, conferring benefits mainly on consumers.

There is also strong evidence of increased state investment in public goods supportive of private-sector investment, especially since the CDF was instituted in 2003. The combination of supportive policy changes in the fertilizer, foreign exchange, and maize markets, coupled with the improved access to markets and services made possible by public good investments, appears to have stimulated investment by the private sector in both maize and fertilizer marketing. These factors have worked synergistically to bring about important gains in maize productivity and benefits to smallholder farmers and consumers in Kenya. Evidence of increased smallholder fertilizer use and maize yields is drawn from data from four nationwide household surveys between 1997 and 2007 collected by Egerton University's Tegemeo Institute. Because the data come from a balanced panel of 1,260 households, the results provide a fairly reliable indicator of the changes in fertilizer use patterns over time, although the survey is not strictly nationally representative. The main findings are as follows:

1. The percentage of sampled smallholders using fertilizer on maize increased from 56 percent in 1996 to 70 percent in 2007.

2. Fertilizer application rates (which include all maize fields regardless of whether they received fertilizer) rose from 34 kilograms per acre in 1997 to 45 kilograms per acre in 2007, a 34 percent increase.

3. There remain wide regional variations in fertilizer use on maize. Continued low fertilizer use in the semiarid areas largely reflects the high risks and limited profitability of using fertilizer in drought-prone rainfed systems.

4. Although the total area under maize remained largely constant over the decade, maize yields increased during 1997–2007 by roughly 18 percent. This yield improvement is not borne out in official government maize production statistics, which do not take into account the shift over time in the proportion of maize area intercropped or the shift over time in the proportion of maize area in relatively semiarid regions, which has been facilitated by the release of improved maize cultivars well suited to mid- and low-altitude areas of the country. After stratifying between hybrid and nonhybrid users and between intercropped and monocropped maize fields, the household survey data show that maize yields on all types of fields have increased over time, which reflects the influence of many factors in addition to fertilizer use.

5. Fertilizer marketing costs declined substantially in constant Ksh between the mid-1990s and 2007. Interviews with key informants in Kenya's fertilizer sector identified four factors responsible for the declining fertilizer marketing costs observed in Kenya: (1) the potential for cheaper backhaul transportation has been exploited, taking greater advantage of trucks transporting cargo from Rwanda and Congo to the port of Mombasa; (2) private importers are increasingly using international connections to obtain credit at lower interest and financing costs than are available in the domestic economy; (3) local and international firms have merged, enabling shared knowledge and economies of scope that save local distribution costs; and (4) increased competition among local importers and wholesalers has expanded the number of firms engaged in fertilizer marketing since the early 1990s. It is likely that the fourth factor—increased competition—has to some extent stimulated firms to exploit the other cost-reducing innovations identified in order to maintain their market position.

Some of these conclusions can be challenged by the fact that Kenya's economy and business environment experienced many changes during this period, both positive and negative, which have also undoubtedly affected the incentives of farmers, consumers, and private marketing agents. However, it is reasonable to assume that these influences outside the agricultural sector are of second-order magnitude, compared with the more direct agricultural policy reforms and investments, in explaining the behavioral responses of farmers and fertilizer and maize marketing agents, as documented in this study.

To assess the robustness of the Tegemeo rural survey findings, we compared the proportion of smallholder households purchasing fertilizer with estimates based on other analyses during the same general period. Based on three other studies that cover a subset of the same districts as the Tegemeo survey, we found that the Tegemeo survey estimates are comparable to and in some cases lower than other estimates of fertilizer purchases and dose rates. The rise in smallholders' use of fertilizer in the Tegemeo survey data is also consistent with official MoA figures (shown in Figure 13.1), which indicate that total fertilizer consumption in Kenya rose 65 percent between 1997 and 2007.

Other signs of improvement in maize markets include farmers' level of satisfaction with the performance of maize markets from their subjective perspective. Over 65 percent of farmers surveyed in the nationwide Tegemeo Institute rural survey indicated that they prefer the current liberalized maize marketing system to the former controlled marketing system, primarily because grain is easier to sell, farmers are paid in cash at the time of sale, and maize is more reliably available for purchase.

However, in 2008 the positive trends in Kenya's maize and fertilizer markets were reversed by civil disruption, drought, and the unprecedented surge in world fertilizer prices. Early 2008 witnessed the destruction of much physical infrastructure in western Kenya, such as petrol stations and grain storage facilities, as well as the closing of many input supply stores. Moreover, the incentives to use fertilizer in Kenya have been adversely affected both by drought and world events as maize–fertilizer price ratios plunged to their lowest level in at least 18 years.

This brings us to consider the implications of these findings for policy options to sustain these achievements. There is room for considerable additional gains in smallholder and consumer welfare if progress can also be made in the following areas:

1. *Considering changes in government actions in the transport arena that could reduce fertilizer and grain distribution costs.* Because of frequent delays in off-loading commodities at the port of Mombasa and because of the erosion of the regional railway system, it is difficult to arrange for upcountry transport of a full shipload of fertilizer, especially given the frequent off-loading delays and inefficiencies at the port. Because of this coordination problem, fertilizer importers have invested in storage facilities near the port, where fertilizer can be temporarily stored to wait until trucks arrive for loading and upcountry distribution. These investments make sense if upland transport constraints and the delays and inefficiency at the port of Mombasa are taken as givens. However, if procedures for streamlining the efficiency of off-loading at the port could be achieved (for example, by privatizing stevedore services and issuing performance contracts or by devolving wider management of port operations to professional firms), thus reducing off-loading time and the storage costs incurred at Mombasa for lack of sufficient transport, fertilizer importing firms could avoid these extra charges. In a competitive marketing environment, these reductions in fertilizer marketing costs would then be passed along in the form of lower farm gate prices.[9]

2. *Reducing the transaction costs associated with a value-added tax (VAT) and port operations.* Currently fertilizer is zero-rated with respect to import duties, yet

a VAT is charged on transport and services such as bagging at the port of Mombasa. Although the VAT is supposed to be refunded, the process is lengthy and is a source of continuing frustration for market participants. In addition, port handling charges, Kenya Bureau of Standards charges, and other taxes account for 17 percent of c.i.f. (Gitonga 2004). Port fees, levies, and accessorial charges need to be rationalized and aggregated. In addition, the numerous documentation procedures need to be reduced and, if possible, provided through electronic means. Interviews with key informants in the fertilizer industry have identified numerous other potential sources of cost savings, many of which require action on the part of the government to improve efficiency.

3. *Investing in rehabilitating the eroded rail, road, and port infrastructure,* which would reduce distribution costs. The farm gate price of fertilizer in western Kenya is roughly twice as high as the landed cost at Mombasa, and transport costs are the major component of this cost difference. The high farm gate prices of fertilizer restrict demand for its use and depress agricultural productivity. Hence, efforts to improve the efficiency of port costs and upland shipping would bring major economywide benefits. In particular, rehabilitating the rail system has the potential to reduce transport costs substantially.

4. *Tailoring fertilizer packages to local demand conditions.* This would increase the demand from smaller farmers who require and are able to purchase only small packets. Repackaging of fertilizers from 50-kilogram packets into 25-, 10-, 2-, and 1-kilogram packets is increasingly taking place, but this is sometimes associated with fertilizer adulteration and counterfeit products. Some of the wide fluctuations in the nitrogen and phosphorous concentration in fertilizers can be accounted for by the absence of effective measurement and calibration facilities. In this context, the Kenya Plant Health Inspectorate Service and the Kenya Pesticide Board should become more effective in monitoring and controlling adulteration and counterfeit products, as well as intensifying farmer and retailer awareness programs to help protect farmers from substandard products.

5. *Increasing fertilizer response rates through agronomic training of farmers.* The profitability of fertilizer use could be enhanced by improving the aggregate crop yield response rates to fertilizer application. This requires making complementary investments in training for farmers on agronomic practices, soil fertility, water management, and efficient use of fertilizer and investments in crop science to generate more fertilizer-responsive seeds.[10] Emerging problems of soil acidity in the maize belt of western Kenya indicate that raising the pH levels of soil may be required to ensure the profitable use of fertilizer in these areas. Survey data commonly indicate that the contribution of fertilizer to foodgrain yields varies

tremendously across farms even within the same villages. Simply bringing fertilizer response rates among the bottom half of the distribution up to the mean would contribute substantially to household and national food security (Nyoro, Kirimi, and Jayne 2004).

6. Finally, *realizing that producer organizations, despite their poor track record, will increasingly be crucial for rural income growth.* Well-managed producer groups have the potential to reduce the transaction costs and risks of private marketing firms dealing with farmers and to develop a production base through the transfer of credit, inputs, and know-how. The Farm Inputs Promotions (FIPS) and the Kenya Market Development Program (KMDP) / Cereal Growers Association farmer training programs are examples of successful attempts by government, development partners, and nongovernmental organizations to assist and train groups to use farm extension knowledge, supply-chain development, and fertilizer technologies.

Although all of these measures can contribute to increased fertilizer use, none is likely to prove effective in isolation. Policymakers should therefore select strategic combinations of supply- and demand-side measures to allow supply and demand to grow in parallel, strengthening the basis for viable private-sector-led commercial fertilizer markets.

The final question is about the role of fertilizer subsidies. The greatest scope for subsidies to promote fertilizer use is in the areas where fertilizer use may be far below its optimal levels after taking into account maize yield response to fertilizer and the risk of applying fertilizer, especially in semiarid regions where crop failure is not unusual. Recent evidence indicates that crops' response to fertilizer application varies widely among smallholder farmers even within the same villages due to differences in management practices, soil quality, timeliness of application, and so forth, and that there is substantial scope for increasing the efficiency of fertilizer use at least for farmers who are currently achieving lower response rates from fertilizer application than their neighbors (Marenya and Barrett 2009; Xu et al. 2009). Moreover, there is little empirical evidence to allow us to determine how prevailing levels of fertilizer application compare to optimal levels, taking these factors into account. Fertilizer use rates are clearly low in the semiarid areas of Kenya, and fertilizer subsidies in these areas would likely increase fertilizer use, but the contribution to yields and smallholder incomes may be quite limited because of the environmental risk and low response rates in such areas. A major question for semiarid areas, therefore, is whether poverty reduction and food security objectives can be best achieved through fertilizer subsidies or other types of public programs and investments. Given that resources are scarce, efforts should be made to identify which types of agricultural expenditures will generate the greatest payoffs.

In the high-potential areas, the large majority of farmers are already purchasing fertilizer and the use rates in 2007 were quite high, although the use rates are likely to have fallen since then due to the adverse conditions mentioned earlier. Fertilizer subsidies are politically attractive because they promise increased fertilizer use and food production, but these outcomes are by no means assured. In 2009 Kenya is facing its lowest maize production level in recent history after having initiated a major fertilizer subsidy program; poor rains in 2009 rendered the fertilizer subsidy program relatively ineffective, and the country has imported more than 1 million tons of maize since early 2009. Moreover, providing subsidized fertilizer in areas of high commercial demand will almost certainly result in a partial crowding out of commercial sales, as shown by findings in Zambia and Malawi where commercial demand for fertilizer is considerably lower than in Kenya (Dorward et al. 2008; Xu et al. 2009). Where purchase levels for commercial fertilizer are high, a ton of subsidized fertilizer distributed by government is unlikely to result in the application of an additional ton of fertilizer to farmers' fields because the farmers previously purchasing fertilizer are no longer likely to buy it if they can acquire the same amount more inexpensively from a government program.

The experience of Kenya demonstrates the role of a supportive policy environment that attracts local and foreign direct investment in improving smallholder farmers' access to input and commodity markets. In Kenya's case, a stable input-marketing policy environment has fostered a private-sector response that supports smallholders' agricultural productivity and poverty alleviation. These goals remain elusive in countries lacking a sustained commitment to the development of viable commercial input delivery systems. Although the government's policy stance toward maize marketing has been prone to vacillation, the operations of the NCPB and the elimination of regional trade barriers since the inception of the EAC Custom Union in January 2005 have both promoted maize price stability (Jayne, Myers, and Nyoro 2008; Chapoto and Jayne 2009). Complementary programs to support small farmers' productivity, such as the FIPS program, the CNFA agrodealer training and credit program, and the organization of farmers into groups to facilitate their access to extension and credit services under the KMDP have also been important factors in increasing fertilizer use in Kenya.

Notes

1. See Kodhek et al. (1998) for details on the sample design. More details about the data can be found in the discussion paper and in Ariga et al. (2008).

2. See Ariga and Jayne (2009) for further details on the reforms related to fertilizer and maize.

3. However, Karanja, Renkow, and Crawford (2003) show that adoption of technologies in high-potential areas is associated with higher gains in yield and profitability compared to those in marginal areas.

4. Because the sample is a stationary set of households that do not change from year to year, changes in distance to markets and services cannot reflect migration or other causes of household relocation.

5. A 2007 Rockefeller Foundation–funded study using different farm survey data in four districts of western Kenya reports either a similar or a higher proportion of small-scale farmers using inorganic fertilizer on maize compared to this study.

6. See Ariga and Jayne (2009) for details.

7. See Ariga and Jayne (2009) for details.

8. This could be due to declining farm size and risk mitigation.

9. Some efficiency improvements in Mombasa port operations have recently been implemented, and more comprehensive reforms are currently under consideration.

10. Research indicates that the highest crop yield response is obtained with a combination of improved seed, fertilizer, and agronomic practices to increase soil organic matter (Kelly 2006; Marenya and Barrett 2008). In some areas, improved management practices may have a greater impact on yields than fertilizer alone (Haggblade, Tembo, and Donovan 2004).

References

Allgood, J. H., and J. Kilungo. 1996. *An appraisal of the fertilizer market in Kenya and recommendations for improving fertilizer use practices by smallholder farmers: A field report.* Muscle Shoals, Ala., U.S.A.: International Fertilizer Development Center.

Ariga, J. M. 2007. Estimation of fertilizer profitability and technical efficiency for maize in Kenya. Tegemeo Institute, Egerton University, Nairobi, Kenya. Photocopy.

Ariga, J. M., and T. S. Jayne. 2008. Maize trade and marketing policy interventions in Kenya. Paper presented at the Conference on Food Marketing and Trade Policies in Eastern and Southern Africa, EST Division, Food and Agriculture Organization of the United Nations, March 1, 2007, in Rome.

———. 2009. Private sector responses to public investments and policy reforms: The case of fertilizer and maize market development in Kenya. IFPRI Discussion Paper 921. Washington, D.C.: International Food Policy Research Institute.

———. Forthcoming. Factors driving the increase in fertilizer use by smallholder farmers in Kenya, 1990–2007. Washington, D.C.: International Food Policy Research Institute.

Ariga, J. M., T. S. Jayne, and J. Nyoro. 2006. Factors driving the growth in fertilizer consumption in Kenya, 1990–2005: Sustaining the momentum in Kenya and lessons for broader replicability in Sub-Saharan Africa. Working Paper 24. Nairobi, Kenya: Tegemeo Institute, Egerton University.

Ariga, J. M., T. S. Jayne, B. Kibaara, and J. K. Nyoro. 2008. Trends and patterns in fertilizer use by smallholder farmers in Kenya, 1997–2007. Working Paper 28. Nairobi, Kenya: Tegemeo Institute, Egerton University.

Bates, R. H. 1981. *Markets and states in tropical Africa: The political basis of agricultural policies.* Berkeley, Calif., U.S.A.: University of California Press.

———. 1989. *Beyond the miracle of the market: The political economy of agrarian development in Kenya.* New York: Cambridge University Press.

Burke, W., T. Jayne, A. Freeman, and P. Kristjanson. 2007. Factors associated with farm households' movement into and out of poverty in Kenya: The rising importance of livestock. International Development Working Paper 90. East Lansing, Mich., U.S.A.: Michigan State University.

Chapoto, A., and T. S. Jayne. 2009. Open versus closed maize border policy: A comparison of maize price instability in East and Southern Africa. International Development Working Paper. East Lansing, Mich., U.S.A.: Department of Agricultural, Food, and Resource Economics, Michigan State University.

Cleveland, W. 1979. Robust locally-weighted regression and smoothing scatterplots. *Journal of the American Statistical Association* 74 (368): 829–836.

de Groote, H., G. Owuor, J. Ouma, L. Mohammed, and K. Danda. 2005. The maize green revolution in Kenya: What happened? *Electronic Journal of Agricultural and Development Economics* 2 (1): 32–49.

De Soto, H. 2000. *The mystery of capital: Why capitalism triumphs in the West and fails everywhere else.* London: Black Swan.

Dorward, A., E. Chirwa, V. Kelly, T. Jayne, R. Slater, and D. Boughton. 2008. Evaluation of the 2006/7 agricultural input supply programme, Malawi. Final report of the School of Oriental and African Studies (SOAS), Wadonda Consult, Michigan State University, and Overseas Development Institute (ODI), undertaken for the Ministry of Agriculture and Food Security, Government of Malawi.

FAO (Food and Agriculture Organization of the United Nations). 2009. FAOSTAT. <http://faostat.fao.org>. Accessed August 20, 2010.

Gitonga. K. T. 2004. *Study on rationalization and harmonization of policies, regulations, procedures, grades, and standards in the fertilizer sub-sector in eastern Africa.* Kenya Report. Kampala, Uganda: Eastern and Central Africa Program for Agricultural Policy Analysis.

Haggblade, S., G. Tembo, and C. Donovan. 2004. Household level financial incentives to adoption of conservation agricultural technologies in Africa. Working Paper 9. Lusaka, Zambia: Food Security Research Project.

Hassan, R., M. Mekuria, and W. Mwangi. 2001. *Maize breeding research in eastern and southern Africa: Current status and impacts of past investments made by the public and private sectors, 1966–1997.* Mexico City: International Maize and Wheat Improvement Center.

IFDC (International Center for Soil Fertility and Agricultural Development). 2001. *An assessment of fertilizer prices in Kenya and Uganda: Domestic prices vis-à-vis international market prices.* Muscle Shoals, Ala., U.S.A. Photocopy.

Jayne, T. S., and G. Argwings-Kodhek. 1997. Consumer response to maize market liberalization. *Food Policy* 22 (5): 447–458.

Jayne, T. S., and A. Chapoto. 2006. *Emerging structural maize deficits in eastern and southern Africa: Implications for national agricultural strategies.* Policy Synthesis 16. Lusaka, Zambia: Food Security Research Project.

Jayne, T. S., and S. Jones. 1997. Food marketing and pricing policy in eastern and southern Africa: A survey. *World Development* 25 (9): 1505–1527.

Jayne, T. S., R. J. Myers, and J. Nyoro. 2008. The effects of government maize marketing policies on maize prices in Kenya. *Agricultural Economics* 38 (3): 313–325.

Karanja, D. 1996. An economic and institutional analysis of maize research in Kenya. MSU International Development Working Paper 57. East Lansing, Mich., U.S.A.: Department of Agricultural Economics, Michigan State University.

Karanja, D. D., M. Renkow, and E. W. Crawford. 2003. Welfare effects of maize technologies in marginal and high potential regions of Kenya. *Agricultural Economics* 29 (3): 331–341.

Kelly, V. 2006. *Factors affecting demand for fertilizer in Sub-Saharan Africa.* Agriculture and Rural Development Discussion Paper 23. Washington, D.C.: World Bank.

Kenya, Ministry of Agriculture. 2008. *Economic review of agriculture: 2008.* Nairobi, Kenya: Central Planning and Project Monitoring Unit, Ministry of Agriculture.

Kibaara, B., J. Ariga, J. Olwande, and T. S. Jayne. 2008. Trends in Kenyan agricultural productivity: 1997–2007. Working Paper 31. Nairobi, Kenya: Tegemeo Institute, Egerton University.

Kimuyu, P. 1994. *Evaluation of the USAID/Kenya fertilizer pricing and marketing reform program.* Nairobi, Kenya: U.S. Agency for International Development.

Kodhek, G. 2004. *Kenya agriculture sector brief.* Rome: Food and Agriculture Organization of the United Nations.

Kodhek, G., T. S. Jayne, G. Nyambane, T. Awuor, and T. Yamano. 1998. *How can micro-level household survey data make a difference for agricultural policy making?* Nairobi, Kenya: Egerton University.

Marenya, P. P., and C. B. Barrett. 2008. *Soil quality and fertilizer use rates among smallholder farmers in western Kenya.* Ithaca, N.Y., U.S.A.: Cornell University.

———. 2009. State-conditional fertilizer yield response on western Kenyan farms. *American Journal of Agricultural Economics* 91 (4): 991–1006.

Mason, N., T. S. Jayne, C. Donovan, and A. Chapoto. 2009. Are staple foods becoming more expensive for urban consumers in Eastern and Southern Africa? Trends in food prices, marketing margins, and wage rates in Kenya, Malawi, Mozambique, and Zambia. MSU International Development Working Paper 98. East Lansing, Mich., U.S.A.: Department of Agricultural Economics, Michigan State University.

Matsumoto, T., and T. Yamano. 2009. *Soil fertility, fertilizer, and the maize green revolution in East Africa.* Policy Research Working Paper 5158. Washington, D.C.: World Bank.

Nyoro, J., M. W. Kiiru, and T. S. Jayne. 1999. Evolution of Kenya's maize marketing systems in the post liberalization era. Working Paper 2. Nairobi, Kenya: Tegemeo Institute, Egerton University.

Obare, G. A., S. W. Omamo, and J. C. Williams. 2003. Smallholder production structure and rural roads in Africa: The case of Nakuru District, Kenya. *Agricultural Economics* 28: 245–254.

Suri, T. 2007. *Selection and comparative advantage in technology adoption.* Yale University Economic Growth Center Discussion Paper 944. New Haven, Conn., U.S.A.: Yale University.

Tegemeo Institute of Agricultural Policy and Development / Michigan State University. Various years. Kenya household surveys. Tegemeo Institute of Agricultural Policy and Development, Egerton University, Nairobi, Kenya; Michigan State University, East Lansing, Mich., U.S.A.

Wanzala, M., T. S. Jayne, and J. Staatz. 2002. Fertilizer markets and agricultural production incentives: Insights from Kenya. Working Paper 4. Nairobi, Kenya: Tegemeo Institute, Egerton University.

Williamson, J. 1997. The Washington consensus revisited. In *Economic and social development in the 21st century,* ed. L. Emmerij. Washington, D.C.: Inter-American Development Bank and Johns Hopkins University Press.

World Bank. 1981. *Accelerated development in Sub-Saharan Africa.* Washington, D.C.

Xu, Z., B. Burke, T. S. Jayne, and J. Govereh. 2009. Do input subsidy programs "crowd in" or "crowd out" commercial market development? Modeling fertilizer demand in a two-channel marketing system. *Agricultural Economics* 40 (1): 79–94.

Chapter 14

The Mungbean Transformation: Diversifying Crops, Defeating Malnutrition

Subramanyam Shanmugasundaram,
J. D. H. Keatinge, and Jacqueline d'Arros Hughes

Mungbeans, *Vigna radiata* var. *radiata* (L.) Wilczek, are one of the major pulse crops supplementing the cereal-based diets of the poor in Asia today. High in protein and easy to digest, mungbeans consumed in combination with cereals can significantly increase the quality of protein in a meal (Thirumaran and Seralathan 1988). Yet 30 years ago, mungbeans were still a crop relegated to marginal lands and cultivated with minimal inputs.

Beginning in the 1960s, cereals were the focus of agricultural research in Asia; the Green Revolution increased cereal production and helped Asia attain sufficiency in food energy (ESCAP 1985). The issue of protein-calorie malnutrition, however, remained unresolved in many developing countries, while the singular emphasis on cereal production generated undesirable outcomes such as depletion of water tables, soil salinization, and overuse of pesticides (FAO 1985; Evenson and Gollin 2003; Sekhon et al. 2007). This emphasis also pushed legumes, including mungbeans, to

We acknowledge the excellent logistical support and editing of the manuscript by Maureen Mecozzi, managing editor / coordinator, Editorial and Library Services at AVRDC, The World Vegetable Center. We thank the following for providing information on short notice to help strengthen the chapter: M. L. Chadha (AVRDC, Regional Center for South Asia, India), T. S. Bains (Punjab Agricultural University, India), and K. D. Joshi (International Maize and Wheat Improvement Center, Nepal). We also recognize the contributions of David J. Spielman, research fellow at IFPRI, and all anonymous reviewers; the final text reflects their comments and guidance.

more marginal environments, leading to stagnant or decreasing production in Asia (Singh 1988; Wallis and Byth 1988) and earning them the label "slow runners" (Borlaug 1973). Traditional mungbean cultivars grown in Asia reached maturity in 90–110 days. These indeterminate cultivars were susceptible to insects and diseases, especially mungbean yellow mosaic virus. Farmers had to harvest multiple times, pod shattering was a problem, and the labor-intensive, low-yielding cultivars produced only about 400 kilograms per hectare of small seed. The World Vegetable Center (AVRDC), an international agricultural research institute based in Taiwan, recognized the potential of mungbeans to supply protein to Asia's hungry, to provide farmers with an income-generating opportunity, and to diversify agroecosystems—if the pulse could fit into high-yielding cereal cropping systems (Shanmugasundaram 2006). As proposed by Borlaug (1973), AVRDC launched a holistic mungbean improvement program to improve productivity and production.

Today, improved mungbeans constitute more than 25 percent of global mungbean production (Shanmugasundaram 2001). The introduction of improved mungbean varieties is the key factor responsible for the 35 percent increase in production in Asia, from 2.3 million tons in 1985 to 3.1 million tons in 2000 (Weinberger 2003a). The transformation of mungbeans from a marginal to a major crop has brought many benefits to Asia, including new income streams for small-scale farmers and new sources of dietary protein and iron for the poor, especially children and women. This chapter describes several key interventions to develop and promote improved mungbean varieties and their impact on food security in Asia.

The Intervention: Improving and Disseminating Mungbeans

Inception

The Green Revolution was a boon to Asia, bringing much-needed food energy from cereals to the region (ESCAP 1985). Per capita yields of wheat increased substantially from 1961 to 1972, but over the same period, per capita yields of legumes decreased in South Asia. Pakistan, for example, spent valuable foreign exchange to import legumes to meet domestic demand (Ali et al. 1997).

Legumes, also referred to as pulses, are the major source of protein in Asia and constitute an important supplement to the predominantly cereal-based diet (Singh 1988; Paroda 1995). Cereals are deficient in the amino acid lysine, which legumes can provide; legumes are low in sulfur-rich amino acids, which cereals can provide (Thirumaran and Seralathan 1988). Legumes are also high in vitamins, minerals, and fiber. Moreover, legume plants are able to fix atmospheric nitrogen in the soil through their symbiotic association with the *Rhizobium* bacteria and are adaptable to a number of cropping systems (Shanmugasundaram 2003). Mungbeans are a warm-season crop (AVRDC 1981) that can grow during hot, wet seasons and in the arid and semiarid tropics (Pandey, Herrera, and Villegas 1988; Pannu and Singh 1988).

Traditional indeterminate mungbean varieties had a long growth duration (90 to 110 days) and required multiple harvests, making them suitable for home gardening but unsuitable for commercial production because of high labor costs. The traditional varieties were susceptible to diseases such as Cercospora leaf spot (*Cercospora cruenta, C. canescens,* and *Cercospora* sp.), powdery mildew (*Erysiphe polygoni*), and mungbean yellow mosaic virus (MYMV) and to insect pests such as bean flies (*Melanagromyza sojae, M. dolichostigma,* and *Ophiomyia centrosematis*), lima bean pod borers (*Etiella zinckenella*), and mungbean weevils (*Callosobruchus chinensis*). Moreover, these traditional varieties did not respond to inputs (AVRDC 1975, 1977).

In a semiwild state, mungbeans were of little value. AVRDC, however, saw an opportunity to domesticate the crop and make substantial improvements. To achieve this objective, it was essential to have a broad, deep pool of genetic diversity upon which breeders could draw. AVRDC obtained mungbean germplasm from various laboratories and other sources; today the center's mungbean collection contains 10,733 accessions of *Vigna* and related species.

A Vision of the Ideal Mungbean for Asia

The initial vision at AVRDC for the mungbean improvement program came from plant breeders, plant physiologists, plant pathologists, entomologists, biochemists, and economists. Plant breeder David R. MacKenzie and plant physiologist Henry Wu had a constructive debate on the type of mungbean plant breeders should develop. Finally they agreed on a short plant type with all the pods on the top of the plant (MacKenzie and Shanmugasundaram 1973). In addition, it was agreed that breeders would aim for a plant that would have the following properties (Fernandez and Shanmugasundaram 1988; Shanmugasundaram and Kim 1996):

- a stable potential yield of more than 2 tons per hectare;

- a maturity duration of around 60–75 days;

- a uniform maturity so that the harvest could be completed in one attempt;

- a bold seed size (50–60 grams for 1,000 seeds instead of the 25–30 grams for 1,000 seeds produced by the local varieties);

- resistance to Cercospora leaf spot, powdery mildew, bean flies, pod borers, and bruchid weevils;

- a compact growth form;

- a favorable harvest index, which measures the ratio of grain weight to total plant weight;

- less sensitivity to photoperiod; and

- a growth habit that was more determinate than those of traditional varieties.

Collaboration to Breed Mungbean Varieties

Having agreed on a blueprint for the mungbean plant type, breeders began screening germplasm for the required traits. Promising lines and superior germplasm accessions were evaluated annually for adaptation and suitability in many countries in different locations and seasons through the International Mungbean Nursery (IMN). The University of Missouri, supported by the U.S. Agency for International Development, conducted the first four IMNs from 1971 to 1975 (Morton, Smith, and Poehlman 1982). The AVRDC carried out the IMN from 1976 onward. Through the mechanism of the IMN, improved mungbean lines were initially spread to Southeast Asia, South Asia, East Asia, and China and then to other parts of the globe (Fernandez and Shanmugasundaram 1988). This intensive, extensive, formal, and informal collaboration led to the release of 112 improved mungbean varieties based on AVRDC breeding lines and germplasm in 27 countries worldwide. The improved varieties have replaced 50–100 percent of local varieties in the region (Ali et al. 1997; Shanmugasundaram 2001; Huijie et al. 2003; Weinberger 2003a).

Development of an Improved Mungbean

Mungbean breeders Ricardo Lantican and Rudy Navarro at the Institute of Plant Breeding, University of the Philippines, developed an early-maturing, compact plant type with uniform maturity and bold seeds. The most prominent lines were CES 55, CES 87, MG 50-10A, and MD 15-2. The Philippine genotypes were susceptible to major diseases, however, such as Cercospora leaf spot and powdery mildew (Lantican and Navarro 1988).

Varieties bred at the Indian Agricultural Research Institute, the All India Coordinated Pulses Program at Kanpur, and other regional research centers did carry resistance to Cercospora leaf spot and powdery mildew and were able to withstand adverse soil and climate conditions with few or no inputs. The Indian varieties, however, were found to be indeterminate, late-maturing, and small-seeded (Lantican and Navarro 1988; Tickoo et al. 1988).

In the mid-1970s, the Institute of Plant Breeding at the University of the Philippines developed improved mungbeans with high yields and disease resistance for the Philippines termed the "Pag-asa" series: Pag-asa 1, 2, and 3 (IPB 1976–78). AVRDC's Philippines Outreach Program and the Bureau of Plant Industry in the Philippines developed a series of varieties for the Philippines based on AVRDC improved lines, which were used in the International Rice Research Institute (IRRI) cropping system network for testing in Southeast Asia (Catipon, Legaspi, and Jarilla 1988).

The AVRDC plant breeder Hyo Guen Park decided to combine the desirable traits of the varieties from the Philippines and India. Data collected for 31 advanced superior breeding lines and promising germplasm accessions from IMNs 5–12 were evaluated for environmental sensitivity. In total, 21 lines were selected and placed into three groups representing diverse environments (Table 14.1).

In an analysis of various breeding lines for photoperiodic response, the following mungbean lines were found to be less sensitive to photoperiod, and they also had high yields and wide adaptation to diverse environments: VC 1973A, VC 1628A, and VC 1158B (AVRDC 1988). Fernandez and Shanmugasundaram (1988) noted that a short-duration mungbean with less sensitivity to photoperiod and temperature would be a promising candidate for crop diversification. A number of varieties were also found to be tolerant to drought, an important factor affecting the stability of mungbean yields in the semiarid and arid tropics (AVRDC 1988).

The potential yield loss due to Cercospora leaf spot and powdery mildew was reported to be 40 and 58 percent, respectively, in susceptible unimproved varieties (Shanmugasundaram and Tschanz 1987). A number of elite lines with high yields, synchronous maturity, and resistance to leaf spot and powdery mildew were developed. By 1987, 16 countries had released 28 varieties to their farmers (Fernandez and Shanmugasundaram 1988).

Development in Thailand. In Thailand, the Department of Agriculture, the Department of Agricultural Extension, regional research stations, several agricultural and vocational colleges, and several international agricultural research institutes cooperated with AVRDC to improve mungbeans. The Department of Agriculture

Table 14.1 Mungbean lines suitable for diverse environments based on environmental sensitivity, 1988

Group	Lines
Group A: High yielding but suitable only for favorable environments	VC 1560D, VC 2565A, VC 2719A, VC 2755A, VC 2764A, VC 2768A, VC 2778A
Group B: High yielding and stable in unfavorable environments	VC 1168B, VC 1209B, VC 1562A, VC 1628A, VC 1973A, VC 1974A, VC 2523A, VC 2582A
Group C: Suitable for low-yielding environments	VC 1089A, VC 1163A, VC 1168A, V 1381, V 1944

Source: Fernandez and Shanmugasundaram (1988).

had been evaluating AVRDC mungbeans since 1974; its active mungbean program evaluated IMN lines from the University of Missouri and received improved lines from AVRDC, the Philippines, and IRRI. In 1982 the results from the 10th IMN, from trials at 218 locations in 14 countries, revealed that VC 2778A and VC 1973A had consistently higher yields than the others (1,189 and 1,145 kilograms per hectare, respectively) and produced the bolder seed preferred by the market (Ahn, Chen, and Chen 1985). From 1982 to 1984 a consolidated testing network involving all the partners in Thailand conducted 28 trials at various experiment stations around the country. The results showed that VC 1973A and VC 2778A outyielded the local Uthong 1 by as much as 37 percent. Kasetsart University successfully released VC 1973A as Kamphaengsaen 1 (KPS1) and VC 2778A as Kamphaengsaen 2 (KPS2) (Fernandez and Shanmugasundaram 1988). The Department of Agriculture released Chai Nat 60 in honor of the 60th birthday of Thailand's king and later another variety, Chai Nat 36. Prince of Songkla University released PSU 1. These varieties now occupy almost 100 percent of the mungbean area in Thailand (Shanmugasundaram, Keatinge, and Hughes 2009).

Development in China. Beginning in 1983, improved mungbean lines from the 11th IMN onward and germplasm from the AVRDC were regularly sent to China for evaluation through AVRDC's Thailand Outreach Program (now the Asian Regional Center). Trial plantings of mungbean lines were undertaken throughout China. With its erect and compact plant type, resistance to lodging, pods set on the upper nodes, early and uniform maturity, high and stable yield, and bold seeds when planted under conditions of medium to above-average fertility and moisture, VC 1973A created excitement among researchers and demand from farmers in China. The Chinese Academy of Agricultural Sciences (CAAS) supported the seed multiplication of VC 1973A, and, beginning with 3.5 kilograms of seed in 1984, it produced enough seed to plant 20,000 hectares in 1987. Analysis showed that VC 1973A was a stable and widely adapted variety, with an average yield of 1,531 kilograms per hectare and a potential yield of about 4,500 kilograms per hectare (Cheng and Lin 1993). Released as Zhong Lu #1, VC 1973A was the first mungbean variety released for nationwide cultivation in China since 1949. In 1989, the total area planted to Zhong Lu #1 reached 267,000 hectares (Cheng 1993; Cheng and Lin 1993; Shanmugasundaram 2003). Other new varieties soon followed (Huanyu and Zhizong 1993).

Development in Myanmar. Following the identification of VC 1973A as a promising breeding line, in 1982 R. K. Palis, the IRRI resident scientist in Myanmar, expressed interest in evaluating mungbeans. Through its Thailand Outreach Program, AVRDC provided the seeds for selected mungbean lines, and Dr. Palis hand carried the seeds to Myanmar, where he conducted trials. AVRDC did not hear from Myanmar for quite some time, however; Dr. Palis had left the country. In the early 1990s, the representative of the United Nations Food and Agriculture Organization (FAO) in Myanmar contacted AVRDC, noting that lines VC 1973A and VC 2768A were doing well and asking for large quantities of the seed for extension to farmers. AVRDC did not produce the quantities of seed required for large-scale distribution, but it agreed to provide the FAO with 100 kilograms of the two lines of seed for multiplication. Since then, the FAO, AVRDC's Asian Regional Center, and the Government of Myanmar have actively produced VC 1973A, VC 2768A, and several other breeding lines, which have been enthusiastically accepted by farmers. The Asian Regional Center provided training for Myanmar scientists in varietal evaluation and seed production, which helped to expand the area and production of mungbeans.

Development in South Asia.   Although improved mungbeans showed spectacular success in Southeast Asia and China (Shanmugasundaram 2001), South Asia was the area most in need of improved varieties. MYMV severely constrained mungbean expansion and production in South Asia, but the disease was not present in Southeast Asia and Taiwan, where AVRDC has its headquarters. Collaborative research with South Asia would be needed to address the problem. Pakistan served as the gateway for AVRDC to initiate breeding for MYMV resistance in South Asia. The Nuclear Institute for Agriculture and Biology (NIAB) in Faisalabad identified MYMV resistance through irradiation of the local variety. Then shuttle breeding—that is, cultivation of two crops a year by sending germplasm back and forth between AVRDC in Taiwan and the NIAB in Pakistan—enabled scientists to develop improved MYMV-resistant varieties for South Asia. Improved varieties now occupy nearly 50 percent of the total mungbean area in Pakistan (Shanmugasundaram 1988; Ali et al. 1997). The improved varieties have a yield potential of up to 1,800 kilograms per hectare compared with the local variety's 1,000 kilograms per hectare; they mature in about 60 days compared with 90 days for the local variety; they mature uniformly, so they can be harvested in one picking. Farmers readily recognized the advantages of the improved variety.

Extending Improved Mungbeans across Asia

To extend the benefits of the new improved mungbeans to all countries in South Asia (the world's major producer of mungbeans, where MYMV is the major factor limiting production), an umbrella agreement was needed to promote collaboration. AVRDC had earlier established the South Asia Vegetable Research Network (SAVERNET), and all South Asian countries were signatories of the network. . SAVERNET committee members agreed to allow a mungbean subnetwork to be organized under the umbrella of SAVERNET. In a project beginning in 1997, supported by the U.K. Department for International Development (DFID), participants agreed to conduct trials in two seasons (rainy and summer) using 17 entries for the rainy season and 16 entries for summer. Seeds for all the agreed varieties were multiplied at AVRDC headquarters in Taiwan, and a trial set was packaged and sent to each of the participating network partners.[1] In total, 64 multilocation trials were conducted in six countries in three years. After four years, scientists had managed to release these improved, virus-resistant varieties, which yield at least 2 tons per hectare and mature in 55 to 65 days, throughout the region to Bangladesh, Bhutan, India, Nepal, Pakistan, and Sri Lanka.

In 2002, DFID approved a two-year project to improve rural household income and employment opportunities, diversify diets, and increase nutritional security through the increased availability of pulses (mungbeans) for the poor and to enhance soil fertility in South Asia through the participation of 500,000 farmers. The project's purpose was to help farmers diversify rice–wheat rotations by adopting improved mungbean cultivars, persuade national agricultural research systems (NARSs) to use improved mungbean germplasm and methodologies for location-specific adaptation, and promote mungbean consumption for balanced diets. Agronomic practice guides were prepared in local languages and distributed to farmers in each country. Efforts to produce large quantities of seed were also undertaken by various agencies and universities in the region.

In India, these efforts were augmented by the Seed Village Program, initiated in 2003. Under this program, a total of 270 farmers, each planting mungbeans on just 0.4 hectare of land, succeeded in producing about 2,700 tons of high-quality seed that was distributed to other farmers in the next season (Table 14.2).

Thanks to this intensive effort, the area under mungbeans during the rainy season increased from 23,000 hectares in 2001 to 65,000 hectares in 2003. The area planted to mungbeans was estimated at 25,000 hectares in the summer season of 2003 and 40,000 hectares in 2004. Farmers from neighboring states, including Rajasthan, Haryana, western Uttar Pradesh, and Bihar, went to Punjab to obtain the seeds of improved varieties.

Table 14.2 Seed production in India through Seed Village Programs, 2003

Production parameter	Measure
Number of villages in each Seed Village Program	6
Number of farmers in each Seed Village Program	45
Area each farmer given seed for planting	0.4 hectares
Seed production from 0.4 hectares	0.4 metric tons
Area planted with 0.4 metric ton in the rainy season, 2003	10 hectares
Seed produced by each farmer in the rainy season, 2003	10 metric tons
Total seed production in one Seed Village Program (10 × 45)	450 metric tons

Source: Bains, Aggarwal, and Barakoti (2006).

In Bangladesh seven agencies multiplied the seed of six mungbean varieties for extension to farmers. A seed exchange program was instituted in which the farmers were given seeds of improved cultivars in exchange for seeds of their local variety. The total quantity of seeds produced was 11,538 tons—sufficient to plant about 330,000 hectares in 2004. And after Bangladeshi scientists observed the Seed Village Program in Punjab, they introduced a similar program in Bangladesh.

In Nepal several research and nongovernmental organizations conducted full-scale on-farm participatory trials in 21 districts in 2003. The improved varieties were well adapted to the spring, summer, and autumn seasons in the low hill and *terai* agro-ecosystems of Nepal and the Indo-Gangetic Plains (Khanal et al. 2004). Improved varieties now occupy about 12,000 hectares in Nepal (Shanmugasundaram 2006).

Thailand's seed exchange program has played a major role in replacing the old local mungbean variety with the new improved varieties. Government policy limited the release of water for the dry season to prevent farmers from growing another rice crop, which forced the farmers to grow a nonrice crop such as mungbeans. This situation in part accounted for the low mungbean yield in the dry season; farmers did not get enough water and inputs for a good yield. However, they received improved mungbean seed from the government because they were growing a non-rice crop (Jansen and Charnnarongkul 1992).

Enhanced Production in Asia

The mungbean research and seed production efforts led to a significant boost in mungbean production in Asia. In China, for example, as of 2003 the Chinese National and Provincial Crop Review Committee had released nearly 20 new improved mungbean varieties and extended them for large-scale planting throughout the country. The two improved varieties selected from AVRDC, Zhong Lu #1 and Zhong Lu #2, were planted on nearly 85 percent of the 772,000 hectares under mungbean cultivation in China (Huijie et al. 2003). Owing to the release of these improved varieties, mungbean production in China increased at an annual rate of 2.4 percent from 1986 to 2000, primarily because of the 1.7 percent annual increase in yield. Annual per capita consumption of mungbeans in China increased from 0.3 to 0.5 kilograms from 1986 to 2000, and during the same period the share of mungbeans in total pulse consumption increased from 14.2 to 28.0 percent. Development and dissemination of improved mungbeans in China also benefited from mungbeans' inclusion in the country's major reforms to the grain distribution system, which helped make production profitable for farmers (Huijie et al. 2003).

In India's Punjab, the area planted to mungbeans increased from 23,000 hectares in 2000–01 (Grover, Weinberger, and Shanmusgasundaram 2006) to more than 75,000 hectares in 2002–03 (Bains et al. 2006) thanks to the improved mungbean variety SML668. More than 95 percent and 60 percent of the mungbean area in the state was planted to SML668 during the 2006 summer and rainy seasons, respectively (Chadha 2009). States neighboring Punjab also benefited from the transfer of mungbean technology. For example, by the end of 2008 the area planted to SML668 in Haryana, Himachal Pradesh, and Rajasthan was estimated to be around 70,000, 50,000, and 60,000 hectares, respectively (M. L. Chadha, personal communication).

In Pakistan, a detailed mungbean production survey was conducted in the Punjab in 1994, with 250 representative farmers selected at random for the study. Data on farm management practices, cropping patterns, input use, varietal adoption, cost and return, and production constraints in mungbean cultivation were collected using a detailed questionnaire. The survey results were striking. They showed that improved mungbean varieties replaced 90 percent of the local varieties in 1994 compared with 20 percent in 1988. Improved varieties produced yields nearly 55 percent higher than yields of the local variety. The cost of producing wheat in a wheat–mungbean production system was 23.5 percent less than in a wheat–other crop rotation. The benefit–cost ratio of improved mungbeans ranged from 1.87 to 2.21, depending on the variety, compared with 1.31 for local varieties. And mungbeans' share of the total pulse area increased from 3 percent in 1980 to 11 percent in 1993–94 (Ali et al. 1997).[2]

In Bangladesh, the impressive growth in mungbean area and production was also attributed to the introduction of improved varieties. In 1980 Bangladesh had a mungbean production area of 15,000 hectares, which produced 7,000 tons annually at 467 kilograms per hectare. In 2000 the production area increased to 55,000 hectares, and production rose to 36,000 tons annually at 654 kilograms per hectare. Improved varieties have replaced local varieties in 70 percent of the mungbean area of Bangladesh. A socioeconomic survey of 320 farmers in the districts of Barisal, Dinajpur, and Jessore conducted in 2002–03 showed that about 50 percent of the farmers cultivated improved mungbeans, and the farmers who grew mungbeans consumed more mungbeans (Weinberger, Karim, and Islam 2006). In addition, the Bangladesh Ministry of Agriculture approved the Lentil, Blackgram, and Mungbean Development Pilot Project (LBMDPP) in collaboration with DFID and AVRDC. The country's seed exchange program was funded primarily through the LBMDPP, which facilitated the rapid area expansion of improved mungbean varieties. Problems arose with seed purity, but the LBMDPP is determined to rectify them.

With only a few grams of test material initially sent for trial in the early 1980s, Myanmar's local researchers and extension agents were able to move the promising selections to farmers and bring about significant change in agriculture and diets. The mungbean area in Myanmar increased from 68,000 hectares in 1980–81 to 650,000 hectares in 1997–2000.

Afghanistan has become the latest country to embrace mungbeans. On April 15, 2009, Deputy Minister Mohammad Sharif formally released two improved mungbean varieties, Maash-2008 (NM 92) and Mai-2008 (NM 94), developed from AVRDC lines. The International Center for Agricultural Research in the Dry Areas, working in collaboration with the Agricultural Research Institute of Afghanistan, the Ministry of Agriculture, Irrigation, and Livestock, and the AVRDC, has been evaluating these varieties for the past four years in nine locations in Afghanistan. Both new varieties are short-duration and high-yielding compared with currently grown varieties in Afghanistan. These varieties, which mature about 10 days earlier than the local variety, yield 50–60 percent more than all local varieties and produce 30 percent more than the highest-yielding existing improved variety.

The Impact of Improved Mungbeans in South Asia

Today the area under improved mungbean cultivation in Asia has expanded dramatically (Table 14.3). Assessments of the impact of improved mungbeans have been successfully conducted in most countries highlighted in this chapter. The number of improved varieties released was the easiest output to assess; it reflects the use of germplasm combined with serendipitous selection, insight, imagination, and sheer hard work (Anderson 1997). Mungbeans have strengthened economies, opened up new market niches, improved health, promoted better agricultural policies, and prompted close networking at many levels to share skills and knowledge, build social capital, and enhance local capacity for research and production.

Table 14.3 Estimated area under mungbeans before and after the intervention in different Asian countries (hectares)

Country	Before intervention	After intervention
Bangladesh	15,000 (1985)	70,000 (2006)
China	547,000 (1984)	776,000 (2000)
India	284,500 (1980)	550,000 (2008)
Myanmar	43,000 (1980–81)	1,000,000 (1998–99)
Pakistan	100,000 (1985)	200,000 (2000)
Sri Lanka	14,000 (1980)	33,200 (1995)
Thailand	308,000 (1984)	335,000 (1995)

Sources: Bangladesh, China, Pakistan, and Sri Lanka: Weinberger (2003a); India: Shanmugasundaram (2006); Myanmar: Bahl (1999); Thailand: Chainuvati, Potan, and Worasan (1988), Srinives (1998).

The impact studies cited in this chapter provide only indicative magnitudes of the effects of the introduction of the improved mungbean varieties, because the analyses include methodological weaknesses and assumptions that can be challenged. To obtain more precise and reliable estimates of the benefits of the widespread adoption of improved mungbean varieties, more rigorous studies need to be conducted. However, the best proof of the success of the mungbean transformation —the increased rate of adoption of the improved varieties along with the rapid decline in the share of traditional varieties—is unquestionable. Farmers would not adopt the improved varieties if they were not significantly superior to the other varieties (Evenson and Gollin 2003).

Economic Benefits

Researchers have documented a range of economic benefits from the diffusion of improved mungbeans in Asia. Some benefits are aggregated at the macro level. For example, between 1986 and 2000 China's mungbean imports decreased from $13.6 million to $1.4 million while its exports increased from $45 million to $50 million. In Pakistan the economic benefits from mungbeans have been estimated at around $20 million, of which the replacement of local varieties with improved varieties accounts for $5.3 million; the increase in area with improved varieties accounts for $3.6 million; the improvement in quality accounts for $4.4 million; and the residual effects of mungbeans on the following wheat crops accounts for $6.4 million. Consumers' share of the benefits was 38 percent, and producers' share was 62 percent (Ali et al. 1997).[3]

Other benefits are measured at the farm level. A farm-level survey of 115 households in six provinces in the northern and central plains of Thailand was conducted in April–May 1992. The key objectives of the survey were to obtain a detailed input–output characterization of mungbean cultivation in Thailand, assess the adoption of AVRDC-related mungbean technologies, and determine their profitability. The results clearly showed that more than 90 percent of the farmers surveyed grew the three improved varieties—KPS1, KPS2, and CN60 (46, 38, and 7 percent, respectively). KPS1 and KPS2 were cultivated during the dry season, whereas KPS1 was preferred for the rainy season. The net return from mungbean cultivation in the dry season was low. In the rainy season, the net return was almost double that of the dry season (Jansen and Charnnarongkul 1992).

The economic performance of improved mungbeans in Bangladesh, demonstrated through farmers' management practices, is shown in Table 14.4.

Health and Nutrition Benefits

An important health and nutrition benefit of mungbean consumption is its role in improving iron intake (AVRDC 1998, 1999; Yang and Tsou 1998). In 2001

Table 14.4 Economic performance of improved mungbean varieties in farmers' fields in Bangladesh, 2002

Indicator	Demonstration	Farmers' management	Difference (percent)
Yield (kilograms per hectare)	1,225	956	28.13
Gross return (US$ per hectare)	554	416	33.23
Gross cost (US$ per hectare)	240	191	26.31
Gross margin (US$ per hectare)	314	225	39.36
Benefit–cost ratio	2.31	2.18	5.36

Source: Afzal and Bakr (2002).

and 2002, Weinberger (2003b) conducted a survey in Lahore, Pakistan, with 200 working women using a seven-day food consumption recall. The results of the study showed that blood hemoglobin levels are a major determinant of workers' productivity. Based on the increase in mungbean production due to improved varieties (Ali et al. 1997) and the higher iron content of the improved mungbean varieties (6.0 milligrams per 100 grams of dry grain compared with 3.5 milligrams for the local varieties) (Vijayalakshmi et al. 2003), the productivity increase of anemic female workers with increased iron availability due to improved mungbean varieties was estimated to be $3.5–$4.2 million annually (Weinberger 2003a, 2003b).

The Avinashilingam Home Science University for Women in Coimbatore, India, and Punjab Agricultural University addressed the health and nutritional benefits of mungbeans, especially to alleviate anemia. Two women scientists worked at the AVRDC Biochemistry and Nutrition Laboratory and developed improved recipes using mungbeans and other ingredients. The recipe books published in Punjabi and English help rural and urban women across northern India prepare nutritious food to improve the bioavailability of iron and alleviate malnutrition in family diets (Subramanian and Yang 1998; Bains, Yang, and Shanmugasundaram 2003). A feeding trial with 225 boys and girls ages 10 to 12 held for one year in Coimbatore showed that supplementing the children's diets with mungbean dishes prepared with improved high-iron recipes did improve their overall physical stamina, although the greater availability of energy and protein to these children may also have contributed to their improved physical stamina. Enhancing iron bioavailability through improved recipes for local dishes is a cost-effective way of improving iron nutrition in population groups at risk (Vijayalakshmi et al. 2003; Bains et al. 2006).

Sustaining Soil Productivity

As legumes, mungbeans have the ability to fix nitrogen in the soil and thus can benefit the succeeding cereal crop. Meelu and Morris (1988) reported that incorporating mungbean residues generated a rice yield increase equivalent to 25 kilograms of nitrogen fertilizer per hectare. Research results in Punjab have shown that the extra-short-duration SML668 mungbean, which fits well between wheat and rice as a catch crop, leaves a residual of 33–37 kilograms of nitrogen per hectare for the succeeding crop after meeting its own requirement. This additional nitrogen provides about 25 percent of the nitrogen requirement of the succeeding crop (Sekhon et al. 2007).

Guaranteed prices for cereals and subsidized inputs mean that farmers have little incentive to diversify the rice–wheat cropping system. Intensive cereal cropping has resulted in response to a decline in the partial factor productivity of nitrogen fertilizer (Hobbs and Morris 1996), but crop diversification is the key to reversing declining crop productivity (Pingali and Shah 1999). In 2002 a survey of 200 farmers was conducted in four ecological regions in Punjab, with 75 adopting the improved mungbean variety SML668. The net return per hectare for the rotation using wheat–mungbeans–rice was $235 compared with rice–fallow–wheat (Weinberger 2003a). The increase was due to the residual nitrogen available to the succeeding crop, which increased its yield. The total net return from the rice–wheat rotation including mungbeans was 27 percent higher than that from the rotations without mungbeans. Assuming that the 1 million hectares currently under rice–wheat rotation can include mungbeans, the total rice yield would increase by 450,000 tons, valued at $50.7 million in additional income for farmers in the Indo-Gangetic Plains in India (Weinberger 2003a).

Mungbeans in the Processing Value Chain

Mungbeans are used in making transparent noodles and vermicelli throughout Southeast Asia (Maneepun 2003). Given that mungbean noodles are a high-value commodity, research to improve the manufacture of mungbean noodles and other processed mungbean products would help increase the value of mungbeans in Thailand and the region. In addition, mungbean sprouts are a popular vegetable in Southeast Asia, China, and Taiwan. The hardness of the seed coat affects the quality and quantity of mungbean sprouts. The hard seed trait is complex and is influenced by a number of factors. There is some evidence that bold-seeded varieties have fewer hard seeds than small-seeded varieties (Sekhon et al. 2006). Further research into this trait would be useful in developing improved mungbean varieties for fresh market sprouts.

In South Asia, mungbeans are processed into *dhal.* In *dhal* processing, the husk and bran are removed by milling, after which about 73.90 percent of the *dhal* is recovered (the husk and bran account for 18.45 percent, and the milling loss is about 7.60 percent) (Amiruzzaman and Shahjahan 2006). Modern milling methods could help increase the maximum recovery of *dhal* to 89.5 percent (Singh 2006). It is not known whether the milling recovery percentage for improved mungbean varieties is better than that for traditional varieties, but the potential appears to be present, given that improved mungbean varieties have a larger grain size than traditional varieties. In the case of other grain legumes, such as pigeon peas and chickpeas, the recovery from varieties with large grain sizes is better than from varieties with small grain sizes. In any case, improvements in milling are likely to increase the marketable yield of mungbeans and add value to the crop, which can help stabilize mungbean prices.

Building Capacity

Knowledge and skills related to mungbean research and development and seed production have been extended to thousands of people across Asia. In China, training and cooperation have given a number of researchers expertise in mungbeans. The CAAS led research on mungbeans in cooperation with institutes from 25 provinces during the country's seventh and eighth Five-Year Plans, from 1986 to 1995. The total number of researchers committed to mungbeans included 30 from the Institute of Crop Germplasm Resources and 70 from other institutions. The CAAS has cooperated with AVRDC since 1983, and four of its researchers were trained at AVRDC's Asian Regional Center in Thailand. The CAAS also founded the Association of Mungbean Research, establishing a strong base for mungbean research and development in China (Institute of Crop Germplasm Resources 1993, 1999; Cheng 1996).

The Seed Village Program launched in Punjab, India, has spread to other Indian states. Beginning with 15 tons of SML668 seed in 2001, Punjab increased its seed production to approximately 810 tons in 2002 and 6,950 tons in 2003. Under the umbrella of the Seed Village Program, 600 seed kits were distributed to 600 farmers to plant 600 acres with mungbeans in Bihar, Punjab, and Rajasthan from 2005 to 2008; the program has generated 17,000 tons of seed to date (Chadha 2009). Since 2005, AVRDC has organized 15 field days and 1,200 demonstrations for more than 4,000 farmers, extension specialists, rural women, and schoolgirls to promote mungbean cultivation, which has raised awareness of improved mungbeans.

To promote the improved mungbean varieties in Bangladesh 11,280 demonstrations were conducted in 47 *upazillas* (subdistricts) in 15 districts. The average yields of demonstration plots (1,110 kilograms per hectare) were 54–56 percent higher than the national average (705 kilograms per hectare) (Afzal et al. 2006). To ensure self-sufficiency in pure, high-quality seed of improved varieties, Bangladesh followed two approaches: (1) a seed exchange program from 2002 to 2004 in which improved seeds were given to farmers in exchange for an equal quantity of the local variety seeds and (2) adoption of the Seed Village Program, initially launched in Punjab, India. In 2000–01 Bangladesh produced 1,074 tons of seed for six improved varieties; in 2002–03 seed production increased to 11,538 tons. Assuming that only 25 percent of the seed is used for planting, the amount produced was sufficient to cover the whole mungbean area of Bangladesh (Afzal et al. 2006).

Worthy of Special Note: Nepal and Sri Lanka

Normal impact assessments and economic studies could not be conducted in Nepal and Sri Lanka because of political strife, lack of personnel, and lack of resources. Nonetheless, demand for mungbeans is rising in Nepal. Collaboration with AVRDC resulted in the release of three varieties, which are planted in an increasing area. The immediate intervention issue in these postconflict countries is availability of and access to good-quality seeds (Erskine and Nesbitt 2009). A recent DFID-funded project in Nepal planned to multiply seed in 18 out of 21 districts of Nepal *terai* and hoped to produce substantial seed in the 2009 summer season, which would be used to scale out mungbean cultivation in subsequent seasons (Joshi 2009). Collaboration with private-sector nongovernmental organizations (NGOs) in Nepal is vital for the success of the program.

In Sri Lanka the total area under mungbeans in 1980 was 14,200 hectares, production was 12,900 tons, and the yield was 908 kilograms per hectare. In 1995 the production area expanded to 33,200 hectares and production increased to 26,400 tons, but the yield declined to 795 kilograms per hectare. The improved varieties have encouraged more farmers to plant mungbeans (Weinberger 2003a), but the latest data on the area planted and the production of mungbeans are unavailable.

Conclusions and Lessons Learned

The rollout of improved mungbeans offers many lessons for partners in the mungbean effort as well as for those seeking to promote other new agricultural technologies:

1. *Start with commitment.* The strong commitment of all parties—international research institutes, national partners, donors, NGOs, and individual farmers— was needed for the intervention to succeed.

2. *Plan with a purpose.* A realistic, definite goal and specific plans based on available resources helped the intervention move forward.

3. *Keep track of time.* Deadlines were set for achieving goals. Progress was monitored semiannually and annually, which allowed the project participants to make midcourse adjustments as needed in the work plan.

4. *Focus on farmers.* Crop management practices varied in each country and season, and NARSs developed farmer-friendly technologies to address farmers' specific constraints in their countries. In addition, the official release of a variety does not guarantee that farmers will cultivate it; several important benchmarks

must be met for successful adoption. Farmers need to be convinced of the variety's superior performance. Pure seed of the improved variety must be available at a reasonable price in sufficient quantities in time for farmers to plant. The management technologies associated with the improved variety in specific locations need to be defined and made available as a package to farmers. And national policies and government institutions must be committed to extend the technology to farmers.

5. *Understand differences.* Progress and success varied among countries and among provinces or states within countries. For instance, Punjab in India was able to make rapid progress in seed production and distribution; Bangladesh was able to adopt and follow the example of Punjab. Within India, Bihar, Haryana, Himachal Pradesh, Jharkhand, and Rajasthan made good progress with assistance from Punjab. Success in one area can serve as a catalyst for success in neighboring states and countries.

6. *Welcome all contributions.* Countries had different strengths and weaknesses. China and India had sufficient resources to maintain large germplasm collections and conduct meaningful research, but smaller countries also made significant contributions to the intervention's success. The Philippines developed high-yield, early-maturing, bold-seeded varieties, and Pakistan developed lines with a high level of resistance to MYMV.

7. *Share across boundaries.* Sharing resources between and within countries is the best way to rapidly spread technology and improve productivity, nutrition, and food security. Partners were able to share their technology through formal networks such as SAVERNET and informal networks such as the ones with China and Pakistan. Sharing without reservations helped to achieve the accomplishments in mungbean production.

8. *Adapt as needed.* In Myanmar researchers generated large gains by evaluating improved technology and adapting it to local situations.

9. *Educate for the future.* Training was essential to help researchers, extension staff, farmers, and economists systematically organize, plan, and implement the activities and collect data.

10. *Realize that seeing is believing.* Demonstration trials, farmers' field trials, and field days enabled farmers to see for themselves the performance of the improved varieties compared to local varieties. When farmers actively participated in

research and development efforts, they felt a sense of ownership of the output. They accepted the responsibilities and did their jobs with dignity and pride.

11. *Involve the community.* The Seed Village Program and other seed exchanges proved to be excellent vehicles by which to rapidly expand mungbean area and production and were adapted for use in several countries.

12. *Set the example.* The success of improved mungbeans in Punjab in India attracted new users in neighboring states such as Bihar and Rajasthan and in neighboring countries such as Nepal.

13. *Let others help.* NGOs played an important role in extending technology to the farmers.

14. *Promote the benefits.* Farmers need to understand why maintaining pure seeds of the improved variety is worthwhile and the returns they will earn on their investments.

15. *Include hands-on training.* To enhance nutritional security, women received training in mungbean preparation to ensure that valuable nutrients would not be lost during cooking.

To further advance the mungbean transformation and ensure its sustainability, the partners in this effort will need to pay particular attention to several areas. For example, more research is needed on resistance to insect pests and powdery mildew disease. Biotechnology can be used to speed up selection for these characteristics. In addition, varieties should be developed that can achieve stable yields even under drought and flood conditions. To ensure that improved varieties perform to their full potential, researchers should refine appropriate seasonal and location-specific agronomic practices. These improved agronomic practices should reduce farmers' costs of production, increase their net profit, sustain soil productivity, and not harm the environment. Research on improving production should go hand in hand with studies in marketing, economics, and nutrition to ensure that traders and consumers accept improved varieties and that these varieties benefit those people most in need. To sustain the long-term investment in mungbean research, scientists should come together once every two to three years to discuss the current status of mungbeans, raise concerns and issues, develop solutions to sustain the progress made thus far, and train new and current staff.

Lack of improved, good-quality, pure seed in sufficient quantity at a reasonable price has been one of the major drawbacks preventing the rapid expansion of mungbeans in area and production. Government policies should promote improved varieties and encourage private seed companies to engage in legume seed production and distribution.

Another area in need of attention is price policy. Governments should develop guidelines and if necessary guarantee mungbean prices to encourage farmers to include the legume in diversified cereal cropping systems. Plant breeders can focus on processing traits to enhance the value of the processed products, which can help stabilize prices.

In evaluating the effects of mungbean interventions, partners should include innovative and indirect forms of impact along with traditional consumer and producer surplus studies and publish them in high-quality journals to explain the intervention's value to policymakers and donors. It is important to sustain donors' funding and technical assistance to mungbean research programs across Asia.

International centers need to be sensitive in working with national partners. NARSs should be appropriately credited for their effort. International agricultural research centers (IARCs) should work behind the scenes, providing technology and other support unavailable to the NARSs. Each country has its own set of rules and regulations, and the IARC scientists should follow them carefully to avoid delays. In some cases it has been possible to convince national agencies to make minor modifications to regulations to facilitate the rapid movement of highly promising materials to farmers.

In conclusion, the mungbean will undoubtedly play a growing role in Asian agriculture and diets—as a nitrogen-fixing, protein- and iron-rich legume—as governments seek to enhance food security and sustain their agricultural base. Through continuing research cooperation among local, national, and international partners to improve and share mungbean germplasm and technical expertise, small-scale farmers can increase their yields, diversify the crop rotations on the more than 25 million hectares under rice–wheat cropping systems, and increase their incomes by growing this nutritious legume for their families and communities.

Notes

1. For details of the varieties, trials, and participating institutions discussed in this section, see Shanmugasundaram, Keatinge, and d'Arros Hughes (2009).

2. The analyses cited by Ali et al. (1997) include biases and assumptions that can be challenged. At the time of the study, the authors lacked the funds to collect panel data, but in the absence of such data, they neglected to use matching techniques to eliminate biases emanating from self-selection. A key parameter needed to estimate the producer/consumer surplus was the difference between growers of improved and traditional varieties in the cost per kilogram produced, calculated from the sample mean for each group. First, this estimate did not take into account systemic (nonrandom) reasons for the difference between adopters of the improved variety and nonadopters, which could

be correlated with performance and would have caused farmers to have lower yields or higher unit costs even if they were to grow improved varieties. Second, the tests of the statistical significance of the differences among the subgroups of growers were done at an 85 percent confidence level, which is not acceptable in economics research. A third deficiency is the attribution of the whole increase in mungbean area in Pakistan between 1986 and 1995 to incentives created by the new variety, a claim that is not substantiated.

3. However, see the caveats elaborated in note 2.

References

Afzal, M. A., and M. A. Bakr. 2002. Experience of legume promotion project in Bangladesh—the LBMDPP. In *Integrated management of Botrytis grey mould of chickpea in Bangladesh and Australia: Summary proceedings of a Project Inception Workshop, 1–2 June 2002,* Bangladesh Agricultural Research Institute (BARI), Joydebpur, Gazipur, Bangladesh, ed. M. A. Bakr, K. H. M. Siddique, and C. Johansen. Joydebpur, Gazipur, Bangladesh and Crawley, Australia: Bangladesh Agricultural Research Institute and Centre for Legumes in Mediterranean Agriculture.

Afzal, M. A., M. A. Bakr, A. Hamid, M. M. Haque, and M. S. Aktar. 2006. Adoption and seed production mechanisms of modern varieties of mungbean in Bangladesh. In *Improving income and nutrition by incorporating mungbean in cereal fallows in the Indo-Gangetic Plains of South Asia, DFID Mungbean Project for 2002–2004.* AVRDC Publication 06-682. Shanhua, Taiwan: World Vegetable Center (AVRDC).

Ahn, C. S., J. H. Chen, and H. K. Chen. 1985. Performance of the ninth (1981) and tenth (1983) IMN. AVRDC Publication 85-242. Asian Vegetable Research and Development Center, Shanhua, Taiwan.

Ali, M., I. A. Malik, H. M. Sahir, and A. Bashir. 1997. *The mungbean green revolution in Pakistan.* Technical Bulletin 24. Shanhua, Taiwan: World Vegetable Center (AVRDC).

Amiruzzaman, M., and M. Shahjahan. 2006. Country report for Bangladesh. In *Processing and utilization of legumes,* ed. S. Shanmugasundaram. Tokyo: Asian Productivity Organization.

Anderson, J. R. 1997. On grappling with the impact of agricultural research. Background paper for a presentation to the International Centers' Week, Washington, D.C. An earlier version, "On measuring the impact of natural resources research," was prepared for the Advisor's Conference, Department for International Development, formerly the Overseas Development Administration, Sparsholt, U.K., July 9, in proceedings circulated on diskette.

AVRDC (World Vegetable Center). 1976. *AVRDC mungbean report 1975.* Shanhua, Taiwan.

———. 1977. *AVRDC progress report 1976.* Shanhua, Taiwan.

———. 1981. *AVRDC progress report 1979.* Shanhua, Taiwan.

———. 1988. *AVRDC progress report 1986.* Shanhua, Taiwan.

———. 1998. *AVRDC report 1997.* ACRDC Publication 98-481. Shanhua, Taiwan.

————. 1999. *AVRDC 1999 progress report.* AVRDC Publication 00-503. Shanhua, Taiwan.

Bahl, P. N. 1999. *The union of Myanmar: Report of the consultancy mission on grain legume production.* Bangkok, Thailand: Food and Agriculture Organization of the United Nations, Regional Office for Asia and the Pacific.

Bains, K., R. Aggarwal, and L. Barakoti. 2006. Development and impact of iron-rich mungbean recipes. In *Improving income and nutrition by incorporating mungbean in cereal fallows in the Indo-Gangetic Plains of South Asia,* ed. S. Shanmugasundaram. AVRDC Publication 06-682. Shanhua, Taiwan: World Vegetable Center (AVRDC).

Bains, K., R. Y. Yang, and S. Shanmugasundaram. 2003. *High-iron mungbean recipes for North India.* AVRDC Publication 03-562. Shanhua, Taiwan: World Vegetable Center (AVRDC).

Bains, T. S., J. S. Brar, G. Singh, H. S. Sekhon, and B. S. Kumar. 2006. Status of production and distribution of mungbean seed in different cropping systems. In *Improving income and nutrition by incorporating mungbean in cereal fallows in the Indo-Gangetic Plains of South Asia,* ed. S. Shanmugasundaram. AVRDC Publication 06-682. Shanhua, Taiwan: World Vegetable Center (AVRDC).

Borlaug, N. E. 1973. Building a protein revolution on grain legumes. In *Nutritional improvement of food legumes by breeding,* ed. M. Milner. Rome: Protein Advisory Group of the United Nations.

Catipon, E. M., B. M. Legaspi, and F. A. Jarilla. 1988. Development of mungbean varieties from AVRDC lines for the Philippines. In *Proceedings of the second international mungbean symposium,* ed. S. Shanmugasundaram and B. T. McLean. Shanhua, Taiwan: World Vegetable Center.

Chadha, M. L. 2009. E-mail to author, April.

Chainuvati, C., N. Potan, and T. Worasan. 1988. Mungbean and blackgram production and development in Thailand. In *Proceedings of the second international mungbean symposium,* ed. S. Shanmugasundaram and B .T. McLean. Shanhua, Taiwan: World Vegetable Center.

Cheng, X. Z. 1993. Research and utilization of AVRDC mungbean lines in China. In *The study and application of AVRDC mungbean in China,* ed. X. Z. Cheng, Y. T. Wang, and C. Y. Yang. Bangkok, Thailand: Asian Regional Center / World Vegetable Center (AVRDC).

————. 1996. *Mungbean.* Beijing: China Agricultural Press.

Cheng, X. Z., and L. F. Lin. 1993. High yielding cultivation technique for mungbean. In *The study and application of AVRDC mungbean in China,* ed. X. Z. Cheng, Y. T. Wang, and C. Y. Yang. Bangkok, Thailand: Asian Regional Center / World Vegetable Center (AVRDC).

Erskine, W., and S. Nesbitt. 2009. How can agricultural research make a difference in countries emerging from conflict? *Experimental Agriculture* 45: 313–321.

ESCAP (United Nations Economic and Social Commission for Asia and the Pacific). 1985. *Economic and social survey of Asia and the Pacific.* Bangkok, Thailand.

Evenson, R. E., and D. Gollin. 2003. Assessing the impact of the green revolution, 1960 to 2000. *Science* 300: 758–762.

FAO (Food and Agriculture Organization of the United Nations). 1985. *The state of food and agriculture.* Rome.

Fernandez, G. C. J., and S. Shanmugasundaram. 1988. The AVRDC Mungbean Improvement Program: The past, present, and future. In *Proceedings of the second international mungbean symposium,* ed. S. Shanmugasundaram and B. T. McLean. Shanhua, Taiwan: World Vegetable Center.

Grover, D. K., K. Weinberger, and S. Shanmugasundaram. 2006. Socio-economic impact of new short-duration mungbean varieties in Punjab. In *Improving income and nutrition by incorporating mungbean in cereal fallows in the Indo-Gangetic Plains of South Asia,* ed. S. Shanmugasundaram. AVRDC Publication 06-682. Shanhua, Taiwan: World Vegetable Center (AVRDC).

Hobbs, P., and R. Morris. 1996. *Meeting South Asia's food requirements from rice–wheat cropping systems: Priority issues facing researchers in the post green revolution era.* Natural Resources Group Paper 96/01. Mexico: International Maize and Wheat Improvement Center.

Huanyu, Z., and W. Zhizong. 1993. High-yielding cultivation of Zhong Lu #1 in the Red Loam Region of Hunan. In *The study and application of AVRDC mungbean in China,* ed. C. Xuzhen, W. Youtian, and C. Y. Yang. Bangkok, Thailand: Asian Regional Center / World Vegetable Center (AVRDC).

Huijie, Z., N. H. Li, X. Z. Cheng, and K. Weinberger. 2003. The impact of mungbean research in China. AVRDC Publication 03-550, Working Paper 14. Shanhua, Taiwan: World Vegetable Center.

Institute of Crop Germplasm Resources. 1993. *Compilation of papers on technology and utilization of AVRDC improved mungbeans in China.* Beijing: Institute of Crop Germplasm Resources, Chinese Academy of Agricultural Sciences, Department of Science and Technology, Ministry of Agriculture, and AVRDC Asian Regional Center, China Agricultural Press.

———. 1999. *Proceedings on technology and utilization of mungbean in China.* China: Institute of Crop Germplasm Resources, Chinese Academy of Agricultural Sciences, Department of Science and Technology, Ministry of Agriculture, and AVRDC Asian Regional Center, China Agricultural Press.

Jansen, H. G. P., and S. Charnnarongkul. 1992. Economic analysis of AVRDC mandate crops through baseline surveys: Mungbean in Thailand. AVRDC Working Paper 6. Shanhua, Taiwan: World Vegetable Center (AVRDC).

Joshi, K. D. 2009. E-mail to author, March 31.

Khanal, N. N., R. K. Giri, L. T. Sherpa, S. Thapa, K. Thapa, R. K. Chaudhari, and B. Rayamajhi. 2004. *Promotion of rainfed rabi cropping in rice fallows of Nepal: Review report of achievements from July 2002–June 2003.* Kathmandu, Nepal: Centre for Arid Zone Studies, CIMMYT SARO (South Asia Regional Office).

Lantican, R. M., and R. S. Navarro. 1988. Breeding improved mungbeans for the Philippines. In *Proceedings of the second international mungbean symposium,* ed. S. Shanmugasundaram and B. T. McClean. Shanhua, Taiwan: World Vegetable Center.

MacKenzie, D. R., and S. Shanmugasundaram. 1973. The AVRDC grain legume improvement programme. In *Proceedings of the first IITA grain legume improvement workshop.* Ibadan, Nigeria: International Institute of Tropical Agriculture.

Maneepun, S. 2003. Traditional processing and utilization of legumes. In *Processing and utilization of legumes,* ed. S. Shanmugasundaram. Report of the Asian Productivity Organization Seminar on Processing and Utilization of Legumes held October 9–14, 2000, in Japan. Tokyo: Asian Productivity Organization.

Meelu, O. P., and R. A. Morris. 1988. Green manure management in rice-based cropping systems. In *Sustainable agriculture: Green manure in rice farming.* Laguna, Philippines: International Rice Research Institute.

Morton, F., R. E. Smith, and J. M. Poehlman. 1982. *The mungbean.* Mayaguez, Puerto Rico: University of Puerto Rico.

Pandey, R. K., W. T. Herrera, and A. N. Villegas. 1988. Drought response of mungbean genotypes under a sprinkler irrigation gradient system. In *Proceedings of the second international mungbean symposium,* ed. S. Shanmugasundaram and B. T. McClean. Shanhua, Taiwan: World Vegetable Center.

Pannu, R. K., and D. P. Singh. 1988. Influence of water deficits on morpho-physiological and yield behavior of mungbean *Vigna radiata* (L.) Wilczek. In *Proceedings of the second international mungbean symposium,* ed. S. Shanmugasundaram and B. T. McClean. Shanhua, Taiwan: World Vegetable Center.

Paroda, R. S. 1995. Production of pulse crops in Asia: Present scenario and future options. In *Production of pulse crops in Asia,* ed. S. K. Sinha and R. S. Paroda. FAO/RAPA Publication 1995/8. Bangkok, Thailand: Food and Agriculture Organization of the United Nations, Regional Office for Asia and the Pacific.

Pingali, P., and M. Shah. 1999. Rice–wheat cropping systems and the Indo-Gangetic Plains: Policy re-directions for sustainable resource use. In *Sustaining rice-wheat production systems: Socio-economic policy issues,* ed. P. Pingali. Rice–Wheat Consortium Paper Series 5. New Delhi, India: Rice–Wheat Consortium for Indo-Gangetic Plains.

Sekhon, H. S., G. Singh, S. Poonam, and S. Pushp. 2006. Agronomic management of mungbean grown under different environments. In *Improving income and nutrition by incorporating mungbean in cereal fallows in the Indo-Gangetic Plains of South Asia,* ed. S. Shanmugasundaram. AVRDC Publication 06-682. Shanhua, Taiwan: World Vegetable Center (AVRDC).

Sekhon, H. S., T. S. Bains, B. S. Kooner, and P. Sharma. 2007. Grow summer mungbean for improving crop sustainability, farm income, and malnutrition. *Acta Horticulturae* 752: 459–464.

Shanmugasundaram, S. 1988. *A catalog of mungbean cultivars released around the world.* Shanhua, Taiwan: World Vegetable Center.

———. 2001. New breakthrough with mungbean. *Centerpoint* 19 (2): 1–2.

————. 2003. Present situation and economic importance of legumes in Asia and Pacific region. In *Processing and utilization of legumes,* ed. S. Shanmugasundaram. Tokyo: Asian Productivity Organization.

————, ed. 2006. *Improving income and nutrition by incorporating mungbean in cereal fallows in the Indo-Gangetic Plains of South Asia, DFID Mungbean Project for 2002–2004.* Proceedings of the final workshop and planning meeting, May 27–31, 2004, at Punjab Agricultural University, Ludhiana, Punjab, India. AVRDC Publication 06-682.342. Shanhua, Taiwan: World Vegetable Center (AVRDC).

Shanmugasundaram, S., and D. H. Kim. 1996. Mungbean. In *Genetics, cytogenetics, and breeding of crop plants,* vol. 1: *Pulses and oilseeds,* ed. P. N. Bahl and P. M. Salimath. New Delhi: Oxford and IBH Publishing.

Shanmugasundaram, S., and A. T. Tschanz. 1987. Breeding for mungbean and soybean disease resistance. In *Varietal improvement of upland crops for rice-based cropping systems.* Laguna, Philippines: International Rice Research Institute, Los Baños.

Shanmugasundaram, S., J. D. H. Keatinge, and J. d'Arros Hughes. 2009. *The mungbean transformation: Diversifying crops, defeating malnutrition.* IFPRI Discussion Paper 922. Washington, D.C.: International Food Policy Research Institute.

Singh, J. 2006. Country report for India (1). In *Processing and utilization of legumes,* ed. S. Shanmugasundaram. Tokyo: Asian Productivity Organization.

Singh, R. B. 1988. Trends and prospects for mungbean production in South and Southeast Asia. In *Proceedings of the second international mungbean symposium,* ed. S. Shanmugasundaram and B. T. McClean. Shanhua, Taiwan: World Vegetable Center.

Srinives, P. 1998. Collaborative mungbean breeding research between AVRDC and its southeast Asia partners with emphasis on Thailand. In *Proceedings of the international consultation workshop on mungbean,* ed. S. Shanmugasundaram. Shanhua, Taiwan: World Vegetable Center (AVRDC).

Subramanian, A. M., and R. Y. Yang. 1998. High-iron mungbean recipes from South Asia. AVRDC Publication 98-480. Shanhua, Taiwan: World Vegetable Center (AVRDC).

Thirumaran, A. S., and M. A. Seralathan. 1988. Utilization of mungbean. In *Proceedings of the second international mungbean symposium,* ed. S. Shanmugasundaram and B. T. McClean. Shanhua, Taiwan: World Vegetable Center.

Tickoo, J. L., C. S. Ahn, H. K. Chen, and S. Shanmugasundaram. 1988. Utilization of the genetic variability from AVRDC mungbean germplasm. In *Proceedings of the second international mungbean symposium,* ed. S. Shanmugasundaram and B. T. McClean. Shanhua, Taiwan: World Vegetable Center.

Vijayalakshmi, P. S., S. Amirthaveni, R. P. Devada, K. Weinberger, S. C. S. Tsou, and S. Shanmugasundaram. 2003. *Enhanced bioavailability of iron from mungbean and its effects on health of school children.* Technical Bulletin 20. AVRDC Publication 03-559. Shanhua, Taiwan: World Vegetable Center (AVRDC).

Wallis, E. S., and D. E. Byth. 1988. Food legumes: "Slow runners forever?" In *Proceedings of the second international mungbean symposium*, ed. S. Shanmugasundaram and B. T. McClean. Shanhua, Taiwan: World Vegetable Center.

Weinberger, K. 2003a. *Impact analysis of mungbean research in South and Southeast Asia*. Final report of the GTZ Project. Shanhua, Taiwan: World Vegetable Center.

————. 2003b. The impact of iron bio-availability-enhanced diets on the health and productivity of school children: Evidence from a mungbean feeding trial in Tamil Nadu, India. In *International conference on impacts of agricultural research and development: Why has impact assessment research not made more of a difference?* ed. D. J. Watson. Mexico City: International Center for Maize and Wheat Improvement.

Weinberger, K., M. R. Karim, and M. N. Islam. 2006. Economics of mungbean cultivation in Bangladesh. In *Improving income and nutrition by incorporating mungbean in cereal fallows in the Indo-Gangetic Plains of South Asia*, ed. S. Shanmugasundaram. AVRDC Publication 06-682. Shanhua, Taiwan: World Vegetable Center (AVRDC).

Yang, R. S., and S. C. S. Tsou. 1998. Mungbean as a potential iron source in South Asian diets. In *Proceedings of international consultation workshop on mungbean*. Shanhua, Taiwan: World Vegetable Center.

Chapter 15

The Global Effort to Eradicate Rinderpest

Peter Roeder and Karl M. Rich

The Origins, Impact, and Eradication of Rinderpest

Origins and Impact

Rinderpest (or cattle plague) is thought to have originated along the Indus River with the domestication of cattle, some 10,000 years ago (Possehl 1996; Diamond 1999). From antiquity to recent decades, its impact on wildlife populations as well as farmed livestock has been severe. Introduced into eastern Africa at the end of the 19th century, the disease rapidly engulfed the whole continent.

Classical cattle plague, with high mortality, occurred most recently in 1994 in the Northern Areas of Pakistan. In areas of endemic maintenance, many animals are protected either through earlier exposure or vaccination, and thus morbidity and mortality

The Global Rinderpest Eradication Program (GREP), the final stage of the global rinderpest eradication initiative, was a concept conceived and developed by a relatively small number of farsighted people. Its success owes much to dedicated individuals who made especially significant inputs into rinderpest control and eradication during the past 50 years, notably Yoshohiro Ozawa, Yves Cheneau, Walter Plowright, Gordon Scott, William Taylor, and Mark Rweyemamu. Many others made seminal contributions to global rinderpest eradication, inter alia Gijs van't Klooster, Jeffrey Mariner, Ahmed Mustaffa Hassan, Bryony Jones, Rafaqat Hussain Raja, Gholam Ali Kiani, Manzoor Hussain, Tim Leyland, Andy Catley, John Anderson, John Crowther, Roland Geiger, Martyn Jeggo, Berhanu Admassu, Paul Rossiter, Richard Kock, Dickens Chibeu, Tom Barrett, Adama Diallo, Mohammad Afzal, and Walter Masiga. This chapter draws on some material prepared for various earlier publications, in collaborative partnership with William Taylor and Mark Rweyemamu, and their collaboration is sincerely appreciated.

rates can be low. Mild strains of the virus may persist in these reservoirs of infection and may periodically assume higher virulence even in their host populations.

Rinderpest is related to (and a precursor of) the human measles virus, caused by one of a group of closely related morbilliviruses (Barnard, Rima, and Barrett 2006). It is a contagious disease characterized by necrosis and erosions throughout the digestive tract; in its severest form it is capable of killing 95 percent of the animals it infects. Affected animals develop fever, discharges from the eyes and nose, erosions of the mucosa from mouth to anus, diarrhea, and dysentery.

The impact of the disease has been catastrophic on communities dependent on their ruminant livestock for draft power, milk, meat, skins, and manure, in ancient times just as today in many developing countries. The disease had an impact on African wildlife that is still discernible in the distribution of some ruminant species. Rinderpest continued to cause serious wildlife losses throughout the 20th century (Taylor and Watson 1967; Kock et al. 1999).

Eradication

Rinderpest was a good candidate for eradication. The severity of its impact on trade and livelihoods ultimately fostered international collaboration and coordination to eliminate it. Moreover, rinderpest viruses all belong to a single serotype, and one vaccine can protect against all existing viruses. Additionally, there is no carrier state in the host, and both vaccinated and recovered animals have lifelong immunity.

The eradication of rinderpest is indeed a remarkable achievement—on a par with the eradication of smallpox from the human population, the only other case of eradication of an infectious disease. The decades-long history of rinderpest eradication entails failures and successes marked by political intrigue, institutional failure, and reluctant institutional changes.[1] Scientific, political, and institutional innovation ultimately achieved dramatic success and indeed changed world history.

In 2010 we can be assured that the rinderpest virus is no longer circulating in domesticated or wild animals anywhere in the world; it has been eradicated from its natural hosts. In the first decade of the 21st century, few farmers and even fewer veterinarians have seen the disease, which was identified for the last time in Kenya in late 2001. Active surveillance programs in key countries confirm the eradication of rinderpest, using serological studies to detect both overt and occult virus circulation. The majority of countries have achieved accreditation of rinderpest freedom by the World Organization for Animal Health (OIE), though the global accreditation task is still incomplete.

For the least developed countries, confidence in rinderpest freedom translates into confidence in food production from cattle and buffaloes and thus, potentially, increased trade in livestock and its by-products.[2] Thus the full benefits of rinderpest freedom can be realized only through systematic accreditation. This chapter highlights some technical and coordination lessons of the successful campaign against rinderpest, and provides an initial, provisional assessment of its potential impacts on the broader economy and on household welfare.

Cattle Plague: A Brief Global History

The written record of rinderpest occurrence in the world is relatively sparse, even in recent years.[3] A broad overview will indicate the former global reach of the disease.

Europe and Russia

Rinderpest was repeatedly spread by military campaigns as early as the 4th century and into the 20th, around and between Asia, Europe, and the Middle East. From the 17th to the 19th century, cattle trade—largely from Russia—repeatedly introduced rinderpest into Europe. Rinderpest pandemics raged throughout Europe, which was actually denuded of cattle from 1857 to 1866.

In Europe and Russia, rinderpest was successfully eliminated in the early 20th century through draconian national legislation that combined immunization programs with zoosanitary procedures (isolation and slaughter of infected animals, quarantines, and movement certification). In the Soviet Union, viral invasion was prevented by border fencing constructed in the 1950s, supported by a vaccinated buffer zone.

Africa

The first African rinderpest pandemic reached southern Africa in 1896, prompting international collaboration in the search for tools to control infection. By 1918 rinderpest was rampant throughout the colonies of West Africa, killing many hundreds of thousands of cattle annually and seriously reducing wild animal populations. An internationally funded campaign in the 1960s to 1970s, Joint Project 15, cleared most of West and Central Africa but left a reservoir of infection in migrating pastoral herds. Lacking follow-up vaccination, West African herd immunity waned.

Eastern Africa lagged the other parts of the continent in the eradication effort, with large nomadic cattle herds that served as reservoirs of viral persistence in southern Sudan and the Somali pastoral ecosystem. In the 1980s the virus entered Nigeria from both the west and the east, creating a national emergency. Rinderpest was last detected in October 2001, near Meru National Park in Kenya.

The Middle East, Arabian Peninsula, and North Africa

A particularly severe pandemic from 1969 to 1973 swept from Afghanistan through Iran to the Mediterranean littoral and into the Arabian Peninsula, invading virtually all countries in the region. Rinderpest again engulfed Iraq beginning in 1985.

Epidemiological studies (employing molecular virus characterization) pointed to repeated introduction of rinderpest from South Asia, but a focus established in Iraq acted as a source of several epidemics in Iran and Turkey into the 1990s, causing alarm in Europe. The disease was eliminated from all three countries in 1996. Outbreaks of rinderpest occurred frequently in Oman, Qatar, Saudi Arabia, the United Arab Emirates, and Yemen up to the mid-1990s. A severe epidemic in Egypt was recorded between 1982 and 1986 and was eventually eliminated by a comprehensive vaccination campaign. The rest of North Africa has never recorded the presence of rinderpest, except for a short outbreak in Libya in 1966.

East Asia
Japan was repeatedly invaded by rinderpest from the 16th into the 20th century, leading to the deaths of many thousands of cattle on each occasion. The last cases in Japan and Korea occurred in 1924 and 1931, respectively. During the war period in China and afterward from 1938 to 1948, rinderpest was widespread and millions of cattle died. In Thailand, as recently as 1957 the government appealed for international food aid after rinderpest lay waste to the buffaloes used in rice farming, causing famine to loom.

South Asia
In the late 18th and the 19th century, South Asia experienced repeated rinderpest epidemics that killed significant proportions of cattle and buffaloes. Despite heroic attempts at mass vaccination, little progress was made in eliminating the infection until the 1990s. After partition, Pakistan remained endemically infected with rinderpest and was a source of infection for local epidemics as well as international spread to Afghanistan and the Arabian Peninsula well into the 1990s. India became rinderpest free in 1995, Pakistan in 2000.

Battling the Plague: Programs and Tools

Campaigns in Asia
The first major campaigns aimed at eliminating rinderpest began in Asia following World War II. A regional program supported by the Food and Agriculture Organization of the United Nations (FAO) reinforced national efforts across Southeast Asia. A major national campaign in China was rapidly successful once a vaccine safe for use in all species became available; an equally intensive national effort on the Indian subcontinent took half a century. Success was achieved by 1957, supported by progressive control of infection in northern India.

Rinderpest was eliminated from Iraq by intensive vaccination campaigns organized by the FAO between 1994 and 1996 using funds from the U.N. Oil for Food Program. These campaigns also ended the pattern of outbreaks spreading from Iraq to Iran and Turkey. In turn, the elimination of rinderpest from India in 1995, along with simultaneous progress in Pakistan, played an important part in reducing the risk of virus transmission by livestock trade to the Middle East.

China. During the war period, 1938–41, more than 1 million cattle died from rinderpest in western China, and the disease spread widely in 1948–49. In 1948 cattle plague was killing millions of cattle, buffaloes, and yaks; the new government decreed that its eradication was to be a priority as an essential prerequisite to agricultural development. This goal was achieved by 1955, but it was only in 2008 that China was accredited by the OIE as free from rinderpest.

With little or no motorized transport and no refrigeration for vaccine, clearing pockets of infection in the Himalayas involved heroic feats. Chinese animal health staff infected live sheep with the vaccine virus and then transported the infected sheep to the sites on the backs of yaks and horses.[4]

India and Pakistan. An Indian goat-adapted vaccine was used alone, without immune serum, from 1931 onward; by 1939 it had had only a modest impact on the total mortality rate from rinderpest. However, 1954 saw the launch of a publicly financed, vaccine-based campaign, the National Project on Rinderpest Eradication (NPRE), to vaccinate 80 percent of the population of cattle and buffaloes over five years and the remaining 20 percent (plus the annual calf crop) over an indefinite follow-up period (Taylor, Roeder, and Rweyemamu 2006).

Each state undertook its own vaccination program, with varying results. Mass vaccination was able to eliminate endemic rinderpest in some states but unable to prevent its reintroduction. Although the program was unable to eliminate infection, it did bring considerable benefits to farmers.

By the mid-1980s it was clear that the NPRE was failing. In 1983 a Government of India task force was convened to revitalize the eradication program. The task force proposed a three-year program to reintroduce mass vaccination in the endemic states, aiming at 90 percent coverage within three years, with focused vaccination as needed in other states. Endemic infection was finally eliminated with program funding from the European Commission (EC). India has remained free from rinderpest since 1995, an achievement that helps neighboring countries (Bangladesh and Nepal) control the infection.

Success in Pakistan was possible only after its national program obtained technical and financial assistance from the FAO and the EC. In 1997 intensive vaccination campaigns conducted by the Government of Pakistan, with assistance from the FAO and the EC, virtually eliminated rinderpest apart from a persisting reservoir of infection in the Indus River buffalo tract.[5] Pakistan has been reliably free from rinderpest since the last cases were detected in 2000, and the OIE awarded Pakistan rinderpest-free accreditation in 2007. Elimination from Pakistan, in turn, reduced the risk of transmission to the Arabian Peninsula.

Campaigns in Africa

National authorities struggled individually to control rinderpest with varying degrees of success beginning in the 1940s, and in the 1960s international collaboration brought about a coordinated regional approach. Three campaigns to control or eliminate rinderpest were organized with donor assistance under the aegis of the Organization of African Unity: the African Union's Joint Project 15 (1961–69), the Pan-African Rinderpest Eradication Campaign (1986–98), and the Pan-African Program for the Control of Epizootics (1999–2006).

Joint Project 15 (JP15). The regional challenge of controlling the disease was explicitly recognized by the African Rinderpest Conference of 1948, which called for the establishment of an African Information Bureau on Rinderpest, which evolved into the Interafrican Bureau of Epizootic Diseases and later into the African Union's Inter-African Bureau of Animal Resources (AU IBAR). In 1961 the Commission for Technical Cooperation in Africa South of the Sahara launched JP15 to create a rinderpest-free zone around the Lake Chad basin, including parts of Cameroon, Chad, Niger, and Nigeria, using tissue culture rinderpest virus (TCRV), a freeze-dried live attenuated virus vaccine. This would be the first intensive use of the new vaccine: vaccinating the entire cattle herd of the zone during the first three years of the campaign (1962–65). The campaign achieved significant progress, as shown in Table 15.1.

During 1962–69, approximately 100 million doses of TCRV were used in JP15, and in West Africa rinderpest seemed to have been consigned to history. However, a focus of infection remained in the pastoral community spanning the border between Mali and Mauritania, reinfecting a number of countries in the late 1970s. Although little noted in published reports, rinderpest virus thus continued to circulate in Sub-Saharan Africa after JP15 (see Roeder and Rich 2009 for geographic representation of rinderpest occurrences in cattle and wildlife between 1968 and 1984).

JP15 was successfully extended across West and Central Africa, but later activities in eastern Africa had limited success. Following the phase of intensive, internationally coordinated vaccination, national veterinary authorities continued vaccination to maintain herd immunity—a largely unnecessary exercise. In Nigeria, for example, the last case of rinderpest occurred in 1976, yet annual vaccination campaigns continued for many years.

Table 15.1 Rinderpest vaccination coverage achieved during the first three years of the African Union's Joint Project 15 vaccination campaigns compared to the annual incidence of outbreaks, 1962–65

	Cameroon	Niger	Nigeria	Chad
Average number of outbreaks per year in previous 10 years	148	196	375	250
1962–63				
Percentage of herd vaccinated	85.9	80.0	71.8	85.8
Number of recorded outbreaks	0	47	91	37
1963–64				
Percentage of herd vaccinated	95.0	98.9	87.8	93.4
Number of recorded outbreaks	2	18	2	6
1964–65				
Percentage of herd vaccinated	87.0	93.7	92.5	92.3
Number of recorded outbreaks	0	4	2	4

Source: Taylor (1997).

JP15 failed to resolve—and even to recognize—the three or four persistent reservoirs of rinderpest infection in West and eastern Africa, allowing a later resurgence of rinderpest. The lack of surveillance systems and epidemiological knowledge, combined with an excessive reliance on institutionalized mass vaccination, were primary factors in the ultimate failure of JP15.

Even though the OIE had received informal reports that the virus had not been totally eliminated, international leadership failed to recognize the problem. Finally, in 1989, the OIE Pathway was developed, precisely to provide a monitoring process for accredited freedom from rinderpest.

The Pan-African Rinderpest Campaign (PARC).   PARC was initiated in 1986 to eliminate rinderpest from up to 34 countries in Africa. Funded primarily by the EC, PARC laid down guidelines for national veterinary services, relying on mass vaccination. A PARC Epidemiology Unit performed epidemiological studies, resulting in a strategy of creating immune belts, and addressed the problem of verifying virus elimination.

Phase 1 of PARC focused on building capacity to administer the vaccine in order to create immune populations. In Phase 2, residual endemic foci would be addressed by establishing sanitary cordons of highly vaccinated populations. The final phase would lift the vaccination and provide intensive disease surveillance.

Finding that eastern Africa was more heavily contaminated than West Africa, in 1988 PARC developed the concept of the West African Wall to create a blockade in the central African states of Cameroon and Chad (Taylor, Roeder, and

Rweyemamu 2006). In practice, even though the central African sanitary cordons were not managed well enough to halt the movement of rinderpest, the virus never moved westward again.

A novel eradication strategy was developed for Ethiopia to deal with persisting rinderpest reservoirs. Attempting to provide nationwide vaccination coverage would have required annual vaccination of 30 million cattle. An alternative strategy sought to contain the rinderpest virus within the reservoirs, establishing vaccination buffer zones to protect vulnerable areas (vaccinating fewer than 3 million cattle per year) and strengthening surveillance and emergency preparedness. A key component was fielding community-based animal health workers in remote and insecure areas; another was the use of a thermostable rinderpest vaccine.

After a major struggle for acceptance, the strategy was implemented in 1993 and rapidly proved successful. At the end of PARC in 1998, the rinderpest virus was circulating only in a few areas: northwestern and northeastern Kenya, southern Somalia, and southern Sudan. This remarkable success had a major impact on the design of future eradication programs in both Africa and Asia.

The Pan-African Control of Epizootics (PACE) Project. The PACE project—also funded by the EC and coordinated by the AU IBAR—ran from 1999 to 2006 in 32 countries. A PACE team based in Nairobi successfully provided epidemiological and strategy expertise to address the problem of persistent rinderpest in the Somali pastoral ecosystem and to design accreditation procedures for all other member countries.[6] Indeed, when the PACE program ended in 2006, rinderpest was no longer circulating in either wild or domesticated hosts in Africa.

Community-based animal health worker programs (developed by nongovernmental organizations in Africa under the aegis of the United Nations Children's Fund) enhanced both the delivery of rinderpest vaccines and the understanding of rinderpest epidemiology (Leyland 1996). "Participatory epidemiology" has also aided disease surveillance for FAO programs in Afghanistan, Pakistan, Tajikistan, and Uzbekistan, using techniques developed in the PARC and PACE programs (Catley 1999; Mariner and Roeder 2003). In southern Sudan, rinderpest was eliminated by 2000 through community-based animal health programs under the aegis of Operation Lifeline Sudan—a remarkable achievement.

The experience of PARC and PACE points to several essential program requirements: adaptive and flexible management, clearly focused objectives, sound technical (especially epidemiological) knowledge, high-caliber technical staff with international credibility, and national and international networking efforts.

Vaccines and Vaccination

Effective rinderpest control began with the introduction of zoosanitary measures. The key to eradication, however, was the discovery in the 1880s, in Russia (Semmer

1983) and in South Africa, that the serum from a recovered animal had protective powers (passive immunity). Systematic vaccination could potentially eliminate rinderpest from a country and even provide international control. However, the search for an effective vaccine would be a long one.[7]

Seminal research demonstrated that the simultaneous administration of immune serum *and* virulent blood—the serum-simultaneous method—could produce an active immunity. However, inoculated animals were potentially infectious and needed to be physically separated from noninoculated animals, and the virus inoculum itself might contain infectious piroplasms. Nevertheless, by 1928 this method had eliminated rinderpest from European Russia and was widely used in Africa and India. Goat-adapted rinderpest virus was soon developed using the serum-simultaneous method and was increasingly used throughout South Asia on its own, successfully eradicating rinderpest from Thailand after World War II. A dessicated goat vaccine was introduced into Egypt in 1945, and six months later the epidemic was eliminated. The goat vaccine was still being used in Bangladesh and Myanmar as late as 1990.[8]

In Japan and Korea, rabbits were used to pass the virus for serum-simultaneous vaccination of ultrasusceptible breeds of cattle (Nakamura, Fukusho, and Kuroda 1943), eventually resulting in the Nakamura III vaccine. After the war the FAO disseminated this vaccine to a number of countries including Egypt, Ethiopia, India, Kenya, Pakistan, and Thailand (Hambidge 1955).

Mariner et al. (1990) developed a more thermostable variant of TCRV called Thermovax, which is stable for up to four weeks if shielded from sunlight and excessive heat. This vaccine was widely and effectively used in Africa within community-based vaccination programs in remote areas, as well as in Afghanistan.

Diagnostics: Seromonitoring and Focused Vaccination. An important component of PARC's design was regular assessment of the efficacy of vaccination through a coordinated seromonitoring program. (The program was maintained as part of PACE until 2004, funded by the EC through the FAO.)[9] The concept was that if overall herd immunity exceeded 80 percent, the rinderpest virus would stop circulating in that population.[10] Populations shown to be inadequately vaccinated (through seromonitoring) would be revaccinated. It proved very difficult to obtain the serological data in time to influence ongoing vaccination programs, but networking significantly increased local expertise.

The Value of Vaccine Quality Assurance. To address the variable quality of rinderpest vaccines being produced by various laboratories, the FAO established the Pan-African Veterinary Vaccine Centre (PANVAC), operating from two sites in West and eastern Africa, to provide rinderpest vaccine quality assurance (QA) for PARC.[11]

Standard operating procedures for vaccine production and QA were defined, published by the FAO, and incorporated into OIE guidelines. PANVAC was highly successful: the quality of vaccines increased rapidly, and PARC managers insisted on the use of PANVAC-certified vaccines. Transfer of PANVAC vaccine production and QA technology to Pakistan in 1995 was a decisive factor in eliminating rinderpest there.

International Coordination

The Global Rinderpest Eradication Program (GREP) was launched in 1993 as an international coordination mechanism under the auspices of the FAO Animal Health Service through its Emergency Prevention System for Transboundary Animal and Plant Pests and Diseases. GREP broadly encompasses all the activities that contributed to the final eradication of rinderpest in 1993.

GREP was designed as a time-bound program, with a deadline of 2010 to achieve accredited global freedom from rinderpest. The initial concept was that control activities would proceed on three fronts, implemented by regional organizations: PARC in Africa (discussed earlier), the West Asian Rinderpest Eradication Campaign (WAREC), and the South Asian Rinderpest Eradication Campaign (SAREC) (Figure 15.1). WAREC, funded by the United Nations Development Program, brought coordination to the Middle East until it was disrupted by the Gulf War. SAREC was never implemented as a regionally coordinated program.[12]

GREP's initial epidemiological studies identified seven potential reservoirs of infection. In 1999 these areas became the focus of a five-year Intensified GREP, although only four of them proved to be sources of infection: the Arabian Peninsula, Pakistan (with Afghanistan), the Somali ecosystem, and southern Sudan.[13] Through the elimination of these reservoirs, the target of the Intensified GREP was achieved by 2001.

The OIE played an important role in GREP, publishing guidelines for rinderpest surveillance and freedom accreditation and also playing an active role in strategy decisions (with the FAO) in PARC and PACE. As GREP evolved, the OIE provided guidelines for the accreditation process through its Scientific Commission. In 2010, both organizations are working toward a long-awaited joint declaration that global rinderpest freedom has been achieved.

Accreditation of Rinderpest Freedom

As mentioned earlier, following mass vaccination there is a tendency to institutionalize annual, pulsed vaccination campaigns, even though they have diminishing effectiveness—even after disease has been eradicated with minimal risk of reinvasion. This tendency reflects an inevitable lack of certainty in the absence of a test

Figure 15.1 The campaigns outlined under the Global Rinderpest Eradication Program (GREP)

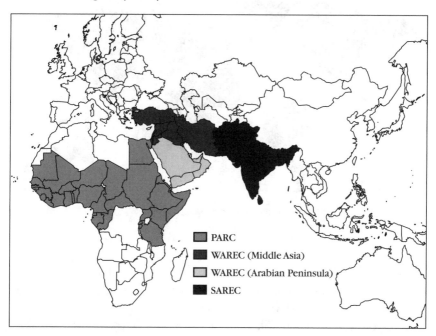

Source: Authors, based on FAO GREP data.
Notes: PARC means Pan-African Rinderpest Eradication Campaign; SAREC means South Asian Rinderpest Eradication Campaign; WAREC means West Asian Rinderpest Eradication Campaign.

that can differentiate between vaccine-induced antibodies and the wild virus infection, coupled with a lack of trust and effective communication among countries.

An expert group on rinderpest surveillance systems, convened in Paris in 1989, produced guidance that became part of the Rinderpest Chapter of the Animal Health Code, later referred to as the OIE Pathway. The Recommended Standards laid out three stages—provisional freedom, freedom from disease, and freedom from infection—as verifiable epidemiological objectives. Significantly, *cessation of vaccination* was integral to the program. It was clear that the verification process, including disease searches and serosurveillance, would be far less costly than continued mass vaccination. This OIE Pathway served as a template until 2007, when the accreditation process was amended from a four-year to a two-year intensive surveillance.

With PARC and GREP, rinderpest disease intelligence improved, providing a more complete global picture of actual risks. Countries now found the courage to

Figure 15.2 Status of OIE accreditation of rinderpest freedom, May 2010

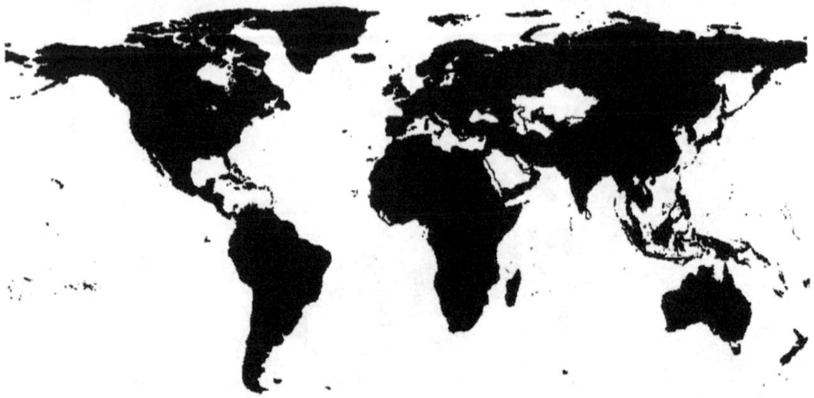

Source: OIE (2010). © 2010 OIE. Reprinted with permission.
Note: Countries accredited as rinderpest free are shaded.

suspend vaccination and launch the freedom verification phase. Innovative programs to demonstrate the absence of infection were established, combining participatory approaches (Jost et al. 2007) with conventional epidemiological techniques.

Progress in accreditation is illustrated in Figure 15.2. There is cause for concern, however, that not all countries will be accredited by the end of 2010 as rinderpest eradication is eclipsed by more pressing disease issues.

The Impact of Rinderpest Eradication

Overview of Economic Impact Studies

Despite the success of the global effort to eradicate rinderpest, there is surprisingly little information on the scope of its benefits, beyond specific outbreaks or control programs. Most information relates to Africa, with little on Asia, and reliable data on livestock production and the economic impacts of disease control are in general difficult to obtain.

The most comprehensive study on the economic impact of rinderpest is that of Tambi et al. (1999), who evaluated the PARC program in a subset of 10 recipient countries. They estimated the cost of the program at ECU 51.6 million (including donor funding and national matching funds).[14] Leslie and McLeod (2001) estimated that the vaccination costs under the program ranged from 5 to 33 percent of the total costs based on a per-unit vaccination cost of $0.09.

Tambi et al. (1999) also estimated that the program resulted in "avoided losses" of 126,000 tons of beef, 39,000 tons of milk, 14,000 tons of manure, and 86,000 hectares of animal traction. The total value of the avoided losses was estimated at ECU 99.2 million, implying a benefit–cost ratio (BCR) of 1.85 for the PARC program (Tambi et al. 1999). Country-specific benefits ranged from a high of ECU 35.4 million in Ethiopia to a low of ECU 0.5 million in Benin. BCRs in all 10 countries were greater than 1 (ranging from a high of 3.84 in Tanzania to a low of 1.06 in Côte d'Ivoire). The authors also computed internal rates of return (IRRs) for selected countries that ranged between 11 and 118 percent, showing a high return to investment from PARC relative to alternative uses.

The study finally attempted to assess the welfare impacts of the PARC program based on changes in producer and consumer surplus: the analysis assumed that the program would increase the supply of livestock (and products), thus lowering prices. A simulation analysis estimated welfare gains of ECU 57.5 million, with 81 percent of this accruing to producers and 19 percent to consumers. The overwhelming majority of gains on the production side went to meat production (92 percent), the remainder to milk production. However, the authors' assumption of price declines from increased livestock production failed to take into account international trade in cases in which meat (or milk) is a tradable product; in such cases, domestic prices would not necessarily fall because the relevant price is the world price, not the domestic price (Sadoulet and de Janvry 1995). Moreover, the analysis neglected to consider additional multimarket impacts, notably on feed and other service markets.

A few other studies also provide insights. Blakeway (1995) evaluated the benefits of rinderpest control in the context of southern Sudan, arriving at a rather high benefit–cost ratio of 34:1. He based that calculation on an estimated 7 percent annual rise in cattle numbers, estimated benefits from avoided losses in livestock of $3.8 million and from avoided food aid of $3 million, and estimated program costs of $200,000.[15] Earlier country-level studies gave much lower benefit–cost ratios. Felton and Ellis (1978) computed a BCR of 2.48 and an IRR of 48 percent for rinderpest control in Nigeria. An ex ante simulation study for Ethiopia found that intensive mass vaccination programs (if feasible) would have a BCR of 5.08, whereas a combination of intensive surveillance and targeted vaccination would have a BCR of 3.68 (Tambi, Maina, and Mariner 2004).

Impacts of Rinderpest

By all accounts, rinderpest control proved an impressive success, with benefit–cost ratios in some instances exceeding 30 (in south Sudan). Given that the majority of beneficiaries are livestock farmers, including poor pastoralists, the eradication of rinderpest would presumably have strong pro-poor effects. But how does disease eradication compare to other potential interventions in its impacts on poverty?

In order to attempt to answer this question, we first attempt to identify the beneficiaries of rinderpest eradication and roughly gauge their benefits.[16] Thornton et al. (2002) provided estimates of populations in South Asia and Sub-Saharan Africa in various production areas, including those in predominately livestock areas, that would be overwhelmingly at risk from rinderpest. In 2000 more than 19 million people resided in livestock-dominated zones in South Asia, with an additional 62 million in similar zones in Sub-Saharan Africa. Globally, nearly 4 percent of the world's population in 2000 resided in areas dominated by rangeland production systems and were particularly at risk from diseases such as rinderpest. Moreover, in South and Southeast Asia, cattle play an important role in plowing, while in eastern Africa, India, and Pakistan, dairy production also contributes to household incomes.

Table 15.2 summarizes the total number of poor livestock keepers in each of the zones provided by Thornton et al. (2002). In Sub-Saharan Africa the total is more than 21 million; in South Asia, the total amounts to approximately 5 million people, representing close to one-fourth of the total population in these areas. Globally, some 41 million poor livestock keepers live in rangeland areas and would be most at risk from rinderpest (denoted in areas LGA, LGH, and LGT in Table 15.2).

A useful indicator of the success of rinderpest eradication is its dynamic impacts on animal stocks and milk production in Africa during the past 25 years. Data from FAOSTAT show a gradual increase in animal stocks and milk production during the 1990s (corresponding to the timing of more intense efforts to control rinderpest); by contrast, the 1980s showed relatively stagnant or even negative growth in livestock numbers (Table 15.3; see Roeder and Rich 2009 for details). The increase in milk production was mainly due to an increase in animals rather than productivity, which was relatively stagnant. Because increases in livestock numbers still trailed population growth in Africa (Table 15.4), rinderpest control did not improve the per capita availability of domestic livestock products.

At the same time, although rinderpest-related animal morbidity and mortality had their greatest impact on the production of milk and meat, they also engendered important downstream impacts in the rural economy.[17] In Asia, a major impact was the loss of animals used for plowing rice and other staple crops. These losses also impacted traders, slaughterhouses, brokers, retailers, and other support structures in the rural economy. For instance, Rich and Wanyoike (2010) found that a recent outbreak of Rift Valley fever in Kenya had large and lasting downstream effects on traders and slaughterhouse workers.

To this end, we focus on five case studies of rinderpest control—in Ethiopia, Kenya, Pakistan, Tanzania, and Uganda—that have successfully eradicated the disease to tease out the potential downstream impacts of control. Our analysis uses

Table 15.2 Poor livestock keepers, numbers (and their percentage of the population), by production area and region

Region	LGA	LGH	LGT	MIA	MIH	MIT	MRA	MRH	MRT	Other	Total
South Asia											
Number of poor livestock keepers	4,831,532	87,651	106,466	47,940,657	18,982,429	122,519	68,872,440	47,637,565	3,196,044	7,354,918	199,132,222
Percentage of population	(25.97)	(26.74)	(27.82)	(9.07)	(9.17)	(10.24)	(23.78)	(24.08)	(24.00)	(9.09)	(14.89)
Sub-Saharan Africa											
Number of poor livestock keepers	15,170,239	5,507,258	1,140,097	471,820	14,572	249,095	48,335,642	53,079,093	27,896,841	14,162,359	166,027,017
Percentage of population	(36.10)	(32.98)	(35.22)	(11.44)	(11.44)	(11.44)	(30.77)	(28.07)	(30.50)	(11.71)	(26.48)
Total world											
Number of poor livestock keepers	28,020,377	7,359,589	5,411,445	55,324,996	35,929,740	11,994,162	136,569,357	160,030,296	68,887,013	46,236,914	55,763,888
Percentage of population	(23.98)	(25.94)	(15.18)	(8.43)	(6.83)	(1.68)	(22.95)	(20.29)	(11.64)	(6.67)	(11.71)

Source: Thornton et al. (2002).

Notes: Numbers are based on World Bank rural poverty rates and typologies from LID 1999. The data do not differentiate among the types of livestock produced in each area. Production systems: LGA means livestock only, rangeland-based arid and semiarid; LGH means livestock only, rangeland-based humid and subhumid; LGT means livestock only, rangeland-based temperate and tropical highland; MIA means mixed irrigated arid and semiarid; MIH means mixed irrigated humid and subhumid; MIT means mixed irrigated temperate and tropical highland; MRA means mixed rainfed humid and subhumid; MRH means mixed rainfed humid and subhumid; MRT means mixed rainfed temperate and tropical highland.

Table 15.3 Cumulative annual growth rates in livestock inventories, selected regions in Africa, 1980–2005 (percent)

Region	1980–85	1985–90	1990–95	1995–2000	2000–2005
East Africa	1.10	0.78	0.31	1.16	1.71
Central Africa	0.87	1.55	0.19	3.22	0.97
West Africa	–0.53	1.75	2.00	2.05	2.45

Source: FAO (2009).

Table 15.4 Annual growth rates for the human population, selected regions in Africa, 1980–2005 (percent)

Region	1980–85	1985–90	1990–95	1995–2000	2000–2005
East Africa	2.91	3.01	2.61	2.78	2.57
Central Africa	2.97	3.08	3.43	2.53	2.87
West Africa	2.8	2.73	2.62	2.6	2.54

Source: United Nations Secretariat, Department of Economic and Social Affairs (2008).

social accounting matrixes (SAMs) developed by the International Food Policy Research Institute (IFPRI) for each country. A SAM is a database of economic trans-actions that take place between different aggregate sectors of an economy. SAMs can be used to calculate "multipliers" that denote the economic activity generated from a unit increase in government spending (including expenditures on rinderpest control) or other sources of final demand (exports, investment). In this manner, multipliers can be used to assess the strength of linkages between the livestock sector and the broader economy and the distribution of household impacts resulting from rinderpest control (Sadoulet and de Janvry 1995). Unlike the analysis of Tambi et al. (1999), SAM multipliers provide a much more complete picture of the chang-es in national output (not just in the livestock sector itself) potentially engendered by disease-control programs and, more saliently, allow us to explicitly tease out the potential poverty impacts of such programs as well.

SAM Multiplier Analysis: Ethiopia. Ethiopia, with the largest cattle herd in Africa, was the recipient of the largest amount of PARC funds for rinderpest control, totaling ECU 14.4 million in 1989–96 (Tambi et al. 1999). Tambi et al. estimated the avoided losses due to the PARC program (in production, manure, animal trac-tion, and milk) at ECU 35.4 million. The computation of SAM multipliers shows slightly larger impacts, summarized in Table 15.5. "Activity multipliers" reflect the impact of a one-unit increase in final demand for a given commodity on total pro-duction in the economy.[18] Livestock has very strong linkages with other productive

Table 15.5 Summary of social accounting
matrix multipliers in study countries

Country	Activity multiplier (rank in country)	Household income multiplier (rank in country)
Ethiopia	3.31 (4)	2.65 (1)
Kenya	2.89 (15)	1.22 (20)
Pakistan	5.18 (5)	2.68 (4)
Tanzania	5.07 (10)	3.13 (10)
Uganda	3.84 (8)	2.96 (7)

Source: Model simulations.

sectors of the economy, with an activity multiplier calculated at 3.31—the fourth highest in the Ethiopia SAM—suggesting that spending in the livestock sector will have large effects in stimulating economic growth relative to spending in other sectors. Given this activity multiplier, the economywide impacts of rinderpest control are estimated at ECU 47.6 million—that is, ECU 12 million higher than the partial equilibrium estimates of Tambi et al. (1999).

The computation of household multipliers highlights the distributional impacts of injections of government spending, such as for rinderpest control. In Table 15.5, total household multipliers show the impact on total household income of a one-unit increase in final demand in a given commodity sector. Livestock has the highest income multiplier among all sectors of the economy: a 1-unit increase in government spending yields a 2.65-unit increase in household income (Roeder and Rich 2009). Moreover, by isolating household income multipliers for the rural poor, along with income multipliers for livestock assets in drought-prone and pastoral areas, we see strong pro-poor impacts from injections of spending into the livestock sector (Roeder and Rich 2009). Rinderpest control would presumably have had strong poverty-reducing impacts compared to alternative interventions in other sectors.

Income distribution from injections into the livestock sector shows that rural households obtained 90 percent of the income generated from an increase in the final demand for livestock (Roeder and Rich 2009). The rural poor did particularly well, gaining 32 percent of the total increase in household incomes. Examining the PARC program in Ethiopia, we find that it raised household income by ECU 38.1 million, of which ECU 34 million accrued to rural groups—the largest amount among the five study countries (Table 15.6).

Kenya. The PARC program's efforts in Kenya focused primarily on an emergency campaign of mass vaccination after a 1996 outbreak. Tambi et al. (1999) estimated the avoided losses attributable to these control efforts at ECU 4.23 million.

Results from the SAM multiplier analysis suggest that the impact on the Kenyan economy was potentially much larger, despite high unit costs. Unlike the Ethiopian cattle sector, that in Kenya is less important than other sectors of SAM: the beef sector ranks 15th out of the 50 productive sectors. But with a sector multiplier of 2.89, an investment in rinderpest control of ECU 3.42 million translates into an economywide benefit of ECU 9.88 million.

Household multipliers in Kenya, by contrast, are relatively small, particularly for the beef sector. As summarized in Table 15.5, overall household multipliers for the cattle sector rank 20th out of the 50 sectors. The total rural household multiplier for the beef sector is just 0.51, less than a quarter of that of Ethiopia. The benefits of an injection into the cattle sector (such as from the PARC emergency campaign) go overwhelmingly to urban areas, particularly to higher-income deciles; thus the potential poverty impacts are thus much smaller than in Ethiopia (Roeder and Rich 2009).

Pakistan. The case of Pakistan is different from those of the four African countries because Pakistan had its own control program. Between November 1999 and June 2005, two separate FAO-led programs spent approximately $1.8 million on rinderpest control activities. Pakistan was provisionally rinderpest free in 2003 (FAO 2003), opening export markets for Pakistani beef products, particularly in the Middle East (Matin and Rafi 2006). Pakistan's beef exports surged from $991,000 in 2002 to $2.26 million in 2003 and $4.32 million in 2006 (FAOSTAT).

The activity multiplier for cattle (5.18) is quite large, ranking fifth out of 33 sectors (see Table 15.5). The $3 million increase in beef exports would thus yield a national benefit of $15.5 million. Household multipliers for cattle are also high, with a 1-unit increase in export demand for cattle engendering an increase in household incomes of 2.68 units, fourth among the 33 productive sectors in the Pakistan SAM (see Table 15.5). Much of the benefit goes to urban areas (42 percent), particularly to the urban nonpoor. However, in rural areas the largest benefits are captured by small farmers, who account for 23 percent of the total household multiplier. Another 14 percent accrues to rural nonfarm (nonpoor and poor) households, ostensibly in wage labor or service industries (Roeder and Rich 2009). Given a household multiplier of 2.68, an increase in export demand of $3 million would raise household incomes by more than $8 million (see Table 15.6).

Tanzania. Tanzania recorded the second highest estimated benefits, at ECU 13.1 million, and had the highest net benefits per animal at ECU 0.88 (Tambi et al. 1999). The multiplier analysis with the 2001 SAM constructed by Thurlow and Wobst (2003) suggests that the livestock sector has multiplier effects that are high compared to Kenya but lower than Ethiopia. As noted in Table 15.5, the multiplier for livestock ranked 10th out of the 43 sectors in the SAM. However, the magnitude of the livestock multiplier is nearly twice that of Ethiopia (5.07), implying that an

Table 15.6 Summary of household income gains from rinderpest interventions in the social accounting matrix multiplier analysis performed in study countries

Country	Household income benefits from rinderpest-related interventions (ECU)
Ethiopia	38,136,716
Kenya	4,160,782
Pakistan	8,037,227
Tanzania	11,493,353
Uganda	16,021,296

Source: Model simulations.

investment of ECU 3.67 million in rinderpest control would generate more than ECU 18.6 million in economywide benefits—higher than the ECU 13.1 million in benefits originally reported by Tambi et al. (1999).

Household multipliers in Tanzania are also quite sizable, particularly for livestock. Table 15.5 shows that the total household multiplier attributed to a 1-unit injection of final demand in livestock would yield an increase in household incomes of 3.13 units, of which 2.34 units would be retained in rural areas. Based on PARC expenditures of ECU 3.67 million, household incomes would rise by almost ECU 11.5 million (see Table 15.6). The poorest rural groups would gain approximately 16 percent of the total rise in household incomes, while nonpoor groups would gain nearly one-half (48 percent).

Uganda.   Uganda spent the second highest amount on PARC-related expenditures, at ECU 5.4 million (Tambi et al. 1999). Tambi et al. estimate the resulting benefit at ECU 10.4 million, just half the benefit estimated using the multiplier analysis. The multiplier for livestock in Uganda of 3.84 (see Table 15.5) suggests that the investment of ECU 5.4 million would increase economywide output by ECU 20.7 million, second only to the amount for Ethiopia among the four African cases. The relative ranking of the livestock sector compared to other sectors (eighth out of 26) suggests important economywide linkages of the livestock sector.

Household income multipliers for Uganda are also sizable, as noted in Table 15.6. A 1-unit increase in final demand increases the total household income by nearly 3 units, of which 2.44 units remain in rural areas (see Table 15.5). Given expenditures of ECU 5.4 million, the total income generated would raise total household incomes by ECU 16 million (see Table 15.6). Twenty-one percent of this income would go to low-potential regions in rural areas (Dorosh and El-Said

2004). An additional 18 percent would go to rural nonfarm households, including wage labor and service providers.

Summary: Economic Impacts of Rinderpest

Investments in livestock, including rinderpest control efforts, have strong, positive economic impacts that spill over into other sectors and that are, to varying degrees, pro-poor in nature. Investments in the livestock sector tend to provide more benefits than other interventions in rural areas. An important consequence of rinderpest control has likely been a general increase in food security in rural areas, including for nonfarmers. Although gains per household are relatively modest, households may benefit further from second-round effects of gains in income that are not captured in this analysis.

Risks to the Sustainability of Rinderpest Eradication

The global eradication of a disease agent should be sustainable by definition: once extinct, it should not return. However, the smallpox eradication experience shows that it is difficult to persuade countries to destroy their officially held virus stocks, even under international conventions. What sources of risk might bring back rinderpest disease?

Reemergence of Infection from a Wildlife Reservoir. The only populations potentially capable of harboring rinderpest today are the wild ruminants of eastern Africa, populations that are so fragmented that they are unlikely to maintain rinderpest for long periods. Neither the rinderpest virus nor seroconversion of wild ruminants has been detected there since 2001; only old animals are seropositive. Moreover, rinderpest does not persist indefinitely in wildlife once the disease is eliminated from cattle (Taylor and Watson 1967). There is thus little risk from this source.

Reversion to Virulence of a Live Vaccine. With live attenuated vaccines there is always the risk of reversion to virulence. Fortunately, the routine use of rinderpest vaccination has reduced dramatically over the past decade, and it has not been used for several years. Vaccine producers have largely ceased to make the rinderpest vaccine, and stocks are dwindling.

Presence of an Occult Reservoir of Infection. Rinderpest virus has not been detected in Asia since 2000 and in Africa since 2001; there is no reason to suspect its presence anywhere. The OIE rinderpest freedom accreditation process provides scientific, surveillance-based assurance for increasing numbers of countries. The few countries that have not formally applied for accreditation have conducted adequate surveillance with negative results. The likelihood that a rinderpest reservoir still exists is vanishingly small. Nevertheless, countries need to be vigilant against the reemergence of rinderpest until a credible global freedom accreditation process is complete.

Agroterrorism and Accident.   As long as stocks of virulent viruses are stored in laboratories around the world, there is a risk of malicious use. Moreover, scientists might conduct experiments using the live virus without adequate biosafety. There is an urgent need to catalogue and sequestrate viruses, and it is currently being addressed by the FAO and the OIE.

Synthesis of Rinderpest Virus de Novo.   In the 20th century, with the spectacular advances of molecular biology, it is no longer necessary to have access to a virus. In theory, only a blueprint is needed, and in this case the full virus genetic sequence information is freely available. How the world will cope with the risk of a virus being created for malicious purposes is not clear.

Reconstruction of Rinderpest Disease in Cattle with Another Morbillivirus.   There are a number of morbilliviruses related to the rinderpest virus that could conceivably enter cattle and evolve to reconstruct a form of rinderpest. *Peste des petits ruminants* virus is the most likely candidate, although there has never been evidence of its evolving into a cattle disease agent. Guarding against this risk requires strengthening the world's capacity to detect epidemiologically significant events and to take rapid action if needed.

Conclusions and Lessons Learned

The eradication of rinderpest has been a remarkable achievement that has brought major benefits to human populations in the developing world (however difficult they may be to quantify). The benefits include strong positive economic impacts that can be judged pro-poor given the general rise in food security in rural areas. Confidence in the eradication of rinderpest has also improved financial returns on trade for a number of countries, despite the continuing risk of other diseases that constrain trade, such as foot-and-mouth disease (FMD).

Programs implemented for rinderpest eradication at the national and regional levels helped to strengthen the surveillance and disease control capacities of many countries, including in marginalized areas. The experience gained has significantly improved the tools available for veterinary disease surveillance and control. Finally, the removal of rinderpest helps safeguard the wildlife heritage of Africa, Asia, and the Middle East.

The eradication of rinderpest was achieved progressively by national control programs organized within a series of intermittent, concerted international efforts during a period exceeding 50 years. The design of any future major disease eradication program must take into account the lessons of that experience. The rinderpest eradication strategy evolved from routine annual, pulsed vaccination to a process of active monitoring and containment based on a sound epidemiological understanding of the disease, followed by confirmation of its absence in a given region. This successful strategy relies on certain essential components.

Organizational and Institutional Elements

Political Support of the International Community. Political support at the national, regional, and global levels is an essential prerequisite of coordinated and sustained action. A compelling socioeconomic appraisal is needed to mobilize political support at the national level and to compete for limited rural development funds. The case for a disease control and eradication program must also be persuasive to residents of the affected areas; animal owners will comply only if they have a personal stake in the outcome.

An International Coordinating Body Hosted by an International Organization. One organization needs to take the lead, providing global coordination as well as technical leadership to regional campaigns. However, donors will often have competing priorities, and coordination will not always work smoothly. When the FAO was mandated to coordinate GREP in 1993, for example, funding by member countries proved insufficient to its evolving responsibilities, especially because some donors were reluctant to cede budgetary control to the FAO.

Programs That Are Time-Bound, Objective-Focused, Realistic, and Clearly Expounded. The preparatory phase of a control program needs to clearly describe the implementing mechanisms that affect disease control, defining a minimal package of capacity development and taking into account the capacity of veterinary services of targeted countries, vaccine delivery systems, and policy issues such as cost recovery and legal provisions. Any technical deficits need to be clearly understood and properly addressed.

The focus on rapid disease elimination can become diffused by crosscutting concerns for infrastructure and capacity development. Some participants may want to exploit the disease control initiative to address broader development concerns such as privatization, decentralization, governance, and human rights. Similarly, a focused program may tend to be expanded to address other disease control issues; for example, PARC evolved into PACE, broadening its focus to include generic "epidemio-surveillance systems" as well as control of contagious bovine pleuropneumonia.[19]

Regional Organizations That Are Committed to Working Closely with the Global Coordinating Body. Global campaigns are best implemented through regional coalitions coordinated by credible regional organizations. GREP was designed to coordinate several regional programs, but only the African region provided an implementation infrastructure (through PARC and the PACE project of the AU

IBAR). The countries not covered by regional campaigns became the direct responsibility of the GREP secretariat, which was ill equipped for this task. Moreover, countries represented on a regional accreditation working group are more likely to develop "ownership" of the accreditation process.

Dynamic, Adaptive, and Technically Oriented Leadership and Management. Management must be output oriented rather than process oriented. In rinderpest control, the vaccination rate represents an obvious and easily monitored performance target. But the aim of a control program needs to be to *curtail* mass vaccination as soon as the virus has ceased to circulate in a given region. These vaccination transition targets cannot be set in advance but depend on real-time monitoring.

Eradication programs require clear initial objectives, constant management, and clear exit strategies designed on a basis of epidemiological understanding. Effective management of rinderpest eradication programs involves taking carefully calculated risks. Unfortunately, when a single team leader is put in charge of both program management and applied population medicine, the planning, financial, logistical, and reporting issues tend to overshadow the innovative technical requirements.

Technical Elements

A Clear and Evolving Understanding of the Epidemiology of the Targeted Disease. Only after the geographic extent of rinderpest infection was established—including the role played by specific, discrete reservoirs of infection—was it possible to set a strategy for its progressive elimination. At the start of any control program there will be deficits in understanding and the need for a mechanism to provide epidemiological information along with laboratory diagnostic support. Epidemiological expertise, with all the available tools of surveillance and data analysis, is fundamental to effective disease control.[20]

Safe, Efficacious, Affordable, and Quality-Assured Vaccines. The seminal research to develop a thermostable rinderpest vaccine made a significant contribution to rinderpest eradication in tropical regions; thermostability is an important attribute for any vaccine to be used in developing countries. Nevertheless, the lack of small-dose vials hampered rural rinderpest vaccination campaigns because the vaccine rapidly loses efficacy once reconstituted.

PANVAC, established by the FAO for PARC, is unique as an independent, dedicated vaccine quality assurance facility. Despite recurring funding deficits, it provided a reliable means to ascertain the suitability of vaccines for the intended environment.

A Set of Robust, Validated Laboratory Diagnostic Tools for Agent Detection. Tests need to comply with OIE standards for disease reporting and freedom accreditation procedures. Ideally the technology should be appropriate for countries without sophisticated diagnostic facilities (filter paper sampling, for example). Support for the development of appropriate diagnostic techniques needs to be built into programs.

Rapid pen-side tests for rinderpest proved valuable in searching for the virus, especially in Pakistan. Such tests are now becoming available for a range of other diseases such as FMD and are especially useful once disease incidence has been reduced.

A serious issue in rinderpest monitoring is the problem of discriminating between antibodies induced by vaccination and those resulting from wild virus infection. The unavailability of validated discriminatory test systems meant that testing of animals could begin only long after the cessation of vaccination. (Similar problems have been encountered with avian influenza, Newcastle disease, *peste des petits ruminants,* and swine fever.)

A Global Reference Laboratory Network to Support Technology Transfer and Information Exchange. The World Reference Laboratory for Rinderpest (now Morbilliviruses) (WRLR) made a significant contribution to supporting national and regional control programs in all technical areas: developing diagnostic tests and technology transfers, providing reference diagnostic services, establishing molecular epidemiology, providing vaccine quality assurance, training operational staff, and developing training materials.[21] Although regional reference laboratories in general lacked sufficient resources, the ideal network would bring together national laboratories supported by a regional laboratory and regional laboratories supported by a global reference laboratory.

Dynamic and Innovative Disease Control and Eradication Strategies Based on Epidemiological Studies, Adapted and Continually Revised for Local Conditions. Control and eradication planning must be a dynamic process that evolves as situations change—especially in the final approach to disease eradication. In particular, livestock owners must be consulted on strategies to accommodate differences in farming systems and socioeconomic status (Mariner, Roeder, and Admassu 2002). In Ethiopia, for example, central planning of vaccination campaigns failed to give due attention to livestock owners' wishes regarding sites and practices for vaccination. When these issues were addressed through community-based animal health worker programs (along with use of the thermostable vaccine), herd immunity was increased and rinderpest quickly eradicated. Vaccine delivery strategies need to be adapted to the needs of livestock owners and specific livestock populations.

A Clearly Defined Disease Freedom Accreditation Process. The goal of global rinderpest eradication by 2010 was selected at the inception of GREP, and this tight schedule proved very valuable in guiding progress along the OIE Pathway. However, although rinderpest apparently ceased to circulate (in both domesticated and wild animals) by 2001, today no declaration of global freedom has been made because a small minority of countries has not officially applied for accreditation as infection free. Because the accreditation process can involve considerable expense, it is of little interest to countries that do not have significant trade in livestock. Proving global freedom from an animal disease is being undertaken for the first time, and the processes are still not fully defined.[22] In any future eradication program, the final accreditation process and the determination of global eradication need to be clearly defined in advance.

A key consideration in embarking on a disease eradication program is whether the needed funding—and political commitment—can be maintained for the required time frame. Differences of opinion underline the problem of international commitment. There is an ongoing debate about whether campaigns to eradicate disease are the most appropriate way to deal with the risks of dangerous pathogens. Western countries, with industrialized livestock production systems that are essentially free from epidemic diseases, are generally not motivated to support disease eradication in developing countries.

An alternative to eradication, for purposes of international trade, is the concept of commodity accreditation (Thomson, Leyland, and Donaldson 2009). Although it does not ameliorate the direct impact of disease on livestock-dependent livelihoods, this approach can have beneficial effects on mobilizing finance from livestock resources. A combination of approaches may prove more effective than global eradication campaigns, including regional control programs targeting the least developed countries (which act as persisting reservoirs of infection), combined with defined processes of risk management in trade. Over time, such a combined approach might well produce global eradication, because the havens for epidemic livestock diseases lie primarily within such target regions, where livestock are essential for food production and rural livelihoods.

In view of the extensive damage caused by rinderpest in the past two decades alone—in Africa, the Arabian Peninsula, and South Asia—it is testimony to the effectiveness of the control effort that the impact of the disease is by now almost forgotten.

Notes

1. In many countries, rinderpest control was the prime reason for the establishment of veterinary services. It was also the impetus for the founding in 1924 of the Office International des Epizooties (OIE)—now the World Organization for Animal Health—and in 1945 of the Food and Agriculture Organization of the United Nations (FAO).

2. Cattle exports from Pakistan into the Middle East and the Gulf States increased dramatically after Pakistan declared provisional freedom from rinderpest (FAO 2003). However, the eradication of rinderpest alone will not suffice to open up the high-value markets of Europe and the Russian Federation, which will remain wary of other cattle diseases such as foot-and-mouth disease (FMD).

3. In general, only exceptional events are reported, especially because rinderpest occurrence is a sensitive issue owing to its impact on international livestock trade. For more details, see Spinage (2003) and Barrett, Pastoret, and Taylor (2006).

4. The early live vaccines were unacceptably virulent when used on some breeds of Chinese cattle and yaks. By passing the Japanese lapinized vaccine virus through goats and sheep, a safe, attenuated vaccine was developed.

5. This was eliminated through a science-based eradication program. Active village monitoring using a participatory disease-searching methodology (Mariner and Roeder 2003) was also instituted throughout the country, with a concentration on Sindh and Punjab.

6. PACE established a Somali Ecosystem Rinderpest Eradication Coordination Unit that would work with the GREP secretariat to resolve questions about whether rinderpest continued to circulate after its last confirmed occurrence in 2001.

7. Still needed is a serological test to discriminate between vaccination and field infection. A marked vaccine with a differentiating test would be of great benefit in the accreditation process.

8. Goat-adapted vaccines were inexpensive and efficacious and induced long immunity, but they produced a clinical reaction in recipients and had to be stored at low temperatures, even when dried and vacuum packed.

9. The U.K. Institute for Animal Health, Pirbright Laboratory, was designated by the FAO in 1994 as the World Reference Laboratory for Rinderpest (WRLR) to provide diagnostic and molecular technology for morbilliviruses. Characterizing viruses from outbreaks proved valuable in determining their origins. The FAO / International Atomic Energy Association Joint division drew heavily on WRLR expertise to establish and run its African and West Asian laboratory networks.

10. The level of herd immunity required to disrupt rinderpest virus transmission is variously quoted as being between 80 and 90 percent, but these figures are empirical and not adduced from epidemiological studies (Roeder and Taylor 2007).

11. PANVAC, hosted by the Government of Ethiopia, continues to function as an autonomous unit within the AU IBAR.

12. National projects in Bangladesh, India, Nepal, and Pakistan, funded by the EC, nevertheless contributed significantly to rinderpest eradication in South Asia.

13. The others were the Kurdish Triangle (Iran, Iraq, and Turkey)—rinderpest free since 1996—and the southern part of peninsular India—rinderpest free since 1995—along with the shared border areas of China, Mongolia, and Russia.

14. Total donor funding of the nonemergency phase of the program (1988–99) is estimated at ECU 57.5 million (Tambi et al. 1999).

15. Blakeway's BCR of 34 takes into account a multitude of household and other nonmarket effects of rinderpest eradication. However, the analysis does not capture the variety of nonmarket costs (transactions costs, organizational costs, etc.) associated with the development and administration of the program, which would result in a figure lower than 34.

16. Full details of the analysis are found in Roeder and Rich (2009).

17. Catley, Leyland, and Bishop (2005) report a number of positive micro-level benefits resulting from rinderpest control, including significant increases in milk production, improved human health, a 40 percent increase in the population of sheep and goat, and a decline in cattle mortality of 39–72 percent.

18. Activity multipliers were calculated based on a 2005–06 SAM for Ethiopia (Ahmed et al. 2009; Roeder and Rich 2009).

19. However, disease control programs often need to address supplementary disease issues. An example from Pakistan concerns rinderpest and peste des petits ruminants (PPR). The same vaccine was in use for both diseases, so in order to cease use of rinderpest vaccination (for rinderpest freedom accreditation), it was necessary to provide an alternative PPR vaccine for use in small ruminants.

20. Mathematical modeling, both deterministic and stochastic (James and Rossiter 1989; Rossiter and James 1989; Tillé et al. 1991; Mariner et al. 2005), contributed significantly to the understanding of virus persistence.

21. Designated and partially funded by the FAO, the WRLR is hosted by the U.K. Institute for Animal Health, Pirbright Laboratory.

22. Opinions differ as to whether a global declaration needs to be linked to the cataloguing and sequestration of viruses.

References

Ahmed, H., A. Amogne, T. Tebekew, B. Teferra, E. Tsehaye, P. Dorosh, S. Robinson, and D. Willenbockel. 2009. *A regionalized social accounting matrix for Ethiopia, 2005/06: Data sources and compilation process.* Addis Ababa, Ethiopia: Ethiopian Development Research Institute.

Barnard, A. C., B. K. Rima, and T. Barrett. 2006. The morbilliviruses. In *Monograph series biology of animal infections, rinderpest and peste des petits ruminants: Virus plagues of large and small ruminants,* ed. T. Barrett, P.-P. Pastoret, and W. Taylor. Amsterdam, the Netherlands: Elsevier.

Barrett, T., P.-P. Pastoret, and W. P. Taylor. 2006. *Rinderpest and peste des petits ruminants: Virus plagues of large and small ruminants.* Amsterdam, the Netherlands: Academic Press.

Blakeway, S. 1995. *Evaluation of the UNICEF Operation Lifeline Sudan, southern sector livestock programme.* Nairobi, Kenya: United Nations Children's Fund–Operation Lifeline Sudan.

Catley, A. 1999. *Methods on the move: A review of veterinary use of participatory approaches and methods focusing on experiences in dryland Africa.* London: International Institute for Environment and Development. <http:/www.participatoryepidemiology.info/userfiles/MethodsontheMove.pdf>. Accessed August 9, 2009.

Catley, A., T. Leyland, and S. Bishop. 2005. *Policies, practice, and participation in complex emergencies: The case of livestock interventions in south Sudan: A case study for the Agriculture and Development Economics Division of the Food and Agriculture Organization.* Medford, Mass., U.S.A.: Tufts University.

Diamond, J. 1999. *Guns, germs, and steel.* New York: Norton.

Dorosh, P., and M. El-Said. 2004. *A 1999 social accounting matrix (SAM) for Uganda.* Washington, D.C.: International Food Policy Research Institute.

FAO (Food and Agriculture Organization of the United Nations). 2003. *Pakistan's victory over rinderpest.* <http://www.fao.org/english/newsroom/news/2003/13841-en.html>. Accessed May 13, 2009.

———. 2009. FAOSTAT. <http://faostat.fao.org>. Accessed August 20, 2010.

Felton, M. R., and P. R. Ellis. 1978. Studies on the control of rinderpest in Nigeria. Study 23. University of Reading, Berkshire, U.K. Photocopy.

Hambidge, G. 1955. *The story of FAO.* New York: D. Von Norstrand Company.

James, A., and P. B. Rossiter. 1989. An epidemiological model of rinderpest: (1) Description of the model. *Tropical Animal Health and Production* 21: 59–68.

Kock, R. A. 2006. Rinderpest and wildlife. In *Monograph series biology of animal infections, rinderpest and peste des petits ruminants: Virus plagues of large and small ruminants,* ed. T. Barrett, P.-P. Pastoret, and W. Taylor. Amsterdam, the Netherlands: Elsevier.

Leslie, J., and A. McLeod. 2001. *Socio-economic impacts of freedom from livestock disease and export promotion in developing countries.* Livestock Policy Discussion Paper 3. Rome: Livestock Information, Sector Analysis and Policy Branch.

Leyland, T. 1996. The world without rinderpest: Outreach to the inaccessible areas—The case for a community-based approach with reference to Southern Sudan. In *FAO animal production and health paper 129.* Proceedings of the FAO Technical Consultation on the Global Rinderpest Eradication Programme. Rome: Food and Agriculture Organization of the United Nations.

Mariner, J. C., and P. L. Roeder. 2003. The use of participatory epidemiology in studies of the persistence of rinderpest in east Africa. *Veterinary Record* 152 (21): 641–647.

Mariner, J., P. Roeder, and B. Admassu. 2002. Community participation and the global eradication of rinderpest. In *Participatory learning and action notes 45: Community-based animal healthcare.* London: International Institute for Environment and Development.

Mariner, J. C., J. A. House, A. E. Sollod, C. Stem, M. C. van den Ende, and C. A. Mebus. 1990. Comparison of the effect of various chemical stabilisers and lyophilisation cycles on the thermostability of a Vero cell-adapted rinderpest vaccine. *Veterinary Microbiology* 21: 195–209.

Mariner, J. C., J. McDermott, J. A. P. Heesterbeck, A. Catley, and P. Roeder. 2005. A model of lineage-1 and lineage-2 rinderpest virus transmission in pastoral areas of east Africa. *Preventive Veterinary Medicine* 69: 245–263.

Matin, M. A., and M. A. Rafi. 2006. Present status of rinderpest diseases in Pakistan. *Journal of Veterinary Medicine B* 53: 26–28.

Nakamura, J., K. Fukusho, and S. Kuroda. 1943. Rinderpest: Laboratory experiments on immunization of chosen cattle by simultaneous inoculation with immune serum and rabbit virus. *Japanese Journal of Veterinary Science* 5 (5): 455–477.

OIE (World Organization for Animal Health). 2010. Map of rinderpest-free countries. Paris. <http://www.oie.int/wahis/public.php?page=disease_status_map&disease_type=Terrestrial&disease_id=194&empty=999999&sta_method=semesterly&selected_start_year=2008&selected_report_period=1&selected_start_month=1&page=disease_status_map.> Accessed October 12, 2010.

Possehl, G. L. 1996. Mehrgarh. In *The Oxford companion to archaeology,* ed. B. Fagan. Oxford, U.K.: Oxford University Press.

Rich, K. M., and F. Wanyoike. 2010. An assessment of the regional and national socioeconomic impacts of the 2007 Rift Valley Fever outbreak in Kenya. *American Journal of Tropical Medicine and Hygiene* 83 (supp. 2): 52–57.

Roeder, P. L., and K. M. Rich. 2009. *The global effort to eradicate rinderpest.* IFPRI Discussion Paper 923. Washington, D.C.: International Food Policy Research Institute.

Roeder, P. L., and W. P. Taylor. 2007. Mass vaccination and herd immunity: Cattle and buffalo. *OIE Scientific and Technical Review* 26 (1): 253–263.

Rossiter, P. B., and A. James. 1989. An epidemiological model of rinderpest: II. Simulations of the behaviour of rinderpest virus in populations. *Tropical Animal Health and Production* 21: 69–84.

Sadoulet, E., and A. de Janvry. 1995. *Quantitative development policy analysis.* Baltimore: Johns Hopkins University Press.

Semmer, E. 1893. Rinderpest-infektion und Immunisierung und Schutzimpfung gegen Rinderpest. *Berliner und Münchener Tierärztliche Wochenschrift* 23: 590–591.

Spinage, C. A. 2003. *Cattle plague: A history.* New York: Kluwer Academic.

Tambi, E. N., O. W. Maina, and J. C. Mariner. 2004. Ex-ante economic analysis of animal disease surveillance. *OIE Scientific and Technical Review* 23 (3): 737–752.

Tambi, E. N., O. W. Maina, A. W. Mukhebi, and T. F. Randolph. 1999. Economic impact assessment of rinderpest control in Africa. *OIE Scientific and Technical Review* 18 (2): 458–477.

Taylor, W. P. 1997. Vaccination against rinderpest. In *Veterinary Vaccinology,* ed. P.-P. Pastoret, J. Blancou, P. Vannier, and C. Verschuren. Amsterdam, the Netherlands: Elsevier.

Taylor, W. P., and R. M. Watson. 1967. Studies on the epizootiology of rinderpest in blue wildebeest and other game species of Northern Tanzania and Southern Kenya, 1965–67. *Journal of Hygiene* 65: 537–545.

Taylor, W., P. L. Roeder, and M. M. Rweyemamu. 2006. Use of rinderpest vaccine in international programmes for the control and eradication of rinderpest. In *Monograph series biology of animal infections, rinderpest and peste des petits ruminants: Virus plagues of large and small ruminants,* ed. T. Barrett, P.-P. Pastoret, and W. Taylor. Amsterdam, the Netherlands: Elsevier.

Tillé, A., C. L. Lefèvre, P.-P. Pastoret, and E. Thiry. 1991. A mathematical model of rinderpest infection in cattle populations. *Epidemiology and Infection* 107: 441–452.

Thomson, G. R., T. J. Leyland, and A. I. Donaldson. 2009. De-boned beef—An example of a commodity for which specific standards could be developed to ensure an appropriate level of protection for international trade. *Transboundary and Emerging Diseases* 56: 9–17.

Thornton, P. K., R. I. Kruska, N. Henninger, P. M. Kristjanson, R. S. Reid, R. Atieno, A. N. Odero, and T. Ndegwa. 2002. Mapping poverty and livestock in the developing world. A report commissioned by the U.K. Department for International Development, on behalf of the Inter-Agency Group of Donors Supporting Research on Livestock Production and Health in the Developing World. International Livestock Research Institute, Nairobi, Kenya.

Thurlow, J., and P. Wobst. 2003. *Poverty-focused social accounting matrices for Tanzania.* Trade and Macroeconomics Discussion Paper 112. Washington, D.C.: International Food Policy Research Institute.

United Nations Secretariat, Department of Economic and Social Affairs. 2008. World population prospects: The 2008 revision. <http://esa.un.org/unpp>. Accessed August 23, 2010.

Chapter 16

Rural and Urban Linkages: Operation Flood's Role in India's Dairy Development

Kenda Cunningham

India, historically a milk-consuming country, has not always produced enough milk to satisfy consumer demand. In the 1950s and 1960s the country depended heavily on milk imports (Aneja 1994). In recent decades, however, a variety of programs and interventions targeting India's dairy industry have made the country one of the world's leading milk producers. Since 1970 India's output of milk and milk products has increased faster (at a rate of 4.5 percent per year between 1970 and 2001) than crop output (Sharma and Gulati 2003). In the early 2000s, milk and milk products accounted for 70.8 percent of the total output value of all livestock products in India and were the largest agricultural commodity category by value (Staal, Pratt, and Jabbar 2008b). These production increases were met between the early 1980s and the late 1990s with a nearly twofold increase in aggregate milk consumption throughout India, a total equivalent to 31 percent of all developing countries' milk consumption and 13 percent of the world's total (Delgado 2003). By 2007, with more than 12 million milk producers, India had become one of the largest producers of buffalo, goat, and cow milk in the world (Figure 16.1), and milk was a greater contributor to the nation's gross domestic product (GDP) than rice.

India's dairy industry is dominated by an unorganized, traditionally informal sector; only 10–15 percent of the market is formalized. In the late 20th century, population growth, economic development, rising incomes, and urbanization presented India with challenges in meeting an increasing dairy demand. National efforts to scale up production and distribution began in 1970 with the implementation of Operation Flood by the National Dairy Development Board. Operation Flood

Figure 16.1 Milk availability, production, and population in India and other major milk-producing nations, 2008

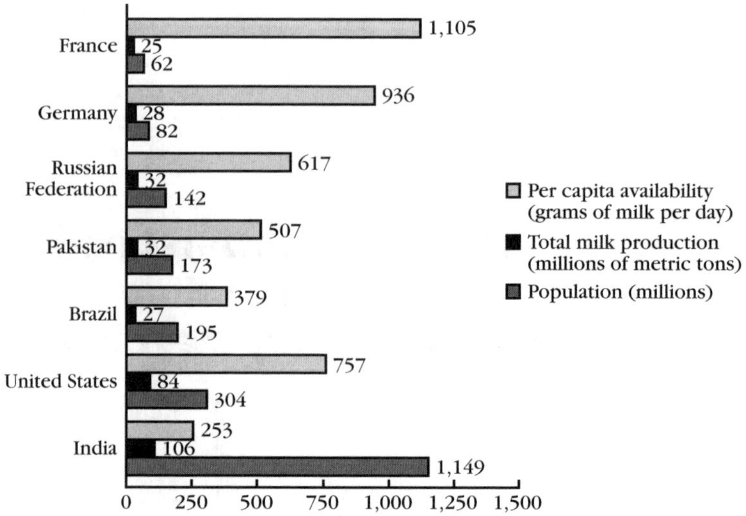

Sources: Calculated by author using data from Population Reference Bureau (2008); FAO (2009).

laid the foundation for a dairy cooperative movement. Although Operation Flood cooperatives actually produced and marketed only a small percentage of India's dairy products (Munshi and Parikh 1994), the program was critical to the industry's overall evolution in India: it transformed the policy environment, brought significant technological advancements to the rural milk sector, established many village cooperatives, and oriented the dairy industry toward markets (World Bank 2006). The intervention led to infrastructural improvements that enabled the production, processing, procurement, and marketing of milk throughout India. It linked India's major cities with dairy cooperatives nationwide. Many Indians, including small-scale dairy farmers, urban and rural consumers, and even landless milk producers, benefited from this large-scale agricultural intervention. Furthermore, by increasing milk production this cooperative scheme has enhanced food security for millions of people throughout India and improved employment, income levels, and the nutritional quality of diets.

The Intervention

The antecedent to the Operation Flood initiative was a milk producers' private cooperative venture in Anand, Gujarat. India's then–Prime Minister Lal Bahadur

Shastri visited the cooperative—the Kaira District Cooperative Milk Producers' Union Ltd. (KDCMPUL)—and was impressed with its effectiveness. In response, the Government of India established the National Dairy Development Board (NDDB) in 1965 and mandated that it replicate the Anand model of dairy cooperatives throughout India. The NDDB launched Operation Flood to flood India with milk, using a sophisticated procurement system to connect rural production with urban demand. This federally sponsored national intervention, which lasted from 1970 to 1996, was designed to ensure that dairy products reached both rural and urban consumers in an efficient and effective manner without seasonal fluctuations in the supply and price of milk.

Verghese Kurien, general manager of KDCMPUL (which was later renamed the Anand Milk Union Limited), was the founding chair of the NDDB and turned the idea of cooperative dairying into a reality. The NDDB set up the cooperatives and provided technical support for planning, farmer extension services, engineering, dairy technology, veterinary services, and nutrition.

The scaling up of production, marketing, and processing was strategically synchronized in three distinct phases over a period of 25 years, with each phase building on the previous phase's achievements. Delays sometimes affected implementation. For instance, the goals of Operation Flood I (OF I) were achieved in double the time that was originally anticipated. Lags in aid delivery, the global dairy crisis, difficulties with absorbing commodities locally, and internal program difficulties all contributed to the delays (Doornbos et al. 1990). OF I, in operation from 1970 to 1981, focused on market policy and initially targeted just four major cities—Mumbai, Kolkata, Delhi, and Chennai, at the time known as Bombay, Calcutta, Delhi, and Madras—for milk distribution. During OF I, 1 million rural milk producers with 1.8 million milch animals were incorporated into the scheme. Operation Flood II (OF II), in operation from 1981 to 1985, incorporated 10 million rural producers with several million head of improved animal stock. During this second phase, the number of milk sheds increased from 18 to 27, and urban milk centers extended to cover all 147 major Indian cities. Operation Flood III (OF III), which lasted through 1996, filled the remaining gaps, targeting nearly 7 million farmer families, building 170 milk sheds, and improving veterinary care (Atkins 1988; Chothani 1989). All three stages received various forms of financial support, including international loans, donations of commodity aid, aid from the Government of India, and internal program resources, as well as finances generated from the intervention.

In 1970 the Indian Dairy Corporation was established to manage the financial aspects of Operation Flood, such as receiving and monetizing donated commodities; the profits generated from selling the donated commodities were used to finance the Operation Flood program. Later the two entities merged, and today the NDDB continues to fund and provide technical assistance to cooperatives through-

Table 16.1   Cooperative growth during and after Operation Flood, 1970–2006

Year	Farmer members (millions)	Village milk cooperatives (thousands)
1970/71	0.3	1.6
1975/76	0.6	4.5
1980/81	1.8	13.3
1985/86	4.5	42.7
1990/91	7.5	63.4
1995/96	9.0	69.6
2000/01	10.7	96.2
2003/04	12.0	108.6
2004/05	12.3	113.2
2005/06	12.4	117.6

Sources: Aneja (1994); Gupta (1997a); National Dairy Development Board (2005).

out India. The Institute of Rural Management in Anand, founded in 1979, was set up to provide professional management and research support to cooperatives (Terhal and Doornbos 1983).

By 1996 Operation Flood included approximately 70,000 dairy cooperatives in 170 milk sheds encompassing 8.4 million milk-producing families (Banerjee 1994), and the system of cooperatives has continued to grow (Table 16.1). As milk production has nearly tripled, India not only has become self-sufficient in milk but also has become an exporter of milk powder; overall, the industry produces enough to almost entirely eliminate India's dairy imports (Perumal, Mohan, and Suresh 2007).

Operation Flood: Approaches to Dairying

Getting Organized

Operation Flood replaced the ad hoc production, marketing, and selling of milk with an organized dairy supply chain using a three-tiered structure. The first tier consisted of the village-level cooperatives, which were responsible for all of the microinputs, including the production and testing of milk. Local farmers, along with an elected management committee, controlled these cooperatives. Over time they were required to include at least one female manager. The second tier, composed of district-level cooperative unions, provided macroinputs, such as the transportation of milk collection and processing equipment. These unions owned and operated the dairy processing plants, managed the cattle feed plants, and provided animal health care at the village level. The third tier was formed by state-level marketing federations, which engaged in marketing and coordinated the logistics of interstate sales (Aneja 1994). The NDDB operated externally as a facilitator, providing guidance and support for setting up the cooperative structure, funding it, and providing technical assistance.

Although this system suffered from some inefficiencies and diseconomies, it created a clear chain for getting milk from the small cattle owners to ordinary urban consumers, and it ensured seasonal continuity. The cooperatives, voluntary groupings of individual economic entities, in turn ensured that dairy development was directly responsive to the evolving needs of smallholder farmers (Narayanaswamy 1996). Cooperatives also created natural channels by which to disseminate information and distribute technologies for dairy production and processing (Munshi and Parikh 1994). In fact, Operation Flood provided the physical and organizational infrastructure for a national milk grid linking the surplus milk in one region to the demand for milk in another region (Figure 16.2).

Operation Flood attempted to address various types of barriers that smallholder producers could face: a lack of assets and technical skills, sociocultural differences, and transaction costs. Although cooperatives were the mechanism through which Operation Flood addressed these obstacles, this mechanism did not always succeed. Sometimes dairy development occurred even faster in places where cooperatives were weak or did not exist. Some challenges for cooperatives included weak management, overstaffing, poor market orientation, and a lack of flexibility in responding to changes in market conditions (World Bank 1996). All cooperatives are not the same, and hence the success rates have varied: success at the village or federal level often has been based on how effectively the cooperative serves the main needs of its members (Shah 1995).

Using Aid for Development

Operation Flood marked the first time that food aid was viewed as a critical investment resource (Banerjee 1994). During Operation Flood, the European Economic Community (EEC) donated surplus dairy commodities to India. Cooperative and farmer-owned processing plants combined the donated commodities—skimmed milk powder and butter oil—with milk produced by Indian farmers and sold the combined product to urban consumers at the prevailing market price. The recombined milk helped meet the domestic demand for milk, as well as financing Operation Flood. This source of funding was combined with substantial loans from the World Bank and some bilateral financial and technical assistance from Australia, Canada, Denmark, Germany, Sweden, the United Kingdom, and the United States. It is also important to note that member equity and profits generated by cooperative business were also a large part of Operation Flood's financing.

Figure 16.2 India's national milk grid, 1993

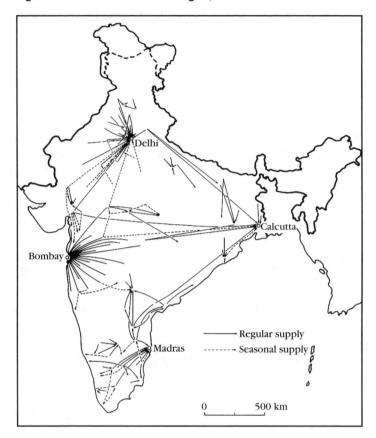

Source: Atkins (1993).

Indian leaders recognized that simply dumping excess European milk onto India's dairy market could discourage local milk production. Operation Flood thus stressed reliance on local production, procurement, marketing, and selling and used the commodities and proceeds to promote overall dairy development instead of dependence. The European surplus was used to create needed dairy supply and eliminate gaps in supply and demand. Once India's production and processing reached a certain scale, Indian farmers had an incentive to produce more milk to supply processing plants. Imported commodity aid was then phased out (Table 16.2). Although some observers claimed that the EEC surpluses offered a disincentive to Indian dairy producers and skewed the market (Doornbos and Gertsch 1994), the continued growth of Indian dairy production since Operation Flood seems to show

Table 16.2 European Economic Community aid for Operation Flood, 1970–96 (metric tons)

Phase	Skimmed milk powder	Butter oil
Operation Flood I (July 1970–March 1981)	126,000	42,000
Operation Flood II (April 1981–March 1985)	240,000	84,700
Operation Flood III (April 1985–March 1996)	75,000	25,000

Sources: Atkins (1988); Kaye (1988); Gupta (1997a).

that India successfully avoided the dangers of dependency on imported milk. The use of food aid as a development investment was anti-inflationary because it helped stabilize market prices by providing a constant buffer stock. Milk was recognized as a marketable commodity due to long-standing market demand, and markets were established in urban areas to give confidence to the suppliers—dairy farmers—to increase their investment (Banerjee 1994).

Several attempts have been made to sum up the total project costs. The total investment, estimated at about $1 billion at the time of the intervention, was divided into three categories of sources: 50 percent loans, 40 percent commodity aid from the EEC, and 10 percent investment of the NDDB's own resources (Gupta 1997a; Tikku 2003). The World Bank's evaluation of Operation Flood reported a higher total project cost: $2.7 billion (1996 U.S. dollars), of which nearly $2 billion were in the form of direct subsidies and $0.7 billion were bank loans and credits (Candler and Kumar 1998).

Using Political Support Strategically

Political support was important in ensuring the smooth implementation of Operation Flood. Prime Minister Shastri's leadership in setting up the NDDB, his appointment of Dr. Verghese Kurien to replicate the model nationally, and his social entrepreneurship in the overall design of Operation Flood were prerequisites to the many achievements of the intervention. Government policies supportive of dairy cooperatives enabled the industry to develop in sync with the evolving needs of dairy farmers.

Once Operation Flood was under way, state interference was minimal. The NDDB's autonomy allowed it to place producers at the forefront of decisions. Located not in New Delhi but in Anand, the NDDB was able to create supporting entities to provide people, technology, veterinary services, equipment, and marketing that the cooperatives needed at an affordable price (Tikku 2003). Some scholars have criticized the minimal state intervention in the program and argue that the lack of government scrutiny led to corruption and conflicts of interest (Alvares 1985).

To be sure, the intervention did not always run smoothly. In some cases local politicians supported government cooperatives at the expense of farmer-managed cooperatives. In some cooperatives the state maintained control over input and output pricing, and some state officials, originally appointed to temporarily serve in management positions, continued to hold onto this power even after Operation Flood ended instead of turning the cooperatives over to the farmers for bottom-up development. Even in these instances, dairy plants and milk routes were developed; however, true development was limited because the capacity-building of nonelites was not prioritized (Kurien 1996). Although Operation Flood was federally sponsored and managed by semistate institutions, it was not a government-run initiative: ownership and management of its day-to-day activities and resources were local. This situation minimized political interference, turf battles, and administrative red tape. The World Food Programme's Terminal Evaluation Report (FAO 1981) notes the lack of political interference in the Operation Flood model and describes its assets as its quality-control standards, its reliable technical assistance, and its multilevel, democratic structure.

Four consecutive prime ministers supported the large-scale dairy intervention (Kaye 1988), highlighting the durability of the cooperative movement. One main lesson is that nonpolitical interventions can be successful and enjoy the backing of executives long past their initial implementation.

Focusing on Local Methods and Markets

Another prerequisite to Operation Flood's success was its focus on the local context. Operation Flood did not impose foreign ideas, processes, and tools but rather used existing indigenous dairy farming techniques. Given the long history and importance of milk products in India, the architects of Operation Flood had to know what approaches dairy farmers would adopt and what types of milk products consumers would purchase. The project promoted labor-intensive but low-capital, energy-intensive production because smallholder farmers rely primarily on family and local labor. It emphasized the Indian approach to dairying, using crop residues and by-products for milk production. Operation Flood was a mechanism for production by the masses but not a switch to mass production (Patel 2007).

Operation Flood resulted in the creation of an indigenous dairy equipment–manufacturing industry; by the late 1990s less than 10 percent of India's dairy equipment was imported. The intervention also led to a remarkable accumulation of indigenous expertise on animal nutrition, animal health, artificial insemination, management information systems, dairy engineering, and food technology (Candler and Kumar 1998).

Another critical element was the local ownership of resources. This approach allowed rural people to use their initiative, insights, and energy to gain better prices as producers, eliminating actors in the middle and helping producers retain profits so they could upgrade production (Kaye 1988). Operation Flood's successful balancing of external support—such as that of the NDDB—with local management prevented tyranny by midlevel players and benefited millions of farmers.

Operation Flood's Impacts and Achievements

Operation Flood's actual share of the dairy sector is fairly limited. Cooperatives—the heart of Operation Flood—account for only about 6–7 percent of India's milk production. And although cooperatives account for 70 percent of the *organized* milk market's share of national milk production, the informal sector still accounts for the lion's share—80 percent—of India's milk market and milk products (Staal, Pratt, and Jabbar 2008b). Nonetheless, the investments in dairying technology and market infrastructure that occurred under Operation Flood have had positive and measurable outcomes on livelihoods and welfare in India.

Markets, Infrastructure, and Rural-Urban Linkages

A major achievement of Operation Flood was its contribution to market and infrastructural development in India. By gradually ensuring rural milk producers an efficient supply chain through which to sell their milk, Operation Flood opened new avenues to employment and income-generating activities for rural households. And by placing quantitative restrictions on dairy imports and introducing licensing requirements that restricted new private-sector entrants to the dairy market, the Government of India was able to protect dairy cooperatives and their members from excessive competition that would have put them out of business (Staal, Pratt, and Jabbar 2008b).

Operation Flood also set up about 175 dairy plants, 45 cattle feed plants, about 15,000 centers for artificial insemination, 100 rail milk tankers, and more than 1,500 road milk tankers (Tikku 2003). Indirectly, Operation Flood contributed to the improvement and expansion of India's wider transportation infrastructure, such as railways and roads (Candler and Kumar 1998). These infrastructural investments helped link rural and urban India, including key markets such as Delhi and Mumbai.

Rural Incomes

In rural areas Operation Flood was India's largest sustainable employment program of its time. In India agriculture is the main means of livelihood and income for millions of farmers, with crop production and dairying the primary and secondary activities, respectively. A study of three districts—Bikaner in Rajasthan, Periyar in Tamil Nadu, and Sabarkantha in Gujarat—showed that in cooperative villages, average household incomes from all income sources are higher, average incomes from milk are generally larger, and average levels of employment are higher than in noncooperative villages (Singh and Das 1984).

Smallholder dairy farmers supplied more than 60 percent of the milk procured by cooperatives (Aneja and Puri 1997). For poorer farmers, dairy may be even more significant. Atkins (1989), for example, argues that Operation Flood was India's most promising large-scale, wealth-generating rural development program. Achaya and Huria (1986a) assert that Operation Flood is a reason that poverty in India dropped from 49 to 38 percent from 1977–78 to 1983–84. Jul (1988) points out that increased productivity in the dairy industry also enhanced economic development. Specifically, the development of a national milk grid, village cooperatives, and district unions increased employment throughout India. Staff were hired to run the thousands of dairy cooperatives and provide animal husbandry services. In the early 2000s, dairy cooperatives had 11 million participating households; Indian households benefited from cooperatives they owned, cooperatives that sold animal feed to them, veterinary coverage for dairy animals, and cooperatives that purchased their milk (Kurien 2004).

Dairy Production and Productivity

Operation Flood led to substantial breakthroughs in dairy production, processing, procurement, and marketing. For example, technological advances in milk drying, storage, and transportation partially alleviated fluctuations in production levels between the lean and flush seasons. Overall production rose dramatically from 1950 to 2008, as did per capita availability, even in the face of rapid population growth (Table 16.3). The productivity of milch animals tripled between 1970 and the early 1990s, and the total milk production in Operation Flood areas increased from 42 million liters per day to 67 million liters per day between 1988–89 and 1995–96 (Aneja 1994; Shukla and Brahmankar 1999). However, it should be noted that the study by Alderman (1987) found no milk yield differences that could be attributed to the presence of Operation Flood cooperatives.

Milk processing and procurement also advanced during Operation Flood. As the sheer amount of processed goods increased, so did the amount of equipment, such as silos, pasteurizers, rail and road storage tanks, and refrigerators that conformed to international standards. Milk testers were designed to weigh, test, and record milk production levels. The capacity for processing milk reached about 15.6 million liters a day, and the capacity for chilling milk reached 6.5 million liters a day (Banerjee 1994). Processing increases were not always uniform, however. For example, in December 1980, raw milk processing in four OF I dairies based in Gujarat was at full capacity (1.6 million liters per day), but it was below capacity in other Operation Flood facilities (Terhal and Doornbos 1983). Throughout Operation Flood, the

Table 16.3 Production and per capita availability of milk in India, 1950–2008

Year	Production (millions of metric tons)	Per capita availability (grams per day)
1950	17	132
1960	20	127
1968	21	113
1973	23	111
1980	32	128
1990–91	54	182
1995–96	66	207
2000–01	81	230
2005–06	97	241
2006–07	101	246
2007–08	105	252

Sources: Nair (1985); Gupta (1997a); India, Ministry of Agriculture (2010).

NDDB devised new storage methods, and milk was converted from a highly perishable commodity into one that can be stored and traded nationwide. To link village producers and city-based consumers, a network was established that included trucks, chilling plants, refrigerated vans, railway wagons, and processing plants.

Milk procurement in Operation Flood areas rose from 28 million liters a day in 1988–89 to 35 million liters a day in 1995–96 (Shukla and Brahmankar 1999). Baviskar and Terhal (1990), arguing that overall increases in milk production are not necessarily attributable to Operation Flood, point out that milk procurement increases among dairy cooperatives do not necessarily mean that milk animals' productivity or even overall milk productivity has increased. Even milk packaging was transformed. Most of the milk is now packaged in plastic bags, and the machines that make the small bags are produced indigenously. Another major innovation by the NDDB was the development of the bulk vending of milk—an indigenous system using gravity milk feeding and a siphon to provide consistent quantities of milk.

Retail sales from the Operation Flood cooperatives increased steadily. In 1970 only about 1 million liters a day were sold, but as OF II was picking up, sales increased to about 5 million liters a day. Toward the end of the 1990s, sales reached nearly 10 million liters a day (Table 16.4) (Candler and Kumar 1998). In urban areas alone, the amount of milk marketed by cooperatives increased more than 50 times between 1970–71 and 1990–91 (Fulton and Bhargava 1994). At the end of Operation Flood, the dairy cooperatives were meeting 60 percent of the urban milk demand and accounted for 22 percent of all milk marketed in India overall (Candler and Kumar 1998).

Table 16.4 Peak procurement and peak marketing of Operation Flood dairying, 1970–96 (million liters per day)

Phase	Peak procurement	Peak marketing
Operation Flood I (1970–81)	3.4	2.8
Operation Flood II (1981–85)	7.9	5.0
Operation Flood III (1985–96)	1.3	9.4

Source: Gupta (1997b).

In isolation, these technological breakthroughs would have been largely irrelevant, but because a year-round market also supported the producers, the technologies were put into practice. Although only a small proportion of the dairy market is dominated by cooperatives, data suggest that dairy cooperatives have affected the supply of milk because they promoted the introduction of new technologies, in particular enhanced dairy cattle (Staal, Pratt, and Jabbar 2008b). To its critics, however, the centralization of dairy has led to homogenization and hence failed to take into account the unique needs, challenges, and opportunities found throughout India (George 1990). All of the increases in milk production, procurement, processing, and marketing cannot be attributed to Operation Flood, but this intervention was instrumental in setting up a new approach to dairying in India. Further data collection and analysis on both micro-level and macro-level impacts would help to clarify Operation Flood's impacts on poverty and food security.

Improved Nutritional Intake

Increased incomes from dairy sales have allowed smallholder dairy farmers to spend more on nutritious foods. In addition, the average per capita consumption of dairy products in India rose from about 132 grams a day in 1951 to around 200 grams by the end of Operation Flood (Table 16.5) (Aneja 1994; Bhide and Chaudhari 1997). The per capita consumption of milk by dairy farmers increased in Operation Flood areas from 290 grams a day in 1988–89 to 339 grams a day in 1995–96 at the aggregate level (Shukla and Brahmankar 1999). Village-level studies have revealed that consumption of milk and other foodstuffs was substantially higher in rural Operation Flood areas than in rural non–Operation Flood areas, an indication that this program improved the dietary diversity and nutritional status of its participants.

Milk is a primary source of animal protein for Indians, so if production increases consistently reach more people, their protein levels are likely to improve. Food expenditure was positively correlated with Operation Flood's introduction into certain areas of India. A study by Mergos and Slade (1987) noted that Operation Flood had a positive impact on the caloric and protein levels of the rural population. However, because the study lacked a baseline, the authors' attribution of the nutri-

Table 16.5 Daily consumption of milk in India, 1950–96 (average grams per day)

Year	Consumption
1950–51	124
1960–61	124
1973–74	112
1981–82	136
1991–92	178
1995–96	197

Source: India, Ministry of Agriculture (2010).

tional impact to Operation Flood is possibly overstated. Another study by Achaya and Huria (1986a) noted that Operation Flood areas had higher levels of vitamin A and C intake, in addition to caloric and protein intake. Furthermore, Singh and Das (1984) found higher per capita protein intake and higher consumption levels in expectant and nursing mothers as well as in children two to six years old. Of the milk produced in the Operation Flood areas, about 65 percent was traded and 35 percent was consumed by the producing household, according to a study by Bhide and Chaudhari (1997). This finding is significant because a poor person can dramatically improve his or her nutritional status by consuming even marginal amounts of milk; consumption of dairy products is vital, especially for children and mothers who are nursing or expectant.

Development of Extension Services

Operation Flood led to broad advances in the development of extension services and technologies. One goal of these services was to improve overall cattle health and nutrition, thus ensuring more productive dairy cattle. Operation Flood brought about new standards for livestock and promoted technological advances such as artificial insemination, crossbreeding, vaccinations, improvements in cattle feed, urea treatment of straw (to improve digestibility), the use of fans and sprinklers to cool cows in the summer, and the construction of biogas plants for production and processing (Candler and Kumar 1998).

As market competition expanded, the producer price of milk shifted upward, resulting in an increased use of concentrate feed and a higher demand for superior animal husbandry practices, such as veterinary services and artificial insemination (Mergos and Slade 1987). Artificial insemination, a major change for rural producers, grew throughout India as a result of Operation Flood: about 18 million artificial inseminations were performed annually at the village level by paraprofessionals who were supported by trained professionals running semen banks and stud stations (Kurien 1996). Through Operation Flood an additional 16,280 dairy cooperative societies were involved in artificial insemination (Gupta 1997a). Extension activities, such as education in cattle breeding and tours of dairy plants, were and remain essential components of milk cooperatives. Because women play a vital role in caring for milch animals, these extension services have significantly increased women's knowledge, confidence, and social status.

To increase milk production without increasing the overall cattle populations, it was important to increase the productivity of milch animals. A notable, and contentious, aspect of Operation Flood was the crossbreeding of exotic cows with indigenous cows to increase their breeding capacity and milk yield. During Operation Flood, 100 indigenous cows provided only about 150 kilograms of milk a day, whereas 100 crossbred cows provided about 400 kilograms per day (Guha 1980). Critics argue that the crossbred cows benefited mainly richer farmers, had higher feed requirements, and were unable to adapt to Indian conditions (Atkins 1988). Achaya and Huria (1986a, 1986b) state, however, that the near doubling of milk production after Operation Flood's inception should have put an end to the speculation about problems associated with adopting crossbreeding. In fact, landless milk producers did acquire crossbred animals, only a small percentage of Operation Flood's strategy was focused on crossbred animals, and the Indian government sought to improve productivity through crossbreeding and cattle upgrading without exterminating well-known Indian cattle breeds.

Economic and Social Development

Operation Flood significantly altered rural India in other positive ways beyond the dairy and food security realms. As already described, the intervention helped extend the national dairy infrastructure and improved transportation systems, such as railways and roads (Tikku 2003). Furthermore, all population sectors participated in the intervention: for example, both men and women milked their animals and marketed their milk at the cooperative twice daily, and at the milk collection centers men and women lined up together. In this way, Operation Flood helped to lower social barriers. Community development was enhanced as gender differences and social class divisions were broken down (FAO 1978). Finally, discussions could be heard among beneficiaries as they shared ideas about sanitation and cleanliness. Despite these broad social achievements, the impacts of Operation Flood were not felt equally by all: gender, caste, class, and landownership status played a role in how people were affected by these changes in dairy farming.

Growth in the smallholder livestock sector, according to Mellor (2003), has made a direct positive contribution to poverty reduction, employment growth, and eventually demand for employment in the rural nonfarm sectors: "In sharp contrast to crop income the Gini coefficient for dairy production, which is very important to the poor in India because of its labor intensity, is 0.11. That is an extraordinarily low Gini coefficient. And, the Gini coefficient for off-farm work in rural areas is a still low 0.22" (Mellor 1999, 3). Investments in dairy seem to promote income equality more than do investments in crop production (Birthal, Taneja, and Thorpe 2006). Although some scholars argue that poor and marginalized farmers (for example, the landless) were underrepresented and that lack of access to credit and fodder prevented many from accruing potential program benefits from Operation Flood (Verhagen 1990), empirical evidence shows that Operation Flood did not merely help rich farmers get their milk to urban consumers but also directly engaged poor people. In 1984, 72 percent of cooperative members were small and marginal farmers or operated less than 5 hectares of land, and the majority of these were also from minority castes and tribes (Atkins 1989). By promoting universal access to a strong milk market, balanced cattle feed, animal health care, and artificial insemination services, Operation Flood must have somewhat reduced the disparity in income distribution between the rich and the poor.

Additionally, the impact of livestock-based programs—such as Operation Flood—on women can vary. Livestock interventions can be an opportunity to generate income, but they can also increase workloads without truly altering women's level of control over resources. Some scholars argue that even if Operation Flood helps poor women farmers, broader gender disparities within the dairy industry, such as division of labor and stereotypes, remain intact (Sharma and Vanjani 1993). Although it is true that Operation Flood cannot dismantle centuries-old patriarchal traditions and structures, data show that women did benefit from this intervention. For example, employment rates, including those of female workers, were higher among Operation Flood beneficiaries than among nonbeneficiaries (Thirunavukkarasu, Prabaharan, and Ramasamy 1991). In addition, Operation Flood uniquely increased employment for landless female dairy farmers in relation to all female dairy farmers (Singh and Das 1984). Extension activities that were essential components of the milk cooperatives have increasingly engaged women, improving their dairy know-how and self-assurance and, in turn, their social status. Women now make up more than 25 percent of cooperative members, and more than 2,700 all-woman cooperatives are functioning. Women continue to play a small role in running the dairy cooperative societies; however, fewer than 3 percent of board members are women (Nehru Ganju 2005; NDDB 2008).

The Intervention's Sustainability

Although Operation Flood ceased operations in 1996, it has had a lasting impact on India's dairy supply chain. Between 1970–71 and 1990–91, when the program was in effect, village milk producers' cooperatives increased almost 40-fold and the number of producers with cooperative membership increased almost 27-fold (Fulton and Bhargava 1994). Even now, more than a decade after the end of Operation Flood, the NDDB, currently chaired by Amrita Patel, continues to expand India's dairy development, and the program's cooperative system gives millions of rural producers the opportunity to use dairying as a way out of poverty and hunger. The dairy cooperative network continues to grow (Figure 16.3), as do production, marketing, and innovation in the milk sector: presently more than 13 million Indian farmers, including 3.7 million women farmers, belong to India's thousands of village-level dairy cooperatives (NDDB 2009). The average daily procurement of dairy products has reached 21.5 million liters, and annual production has reached more than 100 million tons (Figure 16.4). The daily per capita availability of milk is near 250 grams (India, Ministry of Agriculture 2008).

Operation Flood's financial sustainability can be measured in various ways, including its rate of return, its reliance on subsidies, and the durability of the business model. A cost–benefit analysis of dairying interventions showed that investing R 1 in the dairy sector could spawn Rs 3 worth of employment (Shah 2000). A study of two milk sheds by Punjab Agricultural University showed that in 1994–95 the average gross revenue was Rs 9.30 per liter, with a gross margin of Rs 3.61 per liter and a 33 percent net return (Rs 3.06 per liter). The study included labor, capital, and land costs. The rate of return was high (45 percent), and the payback period was low (25 years) (Candler and Kumar 1998).

Figure 16.3 Growth in India's dairy cooperatives, 1970–2006

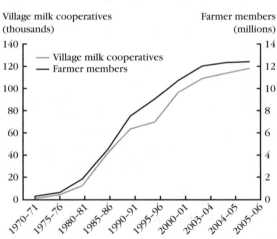

Sources: Aneja (1994); National Dairy Development Board (2007, 2009).

Figure 16.4 India's milk production since Operation Flood, 1997–2007

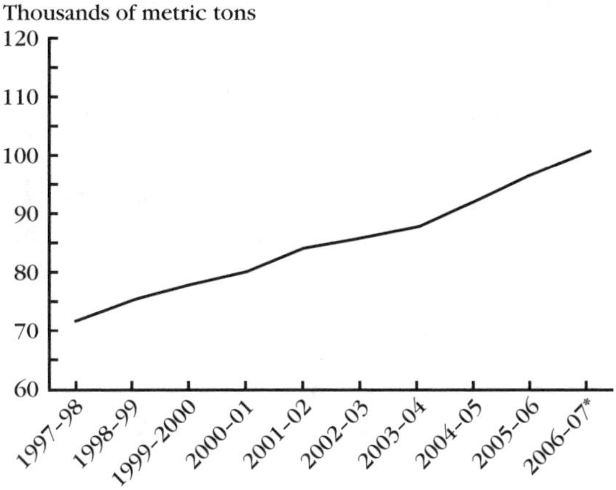

Thousands of metric tons

Source: India, Ministry of Agriculture (2010).
Note: * signifies that the data for this period are projected rather than actual.

Food subsidies, in the form of donated dairy commodities, were an integral part of Operation Flood in all three stages, but the monetization of these commodities only partially financed this program. The donated commodities were marketed to dairies at the then-prevailing market rates, and the funds generated were used to set up cooperatives and other parts of the program. Doornbos, van Stuijvenberg, and Terhal (1987) argue that the difference between the price rural milk producers were willing to accept and the price urban consumers were willing to pay was too small to cover the expenses associated with the processing and marketing of milk; without commodity aid, India's dairy industry would have faced an unpleasant adjustment process. Other financing came from member equity and reserves generated from cooperatives' activities. In the end, Operation Flood achieved its ultimate objective —not financial self-sufficiency per se but self-sufficiency in milk production.

Operation Flood also appeared to have a durable business model. The architects of the program continuously analyzed the rising demand for dairy products in India and devised a dairying program to meet this demand. The project's design included market incentives for cooperatives, infrastructural development, and economies of scale. Finally, Operation Flood focused on ensuring efficiency in supply chain management, quality control by cooperatives, and effective markets for both inputs and services.

The environmental effects of Operation Flood have not been closely analyzed, but some observations can be noted. Anecdotal evidence suggests that the rural production of milk may have helped reduce the cattle population in major cities, thereby improving the environments of urban dwellers, although the extent to which this actually happened is unclear (Jul 1988). Another environmental benefit of traditional Indian dairy production (emphasized under Operation Flood) arises from the symbiotic relationship between crop and livestock production: cattle feed on crop residues and by-products and in turn provide animal traction and fertilizer to replenish the soil (Steinfeld 2003).

Although the cooperative movement has steadily increased over the past several decades in India, its role in the dairy sector is not guaranteed. Since the economic liberalization of the early 1990s, cooperatives, which were long protected, have faced challenges from the private sector. For example, licensing restrictions, which affect the competition faced by cooperatives, have fluctuated over time. Government interference in cooperatives, a lack of strong member equity, inefficiencies and diseconomies resulting from the three-tiered structure, and a lack of professionalism among cooperative managers all present challenges and opportunities for the continued growth of the dairy cooperative movement. Although the program itself followed a sustainable model, the long-term sustainability of the cooperative way of dairying in India is unknown. Various political, social, environmental, and economical factors will determine the future of cooperatives and of dairying more broadly.

Lessons Learned

Operation Flood met its goal of setting the foundation for an enduring cooperative dairy sector that would, in turn, ensure India's long-term self-reliance in milk with maximum benefits to dairy producers. It successfully created a stable, remunerative market for producers, who continued to invest in expanded production and improved productivity to meet the growing demand for dairy products. Some of the underlying principles of Operation Flood are applicable to similar interventions in other settings or with other commodities in India. Imitations have already been spawned. In India alone, the Anand cooperative model is being replicated for other products, including oilseeds, trees, rural electricity, and fish (Shah 1996). The model is also being followed in other Asian countries, such as China, the Philippines, and Sri Lanka (Ali and Bhargava 1998). World Bank President Robert Zoellick recently announced that the World Bank will try to apply some of these best practices in Africa, for example, applying the low-input, low-output Anand model in Tanzania and Uganda. Although the list is not exhaustive, the following are some of the imperative lessons to be learned from Operation Flood.

Use Aid for Development

Although food aid is generally used primarily for humanitarian purposes, Operation Flood used donated commodities to produce dairy products for prime markets and to generate funding to finance the creation of a dairy cooperative infrastructure. This system illustrates how food aid can be used to generate increased levels of domestic production. Moreover, it highlights the foresight of several leaders, who realized what could happen if new avenues for dairy development were not pursued: India's dairy sector could have been overrun by mass quantities of inexpensive dairy imports. The use of food aid as a development investment was also an anti-inflationary measure that helped ensure a stable market and limit price fluctuations.

Invest in Local Markets

Operation Flood was designed to strengthen not only dairy production but also dairy markets. Milk was a highly marketable commodity in India, and the architects of Operation Flood continuously analyzed the demand for livestock products and devised a dairying program to meet this demand. The intervention used market incentives and infrastructural development to establish a successful business model based on cooperatives. By relying on market pull to increase production and by restricting key imports, it strengthened rather than disrupted domestic markets.

Support Collective Action

Cooperatives and the various innovations that took place in Operation Flood also seemed to raise awareness of the power of collective action. The cooperatives' organizational structure minimized petty political differences and strengthened a spirit of cohesion. Local ownership—dairy producers' control over and ownership of the resources throughout the organizational structure—is partly behind this grassroots mobilization.

Adopt Complementary Crop and Livestock Methods

Livestock can convert crop residues and by-products into milk without increasing pressure on the land. Operation Flood showed that significant livestock investments can be made without interfering with land and resource demands for crop production. Milk production was highly integrated into farming in a virtuous circle: an environmentally sustainable use of energy and nutrients.

Envision Creative Structures

Through Operation Flood, the NDDB revolutionized how the dairy industry was perceived and organized. It concentrated on a single primary product and employed a compact, vertically integrated value chain encompassing every stage from primary producer to final consumer. Horizontal integration, which brought inputs, extensions, and services into the same program, helped ensure that the benefits of economies of scale were available to each producer. The cooperative infrastructure, significantly expanded and strengthened under the program, made it easier for participants to adopt and use products and processes. For example, a strong linkage was established between milk production and milk procurement agencies.

Invest in Evaluations

Analyzing Operation Flood's impact is difficult. Data are lacking, and the literature on the intervention abounds with biased evaluations and methodological flaws. Studies must be viewed cautiously because sample sizes may be small or varying, baseline dairy data may be missing, some causal factors may be unaccounted for, and poor research designs may influence the results (Fulton and Bhargava 1994). Although some evaluations of Operation Flood are faulty, they do not necessarily indicate that Operation Flood was also faulty. It is true, however, that the lack of independent metaevaluations and inconsistencies among research methodologies for micro-level impact studies make it difficult to assess the impact of Operation Flood on India's dairy development or food security.

Conclusions

Despite the lack of rigorous evaluations of the intervention, it is certain that Operation Flood laid the foundation for a widespread cooperative movement and ensured cooperative members a regular, remunerative return on their investments in dairy. Although Operation Flood offers a useful model for organizing the dairy sector, any new program must be designed with special attention to the particularities of its environment. Local political, social, economic, and environmental dynamics will all play a role in how an intervention is replicated and how it affects the new beneficiaries. For example, the Anand model succeeded because it provided a market for rural farmers, but the implementation of the model varied slightly from place to place because of differences in demand and in the distance between cooperatives. This experience underlines the importance of synchronizing all managerial and operational processes with marketing capabilities before the Operation Flood model is employed (Ali and Bhargava 1998). Nonetheless, attempts to replicate the cooperative model in Pakistan and Sri Lanka were not successful (Candler and Kumar 1998).

The timing of an intervention can also be an important factor in its success, as it was for Operation Flood. Postindependence India was going through dramatic social changes, and the promotion of animal husbandry complemented landholding and farming at the time. Cooperative farming was not a threat but rather a desired means of creating space for representative democracy to emerge at the village level (Patel 2003). In part, this intervention was also a multiplier effect of the Green Revolution, which developed the crop sector and agricultural infrastructure that set the enabling conditions for dairy development. For example, improvements in irrigation and fertilizer use that resulted from the Green Revolution increased the year-round availability of fodder (Staal, Pratt, and Jabbar 2008a, 2008b). Operation Flood is not necessarily the best approach to developing a dairy industry; however, its positive impact on India's rural development is undeniable. Lessons from Operation Flood are vital: there is a new focus on addressing protein and micronutrient deficiencies in developing countries through livestock, and in many countries the demand for livestock products is rapidly growing. This livestock revolution allows smallholder farmers to benefit economically from expanding markets and provides their families with energy-dense calories and micronutrients (Delgado 2003).

Overall, the lesson of Operation Flood is one of guarded optimism (Hindu Business Line 2009; Scholten and Basu 2009). New challenges, such as rising competition from investor-owned firms, will continue to emerge and must be addressed. Ultimately, however, Operation Flood established a reliable, profit-generating market for smallholder farmers that engendered confidence and increased investment in the dairy sector. The result was greater production and productivity that helped meet the growing demand for dairy products and in turn enhanced the diets of millions of Indians.

References

Achaya, K. T., and V. K. Huria. 1986a. Rural poverty and Operation Flood. *Economic and Political Weekly* 21 (37): 1651–1656.

————. 1986b. Operation Flood: A realistic view. *Economic and Political Weekly* 21 (51): 2241–2242.

Alderman, H. 1987. *Cooperative dairy development in Karnataka, India: An assessment.* Research Report 64. Washington, D.C.: International Food Policy Research Institute.

Ali, A. I., and M. Bhargava. 1998. Marketing capability and performance of dairy cooperatives in India. *ABI/INFORM Global INFOR* 36 (3): 129–141.

Alvares, C. A. 1985. Operation Flood: The white lie. In *Another revolution fails: An investigation into how and why India's Operation Flood Project, touted as the world's largest dairy development programme, funded by EEC, went off the rails.* Jawahar Nagar, New Delhi: Ajanta.

Aneja, R. P. 1994. *Dairying in India: A success story.* APAARI Publication 1994/4. Bangkok, Thailand: Asia-Pacific Association of Agricultural Research Institutions.

Aneja, R. P., and B. P. S. Puri. 1997. Operation Flood ends. In *Dairy India 1997,* ed. P. R. Gupta. New Delhi: Dairy India Yearbook.

Atkins, P. J. 1988. Rejoinder India's dairy development Operation Flood. *Food Policy* 13 (3): 305–312.

————. 1989. Operation Flood: Dairy development in India. *Geography* 74 (324): 259–262.

————. 1993. Towards a macro-spatial interpretation of Operation Flood. In *Geographical studies and Japan,* ed. J. Sargent and R. Wiltshire. Folkestone, Kent, U.K.: Japan Library.

Banerjee, A. 1994. Dairying systems in India. *World Animal Review* 79 (2): 8–15.

Baviskar, S., and P. Terhal. 1990. Internal constraints and external dependence: The EEC and Operation Flood. In *Resources, institutions, and strategies: Operation Flood and Indian dairying*, ed. M. Doornbos and K. N. Nair. New Delhi: Sage.

Bhide, S., and S. K. Chaudhari. 1997. Cooperative producers: Their productivity and income. In *Dairy India 1997*, ed. P. R. Gupta. New Delhi: Dairy India Yearbook.

Birthal, P. S. , V. K. Taneja, and W. Thorpe, eds. 2006. *Smallholder livestock production in India: Opportunities and challenges.* Proceedings of an international workshop held by NCAP-ICAR and ILRI, January 31–February 1, in New Delhi. New Delhi and Nairobi, Kenya: NCAP (National Centre for Agricultural Economics and Policy Research), ICAR (Indian Council of Agricultural Research), and ILRI (International Livestock Research Institute).

Candler, W., and N. Kumar. 1998. *India: The dairy revolution—The impact of dairy development in India and the World Bank's contribution.* Washington, D.C.: World Bank.

Chothani, A. A. 1989. Operation Flood and the national milk grid. *Review of International Cooperation* 82 (2): 23–31.

Delgado, C. 2003. Rising consumption of meat and milk in developing countries has created a new food revolution. *Journal of Nutrition* 133 (11, supp. 2): 3907S–3910S.

Doornbos, M., and L. Gertsch. 1994. Sustainability, technology, and corporate interest: Resource strategies in India's modern dairy sector. *Journal of Development Studies* 30 (3): 916–950.

Doornbos, M., P. van Stuijvenberg, and P. Terhal. 1987. Operation Flood: Impacts and issues. *Food Policy* 12 (4): 376–383.

Doornbos, M., F. van Dorsten, M. Mitra, and P. Terhal. 1990. *Dairy and development: India's Operation Flood.* New Delhi: Sage.

FAO (Food and Agriculture Organization). 1981. Terminal evaluation report on Project India 618— "Milk marketing and dairy development" (Operation Flood I). World Food Programme, Rome.

————. 2009. FAOSTAT. <http://faostat.fao.org>. Accessed August 20, 2010.

FAO Information Division. 1978. *World Food Programme in India: The white revolution.* Rome.

Fulton, J., and M. Bhargava. 1994. The results of a marketing intervention: Dairy cooperatives in India. *Journal of International Food and Agribusiness Marketing* 6 (1): 33–58.

George, S. 1990. Operation Flood and centralised dairy development in India. In *Resources, institutions, and strategies: Operation Flood and Indian dairying*, ed. M. Doornbos and K. N. Nair. New Delhi: Sage.

Guha, H. 1980. Operation Flood II: Some constraints and implications—A comment. *Economic and Political Weekly* 15 (17): 795–796.

Gupta, P. R. 1997a. Operation Flood: The third phase. In *Dairy India 1997*, ed. P. R. Gupta. New Delhi: Dairy India Yearbook.

————, ed. 1997b. *Dairy India.* 5th ed. New Delhi: P. R. Gupta.

Hindu Business Line. 2009. NDDB to provide support for dairy sector in Africa. *Business Daily,* February 11.

India, Ministry of Agriculture. 2008. *Annual report 2007–2008.* New Delhi: Ministry of Agriculture, Department of Animal Husbandry, Dairying, and Fisheries.

————. 2010. National Dairy Development Board website. <http://www.nddb.org/index.html>. Accessed August 23, 2010.

India Times. 1999. The man behind "white revolution." August 20, 51.

Jul, M. 1988. Comments on India's Operation Flood. Dairy Impact Study 3. Department of Food Preservation, Royal Veterinary and Agricultural University, Frederiksberg, Denmark.

Kaye, L. 1988. The white revolution: India's milk production and use soar. *Far Eastern Economic Review* 139 (12): 112–113.

Kurien, V. 1996. The AMUL dairy cooperatives: Putting the means of development into the hands of small producers in India. In *Reasons for hope: Instructive experiences in rural development,* ed. A. Krishna, N. Uphoff, and M. J. Esman. West Hartford, Conn., U.S.A.: Kumarian Press.

————. 2004. India's milk revolution: Investing in rural producer organizations. A case study from Reducing Poverty, Sustaining Growth—What Works, What Doesn't, and Why: A Global Exchange for Scaling Up Success. Presented at the Scaling Up Poverty Reduction: A Global Learning Process and Conference, May 25–27, in Shanghai, China.

Mellor, J. W. 1999. *Faster, more equitable growth—The relation between growth in agriculture and poverty reduction.* Agricultural Policy Development Project Research Report 4. Cambridge, Mass., U.S.A.: Abt Associates.

————. 2003. Agricultural growth and poverty reduction—The rapidly increasing role of smallholder livestock. Keynote address at the FAO Livestock and Livelihoods International Workshop, November 10–12, in Anand, India.

Mergos, G., and R. Slade. 1987. *Dairy development and milk cooperatives: The effects of a dairy project in India.* World Bank Discussion Paper 15. Washington, D.C.: World Bank.

Munshi, K. D., and K. S. Parikh. 1994. Milk supply behavior in India: Data integration, estimation and implications for dairy development. *Journal of Development Economics* 45: 201–223.

Nair, K, N. 1985. White revolution in India: Facts and issues. *Economic and Political Weekly* 20 (June 22–29): A89–A95.

Narayanaswamy, B. K. 1996. Small farmers' development co-operative as a desirable model. In *Rediscovering co-operation,* vol. 1: *Bases of co-operation,* ed. R. Rajagopalan. Anand, India: Institute of Rural Management.

NDDB (National Dairy Development Board). 2005. *Annual report, 2004–2005.* Anand, India.

————. 2007. *Annual report, 2006–2007.* Anand, India.

————. 2008. Achievements of dairy cooperatives: Facts at a glance. <http://www.nddb.org/achievement/ataglance.html>. Accessed April 20, 2009.

————. 2009. National statistics: Production in India. <http://www.nddb.org/statistics/milkproduction.html>. Accessed April 20, 2009.

Nehru, S. 2005. India: Revolutionizing the white revolution. *Women's Feature Service* (September 5).

Patel, A. 2003. Smallholder dairying in India: Challenges ahead. Inaugural address at the FAO Livestock and Livelihoods International Workshop, November 10–12, in Anand, India.

————. 2007. Securing the future of dairying in a rapidly changing environment. J. Raghotham Reddy Memorial Lecture, October 26. Farm and Rural Science Foundation, Acharya N. G. Ranga Agricultural University, Rajendranagar, Hyderabad, Andhra Pradesh, India.

Perumal, M., P. S. Mohan, and M. Suresh. 2007. *Dairy development and income distribution in India.* New Delhi: Abhijet Publications.

Population Reference Bureau. 2008. 2008 world population data sheet. <http://www.prb.org/Publications/Datasheets/2008/2008wpds.aspx>. Accessed August 18, 2009.

Scholten, B., and P. Basu. 2000. *Co-operativization liberalization and dairy industry in India.* Rajsthan, India: ABD.

————. 2009. White counter-revolution? India's dairy cooperatives in a neoliberal era. *Human Geography* 2 (1): 17–28.

Shah, D. 1996. *Catalysing co-operation: Design of self-governing organisations.* New Delhi: Sage.

Shah, T. 1995. *Making farmers' co-operatives work: Design, governance, and management.* New Delhi: Sage.

Sharma, M., and U. Vanjani. 1993. When more means less: Assessing the impact of dairy "development" on the lives and health of women in rural Rajasthan (India). *Social Science and Medicine* 37 (11): 1377–1389.

Sharma, V., and A. Gulati. 2003. *Trade liberalization, market reforms and competitiveness of Indian dairy sector.* MTID Discussion Paper 61. Washington, D.C.: International Food Policy Research Institute.

Shukla, R. K., and S. D. Brahmankar. 1999. *Impact evaluation of Operation Flood on rural dairy sector.* New Delhi: National Council of Applied Economic Research.

Singh, K., and V. M. Das. 1984. *Impact of Operation Flood I at the village level.* Research Report 1. Anand, India: Institute of Rural Management.

————. 2008b. Dairy development for the resource poor—Part 3: Pakistan and India dairy development case studies. PPLPI Working Paper 44-3. Addis Ababa, Ethiopia: International Livestock Research Institute.

Steinfeld, H. 2003. Economic constraints on production and consumption of animal source foods for nutrition in developing countries. *Journal of Nutrition* 133 (11, supp. 2): 4054S–4061S.

Terhal, P., and M. Doornbos. 1983. Operation Flood: Development and commercialization. *Food Policy* 8 (3): 235–239.

Thirunavukkarasu, M., R. Prabaharan, and C. Ramasamy. 1991. Impact of Operation Flood on the income and employment of rural poor—Some micro-level evidences. *Journal of Rural Development* 10 (4): 417–425.

Tikku, D. 2003. Indian dairy sector and the National Dairy Development Board: An overview. Address at the FAO Livestock and Livelihoods International Workshop, November 10–12, in Anand, India.

Verhagen, M. 1990. Operation Flood and the rural poor. In *Resources, institutions and strategies: Operation Flood and Indian dairying,* ed. M. Doornbos and K. N. Nair. New Delhi: Sage.

———. 2006. *Agriculture investment sourcebook.* Washington, D.C.: Agricultural Rural Development, World Bank.

Chapter 17

Rich Food for Poor People: Genetically Improved Tilapia in the Philippines

Sivan Yosef

During the past decade, the overuse of capture fisheries has caused approximately 52 percent of global marine fish stocks to become fully exploited (FAO 2007). In contrast to the decline of capture fisheries, aquaculture has skyrocketed. From 1950 to 2004, aquaculture experienced an 8.8 percent annual growth rate, making it the single fastest-growing food-producing sector in the past three decades (FAO 2006; Acosta and Gupta 2009). In 1987 the Food and Agriculture Organization of the United Nations (FAO) estimated that aquaculture production (excluding seaweed) accounted for a mere 11.1 percent of world fish production; by 1999 this figure had ballooned to 26.2 percent, or more than 32.3 million tons (Ahmed and Lorica 2002). Over the next decade the demand for fish is projected to increase by an additional 37 million tons. Aquaculture may be well poised to meet this demand, with a projected contribution of 41 percent of total fish production by 2020 (Delgado et al. 2003). It is estimated that if just 5 percent of the area deemed suitable for aquaculture in Africa were put to use, enough extra fish could be produced to feed the growing population on the continent until 2020 (WorldFish 2007).

The history of the genetic improvement of tilapia is the story of how a coalition of governments, national and international agricultural research centers, regional networks, and private actors worked together to produce an affordable and hardy fish that could meet the needs of the poor. Improved tilapia is an example of aquaculture's unique ability to meet the projected global demand for fish and, in the process, achieve food security for millions of people around the world. This chapter examines the substantial contribution and impact of tilapia genetic improvement to fish farming in the Philippines.

The Global History of Tilapia

Capture fisheries harvest fish in natural environments, whereas fish culture, or fish farming, is the practice of cultivating fish in a confined water area. Fish culture is similar to agriculture and livestock farming systems in the sense that it often involves applying organic manures and inorganic fertilizers to feed, breed, and care for the health of fish (Kumar 1992).

Fish farming and aquaculture represent one of the most important contributions of the developing world to global food security. Low-income food-deficit countries provide nearly 85 percent of the world's aquaculture production, with 91 percent of all aquaculture originating in Asia (El-Sayed 2006; FAO 2009a).[1] In absolute numbers, China is the leader in aquaculture, producing more than 34 million tons in 2006—a 78 percent increase from 1997 (Table 17.1).

Fish culture can generally be divided into four stages: the production of broodstock, or sexually mature fish; the production of fish eggs; the rearing of juveniles; and growout, the stage at which fish are readied for market. An aquaculture facility, such as a hatchery or fish farm, may include all production stages or focus on just one.

Tilapia, coming from the Tswana word for fish, *thiape,* is the name of a group of warm-water bony fishes originating in Africa and the Middle East's Jordan Valley (Guerrero 2008). The global expansion of tilapia began in earnest in the 1970s but was tempered by a lack of focus on genetics and selective breeding (CGIAR 2006). The former International Center for Living Aquatic Resources Management (ICLARM, now the WorldFish Center) established an internationally recognized selective breeding program. Through this and other similar efforts, 19 of approxi-

Table 17.1 Top aquaculture-producing countries, 1997 and 2006 (metric tons)

Country	1997	2006
China	19,315,623	34,429,122
India	1,864,322	3,123,135
Vietnam	322,378	1,657,727
Thailand	539,817	1,385,801
Indonesia	662,547	1,292,899
Bangladesh	485,864	892,049

Source: FAO (2006).

mately 100 tilapia species have been cultured or cultivated; of these, the ones that hold the most prominence for aquaculture are the Nile tilapia (*Oreochromis niloticus*), the Mozambique tilapia (*Oreochromis mossambicus*), and the blue tilapia (*Oreochromis aureus*) (Guerrero 2008).

Until the 1940s, global tilapia farming was marked by experimental pond cultures in Africa and some commercial activity in Asia. The 1950s and 1960s saw the emergence of subsistence-level farming of *O. niloticus* and other species in Cameroon, French Equatorial Africa, and Nigeria. *O. mossambicus* was brought to the Philippines and the United States, among other countries, and sex reversal (discussed in detail later in this chapter) and sterilization experiments were carried out in Israel, Malaysia, and the United States.

The 1970s were marked by four major developments: the release of hormonal sex reversal technology, the commercialization of tilapia cage culture, the use of *hapas* (floating net enclosures) in breeding, and the rise of Taiwan in the commercial hybrid tilapia market (Guerrero 2008). In the 1980s tilapia farming exploded in Southeast Asia, with three international conferences on tilapia, the commercialization of hormonal sex reversal technology, the development of breeding technologies for *O. niloticus,* and the emergence of the Philippines as the largest tilapia-producing country in the world (Guerrero 2008).

The 1990s established tilapia farming's important role in world aquaculture: tilapia demand skyrocketed in the United States, industrial tilapia farming emerged in Africa and South America, and widespread genetic improvements were made to *O. niloticus* (Guerrero 2008).

Today tilapia is one of the top 10 fish species, contributing more than 1 million metric tons to global fish production (Guerrero 2008). A total of 1.2 million metric tons of tilapia comes from the Association of Southeast Asian Nations (Bartley et al. 2004), although this figure is widely believed to be an underestimate because many tilapia-farming countries do not report their production to the FAO (FAO 2009c). Additionally, an untold amount is consumed within producer households, never entering market chains.

Aquaculture in the Philippines

In 2000 the Philippines' fisheries sector employed more than 1 million people, or 12 percent of the total rural labor force (Garcia, Dey, and Navarez 2005). From 1997 to 2003, while capture fishing experienced low growth rates because of dwindling fish resources, aquaculture saw an 8 percent production surge in the Philippines (Garcia, Dey, and Navarez 2005). Aquaculture's takeoff in the Philippines can be attributed to two main factors. First, environmental degradation has necessitated such development. A study of nine Asian countries, including the Philippines, concluded that in-shore demersal fish (or bottom feeders) stocks had declined by up to 44 percent since the 1970s (Silvestre et al. 2003). Faced with the depletion of fishery resources and population growth, aquaculture was framed as a sustainable alternative to traditional capture fisheries (Acosta et al. 2006). Second, fish are an integral part of the national diet of the Philippines: the average Filipino consumes 28 kilograms of fish every year compared with the world average of 16 kilograms (FAO 2009b). Thus the demand for a reliable source of fish constantly drives new technologies that can increase yields and meet the nutritional demands of a growing population. Aquaculture now makes up 40 percent of the national fisheries sector (Garcia, Dey, and Navarez 2005). The FAO estimated that the 1998 production of all aquaculture fish was 72,000 tons; the 2007 production stood at 2.2 million tons (Philippines Bureau of Agricultural Statistics 2009).

National consumption of fish and fishery products stood at 2.3 million metric tons a year during 1997–2001; during that same period, 0.3 million metric tons of the country's total fish production was exported, making the Philippines a net fish exporter (Garcia, Dey, and Navarez 2005). As of 2001, fishery products contributed to 3.7 percent of the country's gross domestic product.

The Philippines farms freshwater fish in ponds or cages. Whereas ponds boast an average area of 4.91 hectares, cages have an average size of 1.26 hectares. The size of the enclosure is not necessarily an indicator of the operator's wealth. A very small cage farm, for example, can be managed with industrial-level inputs, making it accessible only to farmers with high levels of capital. In 1995–96, some 75–99 percent of pond and cage farms were owned rather than rented, and 71–87 percent were privately operated. Ponds offered net returns of $853 per unit area, while cages offered $263 (Dey et al. 2000b).

Tilapia Breeding in the Philippines

Tilapia is the main freshwater fish species cultured in the Philippines, accounting for 63 percent of total freshwater aquaculture production in 2000 (Dey et al. 2005). Though tilapia production is still surpassed by the production of the traditional milkfish (Figure 17.1), the latter is raised mostly in brackish water, leaving tilapia the leader in freshwater cultivation.

In general, tilapia is considered so versatile across different environments that it has been dubbed the "aquatic chicken" (Acosta and Gupta 2009). Throughout the country, tilapia is reared in freshwater or brackish-water ponds or cages. The practice is characterized as an extensive or semi-intensive monoculture system of farming.[2] Although fish farming can be combined with some other activities, such as rice farming, stand-alone fish culture has always been the tradition in the

Figure 17.1 Milkfish and tilapia production in the Philippines, 1990–2008

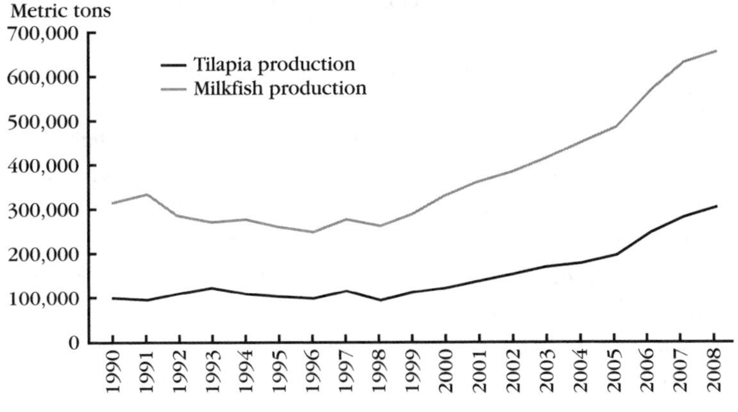

Source: Philippines Bureau of Agricultural Statistics (2009).

Philippines (Guerrero 1994, 1996). The method has returned low production yields (3,599 kilograms per hectare) compared with practices in other countries, but production in the Philippines had a comparatively high value of 3,421 kilograms per hectare (Eknath and Acosta 1998).

Freshwater ponds, located mostly in the central Luzon region, account for approximately 50 percent of tilapia production, while cage culture, mostly associated with the southern Luzon region, represents 36 percent of production. Brackish-water culture accounts for the remainder (Guerrero 1996). Tilapia that are grown in cages have an average size of 175 grams, while those reared in ponds weigh 130 grams (Dey et al. 2000b). Small-scale private hatcheries produce the majority of tilapia seed, with the supply of fingerlings estimated at 600 million in 1996 (Guerrero 1996).

The history of tilapia farming in the Philippines began in the 1950s, when *O. mossambicus* was introduced from Thailand. Publicized as a wonder fish capable of addressing the low supply of animal protein in the region, the Mozambique tilapia was seemingly easy and affordable to breed (Ling 1977; Guerrero 1994). Problems emerged, however, related to inbreeding, overcrowding, invasiveness, and lack of marketability (Yap, Baluyot, and Pavico 1983; Guerrero 1994). Interest in tilapia farming waned and was not revived until a decade later.

In 1974 the Government of the Philippines launched a two-year research program at the Freshwater Aquaculture Center of Central Luzon State University (FAC-CLSU) (Guerrero 1994). Based on early research showing that male tilapia can grow faster than females, the program focused on monosex male culture and sex reversal of females through fry (young fish) hormone treatment. Alongside breeding techniques, other technologies included floating net enclosures for breeding and floating cages for feeding. Upon development of new varieties, the Government of the Philippines transferred the finished products to both resource-poor rural communities and potential large-scale commercial farmers through three main programs (Table 17.2).

These programs provided farmers with technical assistance through technology demonstrations, extension agents, provincewide workshops, and opportunities for collaboration with researchers. Poor rural tilapia farmers working with ponds, cages, and paddy fields were also given bank credit for the first time, enabling them to gain access to tilapia technologies (Guerrero 1994). Efforts to reach the commercial sector, on the other hand, included using private corporations to pilot technology, offering financing and credit through development banks, and providing incentives to farmer nongovernmental organizations and cooperatives for the adoption of new technology (Guerrero 1994).

Meanwhile, an entirely different strain, Nile tilapia, received its first introduction to the Philippines. Native to Africa, Nile tilapia (*O. niloticus*) was introduced to Asian countries in the 1970s to expand small-scale aquaculture (Gupta and Acosta 2004). Even though it showed promise, issues of insufficient fish seed supply, stagnant production, and poor fish growth plagued Nile tilapia farmers (Pullin 1980). Two possible causes were linked to these problems. Some claimed that the Nile tilapia strain introduced to Asia descended from a small number of fish, which had led to inbreeding and the proliferation of undesirable genetic traits (Pullin and Capili 1988). Others reported that the decline in Nile tilapia could be attributed to unintentional hybridization caused by escaped Mozambique tilapia entering ponds and breeding with imported farmed strains of Nile tilapia (Taniguchi, Macaranas, and Pullin 1984; Macaranas et al. 1995).

Table 17.2 Philippine programs with a focus on technology transfer to small-scale farmers, 1971–1980s

Project	Time frame	Partners
Inland Fisheries Project	1971–76	Freshwater Aquaculture Center of Central Luzon State University; Brackish Water Aquaculture Center of the University of the Philippines; National Science Development Board; U.S. Agency for International Development (USAID)
Freshwater Fisheries Development Project	1979–83	Bureau of Fisheries and Aquatic Resources; USAID
National Self-Reliance Movement	1980s	Ministry of Human Settlement

Source: Guerrero (1994).

The Genetic Improvement of Farmed Tilapia Project

In 1988 the Genetic Improvement of Farmed Tilapia (GIFT) project was launched as a starting point for genetic improvement of tropical finfish around the world (Gupta and Acosta 2004). Although part of the impetus for the project was the dissatisfaction surrounding previously introduced strains of tilapia, the real aim of the initiative was to build capacity for genetic breeding by supporting national breeding programs with a high-quality, heterogeneous base stock of fish. The project involved a range of partners, including ICLARM, the Norwegian Institute of Aquaculture Research, and the Philippine national fisheries bureaus and centers, including the Bureau of Fisheries and Aquatic Resources (BFAR), the previously mentioned FAC-CLSU, and the Marine Science Institute of the University of the Philippines. On the donor side, the project was jointly funded by the United Nations Development Programme (UNDP) and the Asian Development Bank (ADB) (El-Sayed 2006). The methodology for GIFT was based on the success of selective breeding programs for salmon and trout established in Norway in the 1970s.[3]

Phase 1: Finding the Right Fit

The project began by comparing the performance of existing Asian *niloticus* farmed strains (which had originally come from Africa) and imported wild fish grown communally in 11 different environments representative of Philippine aquaculture (Eknath et al. 1993). The imported strains were collected from Ghana, Egypt, Kenya, and Senegal and transferred to the Philippines.

In the first GIFT experiment, the wild strains in general outperformed the farmed *O. niloticus* strains (with the exception of the Ghana tilapia). Additionally, the Egyptian Nile and Kenyan Rift Valley strains outperformed the West African strains in most test environments. A second experiment, which evaluated fish in eight different environments, confirmed these results (Bentsen et al. 1998). Although this second study showed that a hybrid between the Egyptian and Kenyan strains would represent a 10 percent improvement over the best pure strain, the logistics associated with a crossbreeding approach would be challenging. ICLARM scientists thus pursued selective breeding of *O. niloticus* rather than crossbreeding—that is, they would choose *O. niloticus* parents for breeding based on certain desirable characteristics rather than trying to breed them with another strain—with the expectation that this approach would improve tilapia performance more than would a crossbreeding program within a few generations (Longalong, Eknath, and Bentsen 1999). *O. niloticus* was chosen for its short generation time, ability to tolerate shallow and turbid waters, high disease resistance, and flexibility for fish culture in many different farming systems (Pullin 1983, 1985; Eknath 1995; Gupta and Acosta 2001). The expectation that GIFT fish would be distributed throughout the region was woven into the original project design.

Scientists constructed a synthetic base population from the 25 best-performing wild and farmed strains experimented with earlier. By 1993 three generations of selection had been completed, and preliminary results showed that the selected fish grew much faster than local tilapia strains and had higher survival rates (Longalong, Eknath, and Bentsen 1999). Tilapia farmers were included as stakeholders in on-farm experiments, which were generally successful (Acosta and Gupta 2009).

Based on successive selective breeding rounds of *O. niloticus,* the GIFT project eventually yielded genetic improvements of 7.1 percent genetic change over nine generations of fish, or a 64 percent cumulative increase in tilapia growth over the base population (Ponzoni et al. 2008). Both higher and lower figures have been reported (Eknath and Acosta 1998), so comparisons between strains differ across time, location, and farming system and must be interpreted in context (ADB 2006).

The first selective breeding program designed for *O. niloticus* at both the national and international levels, GIFT succeeded in overcoming many of the obstacles faced by previous improvement programs for this species (Eknath et al. 1991, 1993). A variety of other GIFT-derived tilapia strains have been released within the Philippines. In 2000, for example, the Government of the Philippines developed GET EXCEL by combining an improved strain of Nile tilapia with a rotational mating scheme—a system of mating that prevents inbreeding (Tayamen and Abella 2004).[4] Two years later BFAR's National Freshwater Fisheries Technology Centre formulated GET EXCEL 2002 by combining strain crosses (such as those from the GIFT, Egypt, and Kenya strains) with rotational mating. Preliminary results have shown that this strain grows faster and has better chances for survival than do other improved market Nile tilapia strains. These results prompted the government-led Nationwide Dissemination of GET EXCEL Tilapia initiative, which seeks to replace old tilapia strains with improved ones (El-Sayed 2006).

Alongside the selective breeding technologies showcased through GIFT and GIFT-derived initiatives, other genetic improvements of tilapia have been made in the Philippines. These mainly involve producing all-male tilapia cultures. Tilapia farmers prefer these types of cultures because males are known to grow faster than females and because such a culture addresses the proclivity of tilapia to reproduce excessively (El-Sayed 2006). The first of these methods for producing all males is sex reversal from female to male—converting female fish to male fish through the use of hormones. At least 10 hatcheries currently produce sex-reversed tilapia, accounting for approximately 15 percent of the fingerling supply (Mair et al. 2002). However, many of these hatcheries have not yet succeeded in producing cultures with a population more than 95 percent male, the level at which the culture can truly be deemed "all male." Sex ratios currently hover at lower than 90 percent, and future improvements seem unlikely (Mair and van Dam 1996; Mair et al. 2002).

The YY male method was developed as an alternative to sex reversal technology and involves combining hormonal feminization and progeny (offspring) testing to breed YY male genotypes, which then produce male tilapia when bred with normal females. The technology has been shown to produce a male–female sex ratio of greater than 95 percent in controlled environments, with increased yields of 30–40 percent compared with normal mixed-sex tilapia (Mair et al. 1995). Because hatcheries cannot produce their own YY males, a network has arisen to deliver fingerlings to growers in each production cycle.

"All-male" production methods have several drawbacks. First, injecting tilapia with hormones raises food safety concerns among consumers. Second, producing YY males takes three generations of breeding, meaning that males lag behind genetically by as much as 20 to 45 percent by the time they are ready for use (Ponzoni et al. 2008). Third, the technology requires a high level of sophistication and advanced laboratory facilities, a less than ideal situation for many developing countries. Finally, the dissemination of fingerlings is a mainly passive process, with the beneficiaries limited to accredited hatcheries, currently numbering only 32 (Mair et al. 2002).

Phase 2: Sharing the Wealth

After developing methods for genetically improving Nile tilapia, WorldFish turned its attention to broader goals: dissemination of the GIFT strain; capacity building of national institutions in aquaculture genetics research; genetic, socioeconomic, and environmental evaluation of GIFT; and the facilitation of national tilapia breeding projects (Eknath 1995).

In the Philippines, throughout the life of the GIFT project, GIFT fish were disseminated to farmers through government agencies. The leading agencies responsible for national dissemination included BFAR and FAC-CLSU (Tayamen, Abella, and Sevilleja 2006). BFAR established the Program for Fish Varietal Regeneration, which was further subdivided into two programs—one to upgrade the production capacity of hatcheries affiliated with the Philippine Department of Agriculture and another to sustain genetic advances and to create a national distribution network for improved tilapia strains. In 1990 the Philippine Council for Aquatic Marine Research and Development established the National Tilapia Production Program to further promote tilapia genetic breeding and to distribute improved fish strains to farmers (Sevilleja 2007).

In 1992, amid worries over biosafety and the capacity of many countries to embark on the widespread dissemination of an improved strain, a meeting on International Concerns in the Use of Aquatic Germplasm was held (Eknath 1995). Experts agreed that GIFT fish should be transferred extremely carefully, adhering to standards set forth by relevant international bodies such as the FAO. Any country wishing to import new fish species would be required to sign a material transfer agreement that set rules for the use of new strains (Acosta and Gupta 2009).

To help spread improved tilapia outside of the Philippines, the UNDP provided $65,000 to the WorldFish Center in 1993 to establish the International Network on Genetics in Aquaculture (INGA) as a forum for the exchange of ideas, research methodology, and genetic materials (CGIAR 2006). Based in Penang, Malaysia, INGA has 12 developing-country members across Africa and Asia (ADB 2006).

In 1994 improved tilapia strains were first disseminated by INGA through trials conducted at stations and farm environments in five member countries (Bangladesh, China, the Philippines, Thailand, and Vietnam) (Acosta and Gupta 2009). Performance evaluations showed third-generation GIFT fish consistently outperforming non-GIFT species. In Bangladesh, for example, GIFT strains showed a 78 percent yield gain (Table 17.3). Following other performance evaluations, 133,494 tilapia germplasm were transferred to national agricultural research centers throughout Asia for use in research, breeding, multiplication, and later dissemination to farmers (ADB 2006).

From 1994 to 2003, INGA facilitated 70,913 transfers of GIFT germplasm among member countries (ADB 2006). The network has also focused on capacity building in breeding and genetics in developing countries. Because of INGA's funding woes, the WorldFish Center and its partners have partially taken over some of these activities (Acosta and Gupta 2009).

GIFT strains do not uniformly show spectacular results. For example, Dan and Little (2000) showed that although GIFT fish obtained a significantly larger individual size in cage and pond environments at final harvest than did competing Thai or Viet strains, the growth difference was less pronounced. Ponzoni et al. (2005) reported, however, that even after a few generations of selective breeding,

Table 17.3 GIFT and non-GIFT yields in on-farm trials in select Asian countries

Country	Production system	Non-GIFT strain yield (kilograms per hectare)	GIFT strain yield (kilograms per hectare)	Yield gain (percent)
Bangladesh	Pond	896	1,593	78
China	Cage	310,967	389,346	25
	Pond	4,275	4,645	9
Philippines	Cage	15,285	23,551	54
	Pond	912	1,361	49
Thailand	Pond	2,044	2,829	38
Vietnam	Pond	558	743	33

Source: Dey et al. (2000a).
Note: GIFT means Genetic Improvement of Farmed Tilapia project.

the population still has additive genetic variance that will allow it to improve even further.

At the close of the GIFT project in 1997, genetic material from the ninth generation of improved tilapia was provided to institutional partners for primarily noncommercial use (Acosta et al. 2006). Donor support for the project ended in the same year, and the public sector was charged with finding a way to continue both breeding and outreach efforts (Acosta et al. 2006). A nonprofit private foundation called the GIFT Foundation International (GFII) was thus established and set about forming seed production partnerships with private hatcheries throughout the Philippines (Acosta et al. 2006). GFII invited privately owned hatcheries that were able to meet certain requirements to enter hatchery agreements (Rodriguez 2006), and it entered into partnerships with seven of them. By the end of 2001, it had disseminated 522,700 GIFT broodstock to these accredited hatcheries (ADB 2006).

The extension system for improved tilapia in the Philippines remained weak, however (Acosta et al. 2006). Surveys conducted with growout farmers and hatchery operators shortly after the transfer revealed that although farmers were able to obtain genetically improved tilapia products, most farmers got technical advice from suppliers of the improved strain rather than from extension agents. Farmers reported that major suppliers focused on selling fingerlings rather than providing information to farmers on fish breeding, nutrition, and health and on water quality (Acosta et al. 2006).

In 2000 WorldFish received a sample of 60 families of the ninth-generation GIFT strain, which it distributed to its partners in 11 countries of Asia (Gupta and Acosta 2004).[5] The 11 recipients established national breeding programs for the further improvement and dissemination of GIFT fish. Vietnam, for one, has produced and disseminated almost 2 million improved tilapia seed, whereas hatcheries in Thailand produce and disseminate 200 million GIFT fry on an annual basis. WorldFish continues selection work in Malaysia on the ninth-generation GIFT, in collaboration with the national Department of Fisheries (Gupta and Acosta 2004). Currently efforts are under way to improve the national breeding programs' genetic improvement and dissemination activities (Acosta and Gupta 2009).

Phase 3: The Uncharted Territory of Public–Private Partnerships

In 1999, seeking to expand its market and improve its earnings, GFII entered into an agreement with GenoMar, a private Norwegian biotechnology company, thus marking the first entry of a foreign private entity into the story of improved tilapia (Acosta et al. 2006). GFII transferred the dissemination rights of GIFT to GenoMar, which in turn rebranded the strain as GenoMar Supreme Tilapia (GST™). GST™ is currently disseminated through GenoMar's private hatcheries in China and the

Philippines. The Chinese hatchery produces about 280 million fingerlings a year, and the Philippines hatchery produces 30 million fingerlings annually. Dissemination in Africa, Asia, and South America occurs through GenoMar's partner hatcheries (GenoMar 2009).

GST™ is currently marketed as using DNA fingerprinting for the genetic tagging of fish, which makes it easier to identify the strains with the optimal characteristics, and a revolving mating scheme, which allows a generation to be completed after only nine monthly batches (Gjoen 2004). According to GenoMar, it has created a tilapia strain with a high salt tolerance, rapid growth rate, high feed conversion efficiency, and improved disease resistance. GST™ is estimated to have an average genetic gain of 20 percent with every generation, representing a 35 percent increase over conventional breeding methods (GenoMar 2009). GenoMar produces a new generation of GST™ every nine months; in China, experiments show that the strain has grown more than twice as fast as local strains (Acosta and Gupta 2009).

It should be noted that many of GenoMar's claims regarding genetic improvement are controversial. Some critics charge that GenoMar's results are not openly published and thus cannot be verified. GenoMar claims to have made genetic gains in an unusually short period of time, pointing to the need for independent verification of these results.

Public–private partnerships in fisheries are still relatively new, and the partnership between GFII and GenoMar has introduced a host of issues. The most prominent issue is the private ownership and dissemination of public goods. GenoMar, as a for-profit private company, holds exclusive commercial rights to GST™ and all subsequent products created from the 10th-generation GIFT strain (Acosta et al. 2006). For example, as of 2003, GenoMar was developing the 14th generation of GIFT-derived strains, while WorldFish had access only to 9th-generation strains (GAIN 2003). Thus, although WorldFish keeps 9th-generation GIFT fish within the public domain, providing them to governments for research and development (R&D), subsequent versions of the improved tilapia are privately held. Moreover, GIFT/GST™ is commercially distributed in the Philippines solely through private channels (Table 17.4) (Tayamen, Abella, and Sevilleja 2006).

Thus the poverty focus that was the original objective of the GIFT project may have suffered from the shift from public-sector to private-sector control. Acosta et al. (2006) point out that the public and private actors involved in improved tilapia breeding have different goals. For example, whereas public-sector GIFT activities focus on small-scale, subsistence-level farmers, GenoMar focuses on medium to large-scale farmers. The relevance and marketability of GIFT strains to the ultimate end users remain serious concerns (Sevilleja 2006).

Another problem plaguing public–private partnerships in the Philippines is a general lack of coordination between private and government actors. The roles of

Table 17.4 Ownership and distribution of improved tilapia in the Philippines, 2006

Strain	Breeding nucleus ownership	Distribution channel
GIFT/GST™	Private	Private
GET EXCEL	Public	Public–private
YY male	Public	Public–private
FaST	Public	Public–private

Source: Sevilleja (2006).
Note: FaST means a breed developed by the Freshwater Aquaculture Center of Central Luzon State University; GET EXCEL means EXcellent strain that has a Comparative advantage over other tilapia strains for Entrepreneurial Livelihoods projects in support of aquaculture for rural development; GIFT means Genetic Improvement of Farmed Tilapia project; GST means GenoMar Supreme Tilapia.

different actors in breeding and dissemination are unclear. Small private hatcheries, for example, find it difficult to compete with government-run breeding centers and hatcheries, which seem to have boundless resources and infrastructure (Tayamen, Abella, and Sevilleja 2006). The complexity of these issues limits the establishment of such partnerships, thus stalling potential advances in tilapia improvement.

Nonetheless, public–private partnerships can be conducive to the development of the tilapia industry. Specific steps can be undertaken to create enabling environments for strategic, win-win public–private partnerships in the improved tilapia sector. Such steps would include ensuring that public-sector institutions have legal protections regarding ownership of improved germplasm, involving the private sector in dissemination in a gradual and thoughtful way, and defining the roles of public and private actors through sound policies (Acosta et al. 2006). The Philippines has already taken some of these steps. In 2002 private and public stakeholders established the Tilapia Science Center, which convenes a biannual National Tilapia Congress to foster collaboration efforts among key players (ADB 2006). The Department of Agriculture, through BFAR, has also led the establishment of the Tilapia Council of the Philippines to coordinate the different programs under the fishery sector that seek to improve the tilapia industry (Sevilleja 2007).

Monitoring and Evaluation

Dissemination and Evaluation of Genetically Improved Tilapia (DEGITA) was an extension of the GIFT project that helped national partners introduce the GIFT strain into their local fish stocks. Funded by the ADB, WorldFish, and five participating countries, the project aimed to evaluate the biological, socioeconomic, and ecological nuances of GIFT strain production; assess the impact of the strain on different income groups; and distribute the strain to smallholders (CGIAR 2006). Many of the impact data on GIFT and other genetically improved tilapia come from DEGITA studies, helping partners tailor, improve, and strengthen their research and outreach activities over time.

The Impact of the "Aquatic Chicken" in the Philippines

GIFT, GIFT-derived, and other improved tilapia have had a great impact in the Philippines. Improved technologies have worked to further lower tilapia prices, making the fish more accessible to low-income population groups. The tilapia industry has also provided tilapia farmers with employment opportunities and increased incomes and has a record of relative environmental and political sustainability.

Production

In the Philippines, sheer tilapia production numbers have steadily increased over the past three decades, with a notable spike at the turn of the century. In 1980 the Philippines produced 18,540 tons of tilapia; by 1990 this figure had increased to 97,424 tons; by 2007 output was nearly 279,000 tons (Figure 17.2). As noted earlier, the increase in tilapia production in the Philippines has not been as dramatic as in other Asian countries, mostly because of problems with nascent management and dissemination techniques.[6] Still, it can be argued that tilapia holds greater importance in the Philippines, where it is one of two main cultured species in a relatively undiversified fish industry, than elsewhere.

Figure 17.2 Tilapia production in the Philippines, 1980–2007

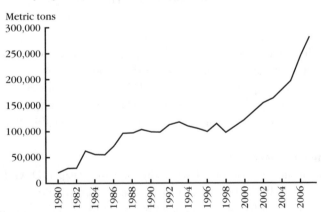

Source: FAO (2009).

The ADB concludes that GIFT and GIFT-derived strains are responsible for most of the increase in tilapia production in the Philippines in the past two decades (Acosta and Gupta 2009). Although data are not available on different strains of tilapia, a 2004 ADB survey of 136 private and public hatcheries showed that the GIFT strain and the GIFT-derived strain GET EXCEL together constituted 67.6 percent of total tilapia seed produced in the country in 2003 (ADB 2006). In terms of tilapia seeds sold, the GIFT and GIFT-derived strains accounted for 66.9 percent of total seed sales, an increase of 3.3 percentage points from 2001. Over this two-year period, the market share of GET EXCEL increased from 44.5 to 48.4 percent, while GST™ remained stable at an average 18.8 percent (Table 17.5).

The Philippines as a Tilapia Exporter

The past two decades have seen impressive growth in global tilapia production (Vannuccini 2001), in response to global demand. Many nonproducing countries and regions, including Central America, Europe, South America, and the United States, have increased their consumption of tilapia. Asia is the main exporter of tilapia, but the Philippines lags behind other Asian competitors (Table 17.6) (El-Sayed 2006). Historically, the national marketable size of 143–200 grams has been much

Table 17.5 Production of different Nile tilapia strains in the Philippines, 2003

Strain	Production[a]	Percentage share
GET EXCEL (GIFT-derived)	421.4	45.7
GST™ (GIFT)	201.6[b]	21.9
FaST (non-GIFT)	187	20.3
GMT (non-GIFT)	59.3	6.4
Local (non-GIFT)	35.6	3.8
Israel (non-GIFT)	17.6	1.9

Source: ADB (2006).
Notes: The survey covered central, southern, and northern Luzon; Bicol; and the provinces of Cotabato, General Santos, Sarangani, South Cotabato, and Sultan Kudarat. FaST means a breed developed by the Freshwater Aquaculture Center of Central Luzon State University; GET EXCEL means EXcellent strain that has a Comparative advantage over other tilapia strains for Entrepreneurial Livelihoods projects in support of aquaculture for rural development; GIFT means Genetic Improvement of Farmed Tilapia project; GMT means genetically male tilapia; GST™ means GenoMar Supreme Tilapia.
[a]In millions of fry and fingerlings.
[b]Hatcheries in Mindanao region used earlier generations of the GIFT strain. Production numbers for the GenoMar Supreme Tilapia strain in southern Luzon and Bicol were not disclosed by respondents.

Table 17.6 Countries' tilapia imports to the U.S. market, 2001 and 2003 (metric tons)

Country	Whole, frozen 2001	Whole, frozen 2003	Fillet, fresh 2001	Fillet, fresh 2003	Fillet, frozen 2001	Fillet, frozen 2003	Total 2001	Total 2003
China	10,870	28,763	191	857	2,529	15,857	13,590	45,477
Taiwan	27,599	19,664	76	286	2,133	2,470	29,809	22,415
Indonesia	39	5.4	n.a.	n.a.	2,179	3,583	2,218	3,588
Thailand	49	121	2	7	209	940	260	1,068
Hong Kong	n.a.	135	n.a.	n.a.	n.a.	n.a.	n.a.	135
Vietnam	7	41	n.a.	17	53	73	60	132
Burma	n.a.	n.a.	n.a.	n.a.	n.a.	19	n.a.	19
Japan	n.a.	n.a.	n.a.	0.5	n.a.	18	n.a.	18.5
Philippines	51	18	n.a.	n.a.	2	n.a.	53	18

Source: El-Sayed (2006).
Note: n.a. means data were not available.

smaller than the international marketable size of 400–500 grams for live fish and 700–1,000 grams for fish that will be filleted (Dey et al. 2000a; World SeaFood Market 2005). Whereas American consumers demand a large fish that can be filleted, poor Philippine consumers traditionally preferred a smaller fish (World SeaFood Market 2005), although these preferences have started to shift. In 1995, 58 percent of Philippine households preferred a larger fish (Dey et al. 2000b). The Government of the Philippines is also taking steps to improve tilapia exports. In 2007 the BFAR projected that tilapia exports would increase by 150–200 tons in one year, with resources focused on the top tilapia-producing region in the country (Sun Star 2007). More attention should also be given to packaging and marketing tilapia as fillets, as well as selecting appropriate pricing and distribution mechanisms (ADB 2006).

Tilapia as a Source of Employment and Income

Aquaculture can be more advantageous than other farm activities such as cash crop or livestock production because of its relatively inexpensive inputs, low capital requirements, and low labor requirements (El-Sayed 2006). Despite the small amount of labor required for tilapia farming, it is estimated that 280,000 people in the Philippines benefit directly or indirectly from employment in the tilapia industry (CGIAR 2006). The tilapia industry provides employment in excavation of ponds, cage and net making, as well as fish feeding and harvesting, sorting and grading, marketing, and transport, among others. Two-thirds of the nation's 604 hatcheries are dedicated to producing GIFT and GIFT-derived seed (ADB 2006).

Although tilapia farming is a male-dominated occupation, females comprise 11 percent of improved tilapia growout farmers and hatchery operators. That number jumps to 33 percent among GIFT users involved in hatchery operations (Sevilleja 2006). Improved tilapia operations are thus relatively inclusive of women.

As mentioned earlier, most tilapia farmers rely on semi-intensive systems that yield substantial profits (Dey and Gupta 2000). Average net returns from improved tilapia farming have been shown to be particularly high. Ninety-seven percent of these farmers culture improved tilapia as a cash crop, and most farmers sell more than 30 percent of their harvested product (Eknath and Acosta 1998). The major costs associated with tilapia farming are the cost of feed and the cost of fry or fingerlings.

Tilapia hatchery farming in the Philippines and Thailand yields average net returns of $5,074 per hectare per year; $1,867–4,241 per hectare per four-month crop cycle for growout operations, not including on-farm household fish consumption; and $390 per cage per cycle for growout cages. Cage farming in central Luzon and Taal Lake has a potential yield of $3,120 annually for a farmer (Acosta and Gupta 2009).[7]

Growing improved strains of tilapia in the Philippines also significantly reduces farmers' production costs. Depending on the production environment, improved strains are 32–35 percent cheaper to produce than nonimproved strains (Table 17.7).

Distribution of Benefits to Producers

Tilapia dissemination strategies in the Philippines were not able to reach as many small-scale farmers as was hoped (Mair et al. 2002). The demographics for tilapia users imply that farmers with a high level of expertise are better able to access

Table 17.7 Yield and variable costs of tilapia pond farming in the Philippines using GIFT and non-GIFT strains, 2002

Item	Cage	Pond
GIFT		
Yield (kilograms per unit area)	236	1,361
Variable cost	168	1,385
Variable cost of fish per kilogram	0.71	1.02
Non-GIFT		
Yield (kilograms per unit area)	153	912
Variable cost	168	1,375
Variable cost of fish per kilogram	1.1	1.51
Percentage difference		
Yield	54.2	49.2
Variable cost	0	0.7
Variable cost of fish per kilograms	−35.5	−32.5

Source: Dey (2002), as presented by Gupta and Acosta (2004).
Note: GIFT means Genetic Improvement of Farmed Tilapia project.

improved tilapia technologies (Sevilleja 2006). Additionally, although most users of improved tilapia are small landowners, those with access to their own sources of capital are better equipped to receive the benefits of improved tilapia through private-sector collaboration (that is, GIFT/GST), suggesting that the current dissemination mechanism fails to reach small and poor tilapia farmers (Sevilleja 2006).

Small-scale fish farmers face numerous obstacles, such as high input costs and little bargaining power in their relationships with traders and midlevel agents (Ahmed and Lorica 2002). This problem is compounded by hatchery market channels; nearly 75 percent of hatcheries sell all or most of their fingerlings to traders instead of directly to growers (Eknath and Acosta 1998). The legal documentation requirements associated with the GIFT hatchery accreditation process tend to limit interested hatchery operators to those with a high level of education, business expertise, and access to capital (Rodriguez 2006).

Yet criticisms of the poverty focus of improved tilapia activities should be slightly tempered. A study of the poverty levels of hatchery and growout farmers showed that fish pond owners represented 10–45 percent of all middle-income village households and 23–55 percent of high-income households. Additionally, most growers did not have access to formal credit. Eknath and Acosta (1998) conclude that there is no evidence that it is the poorer or richer members of communities that enter into aquaculture. Additionally, the monoculture technology that is prevalent in the Philippines has been found suitable for the poor, especially in the context of herbivorous species, because it does not require abundant capital (Dey et al. 2005). The GIFT strain is regarded as a scale-neutral technology in terms of feed and fertilizer use (Acosta and Gupta 2009). Although tilapia farming represents a secondary source of income for some improved strain users, its contribution to their total income is considerable (Table 17.8) (Acosta et al. 2006). Despite challenges, only 3–7 percent of pond operators planned to discontinue tilapia farming, with 20–46 percent planning to expand their operations (Dey et al. 2000b).

Tilapia as a Low-Priced, Nutritious Fish

The genetic strides made in tilapia have helped increase national production, working to keep tilapia prices low for consumers. As a source of protein, tilapia is generally more affordable than pork, beef, chicken, and even other freshwater fish. For example, the average price of tilapia rose by 111 percent from 1990 to 2007, whereas beef prices jumped 148 percent and pork prices rose 157 percent (Figure 17.3). Dey (2000) concluded that adopting improved tilapia species would reduce tilapia prices by 5–16 percent in various Asian countries, including Bangladesh, China, the Philippines, Thailand, and Vietnam.

Tilapia is also a significant source of protein. While eggs, milk, rice, and wheat may contain 3.5–12.0 percent protein, fish contain about 16–20 percent protein,

**Table 17.8 Tilapia farming as a source of income among
improved tilapia users in the Philippines, 2006**

Strain	Tilapia farming as the primary source of income (percent)	Percentage of income from tilapia farming
Growout		
GIFT/GST™	24	39
GET EXCEL	20	38
GMT	16	35
FaST	60	74
Hatchery		
GIFT/GST™	83	64
GET EXCEL	55	56
YY male	25	34
FaST	47	61

Source: Sevilleja (2006).
Note: FaST means a breed developed by the Freshwater Aquaculture Center of Central Luzon State University; GET EXCEL means EXcellent strain that has a Comparative advantage over other tilapia strains for Entrepreneurial Livelihoods projects in support of aquaculture for rural development; GIFT means Genetic Improvement of Farmed Tilapia project; GMT means genetically male tilapia; GST™ means GenoMar Supreme Tilapia.

Figure 17.3 Retail prices of tilapia, pork, beef, and chicken in the Philippines, 1990–2007

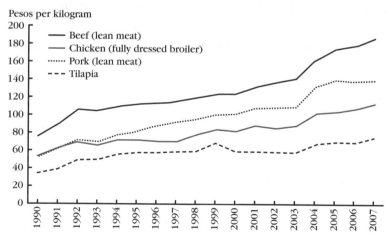

Source: Philippines Bureau of Agricultural Statistics (2009).

as well as a variety of essential minerals, vitamins, and amino acids. In Asia, fish protein accounts for an average 31 percent of total animal protein supply, with this figure jumping as high as 51 percent in Bangladesh, 58 percent in Indonesia, and 75 percent in Cambodia (Acosta and Gupta 2009). For the poorest of the poor, fish is often the only source of animal protein (Kumar 1992). GIFT fish have been shown to have 17–21 percent protein content, representing a 4.7–5.5 percent increase in whole-body protein content over red hybrid tilapia strains, showing that protein efficiency and use are influenced by tilapia genotype (Ng and Hanim 2007; Ponzoni et al. 2008).

A combination of high nutritional value and high consumption is good news for food security. In 1997–2001, Filipino national consumption of fish and fishery products averaged 2.3 million metric tons, increasing 2.2 percent annually (Garcia, Dey, and Navarez 2005). Because multiple macro- and micro-level confounding factors affect consumption, the causality between improved tilapia and increased consumption has not yet been established. Nonetheless, consumption of tilapia in particular has increased recently. Before the development and introduction of genetically improved strains, the average per capita consumption of tilapia in the Philippines was 0.66 kilograms per year. By 2007, this amount had increased by 362 percent to 3.05 kilograms (Figure 17.4) (Philippines Bureau of Agricultural Statistics 2009).

Although no peer-reviewed estimates are readily available for the number of Filipinos who have benefited from these low prices, Falck-Zepeda and Horna (2009) conducted an ex post evaluation of GIFT and GIFT-derived tilapia using an economic surplus approach and concluded that these strains have allowed an addi-

Figure 17.4 Tilapia consumption per capita in the Philippines, 1990–2007

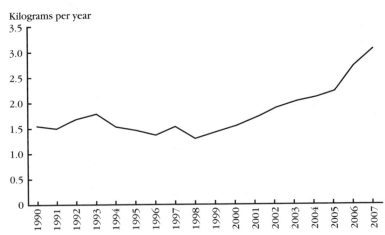

Source: Philippines Bureau of Agricultural Statistics (2009).

tional 18–21 million people to fulfill their daily protein requirements and another 1.3–1.6 million people to fulfill their daily caloric requirements. The quantity and quality of fish consumption vary widely in the Philippines according to income. In 2000 the monetary value of per capita fish consumption among wealthy consumers was 3.4 times that of the poorest consumers (Garcia, Dey, and Navarez 2005).[8] Poor consumers tend to spend their income on less expensive fish, including processed fish (Table 17.9). As their incomes increase, these consumers tend to spend a smaller proportion of their budget on fish and a larger proportion on meat. Poor households thus rely more on fish as their primary source of animal protein, although fish is consumed in all income groups (Garcia, Dey, and Navarez 2005). Figures from 1995 show rural tilapia farmers consuming an average of 39.5 kilograms of fish every year, with rural nonproducers consuming 15.9 kilograms and urban nonproducers 5.8 kilograms (Dey et al. 2000b).

Tilapia follows these income-related consumption patterns, exhibiting a particularly high consumption share among the poorest segment of the national population. Indeed, household survey results confirm the popularity of tilapia among this segment of population: in one, 85–100 percent of respondents reported consuming tilapia one to three times a week; in another, 65 percent of respondents preferred tilapia to other fish species (Dey et al. 2000b; Mair et al. 2002).

Returns

A study conducted by Ponzoni, Nguyen, and Khaw (2007) found that the national economic benefit derived from GIFT's genetic improvement activities was extremely favorable. Even with relatively simple operational systems, such as ponds operating with moderate efficiency, the economic benefit of such activities was valued at more than $4 million and the benefit–cost ratio was 8.5. Upgrading the reproductive efficiency of these contexts by introducing available and inexpensive technologies

Table 17.9 Share of tilapia consumption in Philippine nonproducer households by expenditure class, 2000 (percent)

Expenditure quintile	Share of tilapia in total fish consumption
1	25.52
2	25.18
3	35.29
4	44.95
5	53.33

Source: Dey et al. (2000b).

such as *hapas,* artificial incubation, and good management raised these values to $32 million and $60 million, respectively.

The internal rate of return of the development and dissemination of GIFT was estimated to have been more than 70 percent from 1988 to 2010 (ADB 2006). By comparison, for tilapia in the Philippines, including both GIFT and non-GIFT strains, the rate of return for monoculture of tilapia in cages was 20 percent while that of monoculture in ponds was 30 percent in 1999 (Dey et al. 2005).

The Sustainability of Tilapia Farming

Environmental Sustainability

In developing countries such as the Philippines, the use of antibiotics or chemicals in aquaculture production is generally less intense than in other parts of the world (Charo-Karisa et al. 2008). Even so, with the recent trend toward intensification of tilapia farming, the tendency to apply such artificial farm inputs as prepared feed, hormones, and fuels will only grow, increasing the possible ecological and health impacts (El-Sayed 2006).

Many fish farming operations, especially those raising fish intended for export markets, include carnivorous species with high protein needs. An individual carnivorous fish may consume four to five wild fish to meet its dietary requirements, thus depleting stocks of natural fish populations (Ahmed and Lorica 2002). The expansion of this type of aquaculture has placed pressure on capture fisheries through increased demand for captured fish for use as feed (Dey and Kanagaratnam 2008). Although Nile tilapia is naturally herbivorous, selective breeding programs such as GIFT are often undertaken for tilapia in ponds that receive high-protein supplementary feed consisting of fish protein. Aside from being costly, these supplementary feeds may lead to further depletion of indigenous stocks (Charo-Karisa et al. 2008).

Indeed, depletion of local fish may be the central issue related to GIFT tilapia (Charo-Karisa et al. 2008). As mentioned previously, when Mozambique tilapia and Nile tilapia were introduced in the 1950s and 1970s, they earned a reputation as prolific breeders. The impact of these introductions on biodiversity, compounded with overfishing, pollution, siltation, and water diversion, makes it difficult to assess the impact of Nile tilapia on Philippine lakes (ADB 2006).

It should be noted that WorldFish took extensive precautions with regard to the invasive species issues under the GIFT program. An assessment team brought out to confirm claims that tilapia led to reductions or displacement of other fish did not find any displacement in natural waterways other than lakes and reservoirs (CGIAR 2006). An international workshop on the possible environmental impacts of GIFT strains concluded that "responsible development and dissemination of GIFT would be unlikely to cause serious environmental damage" (CGIAR 2006, 4).

El-Sayed (2006) notes that the predicted future shortage of freshwater will necessitate innovations in tilapia culture. Tilapia will increasingly need to be produced in closed-recirculation systems alongside improvements in aeration, filtration, and feeding techniques; waste settlement and removal; and reuse of water.

Financial Sustainability

The institutions responsible for genetic research in the Philippines are funded through grants, government monies, seedstock sales, and partnerships with private-sector actors. Since 1998 GFII operating revenues have come from fingerling sales, fees associated with issuing hatchery licenses, and grant service income (Rodriguez 2006). A sizable amount of this revenue, PhP 15 million, has gone toward the maintenance and breeding of its own independent nucleus of tilapia. GFII's financial health, as measured by the relationship between operating revenue (generated from fingerling sales and licensing fees) and R&D expenditures (comprising expenditures for personnel, supplies, services, travel, and depreciation of fish stocks), has been somewhat unpredictable over the long term (Figure 17.5).

In terms of external funding, tilapia genetics research has received more than $7 million since 1979 from such international organizations and aid agencies as

Figure 17.5 GIFT Foundation International operating revenues and expenses, 1998–2002

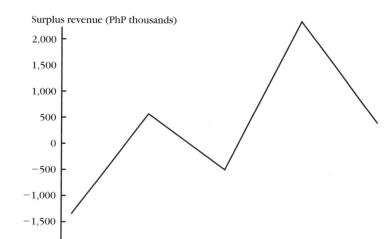

Surplus revenue (PhP thousands)

Source: Rodriguez (2006) as presented in Acosta et al. (2006).

the ADB, the U.K. Department for International Development, UNDP, and ICLARM, among others. National support has totaled more than $1.2 million (ADB 2006). It took approximately 10 years of investment and technical assistance to see GIFT through the entire agricultural value chain, from germplasm collection to selective breeding, on-farm trials, establishment of national breeding programs, and distribution (ADB 2006). This experience makes the case that sustained investment can pay large dividends in the long term.

Political Sustainability

In the past, national fishery services have not always received the resources and management oversight necessary to allow the improved tilapia industry to thrive (Sevilleja 2007). The 1987 Philippine Constitution, however, explicitly recognized the role of the state in managing aquatic resources, saying that the state will support appropriate technologies and research and provide financial backing for the production and marketing of marine resources (Sevilleja 2007). The Government of the Philippines has since emerged as a central player in improved tilapia breeding and dissemination. The government is also building infrastructure vital to the fisheries sector, including seaports, storage facilities, roads connecting farms to markets, and water supply systems, and is supporting the preparation and implementation of a Comprehensive National Fisheries Development Plan (Sevilleja 2007). Local government administrative units also now offer basic fisheries extension services and have created local environmental and natural resources councils comprising government agencies, civil society organizations, and nongovernmental agencies (Sevilleja 2007). As described earlier, the remaining area of weakness lies with the harmonization of public- and private-sector roles in seed supply; policies toward this end have not yet been formally enacted (ADB 2006).

In 2003 the Philippines' then–President Gloria Arroyo declared that tilapia would soon replace the local fish, *galunggong*, as the national staple fish (CGIAR 2006). A year later the country hosted the Sixth International Symposium on Tilapia and Aquaculture in Manila. These developments suggest that there is a sound political environment for the continuation of improved tilapia activities in the future.

Lessons Learned

Farmers Are Open to Changes in Tilapia
Hatchery and Farming Practices

Tilapia seed producers now have greater access to high-quality tilapia broodstock and are able to manage them more carefully by replacing spawners (fish that produce young) every 18–24 months. The availability of a wide array of strains, including

GIFT/GST™, GET EXCEL, YY, and FaST, another line developed by FAC, makes seed producers more willing to try new technologies. For example, a micro-level analysis of hatchery operators in the Philippines showed that 75 percent of hatchery operators were aware of GIFT fish and were willing to take on new technologies, as well as use better tilapia strains when they are available at reasonable prices (Bimbao et al. 2000). Farmers have also been able to more intensively manage ponds, harvesting fish after 3–4 months instead of after 6–7 months, as they did before GIFT (ADB 2006).

We Should Extend the Benefits of Technology Transfers

The GIFT project represented the first-ever systematic collection and transfer of Nile tilapia germplasm from Africa to Southeast Asia. Although this episode shows that technology and knowledge transfers do not originate exclusively in the developed world, Africa has so far received no benefits from the transaction. This situation has arisen mainly because of the WorldFish Center's past policy of not introducing GIFT fish to countries where it is indigenous for fear of contaminating wild germplasm (Gupta and Acosta 2004). Recently WorldFish reversed its decision and made GIFT fish available to any African government that could demonstrate that it had a well-defined maintenance and dissemination strategy for the fish, as well as a plan for how to manage environmental and biodiversity risks (WorldFish Center 2007). Any willing government must still abide by the Convention on Biological Diversity, which seeks to protect ecosystems from alien species (Acosta and Gupta 2009). It has been estimated that introducing GIFT to Africa may improve the growth of the current stock by 64 percent (Ponzoni et al. 2008).

Extension Matters

Although traditional tilapia farming does not currently reach the poorest of the poor in the Philippines on a large-scale basis, genetic technologies may be well equipped to benefit these people, but only when coupled with complementary services. Strengthening the presence and outreach of extension officers would not only enable the diffusion and adoption of a particular technology but also serve as a feedback mechanism to scientists, in turn enabling them to further improve and refine the technologies. Ultimately, a stronger extension system translates into technologies that are relevant to their end users.

Gene Banking, Breeding Technology, and Institutional Capacity Pay Off

Aquaculture has been framed as one of the most viable ways of increasing fish production and food production over the next century. The GIFT program has shown the benefits of applying genetics to aquaculture, especially selective breeding, which is a particularly cost-effective way of achieving genetic improvement.

Before the GIFT project, the Philippines did not systematically bank farmed fish genetic resources. In fact, fish gene banks are relatively rare, especially in tropical developing countries. The GIFT project introduced technology and training for gene banking and now maintains a tilapia gene bank that is of international importance (ADB 2006). GET EXCEL has been the premier product emerging from this gene bank. If not for the ready availability of FaST and GIFT germplasm, researchers would have been forced to reintroduce wild tilapia strains from Africa, thus stalling the creation and dissemination of GET EXCEL.

The GIFT selective breeding technology can also now be applied to the genetic improvement of other species. In the past, cultured fish stocks were considered inferior to wild populations; their prospects have now been improved (Pullin and Capili 1988). GIFT technologies and methodologies are currently being applied to carp, the world's most popular farmed fish, in Bangladesh, China, India, Thailand, and Vietnam. Some experiments have shown genetic improvements of up to 50 percent (Acosta and Gupta 2009).

Multilevel, Multisectoral Cooperation Is Essential

The success of GIFT was buoyed by a vast multilevel network of international and national research institutes, governments, private actors, and donors. In its role as host country, the Philippines and its national fisheries bureaus and centers made many important contributions to the evolution of the project, including acting as the testing ground for the initial rounds of research on and dissemination of improved tilapia. INGA, an international network of multisectoral actors, helped coordinate global linkages and transfer tilapia germplasm, technical expertise, and lessons learned for institutional capacity-building across borders. Scientists from developed and developing countries alike contributed expertise and training, while private hatcheries throughout Asia have been integral to the wide dissemination of improved tilapia strains. International donors provided sustained investments, seeing the project to the end of the agricultural value chain. Finally, the WorldFish Center successfully coordinated all of these activities, in the process becoming renowned as the world leader in tilapia breeding research.

Concluding Remarks

Genetically improved tilapia may contribute significantly to meeting the global food security challenge. Over the past two decades it has positioned itself as a low-cost, high-yielding, and profitable fish. The tilapia industry has offered direct, measurable benefits for nutrition, employment, and income generation, as well as indirect benefits that include increased availability of fish in local rural and urban markets at a lower price that meets booming consumer demand.

This success story has been aided by several factors. First, a strong commitment on the part of the Government of the Philippines to create a favorable policy environment, set up infrastructure, and lead the way in R&D despite past setbacks was key in producing a series of exciting fish strains. Second, strong public–private partnerships in the dissemination phase enabled the public to access improved strains and set the stage for regulation of participating hatcheries. Third, regional networks coordinated technology transfers to other countries, as well as initiating monitoring and evaluation. Finally, a strong initial mandate to apply the lessons learned in tilapia production to the larger aquaculture picture is now coming to fruition.

What remains to be seen is whether improved tilapia can overcome the challenges associated with poor management, dissemination, and limited natural resources, especially freshwater. For all the successes in tropical finfish improvement in Asia, genetically improved fish stocks still account for less than 1 percent of global aquaculture production (Pullin and Capili 1988). Aquaculture only stands to gain from genetic work. Although the improvements achieved through genetic selection may seem limited to a small population of fish, the cumulative gains made in millions of fish as that gain is disseminated to hatcheries and farmers can be a powerful tool in the sector (Ponzoni et al. 2008). If the successes highlighted in this case study can be further scaled up, as they have been so far in Asia, the food security of future generations can be significantly improved.

Notes

1. Low-income food-deficit countries are defined by the FAO as developing countries whose annual per capita net income is below $1,395 and whose imports of basic foodstuffs outweigh their exports.

2. Monoculture is the practice of cultivating only one fish species in one pond. Extensive culture is low-intensity aquaculture, such as that in ponds, as practiced by subsistence farmers. No nutritional inputs, such as manure or feed, are given to the fish, which rely solely on natural food produced in the pond. This form of production offers low yields. Semi-intensive culture uses fertilization (manure) and supplementary feed and offers moderate yields. In intensive culture, all the fishes' nutritional requirements are met by formulated feed, and pond water is replenished, aerated, or recirculated. The yields are high (Rahman, Varga, and Chowdhury 1992).

3. The genetically improved stocks that resulted from the Norwegian programs now account for more than 80 percent of the salmon produced in Norway, which is the world's top producer of Atlantic salmon.

4. GET EXCEL stands for "EXcellent strain that has a Comparative advantage over other tilapia strains for Entrepreneurial Livelihoods projects in support of aquaculture for rural development" (Tayamen, Abella, and Sevilleja 2006).

5. These 11 countries are Bangladesh, China, Fiji, India, Indonesia, Laos, Malaysia, Papua New Guinea, Sri Lanka, Thailand, and Vietnam.

6. For example, in Bangladesh GIFT fish exhibited a 9–25 percent increase in yield compared with non-GIFT fish, representing a 7–20 percent decrease in the variable cost of tilapia farming. In

China and Thailand the yield difference was 9–33 percent, representing a 7–28 percent decrease in farming costs (Dey 2002).

7. For four cages and two harvestings.

8. Wealthy consumers are those with an average income of PhP 91,097, while the poorest consumers have an average income of PhP 7,244 (Family and Income Expenditure Survey as analyzed in Garcia, Dey, and Navarez 2005).

References

Acosta, B. O., R. C. Sevilleja, M. V. Gupta, B. M. Rodriguez Jr., T. Abella, and M. Tayamen. 2006. Public and private partnerships in tilapia research and development: An overview of Philippine experience. In *Public and private partnerships in aquaculture: A case study on tilapia research and development,* ed. B. O. Acosta, R. C. Sevilleja, and M. V. Gupta. Penang, Malaysia: WorldFish Center.

Ahmed, M. M., and M. H. Lorica. 2002. Improving developing country food security through aquaculture development—Lessons from Asia. *Food Policy* 27: 125–141.

Bentsen, H. B., A. E. Eknath, M. S. Palada de Vera, J. C. Danting, H. L. Bolivar, R. A. Reyes, E. E. Dionisio, F. L. Longalong, A. V. Circa, M. M. Tayamen, and B. Gjerde. 1998. Genetic improvement of farmed tilapias: Growth performance in a complete diallel cross experiment with eight strains of *Oreochromis niloticus. Aquaculture* 160 (1–2): 145–173.

Bimbao, G. B., F. J. Paraguas, M. M. Dey, and A. E. Eknath. 2000. Socioeconomics and production efficiency of tilapia hatchery operations in the Philippines. *Aquaculture Economics and Management* 4 (1–2): 49–63.

CGIAR Science Council. 2006. *Improved tilapia benefits Asia.* Science Council / Standing Panel on Impact Assessment Brief 6. Rome: Consultative Group on International Agricultural Research Science Council Secretariat.

Charo-Karisa, H., H. Komen, H. Bovenhuis, M. A. Rezk, and R. W. Ponzoni. 2008. Production of genetically improved organic Nile tilapia. *Dynamic Biochemistry, Process Biotechnology and Molecular Biology* 2 (1, special issue): 50–54.

Delgado, C. L., N. Wada, M. W. Rosegrant, S. Miejer, and M. Ahmed. 2003. *Fish to 2020.* WorldFish Center Technical Report 62. Washington, D.C., and Penang, Malaysia: International Food Policy Research Institute and WorldFish Center.

Dey, M. M. 2000. The impact of genetically improved farmed Nile tilapia in Asia. *Aquaculture Economics and Management* 4 (1–2): 107–124.

———. 2002. Overview of socioeconomics and environmental issues. In *Tilapia farming in the 21st century: Proceedings of the international forum on tilapia farming in the 21st century,* ed. R. D. Guerrero and M. R. Guerrero-del Castillo. Los Baños, the Philippines: Philippine Fisheries Association.

Dey, M. M., and M. V. Gupta. 2000. Socioeconomics of disseminating genetically improved Nile tilapia in Asia: An introduction. *Aquaculture Economics and Management* 4 (1–2): 5–11.

Dey, M. M., and U. Kanagaratnam. 2008. *Community based management of small scale fisheries in Asia: Bridging the gap between fish supply and demand.* WorldFish Center Conference Paper 23. Penang, Malaysia: WorldFish Center.

Dey, M. M., A. E. Eknath, L. Sifa, M. G. Hussain, T. M. Thien, N. Van Hao, S. Aypa, and N. Pongthana. 2000a. Performance and nature of genetically improved farmed tilapia: A bioeconomic analysis. *Aquaculture Economics and Management* 4 (1–2): 85–108

Dey, M. M., G. B. Bimbao, L. Yong, P. Regaspi, A. H. M. Kohinoor, N. Pongthana, and F. J. Paraguas. 2000b. Current status of production and consumption of tilapia in selected Asian countries. *Aquaculture Economics and Management* 4 (1–2): 47–62.

Dey, M. M., M. A. Rab, F. J. Paraguas, R. Bhatta, M. F. Alam, S. Koeshendrajana, and M. Ahmed. 2005. Status and economics of freshwater aquaculture in selected countries of Asia. *Aquaculture Economics and Management* 9: 11–37.

Eknath, A. E. 1995. Managing aquatic genetic resources: Management example 4—The Nile tilapia. In *Conservation of fish and shellfish resources: Managing diversity,* ed. J. E. Thorpe, G. Gall, J. E. Lannan, and C. E. Nash. London: Academic Press, Harcourt Brace Company.

Eknath, A. E., and B. O. Acosta, 1998. *Genetic Improvement of Farmed Tilapias (GIFT) project: Final report, March 1988 to December 1997.* Makati City, the Philippines: International Center for Living Aquatic Resources Management.

Eknath, A. E., J. M. Macaranas, L. Q. Agustin, R. R. Velasco, M. C. A. Ablan, M. J. R. Pante, and R. S. V. Pullin. 1991. Biochemical and morphometric approaches to characterize farmed tilapias. *NAGA—The ICLARM Quarterly* 14 (2): 7–9.

Eknath, A. E., M. M. Tayamen, M. S. Palada de Vera, J. C. Danting, R. A. Reyes, E. E. Dionisio, J. B. Gjedrem, and R. S. V. Pullin. 1993. Genetic improvement of farmed tilapias: The growth performance of eight strains of *Oreochromis niloticus* tested in eleven different environments. *Aquaculture* 111: 171–188.

Falck-Zepeda, J., and D. Horna. 2009. Proven successes in global agriculture: A common quantitative impact assessment platform. International Food Policy Research Institute, Washington, D.C. Photocopy.

FAO (Food and Agriculture Organization of the United Nations). 2006. *FAO yearbook: Fishery and aquaculture statistics, 2006.* Rome.

———. 2007. Concern over the situation of high-seas fish species: Strengthening fisheries management in international waters "a major challenge." FAO report. <http://www.fao.org/newsroom/en/news/2007/1000505/index.html>. Accessed August 1, 2009.

———. 2009a. *FAO yearbook 2007: Fishery and aquaculture statistics.* Rome: Fisheries and Aquaculture Information and Statistics Service.

———. 2009b. FAOSTAT. <http://faostat.fao.org>. Accessed August 20, 2010.

———. 2009c. The introduction and distribution of tilapias in Asia and the Pacific. FAO Corporate

Document Repository. <http://www.fao.org/docrep/007/y5728e/y5728e04.htm#fn1>. Accessed March 25.

GAIN (Gippsland Aquaculture Industry Network). 2003. Genetically altered fish raises ethical concerns. <http://www.growfish.com.au/content.asp?contentid=818>. Accessed August 5, 2009.

Garcia, Y. T., M. M. Dey, and S. M. M. Navarez. 2005. Demand for fish in the Philippines: A disaggregated analysis. *Aquaculture Economics and Management* 9: 141–168.

GenoMar. 2009. Tilapia breeding. <http://www.genomar.no/default.aspx?aid=9078104>. Accessed June 25.

Gjoen, H. M. 2004. A new era: The merging of quantitative and molecular genetics—Prospects for tilapia breeding programs. In *Proceedings of the sixth international symposium on tilapia in aquaculture,* ed. R. B. Bolivar, G. C. Mair, and K. Fitzsimmons. Manila, the Philippines: Bureau of Fisheries and Aquatic Resources.

Guerrero, R. D. 1994. *Tilapia farming in the Philippines: A success story.* Bangkok, Thailand: Asia-Pacific Association of Agricultural Research Institutions.

———. 1996. Aquaculture in the Philippines. *World Aquaculture* 27 (1): 7–13.

———. 2008. Tilapia farming: A global review (1924–2004). *Asian International Journal of Life Sciences* 17 (2): 207–229.

Gupta, M. V., and B. O. Acosta. 2001. Development of global partnerships for fish genetics research—A success story. Paper presented at the Technical Workshop on Methodologies, Organization and Management of Global Partnership Programmes, October 9–10, in Rome.

———. 2004. From drawing board to dining table: The success story of the GIFT project. *NAGA—WorldFish Center Quarterly* 27 (3–4): 4–17.

Kumar, D. 1992. *Fish culture in undrainable ponds: A manual for extension.* FAO Fisheries Technical Paper 325. Rome: Food and Agriculture Organization of the United Nations.

Ling, S. W. 1977. *Aquaculture in southeast Asia: A historical overview.* A Washington Sea Grant Publication. Seattle, Wash., U.S.A.: University of Washington Press.

Longalong, F. M., A. E. Eknath, and H. B. Bentsen. 1999. Response to bi-directional selection for frequency of early maturing females in Nile tilapia (*Oreochromis niloticus*). *Aquaculture* 178 (1–2): 13–25.

Macaranas, J. M., L. Q. Agustin, M. C. Ablan, M. J. Pante, A. E. Eknath, and R. S. V. Pullin. 1995. Genetic improvement of farmed tilapias: Biochemical characterization of strain differences in Nile tilapia. *Aquaculture International* 3: 43–54.

Mair, G. C., and A. A. Van Dam. 1996. The effect of sex ratio at stocking on recruitment in Nile tilapia (*Oreochromis niloticus* L.) ponds. In *The third international symposium on tilapia in aquaculture: ICLARM Conference Proceeding 41,* ed. R. S. V. Pullin, J. Lazard, M. Legendre, J. B. Amon Kothias, and D. Pauly. Makati City, the Philippines: International Center for Living Aquatic Resources Management.

Mair, G. C., J. S. Abucay, J. A. Beardmore, and D. O. F. Skibinski. 1995. Growth performance trials of genetically male tilapia (GMT) derived from YY-males in *Oreochromis niloticus L.*: On station comparisons with mixed sex and sex-reversed male populations. *Aquaculture* 137: 313–322.

Mair, G. C., G. J. C. Clarke, E. J. Morales, and R. C. Sevilleja. 2002. Genetic technologies focused on poverty? A case study of genetically improved tilapia (GMT) in the Philippines. In *Rural aquaculture,* ed. P. Edwards, D. Little, and H. Demaine. Wallingford, Oxon, U.K.: CABI.

Ng, W.-K., and R. Hanim. 2007. Performance of genetically improved Nile tilapia compared with red hybrid tilapia fed diets containing two protein levels. *Aquaculture Research* 38: 965–972.

Philippines Bureau of Agricultural Statistics. 2009. CountrySTAT tables. <http://countrystat.bas .gov.ph/>. Accessed March 25.

Pillay, T. V. R. 2004. *Aquaculture and the environment.* Oxford, U.K.: Blackwell.

Ponzoni, R. W., N. H. Nguyen, and H. L. Khaw. 2007. Investment appraisal of genetic improvement programs in Nile tilapia (*Oreochromis niloticus*). *Aquaculture* 269: 187–199.

Ponzoni, R. W., A. Hamzah, S. Tan, and N. Kamaruzzaman. 2005. Genetic parameters and response to selection for live weight in the GIFT strain of Nile tilapia (*Oreochromis niloticus*). *Aquaculture* 247: 203–210.

Ponzoni, R. W., N. H. Nguyen, H. L. Khaw, N. Kamaruzzaman, A. Hamzah, K. R. A. Bakar, and H. Y. Yee. 2008. *Genetic improvement of Nile tilapia (Oreochromis niloticus)—Present and future.* Paper presented at the 8th International Symposium on Tilapia in Aquaculture, October 12–14, in Cairo, Egypt.

Pullin, R. S. V. 1980. Philippine tilapia broodstock project. *ICLARM Newsletter* 3 (1): 8–9.

———. 1983. Choice of tilapia species for aquaculture. In *International symposium on tilapia in aquaculture,* ed. L. Fishelson and Z. Yaron. Tel Aviv, Israel: Tel Aviv University.

———. 1985. Tilapia: "Everyman's fish." *Biologist* 32: 84–88.

Pullin, R. S. V., and J. B. Capili. 1988. Genetic improvement of tilapias: Problems and prospects. In *The second international symposium on tilapia in aquaculture,* ed. R. S. V. Pullin, T. Bhukaswan, K. Tonguthai, and J. L. Maclean. ICLARM Conference Proceedings 15. Bangkok, Thailand, and Manila, the Philippines: Department of Fisheries and the International Center for Living Aquatic Resources Management.

Rahman, M. M., I. Varga, and S. N. Chowdhury. 1992. *Manual on polyculture and integrated fish farming in Bangladesh.* Field Document BGD/87/045/91/11. Rome: Food and Agriculture Organization of the United Nations.

Rodriguez, B. M. 2006. Disseminating genetically improved tilapia fingerlings through the GIFT licensing program. In *Public and private partnerships in aquaculture: A case study on tilapia research and development,* ed. B. O. Acosta, R. C. Sevilleja, and M. V. Gupta. Penang, Malaysia: WorldFish Center.

Sevilleja, R. C. 2006. The effects of evolving partnerships on access to and uptake of tilapia genetic improvement technologies and their products: Results of survey and policy implications. In *Public and private partnerships in aquaculture: A case study on tilapia research and development*, ed. B. O. Acosta, R. C. Sevilleja, and M. V. Gupta. Penang, Malaysia: WorldFish Center.

————. 2007. Genetics-based technologies for sustainable development in Philippine aquaculture: The case of tilapia. In *Species and system selection for sustainable aquaculture*, ed. P. Leung, C.-S. Lee, and P. J. O'Bryen. Ames, Iowa, U.S.A.: Wiley-Blackwell.

Silvestre, G. T., L. R. Garces, I. C. Sotbutzki, M. Ahmed, R. A. Valmonte-Santos, C. Z. Luna, L. Lachika-Aliño, P. Munro, V. Christensen, and D. Pauly. 2003. *Assessment management and future directions for coastal fisheries in Asian Countries.* WorldFish Center Conference Proceedings 67. Penang, Malaysia: WorldFish Center.

Sun Star Davao. 2007. Tilapia export seen to increase next year. April 25. <http://www.sunstar .com.ph/static/dav/2007/04/25/bus/tilapia.export.seen.to.increase.next.year.html>. Accessed July 22, 2009.

Taniguchi, N., J. M. Macaranas, and R. S. V. Pullin. 1985. Introgressive hybridization in cultured tilapia stocks in the Philippines. *Bulletin of the Japanese Society of Scientific Fisheries* 51 (8): 1219–1224.

Tayamen, M. M., and T. A. Abella. 2004. Role of public sector in dissemination of tilapia genetic research outputs and links with private sector. Paper presented at the Workshop on Public–Private Partnerships in Tilapia Genetics and Dissemination of Research Outputs, January 21–23, in Tagaytay City, the Philippines.

Tayamen, M. M., T. A. Abella, and R. C. Sevilleja. 2006. Role of public sector in dissemination of tilapia genetic research outputs and links with private sector. In *Public and private partnerships in aquaculture: A case study on tilapia research and development*, ed. B. O. Acosta, R. C. Sevilleja, and M. V. Gupta. Penang, Malaysia: WorldFish Center.

Vannuccini, S. 2001. Global markets for tilapia. In *Tilapia: Production, marketing, and technical developments—Proceedings of the tilapia 2001 international technical and trade conference on tilapia,* ed. S. Subasinghe and T. Singh. Kuala Lumpur, Malaysia: Infofish.

WorldFish Center. 2007. *Annual report.* Penang, Malaysia: WorldFish Center.

World SeaFood Market. 2005. A fish called tilapia. <http://www.thefishsite.com/articles/68/a-fish-called-tilapia>. Accessed August 20, 2009.

Yap, W., E. A. Baluyot, and J. F. Pavico. 1983. Limnological features of Lake Buluan: Preliminary findings and observations. *Fisheries Research Journal of the Philippines* 8 (1): 18–25.

Chapter 18

"Crossing the River while Feeling the Rocks": Incremental Land Reform and Its Impact on Rural Welfare in China

John W. Bruce and Zongmin Li

Land Reform in China

During 1978–84, when Chinese communities implemented a decollectiviza-tion reform of stunning scope, the management of about 95 percent of farmland in China was returned to more than 160 million farm households. The gains produced were as impressive as the scope of implementation. Production and productivity per hectare increased markedly, with readily discernible impacts on the incomes, food security, and nutritional levels of Chinese families. The efficiency gains in agriculture released a large amount of surplus labor that had existed under collective cultivation, and this, in turn, fueled the development of China's equally remarkable rural industrialization, accomplished through the development of township and village enterprises (TVEs). In more recent years, the government has turned its attention to gradually strengthening the property rights of the households that were reform beneficiaries.

What was the basis in economic theory for this land reform and its success? The economic literature recognizes two fundamental institutional reforms in the land sector: redistributive land reform and land tenure reform. There is a consensus among economists that where labor is relatively cheap, household farming enjoys advantages in labor supervision that make it more efficient in using land than are larger-scale operations, even though the latter may be more highly mechanized. Few larger-scale economies exist, and those that do are not in production itself but in primary processing and marketing. Most often, historically, large agricultural holdings have not been the product of market forces and have had less to do with efficiency than with power, domination, and ideology. Land reform is the scaling down of those large units, usually through a process of subdivision, to smaller, more efficient farm sizes. When these are given to families in secure tenure, there is a significant distribution of wealth and economic opportunity in the society. The principle is the same regardless of whether one is reforming Latin American *latifundia* or collective farms in socialist contexts (Binswanger, Deininger, and Feder 1995).

The economic literature also asserts that the quality of land rights affects the landholder's investment incentives. Tenure security, the confident expectation of being able to continue in possession of one's land long enough to recoup investments in land, and even to leave the land to one's heirs, substantially strengthens the farmer's incentive to invest. This is especially important to investments in activities such as soil building, tree planting, or terracing, which enhance the long-term productivity of the land. The rights to transfer and mortgage land are again asserted to affect both investment incentives (because the investment is reflected in the market value of the land) and credit access (because the land can be used to secure loans). Land tenure reform involves increasing the robustness of the property rights of the producer, thereby increasing incentives to invest and thus the productivity of the land. This is a core understanding, reflected in policy prescriptions of the international development community and in particular the World Bank (Feder and Feeney 1991; Deininger 2003).

The literature on rural reform in China typically breaks the rural reform into the "incentive reforms" that dominated in 1978–84 and the gradual market liberalization that began in 1985 and extended through the 1990s (see, for example, Huang, Otsuka, and Rozelle 2007). This chapter instead focuses on the land reforms. The first reform, the return to household farming from collective agriculture, was, together with price reforms, part of the "incentive reforms" that took place during 1978–84. The second, a gradual enhancement of property rights, paralleled the market liberation that began around 1989 and is ongoing. The first is considered a spectacular success, whereas the second remains incomplete, and its effects are still being assessed.

The constant throughout these two reforms has been what Chinese documents refer to as the "two-tier system": public (collective or state) ownership of land and private-use rights to land rather than full private ownership for users. This conceptualization, with its retention of public ownership of land, has made it possible for a communist state to embrace a reform process that has moved China ever closer to de facto private property in land.

This chapter focuses primarily on the first reform, known as the Household Responsibility System (HRS) reform. It explores its impacts on production, investment, food availability, and welfare, but it also covers the ongoing process of strengthening producers' property rights, with some limited evidence regarding the actual and potential impacts of this second reform. Table 18.1 summarizes the changes that took place during the reform period.

Exiting Collective Agriculture

The collectivization of agriculture in China after 1956 was expected to benefit from the economies of scale predicted by Marx and also to provide a base for the development of rural industries. It was also intended, suggests Lin (2003, 14), "to

Table 18.1 Summary statistics of changes in rural China, 1978 and 1984

	1978	1984	Percent	Source
Rural households affected by the land reform (millions)	160	n.a.	n.a.	
Changes in grain yield (kilograms per mu)	169	239[a]	41.42	Chen, Chen, and Yang (1992, 536).
Changes in food production (100,000 metric tons)				State Statistical Bureau (various years).
Grain	3,047.5	4,070.0	33.55	
Cotton	21.7	62.5	188.02	
Oil-bearing seeds	52.2	118.5	127.01	
Tea	2.7	4.1	51.85	
Sugar	211.2	396.6	87.48	
Meat (pork, beef, lamb)	85.6	152.5	78.15	
Changes in per capita food consumption (kilograms)				Zweig (1997, 14).
Grain	195.46	249.65	27.72	
Edible oils	1.6	4.66	191.25	
Pork	7.67	12.93	68.58	
Beef and lamb	0.75	1.23	64	
Poultry	0.44	1.35	206.82	
Eggs	1.97	3.88	96.95	
Aquatic products	3.50	4.32	23.43	
Changes in rural per capita income (yuan)	220	522	137.27	State Statistical Bureau (various years). Cited in Fan, Zhang, and Zhang (2004, 397).
Changes in calories consumed (kilocalories per day)	2,226.9	2,450.0[b]	45.5	Wakashiro (1989, 18).
Changes in rural poverty rate (percent)	33	11	–22	Fan, Zhang, and Zhang (2004, 396).
Changes in rural nonagricultural labor force (10,000)	3149.5	5169.6	47.04	Chen, Chen, and Yang (1992, 603).

Notes: n.a. means not applicable. 1 mu equals 15 hectares.
[a]This figure is for 1988.
[b]This figure is for 1980.

serve industrialization by affecting a mandatory drain on rural surplus. Taxation was carried out through the control of basic rural production factors, the monopoly in the circulation of agricultural products and especially price scissors between industrial and agricultural products." He notes that Chinese economists have assessed the total value drained from agriculture through scissor pricing from the 1950s to the 1980s at ¥ 600–800 billion.

The impacts of collectivization on production were disappointing. After 1959, grain production declined. The country suffered serious famine in 1960–63. Du Runsheng, the party's director of rural policy from 1978 to 1989, recently summed up the situation: "Per capita grain production never averaged much more than 300 kilograms. Of the 800 million peasants, 250 million were impoverished. The nation as a whole could not achieve self-sufficiency in grain and required massive imports" (Du 2006, 2). Collective agriculture proved incapable of fostering agricultural productivity comparable to that of the "Asian tigers" such as South Korea and Taiwan. Distorted incentives were at the base of the failure. In spite of the system's highly egalitarian aspirations, a work point (*gongfen*) system was introduced in an attempt to create incentives for harder work, but supervision of labor in the brigade proved extremely difficult. Ultimately each worker received the same number of work points each day regardless of the quality or the quantity of his or her labor. With little incentive for individuals to invest their labor, productivity of labor and productivity per land unit declined. Shared tasks without accountability resulted in extensive disguised underemployment (Lin 1988; Lin 1990; Lin 2003).

The reform process began with local experimentation in Fengyang County in poverty-ridden Anhui Province, a poor region plagued by flood and famine, driven by famine and the collapse of confidence in collective agriculture.[1] In 1978 a few production brigades in Anhui Province secretly distributed their land to their member households to farm. The year's productivity increases were impressive. Some brigades in Anhui that returned to household farming had production increases two to five times larger than those in unconverted brigades (Lin 1988). Local officials embraced the reform, which was then carried out under the protection of Wan Li, the provincial governor of Anhui (Ho 2005).

By 1976, Chairman Mao had died and the Cultural Revolution had come to an end. The country was in chaos, and there were grain failures and famine in parts of China. The time was right for the party to reconsider its options. As late as 1977, a return to household farming was specifically forbidden by the Central Committee of the party. Because the performance of collective agriculture had been better in the more fertile coastal areas, opposition to reform was considerable in those provinces. Gradually, however, the party came to accept the idea of contracting land to households. Wan Li was promoted from Anhui to the politburo, and the Fengyang experiments received support from influential leaders such as Chen Yun and Hu

Yaobang. In 1978 the Plenum of the 11th Central Committee of the Communist Party allowed the option of breaking up the communal lands into household holdings (Ho 2005).

The new system spread by voluntary decision of the local collectives.[2] In January 1980 only 1.02 percent of all production teams in China had changed over to household farming, but by December 1980 the figure was 14.40 percent, then 28.20 percent by July 1981 and 45.10 percent by October 1981. By 1981, when the government finally recognized that the HRS reform was broadly applicable, 45 percent of the production teams in China had already been dismantled (Lin 1988; Lin 1992; Lin 2003).[3] By the end of 1983, about 97.7 percent of the production teams and 94.2 percent of the farm households in China were farming under the new system (Lin 1988). In a few years, collective land was contracted out to more than 160 million households (Ho 2005).

The Household Responsibility System

Under HRS, the new system of contracting land to the household, land continued to be "owned" by the old production brigade under the collective system, but these units began the transition from production organizations to communities.[4] A community's rural land under HRS typically was classified in one of four categories: residential land, construction land, responsibility farmland, and household food plots.[5] The reform was focused on responsibility farmland, the collective farmland from the commune period, which was distributed to households to cultivate,[6] usually with some services such as tractor plowing or irrigation maintenance provided by the production teams. It was distributed strictly according to household size, without reference to the size of the household labor force. The distribution was highly egalitarian, and as a result each household's holding was fragmented into an average of nine tracts, even though the size of a holding was only about 1.2 acres (Lin 1989). Regulations prohibited sales of use rights, rentals, and use of land as collateral for both privately and collectively managed land (Li, Rozelle, and Huang 2000).

Tenure was provided to users of farmland through contracts from the brigade's executive committee. These were one- to three-year contracts with households and included an obligation on the part of farmers to produce specific amounts of staple crops (rice, wheat, and some other crops) for sale to the state at fixed prices (hence the emphasis on responsibility in the term *Household Responsibility System*). Crops produced over these quotas could be sold on the open market, as could crops not covered by quotas. The use contracts could involve relatively modest use charges, could be inherited by resident heirs, and could be subleased but not otherwise transferred. One finds variations among communities, because the contracts were produced locally (there was no national model contract), but this was the broad pattern (Wang 2005).

The highly egalitarian ethos of the villages led to frequent redistribution of the responsibility land (referred to as "adjustments") to ensure the equality of holdings as families changed over time. Redistributions were often made at the initiative of the collective but also took place to accommodate government projects that needed land. Other factors were in play as well: a desire on the part of the government to prevent the development of too strong a sense of land proprietorship on the part of farmers and a desire on the part of collective cadres to take advantage of the considerable rent-seeking opportunities such redistributions offered (Rozelle et al. 2005; Wang 2005). Land redistribution frequencies varied considerably. Li, Rozelle, and Huang (2000) found that more than 90 percent of the villages in Hubei had readjusted in the past 15 years, but in Sichuan, the percentage with readjustments was only 22 percent.

Although the security of tenure left something to be desired under the early HRS, the HRS still had manifest advantages. Ling (1991) summarizes them: the flexibility to use family resources to deal with temporary labor shortages, the ability of the family to adjust expenditures, the low management costs of self-management, the fact that the production unit was conterminous with the consumption unit, and the fact that households' income and living standards now depended on themselves and how they managed their farms. The farmers in Ling's study pursued their comparative advantage, reducing areas under grain and shifting to oil-bearing seeds, tobacco, and vegetables and moving into animal husbandry (Ling 1991).

The new system had one important advantage over many land reforms for its beneficiaries, who had previously often found their input supply and marketing chains badly disrupted. The beneficiaries of the HRS reform had a guaranteed market for a quota amount of their major crops at state-set prices and the opportunity to sell their overquota production and nonquota products to the state at market prices.

Marketing Reforms and Land Reform

The sad state of Chinese agriculture in 1977 led the government to introduce a significant reform of the system of state procurement of agricultural products and of procurement prices at the same time that the HRS reform was getting under way. This was in fact the primary government response to the food crisis. The HRS was a more spontaneous development later embraced by the government.

In 1977, the Chinese government was the only legal purchaser of many key commodities, including rice, wheat, maize, oilseeds, and cotton. Provinces were assigned quotas by the central government, and these were broken down geographically, so that production teams, at the base of the pyramid within the communes, were assigned quotas. Quota production had to be marketed to the state at prices set by the state. With the division of communes into household farms, these quotas became the responsibility of households, assigned by the same contract that allocated them their farmland. In these years the state was effectively a monopoly purchaser of quota production, of grain and key agricultural products produced in excess of quotas, and often even of nonquota crop production.

After 1977, the state procurement prices were raised to increase farmers' incentives. In 1979 alone, state procurement prices for major crops increased, on average, by 22.1 percent. The government also began to increase grain imports and loosened restrictions on private interregional trade in agricultural products (Lin 1992). Government purchasing prices rose greatly in the years after 1979, those for grain by 100 percent and those for many other crops by 40–50 percent, with free market prices still higher (Ling 1991). Huang (1998) provides procurement prices for 1952–90.

At the same time, the state introduced important reforms in the state procurement system. From 1977 onward, farmers were allowed to trade grain on free markets once they fulfilled their delivery quotas to the state procurement system. Bans that prohibited peasants from growing cash crops were terminated. Farmers regained the right to grow vegetables or other nonquota cash crops and to sell their products in the open markets. Trade in urban areas was still restricted in the initial stage, however, and most markets were still substantially controlled by the government (Jia and Fock 2007).

Since 1988, a series of reforms have sought to further change the incentive structures of grain production, consumption, and exchange. To enhance farmers' incentives to undertake grain production and exchange, either the state purchase price has been raised or the quantity purchased at the lower state price has been reduced, depending on the measures adopted in the marketing reforms (Kung 1992). To secure low consumption prices in favor of urban areas and industry, the domestic grain market was rationed. The state raised grain prices a number of times after 1979. Grain rationing was abolished in the early 1990s. Although the Chinese government continued to use price controls to keep grain prices low and the supply stable for urban areas, what is important for our purposes is how the procurement and market reforms were coordinated. In contrast to many postcommunist countries, China opted for a two-track approach, maintaining quotas and set prices for quota production while liberalizing markets for nonquota production and allowing markets to control prices for above-quota production of quota crops. Lin (2003, 332–333) believes that it is this "crossing the chasm" between state prices and the market "in two steps" that constitutes the genius of China's agricultural pricing reforms. The approach, he argues, assured the state of sustained grain production and assured farmers of a predictable if modest farm income during a period of uncertainty as China returned to household farming.

Land Reform and Rural Industrialization

With the transition to household farming, it became evident that substantial disguised surplus labor had existed in collective agriculture. In 1985 the main foodgrain crops required fewer labor days per hectare than in 1978—down 22 percent for rice, 48 percent for corn, and 53 percent for wheat (Vermeer 1989). The labor freed up had to find other local employment. Those thrown off by agriculture could not migrate elsewhere to work because of the *hukou* permit system, a remarkably effective system of strict residence control that kept laborers in their existing rural and urban areas (Zhang, Li, and Shao 2006).

The solution to the problem of the rural labor surplus was rural industrialization, reflected in the slogan "Leave the farm but not the village, move to the factory but not the city." The rapid development of the agricultural sector under the HRS reforms created not only a labor surplus but also a surplus of funds for local enterprise development (Lin 2003). Township and village governments inherited the collective factories on the break-up of the communes, and they seized on the opportunity presented by inexpensive local labor. Rural industrialization was only later embraced by the government as policy (Pei 2005). The TVEs became the success story of Chinese development in the 1990s. Like the HRS, the TVEs were not the brainchildren of planners but were formed as a spontaneous reaction to market opportunities by local governments. They were public but autonomous, and their production and other decisions were market driven.

The rapid launch of rural industries was facilitated by continued collective ownership of land by the successors to the communes (Pei 2005). In 1978 the gross output value of the TVEs was just 7.2 percent of China's total output value. In the 1980s rural industry became the leading force behind rural economic development, with 60–70 percent of rural output value produced by TVEs (Lin 2003). Zweig (1997) shows the dramatic growth, with the number of firms increasing from 1.5 million in 1978 to 17.5 million in 1987 to nearly 25.0 million in 1994, while the number of employees increased by a factor of 4.5 and total output value by a factor of 80. By 1996, China's rural enterprises employed more than 135 million people, about one-third of the rural labor force. These laborers had been released by the decollectivization of agriculture, and yet agricultural production continued to grow (Oi 1999). Lin (2003, 312) notes that TVE development, sparked by the HRS reform, has driven subsequent reforms: "TVE development was a market driven process, and greatly pushed the economic reform toward a market economy, corrected the distorted industrial structure, and became part of a dual track system in areas of both resource allocation and price formation which placed the traditional system under growing pressure."

The Property Rights Reform

The initial impact of the HRS reforms seems to have played out by the end of the 1980s. Fan, Zhang, and Zhang (2004) explain that during 1985–89, rural income continued to increase but at the much slower pace of 3 percent per year. Although there were calls for a return to larger-scale farm production, the predominant response among Chinese land policy analysts was to urge a strengthening of farmer property rights. How and when those rights should be strengthened remained topics of debate. The initial HRS contracts were annual but were often extended. The major threat to farmers' security of tenure was the periodic reallocations of land, which were urged for their significant disincentive effects on investment in the holding. Such reallocations could override the terms of contracts with farmers. Reallocations involved high transaction costs (Zhu and Jiang 1993). A 2001 survey of more than 1,600 households in 17 provinces conducted under the auspices of the Research Center for Rural Development found that four-fifths of the villages had conducted at least one land readjustment since the first allocation after the HRS was instituted. Legal reform came gradually. At the outset, there was no law regarding the HRS, and reform was accomplished through a series of rapidly evolving party pronouncements and instructions. The approach to normative innovation was ad hoc experimentation, tweaked through a succession of party edicts. A legal basis for the new institution would be provided only once the new model had proved itself. In China, a new law was not associated with innovation but was seen as evidence that a new arrangement had succeeded and was to be enshrined in law (Du 2006).

Rural Work Document 1 of 1984 urged local officials to extend the duration of farmland contracts to 15 years and for longer periods for special uses such as reforestation and fruit tree plantations. The politburo's Document 5 of January 1987 ("Deepening the Rural Reforms") encouraged farmers who had moved into nonagricultural employment to transfer their use rights to others. To this end, the Constitution's prohibition of all transactions in rural land (Article 10) was amended by the National People's Congress on April 12, 1988 (Bruce and Harrell 1989). The basic land administration arrangements in the wake of the HRS were first set out in law as the Land Administration Law of 1987, a decade after the reform began. The 1987 law reaffirms collective ownership of rural land (Article 8) and allocation to households for use under contracts (Article 12). It does not include some of the tenure-strengthening measures recommended in party pronouncements, which had been permissive rather than mandatory, but in 1999 amendments to the law for the first time *required* a minimum term of 30 years for farmland tenure. Longer terms, up to 70 years, are available for specialized uses such as forestry or agroforestry on hillside lands and for some construction projects. During Bruce's fieldwork in Fujian in the mid-1990s, he realized that although some responsibility contracts were being made for longer periods, they still contained clauses providing for reallocations at the discretion of the collective, nullifying the greater security of tenure provided to farmers by the longer terms (Bruce and Muo 1998). Change came slowly, with local cadres often ignoring the legal changes.

In 2002, after several years of intensive debate in party circles, a full legal framework for the HRS was provided. The Rural Land Contracting Law, which became effective in 2003, does not replace the Land Administration Law but rather supplements it. Now all land use contracts are required to be for 30 years for arable land, 30–50 years for grassland, and 30–70 years for forest land. They must be in writing and must be signed by both parties. Readjustments are restricted. Land is not to be readjusted (redistributed by the collective) during the contract term, except as required by a natural disaster and/or unspecified "other special circumstances." Any readjustment during the term of the contract must be approved by two-thirds of the members of the village assembly or two-thirds of the village representatives. The village may maintain a part of its land in a flexible reserve to adjust landholdings for newly added village residents, and it may also use reclaimed land and land returned voluntarily by contracting parties for this purpose. The contracted land use rights are now clearly inheritable during their terms, an important change. The law confirms the right of the holder to transfer (assign), lease, exchange, or otherwise engage in transactions regarding the use right. For assignments and exchanges, the permission of the collective must be sought, but for other transactions the collective must only be notified.

The new law went into effect on March 1, 2003. Studies suggest that implementation of the reforms under the 1999 Land Administration Law and the 2003 Rural Land Contracting Law has been very uneven (Prosterman, Schwartzwalder, and Jianping 2000; Prosterman, Ping, and Zhu 2004). But the impetus for further property rights reform prevailed. In 2007 China launched a comprehensive new Property Law. Although this law is beyond the scope of this chapter, it does demonstrate the government's commitment to a long-term strengthening of property rights.

Land Reform Impacts

The impacts of the first land reform, the HRS reform, have been carefully studied and analyzed. A remarkable degree of consensus exists regarding both the positive impacts of the reform and the major role it played in initiating broader rural reforms. Those studies focus on the key years 1978–84.

HRS Reform Impacts

A number of indicators have been used to assess the impact of the HRS reforms. These include investment in farms, farm productivity, farm income, and, to a lesser extent, food availability and nutrition.

Investment Impacts. From 1979, when the HRS had begun to spread rapidly, the government's investment in agriculture declined from ¥ 6.24 billion, reaching ¥ 3.43 billion in 1983, then partially recovered to ¥ 4.68 billion in 1987. Meanwhile the collectives' investment decreased from ¥ 8.71 billion in 1979 to ¥ 2.07 billion in 1985. These declines were offset, however, by an investment of ¥ 34.49 per household in 1981, which increased to ¥ 100.13 in 1983 and remained at an annual level of ¥ 80–100 thereafter. Between 1983 and 1984, annual household investments averaged ¥ 18.10 billion, more than twice the annual combined state and collective investments during the same period and more than the annual combined state and collective investments before the reforms (Feder et al. 1992).

Cumulatively, although farmers' investments in agriculture were substantial, they were dwarfed by farmers' investments in residential construction. The housing investment was 85 percent of farmers' total investment in 1982, remaining high throughout the 1980s and still reaching 70 percent in 1992. Feder et al. (1992) found that the ratio of housing investment to productive investment in the counties they studied ranged from 2.71 to 10.88 during the period 1983/84–1987/88. This rush to invest in housing reflects the insecurity of tenure on productive land, which was subject to periodic land reallocations, compared with the more secure tenure on residential land. Farmers' investments in land were for the most part not investments in land itself but in capital stocks such as farm equipment and livestock (Feder et al. 1992).[7]

Production Impacts.   Research on the effects of the HRS has focused primarily on its impact on productivity, especially on teasing out the relative importance of institutional reforms, pricing reforms, technological change, and public investment in achieving the major productivity gains of 1978–90.[8]

In his analysis of provincial-level panel data, Lin (1988, 1989) found 42.2 percent output growth in the cropping sector in 1978–84. Between 1978 and 1984, the key years for HRS implementation, the annual growth rates for the three most important crops, grain, cotton, and oil-bearing crops, averaged 4.8, 17.7, and 13.8 percent, respectively. These rates compared favorably with the average growth rates of 2.4, 1.0, and 0.8 percent per year, respectively, for those crops in the preceding 26 years, 1952–78. Zhang, Li, and Shao (2006) estimated that the grain area output per unit under the HRS was nearly 25 percent higher than that under collective farming. They also found that national grain output rose about 300 million tons in 1978 and increased to 407 million tons in 1984.

Lin (1992) found that 45.8 percent of this output growth came from increases in inputs, the most important of which were fertilizer applications, which alone accounted for about one-third of the output growth. The HRS reform accounted for 48.6 percent of output growth, as much as the combined effects of the various input increases. Changes in market prices and state procurement prices did not affect productivity, but they did contribute nearly 16.0 percent of output growth, probably through input use, cropping intension, or crop mix. This finding is roughly in line with those of the two other serious econometric analyses. MacMillan, Whalley, and

Jing (1989) found that of the total farm productivity increase in 1978–84, 41 percent of the increase in the cropping and animal husbandry sectors could be attributed to total factor productivity growth; of that, 78 percent was attributable to the farm institutional reform (HRS) and 22 percent to price increases. Wen (1989) found that farm output increased by 56 percent due to the institution of the household-based farming system. Zhang, Li, and Shao (2006) reached similar conclusions.

Other researchers have urged greater attention to the roles of technological change (Fan 1991; Huang and Rozelle 1996; Huang, Otsuka, and Rozelle 2007) and the contribution of decades of heavy public investment in infrastructure prior to the reforms (Fan, Zhang, and Zhang 2004). They concur, however, that the majority of the production growth in Chinese agriculture between 1978 and 1984 was due to the rural reforms. There is thus a broad consensus, built on solid survey data and sophisticated econometric analysis, that the institutional reform creating the HRS was the primary factor responsible for the remarkable output growth between 1978 and 1984, accounting for 40–60 percent of that growth.

Income and Poverty Impacts. Per capita income increased to ¥ 522 in 1984 from ¥ 220 in 1978, a growth of 15.0 percent per year in real income per capita at 1990 prices, contrasting sharply with the pace of growth during the pre-reform period, 2.3 percent per year (Fan, Zhang, and Zhang 2004; Gulati, Fan, and Dafali 2005). Lin (2003), deducting for price factors, estimates that productive net income per capita increased from ¥ 166.39 to 291.10, while per capita cash in hand plus the outstanding amount of the savings deposit in rural areas (year-end) increased from ¥ 26.6 to 85.3. Between 1978 and 1983, per capita rural income more than doubled, rising from ¥ 133.6 in 1978 to ¥ 310.0 in 1983, significantly improving in relation to urban incomes (Renwei 1993).

Ling (1991) notes that the rate of growth in peasants' average per capita income (at constant prices) still increased by 2.1, 9.8, and 7.1 percent in 1985, 1986, and 1987, respectively. Ling makes an important point: those increases were due not just to increases in farm output and rises in prices of farm products but to an increase in the number of income sources, notably from employment in rural industries or housing construction. Still, even in 1998, agriculture accounted for 60 percent of rural household income (Oi 1999).

How did these increases in income affect poverty levels in China? In the 20-year period after 1981, the proportion of the population living below the poverty line fell from 53 to 8 percent. Half the decline in poverty came in the first few years of the 1980s. In 1980, a staggering 98 percent of China's poor lived in rural areas, and the bulk of the dramatic reduction in poverty during the first years of that decade came from these areas. The rural poverty rate as calculated by the Chinese government fell from 76 percent in 1980 to 23 percent in 1985. Between 1980 and 1985, the relative inequality between rural mean income and urban mean income fell to less than half of its 1980 level, flattening out in 1985–90 and rising again to 1980 levels during in the 1990s. Agricultural growth did more to reduce poverty and inequality than did growth in either the secondary or tertiary sector. However, the share of the urban population in poverty rose from 19 percent in 1980 to 39 percent in 2002 (Ravallion and Chen 2004).

How were the increases in income distributed? Working off such a radically egalitarian and relatively uniform livelihood base as that in rural China, it seems likely that the reforms would have increased inequality. In fact, the distribution of income improved by most measures during the early part of the reform period as average incomes rose substantially with only a modest increase in inequality. The Gini coefficient of income distribution in the rural areas declined steadily, from 0.32 in 1978 to 0.22 in 1982, but thereafter inequality began to increase and the Gini coefficient rose to 0.34 in 1988 and on to 0.42 in 1995 (Zong 1993; Griffen, Khan, and Ickowitz 2002).

Griffen, Khan, and Ickowitz (2002) ask whether the rise in income inequality after 1988 was one of the consequences of restoring a household farming system. Working with landholdings adjusted or unadjusted for irrigation, they recalculate the Gini coefficient for 1988 and 1995. They find that the distribution of land became more equal between 1988 and 1995. Therefore, it was not the HRS reform, they conclude, that was the source of growing inequality in rural incomes after 1988 but rather the growing inequality in nonfarm sources of rural incomes. Fan, Zhang, and Zhang (2004) attribute this worsening inequality to a growing differential in rural nonfarm opportunities among regions. Writing more recently, Khan and Riskin (2005) and Yingying, Hua, and Harrel (2008) reach the same conclusion: the greatest factor contributing to the development of regional inequality in rural incomes was rural enterprise. In earlier stages, wages from rural industry were the key factor, but by 1995 ownership of rural small businesses emerged as the greatest source of rural income inequality. Assets, these authors note, have replaced labor as the source of income inequality. Analyses by Fan, Zhang, and Zhang (2004), while suggesting that the impact of the HRS on agricultural growth may have been overstated in some studies due to the neglect of other contributing factors, still confirm that more than 51 percent of rural poverty reduction can be attributed to these reforms.

Food and Nutrition Impacts. Hardly any improvement took place in average Chinese per capita food consumption between 1957 and 1974. An official estimate puts the number of people who "suffered from a lack of grain" at more than 100 million in 1977 (Smil 1986, 25). Piazza (1983) provides data from 1950–81, a few years into the HRS. Working with a 1979 estimated energy requirement of 2,160 kilocalories per day, he finds that energy as a percentage of requirements reached 100 percent for the first time in 1972, increasing by 4 percent in the following four years, then by 12 percent in the four years of the HRS reforms for which he has data (1978–81). Working with an estimate of net protein use of 2.8 grams per day, he finds that, whereas in 1972 it ranged between 104 and 123 percent and was 122 percent in 1978, in 1979–81 it was between 140 and 143 percent.

By 1984, per capita availability of rice was 30 percent above the 1977 level. A 1975 survey gave an average of 2,188 kilocalories a day, and by 1983 Smil's estimate of per capita availability was 2,450 kilocalories in the villages and 2,150 kilocalories in the cities, for a national average of 2,380 kilocalories a day, though Smil (1986) notes that there were major regional differences. He estimated that in 1983 some 100 million peasants (11 percent of the population) were consuming fewer than 2,100 kilocalories per day, which he considers a minimum; another 90–100 million were only slightly above that minimum, at about 2,200 kilocalories per day (Smil 1986).

At the same time, the variety in the Chinese diet was improving. After 1978, consumption of secondary foods (foods other than grain) rose rapidly (Wakashiro 1989; Harrold 1992). Areas sown to various crops reflected this change. In the years immediately following 1978, the area sown to grains declined by 8 percent, compared with an increase in the area devoted to economic (nonquota) crops of 40 percent (Vermeer 1989). Looking at the bigger picture, Burgess (1997) points out that the universal and egalitarian access to land in rural China was critical to these mass increases in caloric intake, especially the large areas of China in which residents faced food markets characterized by high transaction costs.

Property Rights Reform Impacts

A "Vitriolic" Debate over Property Rights. The primary thrusts of the property rights reforms have been to reduce periodic reallocation of holdings, to extend the terms of use rights, and to enhance the marketability of the use right. In contrast to the HRS reforms, no consensus has emerged among researchers as to the benefits of these reforms. Ideology and policy preferences for a more or less egalitarian rural society drive an ongoing debate, which Li, Rozelle, and Brandt (1998, 64) correctly describe as "vitriolic."

Some argue that incomplete land rights due to reallocations and restrictions on rentals and other transactions weaken incentives for investments and prevent consolidation of fragmented holdings. They urge greater strengthening of land rights (Feder et al. 1992; Zhou 1994; Wen 1995; Carter and Yao 2004; Prosterman, Ping, and Zhu 2004). Similarly, Zhang, Li, and Shao (2006) are concerned that the HRS has become an obstacle to the development of larger-scale, mechanized grain production. As labor becomes more expensive, they urge increased transferability of the farmland use right to allow specialized grain-producing households to enlarge the sizes of their farms. Zhu and Prosterman (2007) make a strong case for moving forward on property rights reforms.

Others (Kung 1995; Dong 1996; Kung and Liu 1996) suggest that gains from greater liberalization would not be great and that farmers themselves do not favor privatization because they value the assured land access they enjoy under collective ownership. They argue that administrative allocation is often economically rational, favoring efficiency, and believe that China is not ready for full liberalization of the market in land rights, given that credit markets are undeveloped, there is no land registration system, and the legal system is incomplete.

The debate can be seen as one over the balance between the production or social security functions of land, both legitimate functions, and how quickly or slowly that balance should shift. What empirical studies have addressed these issues?

Estimating Economic Impacts. Studies of the impact of these property rights reforms are complicated by the fact that the reforms have been highly incremental and that implementation has lagged well behind party pronouncements and laws. A 17-province survey studied implementation of the 30-year land use rights under the 1998 Land Management Law. It found that in spite of the absolute prohibition on land readjustments during the 30-year term for land use rights, just under half of the respondents had received the land use right contracts mandated by the 1998 law, and many of the provisions in these contracts were inconsistent with or even violated national land readjustment laws and policies. One in five farmers reported that their village had indeed conducted a land readjustment since the implementation of the 30-year rights (Prosterman, Schwartzwalder, and Jianping 2000). These differences reflect the fact that in what is, de facto, a highly decentralized system, decisionmaking on implementation resides largely with local and village authorities rather than the central government (Carter and Yao 2005). The studies therefore seek to explore the impact of property rights by searching out contrasting levels of such rights in different types of plots or in different villages, given that implementation has gone further in some areas than others, and comparing behavior.

Deininger and Jin (2002) examine the impact of stronger land tenure rights in Guizhou Province in the south, which resulted from the provincewide replication of elements in the 1987 Meitan pilot reform between 1994 and 1997. They find that more secure land rights had a significant and positive impact on farmer investment on upland fields, though not on paddy land. They also found an active rental market in use rights (subleases) and suggest that being able to transfer land could have a major impact on agricultural investment (Deininger and Jin 2002; Deininger and Jin 2007).

Other studies tend to confirm that more secure and more transferable property rights would have a positive impact, but the issue of the extent of that impact and the appropriate timing remains. A study of 130 farmers in Fengning County of Hebei Province in northern China found that the right to use land for long periods encouraged land-saving investments (Li, Rozelle, and Brandt 1998). Private plots (household food plots) had 13 percent higher yields and received 18 percent more labor and 14–32 percent more inputs (depending on the input) than did collective parcels. Although the results show that land tenure affects agricultural production decisions, the difference between collective and private plots is small compared to the productivity gap that existed between private plots and communal land in the pre-reform period. Overall, the authors conclude, the differentials are small and may be outweighed by the social insurance role played by collective ownership. Other studies from northeast China show more input use and higher yields on plots held for longer periods of time (Li, Rozelle, and Huang 2000) and greater use of organic fertilizer (a useful indicator because of its long-term impact on fertility) where risks of loss of land were reduced (Jacoby, Li, and Rozelle 2002). The researchers' welfare analysis suggests that guaranteeing land tenure in this part of China would yield only small efficiency gains, however (Rozelle et al. 2005). Citing Dong (1997), they suggest that this may be because most capital-intensive agricultural investments, such as those in canal irrigation, drainage, and terracing projects, are undertaken at the village level, not the individual level (Jacoby, Li, and Rozelle 2002).

Rozelle et al. (2005) describe forgoing a modest efficiency gain from enhanced property rights as an implicit premium for the social insurance provided by collective ownership, with its highly egalitarian distribution. Concurring with Kung (1995), they conclude that tenure strengthening is not urgent and suggest that alternatives to the social security function of egalitarian land distribution be found first. Huang, Otsuka, and Rozelle (2007) urge both implementation of an alternative rural social security system and continued progress toward privatization of rural land. The two are in fact necessary complementary measures, and both can be undertaken if China is prepared to make the needed investment in its rural areas.[9]

Gendered Impacts.   In the meantime, some other, unanticipated effects of the property rights reforms have become clear. Although male and female family members may both benefit from land reforms, it is not unusual for land reforms to have gendered impacts as well. They often place women at a disadvantage in some respects. In much of the postsocialist world, the land reforms vesting land in households have in fact caused land to be vested in the male household heads, thus strengthening patriarchal family control (Meinzen-Dick et al. 1997). To what extent has this been the case in China, which prides itself on legal guarantees of gender equality?

Li (1993) explains that women had in theory participated as equals in collective agriculture under the commune system, though women were typically employed in more menial positions and largely excluded from management. In the conversion to the HRS, households received land according to their labor supply. Wives and adult daughters were counted for this purpose, though sometimes given a lesser value than male laborers. In the first decade of the system, one of the adjustments made in periodic reallocations was to adjust family holdings to reflect changes in family size and marriages. Rural people follow the *cong fu ju* marriage custom, prevalent throughout China, in which a young man marries a girl from outside his home village and his bride comes to live with his family in his village. When the new wife arrived, the land allocation of the family she joined would be increased accordingly at the next opportunity. She would not receive a discrete parcel of land, but her presence would be counted in determining the overall landholding of the family. Wives and daughters, like other family members, shared in the many benefits of the HRS.

Rural industrialization, made possible by the labor surplus revealed by the HRS, began to change the roles of women in their households. In Hebei Province, the number of part-time agricultural households grew as male laborers moved into local jobs in TVEs. By 1992, 25 percent of village families farmed part time. In those families, women were almost entirely responsible for agricultural production. Tasks formerly done only by husbands (storing grains, processing food, trading produce in farmers' or state markets, building houses and walls, and transporting produce by bike and tractor) became shared tasks. Some jobs formerly considered "skilled" (managing pumps, driving a tractor, making adobe bricks, and feeding draft animals) came to be considered "unskilled" and thus appropriate for women (Li 1993, 1997).

At the same time that the role of women in agriculture was growing, the property rights reforms introduced after 1988 to limit periodic land reallocations negatively affected allocations of land to wives moving to their husbands' villages. Women had to wait longer and longer for the family they joined to receive this recognition of their labor contribution, and many villages ceased to provide for it altogether. Failure to receive such an allocation became an acute disadvantage later in cases of divorce and widowhood; the divorcee or widow was then seen as having no land entitlement in the village and was often left landless by the husband's family and the village. Organizations such as the All-China Women's Federation lobbied for more frequent reallocations, recognizing that women were being disadvantaged in the reform (All-China Women's Federation 1999).

This issue became a topic of public debate and was tackled directly in Article 30 of the Rural Land Contracting Law, which provides: "When a woman marries during the contract term, the contract-issuing party cannot take back her original contracted land unless she receives land where she moves. When a woman is divorced or widowed, the contract-issuing party cannot take her land back if she still lives at her current place of residence or moves to a new place of residence where she cannot get land."

Li and Bruce (2005) note that although Article 30 does not help a young woman obtain land rights in her husband's village, it does purport to protect her land rights in her parental village when she marries and moves away (at least until she receives land in the new village, an increasingly unlikely eventuality). The benefits that women derive from this provision depend on the extent to which families recognize that their daughters retain an interest in the family land and accept that they should be able to derive benefits from it. One authority on women's land rights (cited in Li and Bruce 2005, 276) worried that "women's claim to their land rights runs against the interest of their family, including father, brother, and mother, and means a rebellion against the patriarchal system, a cut-off of the kinship ties . . . [so] women who are ready to do so are unlikely to succeed in practice, because it is unlikely that they will get any legal support." On the other hand, Article 30 is potentially helpful to widows and divorcees, though it is short on details such as how land is to be partitioned in such situations. Enforcement, as in the case of all legal rights in China, is likely to be problematic.

The Reform Process and Success in China's Land Reform

The Chinese land reform experience has been positive, and it is important to understand the factors that have contributed to its success. The substance of the reforms is central, of course, but few other countries are emerging from collective agriculture, and the more interesting lessons from China for other countries may have more to do with the reform process. What are the distinctive characteristics of these reforms?

At key transition points, China has provided opportunities for and encouraged community-based pilots to test reform ideas, often waiving or simply ignoring current law to do so. It has been well served by these experiments. Experiments at county and township levels with different models for postcollective agriculture continued through the 1980s (Bruce and Harrell 1989) and even into the 1990s (Prosterman and Bledsoe 2000). The learning from such experiments was developed and conveyed to officials by national research institutions focused on rural issues. Although these institutions were attached to the highest level of government, they were nonetheless allowed a fair degree of autonomy in pursing their studies and drawing their conclusions. Notable in years of HRS implementation and the decade following was the Research Center for Rural Development, led by the renowned economist Du Runsheng. The existence of these research institutions and the degree of autonomy given them (and increasingly, to university-based researchers) to reach their conclusions independently has been critical to transmitting lessons from local experiments into national policymaking.

A profile of a distinctive reform process emerges. It includes (1) a permissive attitude by the central government toward local experimentation, even experiments that violated existing law and policy and were seriously heterodox in terms of ideology; (2) analysis and transmittal of lessons from those experiments into the party and the government by state-sponsored research institutions with a degree of autonomy; (3) extended dialogue on reform issues within the party and the government; (4) reforms initially promulgated through party policy declarations, often with frequent follow-up instructions adjusting the direction of reform and urging implementation; and (5) enactment of a law or laws consolidating the changes once they were considered to have proved themselves. From a rule-of-law standpoint, this seems problematic, and some legal commentators have struggled with China's development successes with so little law. Dam (2006), looking at these issues some years down the line, concludes that there may be more law in China than some commentators imagine. The problem is resolved if we realize that party declarations did in fact have the force of law in China and in this period may well have had more force than law.

Gulati, Fan, and Dafali (2005) speak of a "learning by doing" approach to reform and argue that the adoption of new measures through experimentation rather than following a predetermined blueprint increased the likelihood of the success of reforms. Lin (2003) notes that although the reforms needed to be nonradical for ideological reasons, gradual reform had many advantages. He observes that most of China's economic reforms have not been implemented at the same time across the country. Reform has been local in nature, in the case of both spontaneous reforms, such as the HRS, and reforms imposed by government. He notes that this approach minimizes risks and that the combination of experimental spontaneous reform and incremental reform can provide timely signals of how and when further reform can maximize gains. In China it allowed the market to be established and fostered gradually. At the same time, incremental reform could make full use of existing organization resources, maintain relative stability, and ensure a smooth transition in institutions. Chen, Jefferson, and Singh (1992) provide a similar analysis, emphasizing the benefit of the opportunities for course corrections provided by incremental reform.

The reform of the HRS and the other reforms it sparked have had a profound positive influence on China's growth and the welfare of its people. The reform debate has become ever more public. The reform remains incomplete, however, in that rural property rights remain limited compared to those related to urban land. The prognosis for further reform is good, because the reforms to date have raised popular expectations of further reform, and there is genuine excitement among members of the new middle class about enhanced property rights and their effective protection under a rule of law. As China's economy continues to develop, land tenure institutions will again need to be adjusted as well to meet new needs and challenges. To paraphrase Mao, there appears to be a need not for permanent revolution but for permanent reform.

Notes

1. See Yang and Su (1998) for an account of the politics of famine and reform in China.

2. Lin (1987) analyzes the reform as farmers' institutional choice, using the model of induced institutional innovation developed by Hayami and Ruttan.

3. For a discussion of the politics and ideological debates around these changes, see Zweig (1997, 62–67).

4. Ho (2001) analyses what he describes as the "deliberate institutional ambiguity" involved in the statement that the rural collective owns the land. The "administrative village" (the successor to the brigade) most often managed the contracting of land to farm households, and it was often not clear whether it was this administrative village or the "natural village" (the successor to the production team) that owned the land. Inconsistent use of terms such as "farmers' collective," "collective economic organization," and "villagers' committees" in instructions related to land confounded cadres responsible for implementing them.

5. Rural homes were privately owned by households, though the land on which they stood was still owned by the collective. Housing sites were obtained without charge by residents and could not be transacted, though they could be inherited with a house if the heir was a resident of the collective. The house plot and the household food plot have been the most secure tenure niches in rural China, with no term limits. Construction land was allocated to public facilities and enterprises based on a contract for building. In this case there was considerable variability of terms among the contracts within and among communities, depending on the enterprise and the use.

6. Li, Rozelle, and Huang (2000) indicate that at the outset of the reform, local leaders allocated the collective land to peasants in three tenure types: ration land (*kouliang tian*), to meet household subsistence requirements; responsibility land (*zeren tian*), given on the condition that farmers deliver a low-price grain or cotton quota to the state; and contract land (*chengbao tian*), auctioned off by village leaders for a fee. Not all villages had all three categories of land, and land tenure types differed sharply among the villages: although all villages had responsibility land, a minority had contract land and fewer than 20 percent had ration land.

7. Investments in capital stocks (livestock and equipment) more than doubled in the five-year period covered by the study (1983/84–1987/88), with average annual capital growth of 15 percent over this period. Fewer than 10 percent of households owned tractors, but a large number of households invested in livestock for food (pigs and poultry) and draft power, and one site with extensive canal networks even invested in boats (Feder et al. 1992).

8. The analysis here looks at the contracted responsibility land as a whole, but there were important subcategories, in particular hillside land. This land, often neglected and denuded in the collective period, has been developed aggressively under agroforestry and tree farming since the advent of the HRS. Initial allocations of forest land to households on the same model as farmland unfortunately resulted in deforestation through unsustainable cutting. Later the emphasis shifted to reforestation under long-term contracts, with an emphasis on agroforestry. This has been far more successful and is an important environmental success of the HRS reform. See Bruce, Rudrappa, and Li (1995) and Zhang and Kant (2005).

9. For a strong brief by a legal scholar advocating privatization, highlighting its financial implications, see Palomar (2002).

References

All-China Women's Federation. 1999. Report on violation of women's land rights under the second land adjustment: A working paper. Beijing: Department of Women's Rights and Interests.

Binswanger, H., K. Deininger, and G. Feder. 1995. Power, distortion, revolt, and reform in agricultural land relations. In *Handbook of development economics,* vol. 3, ed. J. Behrman and T. N. Srinivasan. Amsterdam, the Netherlands: Elsevier Science.

Bruce, J. W., and P. Harrell. 1989. *Land tenure reform in the People's Republic of China, 1978–1988.* Land Tenure Center Research Paper 100. Madison, Wis., U.S.A.: Land Tenure Center, University of Wisconsin–Madison.

Bruce, J. W., and L. J. Muo. 1998. *Legal findings: Possible impediments in the law on land use and land tenure in Fujian Province.* Christchurch, New Zealand: Landcare Research NZ.

Bruce, J. W., S. Rudrappa, and Z. Li. 1995. China: The legal basis for community forestry. *Unasylva* 46 (1/180): 44–49.

Burgess, R. 1997. *Land, welfare, and efficiency in rural China.* London: London School of Economics.

Carter, M., and Y. Yao. 2004. Specialization without regret: Land transfer rights, agricultural productivity, and investment in an industrializing economy. World Bank Policy Research Working Paper 2202. Washington D.C.: World Bank.

———. 2005. Market versus administrative reallocation of land: An econometric analysis. In *Developmental dilemmas: Land reform and institutional change in China,* ed. P. Ho. New York: Routledge.

Chen, K., G. Jefferson, and I. J. Singh. 1992. Lessons from China's economic reform. *Journal of Comparative Economics* 16 (2): 201–225.

Chen, J. Y., J. J. Chen, and X. Yang. 1992. Rural economy before and after 1995. In *Chinese rural social and economic changes (1949–1989).* Taiyuan, China: Shanxi Economic Publishing House.

Dam, K. W. 2006. *The law-growth nexus: The rule of law and economic development.* Washington, D.C.: Brookings Institution Press.

Deininger, K. 2003. *Land policies for growth and poverty reduction.* World Bank Policy Research Report. Washington, D.C., and Oxford, U.K.: World Bank and Oxford University Press.

Deininger, K., and S. Jin. 2002. The impact of property rights on households' investment, risk coping, and policy preferences: Evidence from China. World Bank Policy Research Working Paper 2931. Washington, D.C.: World Bank.

———. 2007. Securing property rights in transition: Lessons from implementation of China's Rural Land Contracting Law. World Bank Policy Research Working Paper 4447. Washington, D.C.: World Bank.

Dong, X. 1996. Two-tier land system and sustained economic growth in post-1978 rural China. *World Development* 24 (5): 915–928.

———. 1997. Public investment, social services, and productivity of Chinese household farms. A Stochastic Frontier Analysis Working Paper. Manitoba, Canada: Department of Economics, University of Winnipeg..

Du, R. 2006. *The course of China's rural reform.* Washington, D.C.: International Food Policy Research Institute.

Fan, S. 1991. Effects of technological change and institutional reform on production growth in Chinese agriculture. *American Journal of Agricultural Economics* 73: 266–275.

Fan, S., L. Zhang, and X. Zhang. 2004. Reforms, investment, and poverty in rural China. *Economic Development and Cultural Change* 52 (2): 395–421.

Feder, G., and D. Feeny. 1991. Land tenure and property rights: Theory and implications for development policy. *World Bank Economic Review* 5 (1): 135–153.

Feder, G., L. J. Lau, J. Y. Lin, and X. Luo. 1992. The determinants of farm investment and residential construction in post-reform China. *Economic Development and Cultural Change* 41 (1): 1–26.

Griffen, K., A. R. Khan, and A. Ickowitz. 2002. Poverty and the distribution of land. *Journal of Agrarian Change* 2 (3): 279–330.

Gulati, A., S. Fan, and S. Dafali. 2005. *The dragon and the elephant: Agricultural and rural reforms in China and India.* Markets, Trade, and Institutions Division Discussion Paper 87 / Development Strategy and Governance Division Discussion Paper 22. Washington, D.C.: International Food Policy Research Institute.

Harrold, P. 1992. *China's reform experience to date.* World Bank Discussion Paper 180, China and Mongolia Department. Washington, D.C.: World Bank.

Ho, P. 2001. Who owns China's land? Policies, property rights, and deliberate institutional ambiguity. *China Quarterly* 166 (June): 387–414.

———. 2005. Introduction: The chicken of institutions or the egg of reforms? In *Developmental dilemmas: Land reform and institutional change in China,* ed. P. Ho. New York: Routledge.

Huang, J., and S. Rozelle. 1996. Technological change: Rediscovering the engine of productivity growth in China's rural economy. *Journal of Development Economics* 49 (2): 337–367.

Huang, J., K. Otsuka, and S. Rozelle. 2007. The role of agriculture in China's development: Past failures, present successes, and future challenges. Draft chapter for a forthcoming book edited by Brandt and Rawski. Stanford University, Stanford, Calif., U.S.A.

Huang, Y. 1998. *Agricultural reform in China: Getting institutions right.* Cambridge, U.K.: Cambridge University Press.

Jacoby, H. G., G. Li, and S. Rozelle. 2002. Hazards of expropriation: Tenure insecurity and investment in rural China. *American Economic Review* 92 (5): 1420–1447.

Jia, X., and A. Fock. 2007. Thirty years of agricultural transition in China (1977–2007) and the "New Rural Campaign." Paper prepared for the 106th Seminar of the European Association of Agricultural Economists, October 25–27, in Montpellier, France.

Khan, A. R., and C. Riskin. 2005. China's household income and its distribution, 1995 and 2002. *China Quarterly* 182: 356–384.

Kung, J. 1992. Food and agriculture in post-reform China: The marketable surplus problem revisited. *Modern China* 18: 138–170.

Kung, J. K. 1995. Equal entitlement versus tenure security under a regime of collective property rights: Peasants' preference for institutions in post-reform Chinese agriculture. *Journal of Comparative Economics* 21: 82–111.

Kung, J. K., and S. Liu. 1996. Land tenure systems in post-reform rural China: A tale of six counties. Working paper, Division of Social Sciences. Hong Kong: Hong Kong University of Science and Technology.

Li, G., S. Rozelle, and L. Brandt. 1998. Tenure, land rights, and farmer investment incentives in China. *Agricultural Economics* 19: 63–71.

Li, G., S. D. Rozelle, and J. Huang. 2000. Land rights, farmer investment incentives, and agricultural production in China. Agricultural and Resource Economics Working Paper, Department of Agricultural and Resource Economics. Davis, Calif., U.S.A.: University of California

Li, Z. 1993. *Changes in the role of rural women under the household responsibility system: A case study of the impact of agrarian reform and rural industrialization in Dongyao Village, Hebei Province, North China.* Land Tenure Center Research Paper 113. Madison, Wis., U.S.A.: Land Tenure Center, University of Wisconsin.

Li, Z., and J. W. Bruce. 2005. Gender, landlessness, and equity in rural China. In *Developmental dilemmas: Land reform and institutional change in China.* New York: Routledge.

Lin, J. 1987. The household responsibility system reform in China: A peasant's institutional choice. *American Journal of Agricultural Economics* 69 (2): 410–415.

Lin, J. Y. 1988. The household responsibility system in China's agricultural reform: A theoretical and empirical study. *Economic Development and Cultural Change* 36 (3): S199–S224.

———. 1989. Rural reforms and agricultural productivity growth in China. UCLA Working Paper 576, Department of Economics. Los Angeles: University of California at Los Angeles.

———. 1990. *Collectivization and China's agricultural crisis in 1959–1961.* UCLA Economics Working Paper 579, Department of Economics. Los Angeles: University of California.

———. 1992. Rural reforms and agricultural growth in China. *American Economic Review* 82 (1): 34–51.

———. 2003. *The China miracle: Development strategy and economic reform.* Hong Kong: Chinese University Press.

Ling, Z. 1991. *Rural reform and peasant income in China: The impact of China's post-Mao rural reforms in selected areas.* New York: St. Martin's Press.

McMillan, J., J. Whalley, and Z. L. Jing. 1989. The impact of China's economic reforms on agricultural productivity growth. *Journal of Political Economy* 97 (4): 781–807.

Meinzen-Dick, R. S., L. R. Brown, H. S. Feldstein, and A. S. Quisumbing. 1997. Gender, property rights, and natural resources. *World Development* 25 (8): 1303.

Oi, J. C. 1999 Two decades of rural reform in China: An overview and assessment. *China Quarterly* 159 (special issue): 616–628.

Palomar, J. 2002. Land tenure security as a market stimulator in China. *Duke Journal of Comparative and International Law* 12 (1): 7–74.

Pei, X. 2005. Collective landownership and its role in rural industrialization. In *Developmental dilemmas: Land reform and institutional change in China.* New York: Routledge.

Piazza, A. 1983. Trends in food and nutrient availability in China, 1950–81. World Bank Staff Working Paper 607. Washington, D.C.: World Bank.

Prosterman, R., and D. Bledsoe. 2000. *The joint stock share system in China's Nanhai County.* Seattle: Rural Development Institute.

Prosterman, R., L. Ping, and K. Zhu. 2004. Ensuring sustainable rural income growth by fully implementing the RLCL. Memo to Chen Xiwen, Deputy Director of Central Finance Leadership Group, Rural Development Institute, Beijing.

Prosterman, R., B. Schwartzwalder, and Y. Jianping. 2000. Implementation of 30-year land use rights for farmers under China's 1998 Land Management Law: An analysis and recommendations based on a 17-province survey. *Pacific Rim Law and Policy Journal* 9 (3): 507–568.

Ravallion, M., and S. Chen. 2004. *China's (uneven) progress against poverty.* Development Research Group. Washington, D.C.: World Bank.

Renwei, Z. 1993. Three features of the distribution of income during the transition to reform. In *The distribution of income in China,* ed. K. Griffen and Z. Renwei. London: Macmillan.

Rozelle, S., L. Brandt, G. Li, and J. Huang. 2005. Land tenure in China: Facts, fictions and issues. In *Developmental dilemmas: Land reform and institutional change in China.* New York: Routledge.

Smil, V. 1986. Food production and quality of diet in China. *Population and Development Review* 12: 25–45.

State Statistical Bureau. Various years. *China statistical yearbook.* Beijing: China Statistical Publishing House.

Vermeer, E. B. 1989. Food and agriculture in China during the post-Mao era. *China Information* 3: 83–88.

Wakashiro, N. 1989. Trends in the structure of food consumption and recent grain shortages. *China Newsletter* 83: 17–22.

Wang, W. 2005. Land use rights: Legal perspectives and pitfalls for land reform. In *Developmental dilemmas: Land reform and institutional change in China,* ed. P. Ho. New York: Routledge.

Wen, G. J. 1989. The current land tenure and its impact on long-term performance of the farming sector: The case of modern China. Ph.D. dissertation, Department of Economics, University of Chicago, Chicago.

———. 1995. The land tenure system and the saving and investment mechanism: The case of modern China. *Asian Economic Journal* 9 (3): 233–260.

Yang, D. L., and F. Su. 1998. The politics of famine and reform in rural China. *China Economic Review* 9 (2): 141–156.

Yingying, Z., H. Hua, and S. Harrel. 2008. From labour to capital: Intra-village inequality in rural China, 1988–2006. *China Quarterly* 195: 515–534.

Zhang, H., X. Li, and X. Shao. 2006. Impacts of China's rural land policy and administration on rural economy and grain production. *Review of Policy Research* 23 (2): 607–624.

Zhang, Y., and S. Kant. 2005. Collective forests and forestlands: Physical asset rights versus economic rights. In *Developmental dilemmas: Land reform and institutional change in China*, ed. P. Ho. New York: Routledge.

Zhou, Q. 1994. Land system in rural China: The case in Meitan County of Guizhou Province. In *The land system in contemporary China*, ed. G. J. Wen. Changsha, China: Hunan Science and Technology Press.

Zhu, K., and R. Prosterman. 2007. *Securing land rights for Chinese farmers: A leap forward for stability and growth*. Washington, D.C.: Cato Institute.

Zhu, L., and Z. Jiang. 1993. From brigade to village community: The land tenure system and rural development in China. *Cambridge Journal of Economics* 17 (4): 441–461.

Zong, L. 1993. Agricultural reform and its impact on Chinese rural families. *Journal of Comparative Family Studies* 24: 1978–1989.

Zweig, D. 1997. *Freeing China's farmers: Rural restructuring in the reform era*. Armonk, N.Y., U.S.A.: M. E. Sharp.

Chapter 19

Land Tenure Policy Reforms: Decollectivization and the *Doi Moi* System in Vietnam

Michael Kirk and Nguyen Do Anh Tuan

Following periods of slow growth, declining food production, and famine that resulted primarily from the collectivization of agriculture, in 1986 the Vietnamese government enacted a series of reforms to transform the country from a centrally planned to a market-oriented economy. This *doi moi* reform not only dismantled the rural collectives but also assigned land rights to farmers, liberalized agricultural markets, and introduced other reforms to the wider economy. These reforms unleashed a new entrepreneurial spirit in the agricultural sector that resulted in intensified rice production and diversification into new crops such as coffee. Largely as a result of these reforms, poverty incidence dropped from 58.0 to 29.0 percent between 1993 and 2002, the growth of agricultural gross domestic product (GDP) rose to 3.8 percent a year between 1989 and 1992, and GDP growth averaged 7.0 percent a year from 1993 to 2000. By 1989, Vietnam had not only alleviated national food shortages but also become the world's third-largest exporter of rice. In effect, agriculture had become a driver of overall economic growth in Vietnam.

Decollectivization and land titling have generated powerful incentives to invest in agriculture. The resulting rural growth has raised households' incomes and standards of living and led to greater food security and better nutritional status throughout much of Vietnam. Vietnamese farmers now work under more secure land rights, diversify into different types of agricultural activities beyond staple food production, and participate in active markets that allow the buying, selling, and renting of land. This chapter reviews Vietnam's experience with land reform, focusing on not only the process of decollectivization and related economic reforms but also their impact on productivity, poverty, equity, and the environment.

The Process of Decollectivization and the *Doi Moi* Reform in Vietnam

Land reform and decollectivization in Vietnam were implemented as part of a general economic reform (*doi moi*). This reform process goes back as far as 1954, when the Vietnamese Communists defeated the French, and it continues into the present. This section reviews the various stages of reform, focusing on approaches, processes, actors, and outcomes of interventions that contributed to the success of decollectivization and land reform in Vietnam.

1954–80: Socialist Land Reform and Collectivization

Under the 1954 Geneva Accords, Vietnam was divided into two countries with opposing ideologies: the Democratic Republic of Vietnam in the north—influenced by the former Soviet Union and China with their socialist ideology—and the Republic of Vietnam in the south—dominated by the United States with its capitalist economic system. Although both governments conducted land reforms to redistribute large landholdings soon after 1954, they went about the process in different ways (Do and Iyer 2003).

In North Vietnam, where agriculture was characterized by fragmented landholdings and small-scale petty commodity production, land reforms in the late 1950s led to the collectivization of agriculture as part of a wider centrally planned model for the economy (Fforde and Pain 1987). Although collectivization was incomplete in these early years, it advanced during 1965–68, when the Vietnam War motivated farmers to form communes, for safety more than anything else (Beresford 1985; Chu et al. 1992). Between the effects of war and the poor incentives provided by collectivized production, agricultural productivity in the north declined during the subsequent decades, while recurrent food shortages forced North Vietnam to live off the support of foreign aid from its socialist allies.

In South Vietnam, agriculture was highly commercialized and more export oriented. The production system was based on a tenant sharecropper relationship between the landlord class and what at that time was called the rural proletariat (Watts 1998). A land-to-the-tiller program introduced in 1970 set a land ownership ceiling of 20 hectares per family, thus creating a more middle-class peasantry and favoring further agricultural commercialization.

The war with the United States had left Vietnam severely damaged; entire provinces and cities had been destroyed outright. After reunification in 1975, the Vietnam Communist Party (VCP) attempted to strengthen the centrally planned system—and large-scale agricultural collectivization in particular—as a model for the whole country (VCP 1977; White 1982; Fforde and de Vylder 1996). In the north, the 1974 Thai Binh Agricultural Conference led to the transformation of small-scale cooperatives into large-scale "advanced" cooperatives that received support from the state in the form of mechanization and agricultural infrastructure systems. Cooperatives grew from hamlets with dozens of households to entire villages with hundreds of households. In 1979, cooperatives in northern Vietnam held an average of 202 hectares of land on which an average of 378 households lived and worked. Almost 97 percent of rural northern Vietnamese households belonged to the 4,151 cooperatives (Do and Iyer 2003).

In the south, the socialist transformation forced peasant households to participate in so-called low-level cooperatives, or pre-cooperatives, that were based on labor exchange or "solidarity teams." Collectivization in the south, however, was not particularly successful. In 1979 the south had only 272 cooperatives, and in 1980 only 24.5 percent of farm households belonged to a cooperative. In many cases, southern cooperatives were notional cooperatives rather than cooperatives that actually functioned in practice (Kerkvliet 1995; Tran 2005).

In general, cooperatives throughout reunified Vietnam managed resource allocation decisions about production and distribution in accordance with the material targets of the State Planning Committee while also providing social services to their members. The basis of production was the brigade, which was responsible for directing field labor; a cooperative provided planned targets to the brigades, which then delivered the output upward to the cooperative.

Within cooperatives, peasants were allowed to retain small personal plots, which altogether amounted to no more than 5 percent of the total area of the cooperative. Some of the output of these plots, along with the surplus production of the cooperative, entered the heavily regulated public and private markets either at "negotiated prices" offered by state trading agencies or at free market prices in the "unorganized" markets (Chu et al. 1992; Fforde and de Vylder 1996; Dat 1997).

Not surprisingly, collectivization in the postunification period generated poor agricultural performance (Kirk and Nguyen 2009). The annual rate of agricultural growth was only 0.7 percent during 1976–80, and in 1976 and 1977, agricultural growth actually declined by 0.5 and 6.6 percent, respectively. Foodgrain availability fell by 1.5 percent during 1976–79, despite the sharp rise in foodgrain imports. Food procurement by the state fell from 2 million tons to 1.4 million tons during 1976–79 as farmers tried to avoid the low prices offered by the state procurement system, instead selling through private "unorganized" markets (Akram-Lodhi 2001). It was reported that private markets offered prices 10 times higher than those offered by the public sector (Fforde and de Vylder 1996).

1981–88: The Pre-reform Period

In the early 1980s Vietnam faced an economic crisis due to an impending food shortage, decreasing Western and Chinese aid, and the imminent collapse of the state commodity funds, the state-owned entities that were key to mediating production and consumption in the centralized economic system. Local experiments in more market-based production systems led the Communist leadership for the first time to officially acknowledge the legitimacy of individual interests alongside those of collective interests.[1]

The most radical changes occurred in the agricultural sector. Efforts to further collectivize peasant households in the south were suspended. More important, Directive 100 of the VCP, famously known as Contract 100 (*Khoan 100*) or "output contract" (*khoan san pham*), was issued on January 13, 1981. The output contract system allowed cooperatives to distribute land to households, shifted responsibility for production to the households, and assigned them production quotas. Under this system, households were allowed to keep or sell surpluses on the private market or to the state trading agencies at "negotiated prices" (Vo 1990). In mid-1981, along with the imposition of drastic price regulation measures, procurement prices for agricultural goods were increased to the same level as market prices (Beresford and Fforde 1996). In addition, control stations were established along main transportation routes in order to check illegal circulation of state-controlled goods.

At first this partial reform was successful. Agricultural growth averaged 6.6 percent a year during 1981–84, reaching its peak of 10.6 percent in 1982. Per capita food production increased from 260 kilograms in 1976–80 to 293 kilograms in 1981–84. Nevertheless, such reform measures arguably had only an ad hoc effect on output growth. Agricultural growth started to decelerate in 1983. In particular, food production increased by less than 1 percent in 1983, leading per capita food production to fall below the minimum subsistence level of 300 kilograms. Inflation accelerated at more than 60 percent a year during 1980–84. The gap between free market and official prices, which had narrowed in 1981, widened again to 10 times or more in the mid-1980s.

The failure of price, wage, and currency reforms in 1985 worsened the situation. During 1985–88, GDP growth fell to an annual average of 4.6 percent. Agricultural growth decelerated to only 2.2 percent a year during 1985–88 and even fell by 1.8 percent in 1987, when food production decreased by 4.4 percent. This decline led to famine in some provinces because per capita food production was only 281 kilograms. Inflation started at 91.6 percent in 1985 and accelerated to three-digit percentages during 1986–88. Inflation hit a staggering 774.7 percent in 1986.

The poor agricultural performance, food shortages, high inflation, and low economic growth of 1984–87 were attributed to three factors. First, the output contract system did nothing to move the rural economy toward market-based prices for either inputs or output. The state trading agencies still tried to dominate the marketable surplus of food staples (Fforde and de Vylder 1996). Second, agricultural cooperatives retained the ability to increase or decrease the cooperative members' share of the contracted amount by altering the system of payments for the inputs assigned to both themselves and the members.[2] In many cases, peasants returned land to the cooperatives and put more effort into their personal 5 percent plots of land. Third, agricultural stagnation was further exacerbated by insufficient state investment.

1987–93: Market-Oriented Reforms and Decollectivization

During 1987–89, Vietnam implemented a package of measures that fundamentally changed the nature of the country's economy from a centrally planned to a market-oriented system. The *doi moi* reform process originated from the Sixth National Party Congress of the VCP, held in December 1986. Confronting high inflation, the erosion of state institutions, and severe shortages in the economy, the VCP saw *doi moi* as essential to its own survival. The objectives of this reform were to

1. stabilize the economy, which suffered from high inflation and serious economic imbalances;

2. develop the private sector;

3. increase and stabilize agricultural output;

4. shift the focus of investment from heavy to light industry;

5. focus on export-led growth, based on the experience of Vietnam's dynamic regional neighbors; and

6. attract foreign direct investment (FDI), seen as essential for economic development.

In agriculture, Resolution 10 (popularly known as Contract 10 or *khoan muoi*), issued by the VCP in 1988, initiated the process of decollectivization and revived the development of the peasant household economy in rural areas. Resolution 10 obliged the agricultural cooperatives to contract land to peasant households for 15 years for annual crops and 40 years for perennial crops. Land was generally allocated on the basis of family size. The cooperatives retained ownership of capital stock, working capital, and other means of production but were obliged to rent them out to peasant households. Moreover, peasant households were allowed to buy their own capital stock and working capital irrespective of the supply available from the cooperatives. They could thus buy and sell animals, equipment, and machinery.

This resolution also stated that the proportion of the contracted amount to be left to cooperative members should not be less than 40 percent of production. Contract quotas and unit prices were fixed for five years, bringing a new degree of certainty to peasant households. Households that did not meet the quota had to compensate the cooperatives in cash or in kind, at the market price. Peasant households also had to pay agricultural taxes equivalent to an average of 10 percent of their annual output. Finally, the state accepted private-sector food marketing.

During 1987–91, price liberalization was carried out and the ration system was abolished. In 1987, internal control posts were abolished, accelerating trade within Vietnam. In 1988, foreign trade reform was initiated: tariffs began to replace quantitative restrictions, and the government ceased its exclusive control of foreign trade. Finally, in December 1989, the official exchange rate was sharply devalued and brought near equality with the free market rate (Sepehri and Akram-Lodhi 2002).

The reforms generated positive economic results. Inflation, which had soared during 1986–88, fell to just 36 percent in 1989. Growth accelerated first in agriculture, leading to high demand in the construction and services sectors. During 1989–92, GDP growth averaged 6.1 percent a year, with agriculture and nonagriculture growing at 3.8 and 6.9 percent, respectively. Despite the negative impacts of foreign aid cuts due to the collapse of the Eastern European socialist system during 1990–91, Vietnam maintained a GDP growth of 5.1 and 5.8 percent in 1990 and 1991, respectively. By 1992, Vietnam had fully recovered from the shock caused by the collapse of the socialist system, and growth climbed to 8.7 percent (Dollar 1994).

The improved economic performance during 1989–92 was attributed first to the leading role of the agricultural sector. After importing about 400 tons of food during 1987–88, Vietnam became the world's third-largest exporter of rice in 1989. In addition, rice exports played an important role as the main source of foreign exchange earnings during this period as imports financed by foreign aid from the socialist bloc ended.

Nonetheless, domestic saving and investment rates were still too low, and further economic growth could not be sustained. More important, the legal and institutional framework for efficient market operation had not been firmly established (Fforde and de Vylder 1996). In agriculture, the impacts of reform measures were limited to subsistence crops. Four types of limitations related to land policy impeded the commercialization of agriculture. First, the duration of land use rights was not sufficient to encourage households to invest in agricultural production. Second, because land transfer was not allowed, land acquisition by peasant households, crop specialization, and commodity production in agriculture were discouraged. Third, land use rights were not accepted as collateral by financial institutions, so households were prevented from raising loan funds for agricultural investment. Fourth, local governments still played a dominant role in deciding crop patterns for specific types of land. Most of the land continued to be used for food production because of the emphasis placed on food security, while agricultural diversification and commercialization were not sufficiently encouraged.

1993–2001: Building Market Institutions and the 1993 Land Law

This period witnessed a series of efforts for international integration, starting with the resumption of lending from the International Monetary Fund and the World Bank and large inflows of FDI and ending with trade agreements with the Association of Southeast Asia Nations and the United States. In 1993–2001, reforms continued, mainly to consolidate the previous policy initiatives, but at a slower pace.[3] After a long history of war, poverty, hunger, and international isolation, both the VCP and the Vietnamese people sought economic growth, prosperity, and international integration (VCP 1996). The state actively opened the economy to export expansion and foreign investments, achieving high economic growth. The state also pushed to get prices right by pursuing further gradual market liberalization, particularly in the credit and labor markets, and developing new institutions to replace the previous state administrative system.

In agriculture, the 1993 Land Law built on Resolution 10 by extending land tenure to 20 years for annual crops and 50 years for perennial crops. Although households were limited to 3 hectares of land per farm for annual crops in the Red River Delta and 5 hectares per farm for annual crops in the Mekong Delta, for the first time the exchange, transfer, lease, inheritance, and mortgaging of land use rights was permitted (Sepehri and Akram-Lohdi 2002). To help develop the land market, the government began issuing land use certificates to peasant households. By 1999, more than 10 million households had received certificates for agricultural land, representing about 87 percent of peasant households and 78 percent of the agricultural land in Vietnam (ANZDEC 2000).

Other measures accompanied the land tenure reforms. The 1993 Land Law also reduced the agricultural land use tax from an average of 10 percent of annual output to 7 percent of annual output. Perennial crops farmed on newly reclaimed land were exempted from the tax. Also in 1993, the government promulgated decrees aimed at reforming institutions such as state-owned agricultural enterprises and improving investment and technological innovation. In March 1996 Vietnam's National Assembly approved the Law on Cooperatives, clarifying the cooperatives' role as service providers and establishing a new legal framework for the cooperatives within a multisectoral commercial economy. Under this new framework, cooperatives integrated trade and service supply, and sector-focused and professional cooperatives were established in areas such as pig raising and rice seedling supply.

The period 1993–2000 was considered the golden age of the market economy in Vietnam. GDP growth reached 7 percent a year, with agricultural growth of

4 percent and nonagricultural growth of more than 8 percent.[4] Despite the Asian financial crisis of 1997–2001, Vietnam maintained a strong economic performance, with the agricultural sector growing at more than 5.0 percent a year,[5] although nonagricultural growth slowed to 6.8 percent.

Researchers questioned the sustainability of agricultural growth, however, because it depended mostly on the incentives created by the previous land reform, decollectivization, and price liberalization. The focus was still on quantitative growth rather than on the quality and competitiveness of agricultural production. Food security still dominated approaches to export-oriented and commercialized agriculture. The transformation of the crop structure occurred slowly. The rural–urban income gap was widening, and more than 70 percent of the poor were concentrated in the rural sector (World Bank 1998, 1999). Land fragmentation, which hampered efficiency and productivity, was the major constraint to agricultural commercialization. Land was supposed to be allocated on the basis of need (usually indicated by the size of households) and ability to farm the land (the number of household members who could work on the land). Moreover, land markets failed to develop strongly without a clear regulatory and institutional framework (Akram-Lohdi 2001; Dang et al. 2005; Nguyen 2006). The efforts to develop a new style of agricultural cooperative were not particularly successful. In some areas, relatively few cooperatives survived longer than a year. Moreover, among those considered successful, the majority were very small, with average capital of about VND 1 billion or $63,000. Agricultural tax reform led to the imposition of new categories of local fees on rural households. Local governments were forced to charge fees for rural services, such as health care and education, that cooperatives had previously funded, and these fees constituted the heaviest form of taxation on the rural population. Finally, food security was still a central policy objective, and the state kept tight control on the volume of rice exports, partially limiting incentives to rice farmers.

2001–Present: Land Market Liberalization, Agricultural Commercialization, and International Integration

Since 2001, reform efforts have focused on state-owned enterprises, financial reforms, the development of factor markets after the Asian financial crisis, and trade liberalization in the effort to become a member of the World Trade Organization in 2007. Policy reform measures were influenced by international donors that supported the Comprehensive Poverty Reduction and Growth Strategy (CPRGS) proposed by the World Bank. This document was a major turning point in Vietnam's planning processes. Previous plans and strategies had been heavily based on a command view of the economy. By contrast, CPRGS spelled out clear development goals (such as the Vietnam Development Goals), using empirical evidence and consultation to identify the policies best suited to attainment of those goals, aligning resources behind those policies, and setting up appropriate monitoring and evaluation mechanisms.

In agriculture, policy measures focused on three pillars of rural development: (1) creating opportunity by accelerating market orientation, (2) managing natural assets for broad-based growth, and (3) mainstreaming poverty reduction through inclusion and empowerment (World Bank 1998). The goal of these policy measures was to push up agricultural commercialization, increase competitiveness and the value-added of agricultural production, improve rural livelihoods, narrow the rural–urban income gap, and promote sustainable agricultural growth.

Agricultural market-based growth was further enhanced by amendments to the Land Law in 2001, 2003, and 2004 that permitted foreign investors to acquire land use rights, promoted land consolidation in agriculture, simplified buying and selling procedures for land, and allowed land use certificate holders to buy and sell their usufruct rights or change the functional assignation of their land. A 2004 revision to the Land Law made an important contribution to gender balance by naming both the husband and the wife on each land use certificate.

In 2001, the state scrapped quotas for rice exports and fertilizer imports, allowing all firms engaged in the domestic trade of these commodities to enter international arrangements. The agricultural tax was gradually phased out. By 2003, the agricultural tax raised only 1.3 percent of tax revenue and 0.1 percent of total government revenue (World Bank 2004).

To speed up rural transformation, policy measures issued during 2000–05 supported private investment in the rural sector, the application of modern technologies, development of rural nonfarm activities, and improvements in rural physical and social infrastructure, with a focus on education and health care.

Since 2001 Vietnam's agricultural growth has averaged 3.7 percent a year despite low commodity prices on world markets. Its per capita food production increased from 420 kilograms in 2001 to 470 kilograms in 2007, ensuring national food security and abundant exports. Aggressive agricultural diversification and commercialization took place. Exports of agricultural, forestry, and fishery products grew an average of 16.8 percent annually from 2001 to 2007. Vietnam has become the world's second-largest coffee exporter and the largest exporter of robusta coffee. Within a decade after 1995, coffee production tripled, and in 2007 Vietnam earned more than $1 billion from coffee exports. Similarly, Vietnam has become the world's largest exporter of cashew nuts and pepper.

Since 2007 the state has reconsidered its agricultural and rural development strategies, partly because of the global economic recession, declining agricultural growth, and lessons on rural social instability learned from China. To integrate agricultural and rural development into ongoing industrialization and modernization, new policies focused on improving peasant incomes, rural livelihoods, and infrastructure and further liberalizing land markets.

Impacts of Land Tenure Policy Reforms in Vietnam

This section broadly examines the impact of Vietnam's comprehensive reforms on agriculture, the rural economy, and Vietnamese society as a whole.

Growth and Productivity Increases

Land tenure reforms in Vietnam have—together with complementary productivity-enhancing instruments—generated powerful incentives to invest in farming. In turn, these reforms have induced a strong rural-based growth process, raising households' incomes and their standards of living (Beckman 2001; Henin 2002; Rozelle and Swinnen 2004). Starting in the early 1990s, Vietnam maintained an annual GDP growth rate of 7.6 percent (Fritzen 2002; Minot 2003); at the turn of the millennium these rates slowed to 5 percent (1998–2001), but they sped up again during the past few years. Agricultural growth achieved its peak in 1996–2000, with an annual rate of 4.7 percent; in the longer term (1986–2005), agricultural growth averaged 3.8 percent a year.

Table 19.1 underlines the high rate of growth in total factor productivity (TFP) from 1986 to 1990, thanks to new incentive structures for smallholders generated by Resolution 10. The increasing use of farm inputs such as fertilizers and pesticides between 1991 and 1995 is reflected in price increases for agricultural products and is partly the result of the better availability of rural credit. From 1996 to 2005, the role of land expansion became clear, enabled by increasing production efficiency, more capital-intensive inputs, and improvements in irrigation systems.

Major components of this growth, which led to reduced rural poverty, are attributable to higher per hectare yields in rice and other crops. These yield increases went hand in hand with the diversification into new or intensified non-crop endeavors, such as aquaculture, livestock breeding, and nonfarm activities. The sharply different growth rates in the acreage of annual crops (1.7 percent) compared to perennial crops (6.3 percent) (Nguyen 2008) indicate the impact that strengthened incentives such as secured land use rights for tree and shrub cultivation had on long-term investments in land. Intensification, driven by a growth in labor, fertilizer application, and irrigation, contributed most to agricultural growth during 1986–2005. Interestingly, TFP as a residual proved irrelevant and showed even slightly negative rates (−0.3 percent) during 1986–2005, meaning that the improved factor quality arising from the reform process does not seem to be important, except in the context of organizational innovations during the early period of reform (1986–90).

Table 19.1 Estimates of agricultural growth in Vietnam, 1986–2005 (percent)

Period	Agricultural GDP	Labor	Tractor	Pump	Fertilizer	Area of annual crops	Area of perennial crops	TFP
Growth								
1986–90	2.7	2.5	–4.7	–1.2	–3.4	0.8	5.4	
1991–95	4.7	6.9	28.6	25.6	13.4	2.0	7.8	
1996–2000	4.9	1.3	10.5	10.3	4.0	2.7	9.8	
2001–05	3.7	1.0	5.7	4.4	1.4	1.0	3.0	
1986–2005	3.8	2.7	10.5	9.8	8.0	1.7	6.3	
Factors contributing to growth, based on their elasticities								
1986–90	2.7	0.8	–0.2	–0.1	–0.5	0.3	0.2	2.0
1991–95	4.7	2.2	0.9	1.4	1.8	0.8	0.4	–2.8
1996–2000	4.9	0.4	0.3	0.6	0.5	1.1	0.4	1.5
2001–05	3.7	0.3	0.2	0.2	0.2	0.4	0.1	2.2
1986–2005	3.8	0.9	0.3	0.5	1.1	0.7	0.3	0.0
Factors contributing to 1 percent agricultural growth								
1986–90	100.0	29.9	–5.9	–2.4	–17.1	12.4	9.2	73.9
1991–95	100.0	47.5	19.9	29.4	37.9	17.6	7.4	–59.7
1996–2000	100.0	8.9	7.0	11.4	10.9	22.3	9.1	30.3
2001–05	100.0	8.5	5.1	6.5	5.2	11.4	3.7	59.5
1986–2005	100.0	23.2	9.1	14.0	28.3	18.2	7.5	–0.3
Elasticities	32.5	3.3	5.4	13.4	40.9	4.5		

Source: Nguyen (2008).
Note: TFP (total factor productivity) is the residual of agricultural growth minus the sum of factor growth weighted by elasticities.

Enhanced Tenure Security through Land Use
Certificates and Investment

Beyond decollectivization, issuing land titles is considered critical to enhancing tenure security, promoting investment, and triggering the emergence of tenancy and land sales markets (Deininger and Jin 2008; Do and Iyer 2008). By 2000, nearly 11 million land certificates had been issued to rural households. Starting with a coverage rate of 24 percent in 1994, the titling program had already reached 90 percent of the rural population by 2000 (Do and Iyer 2008), but with variations among the provinces. By 2003, 91 percent of rural households had been issued land use certificates (LUCs) (World Bank 2004).

At the community level, LUCs have been found to be associated with higher levels in the share of total area devoted to perennial crops and increased investment in irrigation (Deininger and Jin 2008). In addition, titling has had a significant impact on labor input in nonfarm activities. Cultivating perennials is normally a labor-saving practice. Consequently, households in provinces with high rates of LUC issuance, which were likely to increase their cultivation of perennials, increased their nonfarm activities by 2.7 weeks per active household member (Do and Iyer 2008).

New incentives were thus given to marketing, food processing, woodworking, and the local garment industry, thereby diversifying rural incomes and safety nets.

It would be naïve, however, to think that the Land Law alone would ensure tenure security (Ravallion and van de Walle 2003). Power struggles at the local level and the capture of rents by local elites may lead either to worse distributional outcomes than before the reform or to an increased desire to protect the poor. In concrete terms, uncertainty persists in some provinces about whether these rights are really inheritable rights or whether they will be reallocated at the end of the 20-year use period, despite repeated government statements to the contrary (Saint-Macary et al. 2008).

From the beginning of the reform, the local authorities' frequent reassignment of land to maintain land productivity resulted in underinvestment (Pingali and Xuan 1992). Case studies in Vietnam's northern mountain region have shown that within the guaranteed 20-year leasehold, two phases of land reallocations have already occurred to accommodate new settlers in the villages. Therefore, there is insufficient evidence (Saint-Macary et al. 2008) that tenancy contracts under titling, from the tenants' viewpoint, are always more secure than the rural collectives' practices.

Secured Tenancy and Rural Factor Markets

Because land rental and land sales markets respond differently to measures that increase resource access for the rural poor and enhance farm efficiency, short-term rental and permanent land transfers will be treated separately (Deininger and Jin 2008).

Land Rental Markets. Between 1993 and 1998, rental market participation more than quadrupled, from 3.8 to 15.5 percent. Despite this promising start, rental markets have not increased in importance over time (Brandt 2006). In 2004, only 3.6 percent of all agricultural land was rented, with 10.7 percent renting in and 6.0 percent renting out. For households renting in, however, tenancy plots are the source of nearly one-third of their managed lands.

Who is active in rental markets? Deininger and Jin (2008) state that agricultural plots are transferred preferentially to those households with limited asset endowments but a high level of agricultural ability (Deininger and Jin 2008). Consequently, rental markets have increased the productivity of land by transferring resources to producers who can make better use of them. At the same time, rental markets allow those whose comparative advantage is not in agriculture to provide land rentals and join nonagricultural sectors, where they gain greater remuneration for their labor. Transfers positively affect not only efficiency but also equity. Neither female-headed households nor those being threatened by adverse shocks are discriminated against in local rental markets. Some elements of discrimination do exist, however; smallholders of the most productive ages are preferred to older people.

It is impossible to generalize regarding the scope and performance of rental markets, because strong regional differences exist. Motives for renting out land are manifold, but there is no evidence that they are dominated by the perceived low abilities of landlords or by excessively large landholdings.

Land Sales Markets. The rapid development of rental markets was accompanied by a reemergence of land sales transactions. In the 1990s, informal, illegal land markets developed in which land was not only mortgaged and rented but also bought and sold (Henin 2002). In the beginning, permanent transactions were underreported because land sales were forbidden and a tax was imposed on all land transactions (Do and Iyer 2003). By circumventing these regulations, more and more people bought land without being impeded by credit market imperfections (Deininger and Jin 2003). The expansion of land sales markets developed even more than rental markets, initiated by households that remained active in agriculture. In fact, market activities increased from involving 1.0 percent of producers in 1993 to 7.2 percent in 1998.

The positive impact of land sales markets on rural poverty reduction remains limited, however, because resource-poor producers, especially female-headed households, can rent land but not buy it. At the same time, those who opt to leave agriculture rent out land but do not necessarily sell their rural holdings, which serve as a safety net. Therefore, attained farming abilities, the level of local nonfarm development, and the security of land rights strongly determine the operation of rural land markets, with rental markets having a pioneering impact on both equity and productive efficiency.

Because of a lively transfer of plots without titles, a quasi market largely independent of state control has emerged. On the other hand, the local state continues to play an active role in setting the terms of land transactions as a measure to prevent landlessness. These ad hoc interventions may make a village even worse off (Ravallion and van de Walle 2008). Prohibitions could lead to illegal transactions that offer less favorable contract terms to vulnerable households that must sell after adverse shocks. As a consequence of emergency sales, wealthier households can accumulate land. New landlord–tenant relations may thus develop, with land concentration and the hiring of agricultural labor reemerging as issues.

Rural Credit Markets. Possession of an LUC regarded as legitimate should improve a rural household's access to credit, particularly from formal banking institutions (Do and Iyer 2008; Ravallion and van de Walle 2008). There is insufficient evidence, however, that recent credit market development will complement ongoing land tenure reforms, facilitate permanent land transfers, and contribute to greater rural investment (Do and Iyer 2008). The volume of credit did not increase significantly after land reforms, and the ability of households to borrow is still low (Do and Iyer 2008). Interestingly, the probability of receiving a loan goes down by

11 percentage points when a province implements the land reform fully. The amounts borrowed, as a fraction of total household expenditure, are lower for households within highly registered provinces. Additionally, there is no change in the fraction of loans from formal sources after tenure reform is implemented. Thus, the Land Law has not been very effective in alleviating credit constraints for rural households (Do and Iyer 2008). Even after obtaining LUCs, rural households remain excluded from formal lending institutions because land is not yet well acknowledged as collateral (Henin 2002). Moreover, a group of landless poor has emerged that is not well served by either formal or informal credit institutions (Ravallion and van de Walle 2008; Kemper and Klump 2009).

Food Security and Nutrition

Even in the early 1990s, the family contract system had a significant effect on rice productivity, transforming Vietnam from a rice importer to one of the world's largest exporters (Pingali and Xuan 1992). From 1980 to 1984, the annual rate of growth in rice yields per hectare of land rose by about 32 percent in northern Vietnam and 24 percent in southern Vietnam (Pingali and Xuan 1992). Most of this growth was attributed to increases in yield per hectare rather than to expansion in cultivated area. During this period, the annual rice output per capita increased by about 40 kilograms for the whole country.

Rice is by far the most important staple food in the Vietnamese diet, accounting for about 75 percent of the total caloric intake of a typical Vietnamese household. Therefore, the continuing increase in per capita rice production in the 1990s contributed to increased food security (Minot and Goletti 2000). As reforms stabilized the economy, rice prices became less volatile and substantially increased during the 1990s, directly affecting Vietnamese households, 72 percent of which both produced and consumed rice. Surplus rice-producing regions thus realized a net benefit from rising prices, whereas deficit regions lost (Minot and Goletti 2000). On average, higher rice prices would benefit rural households at the expense of urban households. Because poverty in Vietnam is still rural, the reforms had the potential to reduce food insecurity and improve nutritional status by attacking some roots of rural poverty.

Changes in rice production and prices are thus likely to have a significant impact on the welfare of Vietnamese households and on rural and urban poverty levels (Niimi, Vasudeva-Dutta, and Winters 2004). At low income levels, rice is a normal good, but beyond certain income levels, rice consumption begins to fall and is replaced with consumption of meat, eggs, and dairy products (Minot and Goletti 2000), so rising incomes in Vietnam led to more diverse diets.

In concrete terms, some nutritional indicators improved rapidly between 1993 and 1998; stunting rates for children younger than age five, for example, fell from

53.0 percent to only 33.4 percent. Rates of underweight children younger than age five declined much more sluggishly (Fritzen 2002). These observed changes in nutritional patterns—more diverse food intake in particular—have had a positive impact on the average nutritional status of the Vietnamese.

Changes in Farming Systems and Agrarian Structures

Market reforms have radically transformed agricultural production and rural structures, not only in the rice-growing plains but also in the more remote upland villages (Henin 2002). In both regions, state cooperatives have been dissolved and replaced with family farms (Beckman 2001). Other means of production, such as capital goods, have also been handed over to the new family farms.

In the first reform phase in the 1990s, per household land endowments remained small but variable across regions, with an average of 0.30 hectare of annual crops and 0.06 hectare of perennial crops (Deininger and Jin 2008). Land distribution did not lead to a significant increase in land inequality during the 1990s; on the contrary, the national Gini coefficient for per capita land endowments declined slightly, from 0.53 to 0.50. Regional differences persist, however. Extending the survey period from 1993 to 2004, the Gini coefficient for rural households' per capita land rose from 0.49 to 0.64, which is a remarkable change (Brandt 2006).

Although land assignment to households was initially equitable, some households ended up with less land than would have been the case in a competitive market allocation (Ravallion and van de Walle 2006). This process also led to the fragmentation of landholdings, generating some efficiency losses. In 2004, the average number of plots was 4.3 per rural household, with a mean of 4.9 in the Red River Delta and only 2.0 in the Mekong River Delta (Brandt 2006). Since land consolidation projects have started, the number of plots is again declining.

With these impacts on agrarian structures, farming systems and cropping patterns changed considerably in only a few years. Besides increasing their investments in irrigation systems and rice production, farmers devoted a higher share of total agricultural area to perennial crops, which were subject to enhanced tenure security (Deininger and Jin 2008). Plots dedicated to multiyear cropping significantly increased at the expense of annual crops, with the coffee boom in Vietnam one indicator.

Poverty Reduction and Equity

The Impact on Rural Poverty. There is extensive evidence that agricultural growth can play a central role in reducing poverty (Ravallion and van de Walle 2008). Vietnam, with more than 50 percent of the labor force employed in agriculture and the major share of consumer income spent on food, is a case in point (Rozelle and Swinnen 2004). And although a range of factors have contributed to this growth and to the consequent decline in rural poverty, land reform is an important one (Beckmann 2001). Household survey data show that the fraction of households living below the poverty line in Vietnam fell from 58.1 percent in 1993, the year the Land Law was issued, to 28.9 percent in 2002 and 16.0 percent in 2006.

Reform came in two stages: (1) decollectivization and the privatizing of land use rights and (2) introduction of markets in privatized land use rights. In the first stage, researchers found that more efficient and equal land allocation together hypothetically would have resulted in slightly lower poverty rates (2 percentage points) than the actual rates (Ravallion and van de Walle 2008). If, however, the poverty lines used at that time (the early 1990s) were applied—and if a purely market-based land allocation were implemented instead of the ones used in the model—poverty incidence would have been higher. They conclude that "it seems that an effort was made to protect the poorest and reduce overall inequality at the expense of overall consumption" (Ravallion and van de Walle 2008, 97).

In the second reform stage, households, having started with inefficiently low amounts of cropland in the first phase, increased their holdings over time. Market forces tended to favor the "land poor" and those who were well rooted in rural society—namely male-headed households and the better educated. Did this reform lead to poverty-increasing landlessness? During 1993–2004, not only did landlessness among the poorest fall but the trend rate of poverty reduction between 1993 and 2004 was slightly higher for the landless than for the rest of the rural population, with the exception of the Mekong Delta, which has a unique history of landownership (Ravallion and van de Walle 2008). Overall, rising landlessness in Vietnam has been poverty reducing, an effect that is statistically significant. Thus, land market development and titling strongly contribute to overall poverty reduction in rural areas.

Regional differences in poverty reduction can be observed, however: improvements in rural standards of living occurred mainly in the lowland rice-growing areas, but poverty persisted in the highlands, mostly because of a lack of access to irrigation water, agricultural credit, and markets, as well as land (Henin 2002).

Asset Distribution and Landlessness.   Although the efficiency gains from land tenure reforms are undisputed, the effects on income, wealth distribution, and equity issues are controversial. From a macroeconomic perspective, considerable equity benefits dominate after nearly two decades of reforms (Deininger and Jin 2003). Land allocation—the Vietnamese approach of titling through LUCs and securing land rights—at least partially initiated a wave of investment in agriculture, particularly in irrigated rice production in northern Vietnam, and thus contributed to more equal asset distribution even in mountainous regions (Neef et al. 2007).

Nevertheless, there is some dispute about whether formalizing land rights and land market development has increased landlessness,[6] but data based on the Vietnam Living Standard Survey from 2004 show that 85.6 percent of all households report having agriculture-related land, with about 94.0 percent in the Red River Delta in northern Vietnam and only 61.2 percent in southern Vietnam (Brandt 2006). The percentage of rural households with land fell from 92.2 percent in 1993 to 89.7 percent in 1998 and 85.6 percent in 2004.

How should we interpret rising landlessness in the aftermath of the agrarian reforms? One line of argument holds that starting from relatively equitable land distribution, land market reforms have created new rural class structures by allowing rich farmers to buy land from poor farmers, who then became poor landless laborers. The other line holds that the more affluent have become landless as they shift partially out of farming into occupations with higher labor remuneration. Rising landlessness and falling poverty thus occur together as part of a wider process of economic transition (Ravallion and van de Walle 2008). If consumption is taken as a proxy for wealth, models show that 10 years after the implementation of the Land Law the poorest tend to be the least likely to be landless in Vietnam.

Land Use Rights and Gender Equality.   Enforceable rights to real property enhance the potential of both women and men to use property for economic purposes in the most efficient way. Women's access to land, however, often depends on their marital status, and unmarried and divorced women who contribute significant agricultural labor are rarely named on land titles. Similarly, land certificates bear only husbands' names (World Bank 2002) because certificates had space for only one name per family, resulting in systematic discrimination against women and disregard of their contribution to productivity increases and welfare generation.

Through pilot project initiatives, this imbalance is being addressed with revised certificates that name both women and men as rights holders, creating incentives for women in particular to invest in land and reap its benefits in case of divorce or widowhood. This change is possible only with the support of local government representatives, because it puts existing informal rules and power relations into question.

Governance and Power Relations

The impact of a new legal framework for land access and transactions cannot be underestimated: securing property rights matters, and it is up to government organizations at different levels to support and finance institutions that ensure security (Deininger 2003). In Vietnam national legislation is transformed into rules and regulations applicable at the local level; therefore, devolution and governance issues are relevant in fully assessing the impact of reforms for local users and stakeholders.

In northern Vietnam, the national government is aiming at large-scale devolution of the use, management, and governance of land and forest resources (Neef et al. 2007). It is not surprising that devolution of responsibilities has created new problems: case studies show that devolution resulted in discrepancies among formal legal rights, actual interpreted rights, and forest use practices. The scope and reach of actual rights has become an issue of intense negotiations or even conflicts among actors involved in local resource use (Thanh and Sikor 2006).

There is evidence that land market development is restricted as long as local political authorities retain their power over land. Local party cadres still oversee titling, define land use restrictions, declare land appropriation for infrastructure, and periodically reallocate land in response to demographic changes and new family formations (Ravallion and van de Walle 2003, 2006). Poor farmers have protested against inadequate compensation in cases of redistribution and misconduct by local officials. Yet the same authorities often defend villagers' interests against national ones, specifically regarding public infrastructure. Concerns about the institutional quality and sustainability of the reforms arise because, for example, perceptions of corruption in Vietnamese officials already rank among the highest in the world, with an upward trend (Fritzen 2002).

Impact on the Natural Environment

The impact of land tenure reforms on resource protection is multidimensional. On the one hand, the issuing of LUCs has led to a significant increase in the adoption of agroforestry practices, including the development of ditches (Saint-Macary et al. 2008). In northern Vietnam, increasing investment in rice terraces due to higher tenure security has led to stronger protection against erosion on steep slopes. Enhanced tenure security on agricultural lands thus supports solutions for environmental concerns (Neef et al. 2007).

On the other hand, for farming systems other than intensive irrigated rice cultivation, case studies show that individual property rights may have negative effects on endangered ecosystems: allocating land rights to families leads to a conversion of rural wetlands into agricultural land. Because use rights depend on using the plots in question for agricultural purposes or allocating them to new settlers, the common pool management of the wetlands is in danger. Those areas with the highest agricultural potential become privatized, while others shift to open access (Adger and Luttrell 2000). Additionally, concerns arise that market reforms have undermined collective action to ameliorate flooding hazards (Adger 1999, 2000). Consequently, land privatization, together with reduced public expenditure, might exacerbate the vulnerability of land to coastal flooding. These risks are aggravated by the diminishing role of former rural cooperatives to provide local public goods such as protection against floods or other hazards.

Conclusions and Lessons Learned

Land tenure policy reform in Vietnam does not stand alone as a single reform instrument. Rather it was a key element in a broader reform process that includes complementary incentives to develop rural product and factor markets (Rozelle and Swinnen 2004). Enabled through land tenure reform, land rental and sales markets had a positive impact on agricultural productivity as well as overall growth. As for economic and social sustainability, creating private land rights strongly contributed to making the transition toward a market economy in Vietnam irreversible. It also generated the necessary economic incentives for family farms, made the agricultural sector an engine of economic growth, and changed social structures in rural areas.

How sustainable are the reforms? Elements that favor the social and political sustainability of the intervention include relatively egalitarian land ownership and the rapid growth of off-farm opportunities (Deininger and Jin 2008). Moreover, nonagricultural employment will be an additional trigger for further development of land rental markets because those leaving agriculture will contribute to the total supply of land. Land tenure reforms have triggered a process that will relieve the country from food insecurity, combat hunger, and improve the nutritional status of the rural and urban populations. Due to the improved health status and the decreased vulnerability of households in this specific field, the conditions for forming human capital and providing better education for youth have become more favorable. With regard to financial sustainability, no discussions have yet taken place about the degree to which a partial recuperation of public funds is possible through land transfer taxes, which are important sources of local revenue in other countries.

In spite of the many positive outcomes, new obstacles have arisen that may undermine the sustainability of the reform process: the new system does not necessarily work to everyone's advantage. Some rural poor may sell and therefore lose their land in times of emergency; thus the emergence of landlessness must be monitored carefully. Also, land tenure reforms have not yet increased rural people's access to credit as much as expected because formal banking institutions seem reluctant to accept LUCs as collateral. Finally, the reforms have had mixed results on the environment. On the one hand, more secure property rights have led farmers to adopt agroforestry and other antierosion measures. On the other hand, strengthening individual rights to land can put fragile lands at risk when land reform allocates rural wetlands to households that then convert them to farmland or aquaculture.

Although several negative feedback mechanisms can be expected as secondary effects of land reform, particularly related to equity and income distribution as well as environmental issues, the positive effects on economic, social, and ecological sustainability dominate.

Vietnam's experiences point to a number of lessons. First, land tenure reforms that enable a peaceful transition from a centrally planned economy to a market economy are unlikely to achieve their objectives if they are not interlinked with other components, such as reform of agricultural product markets, the factor input supply, and complementary rural factor markets, including labor and credit.

Second, as part of the land tenure reform package, generating stronger economic incentives for rural producers is critical, as illustrated in Vietnam's household responsibility model based on the decollectivization of cooperatives, the distribution of land use rights, and market liberalization.

Third, these incentives are unlikely to materialize as long as complementary agricultural policy reforms are not successfully implemented. These complementary policies imply accepting price as an indicator of scarcity and liberalizing marketing channels. This process is often accompanied, at least temporarily, by government intervention in these markets, as demonstrated by Vietnam's rice export quota.

Fourth, strong instruments and clear signals for enhanced tenure security are critically important, especially regarding the duration of leaseholder rights for annual and perennial crops. A high level of investment in sustainable agriculture depends on according long-term, inheritable tenure rights to families.

Fifth, a flexible, incremental approach to designing, sequencing, and implementing the reform has advantages, because all steps must be legitimized at both the national and local levels. Considerable information and communication are necessary to break resistance against reforms and to win over various stakeholders.

Finally, land tenure reforms must go hand in hand with organizational reforms at lower levels, for a local administration can play either a strongly supportive or an extremely negative role in reform implementation. To limit fiscal burdens on local governments, a constructive discussion on the merits and limitations of land taxes must be initiated, even if such taxes may create distributional problems and burdens for the rural poor.

Notes

1. In 1977 and 1978 in the Do Son District near Haiphong, a cooperative introduced household contracts. Under these contracts, households received land from the cooperative for rice cultivation. Once households fulfilled their quota obligations to the cooperative, they were allowed to retain the surplus as either a use-value or an exchange-value. Households were also encouraged to reclaim wasteland and work this land for themselves, retaining the entire output. The results were so impressive that in 1980 authorities in Haiphong instructed all agricultural cooperatives to adopt the new production scheme, a reform that served as a prelude to countrywide reform in the 1980s.

2. Cooperatives still performed certain tasks—usually caring for water supplies and seeds and preparing land—that were paid for by the contracted amount of output. Taxes and deductions for local schools as well as other less popular fees imposed by the local authorities also had to come out of the contracted amount (Fforde and de Vylder 1996).

3. Perhaps the most significant structural reform of the period was in the area of trade policy. With the exception of rice, export quotas ceased. Import quotas were reduced to apply to only seven items, and import permits were introduced for most remaining controlled items.

4. From the second half of the 1990s, growth was led by the industrial sector, whose growth averaged 14 percent annually. Growth was driven by unprecedented levels of investment, which

reached a high level of 30 percent as a result of sharp increases in FDI that accounted for one-third of total investment.

5. In addition, agricultural export expansion made a great contribution to financing the increasing growth of industrial imports.

6. A household is landless if it has no land other than the land it rents or resides on or if it follows shifting cultivation. Households that offer land rentals are "noncultivating" ones (Ravallion and van de Walle 2008, 53).

References

Adger, W. N. 1999. Evolution of economy and environment: An application to land use in lowland Vietnam. *Ecological Economics* 31 (3): 365–379.

———. 2000. Institutional adaptation to environmental risks under the transition in Vietnam. *Analyses of the Association of American Geographers* 90 (4): 738–758.

Adger, W. N., and C. Luttrell. 2000. Property rights and the utilisation of wetlands. *Ecological Economics* 35 (1): 75–89.

Akram-Lodhi, A. H. 2001. "Landlords are taking back the land": The agrarian transition in Vietnam. Working Paper 353. The Hague, the Netherlands: Institute of Social Studies.

ANZDEC, Limited. 2000. *Vietnam agricultural sector program (ADB TA 3223-VIE): Phase I technical report.* Washington, D.C.: International Food Policy Research Institute.

Beckman, M. 2001. Extension, poverty and vulnerability in Vietnam—Country study for the Neuchatel initiative. Working Paper 152. London: Overseas Development Institute.

Benjamin, D., and L. Brandt. 2002. Agriculture and income distribution in rural Vietnam under economic reforms: A tale of two regions. World Bank, Washington, D.C. Photocopy.

Beresford, M. 1985. Household and collective in Vietnamese agriculture. *Journal of Contemporary Asia* 15 (1): 5–36.

Beresford, M., and A. Fforde. 1996. A methodology for analyzing the process of economic reform in Vietnam: The case of domestic trade. Australian Vietnamese Research Project, Working Paper Series 2. Macquarie University, Sydney, Australia.

Brandt, L. 2006. Land access, land markets, and their distributive implications in rural Vietnam. Department of Economics, University of Toronto, Toronto. Photocopy.

Chu, V. L., N. T. Nguyen, P. H. Phy, T. Q. Toan, and D. T. Zuong. 1992. *Hop tac hoa nong nghiep Viet nam, lich su-van de-trien vong* (Agricultural cooperatives in Vietnam: History, problems, perspective). Hanoi, Vietnam: Su That.

Dang, K. S., N. Q. Nguyen, Q. D. Pham, T. T. T. Truong, and M. Beresford. 2005. *Policy reform and the transformation of Vietnamese agriculture.* Rome: Food and Agriculture Organization of the United Nations.

Dat, T. 1997. *Tien trinh doi moi quan ly nen kinh te quoc dan cua Viet Nam* (The process of renovating the management of the national economy in Vietnam). Hanoi, Vietnam: NXB Ha Noi.

Deininger, K. 2003. Land markets in developing and transition economies: Impact of liberalization and implications for future reform. *American Journal of Agricultural Economics* 85 (5): 1217–1222.

Deininger, K., and S. Jin. 2003. The impact of property rights on investment, risk coping, and policy preferences: Evidence from China. *Economic Development and Cultural Changes* 51: 851–882.

————. 2008. Land sales and rental markets in transition: Evidence from rural Vietnam. *Oxford Bulletin of Economics and Statistics* 70 (1): 67–101.

Do, Q. T., and L. Iyer. 2003. Land rights and economic development: Evidence from Vietnam. World Bank Policy Research Working Paper 3120. Washington, D.C.: World Bank.

————. 2008. Land titling and rural transition in Vietnam. *Economic Development and Cultural Change* 56 (3): 531–579.

Dollar, D. 1994. Macroeconomic management and the transition to the market in Vietnam. *Journal of Comparative Economics* 18 (3): 357–375.

Fforde, A., and S. Pain 1987. *The limit of national liberalization*. New York: Croom Helm.

Fforde, A., and S. de Vylder. 1996. *From plan to market: The economic transition in Vietnam*. Boulder, Colo., U.S.A.: Westview.

Fritzen, S. 2002. Growth, inequality and the future of poverty reduction in Vietnam. *Journal of Asian Economics* 13: 635–657.

Henin, B. 2002. Agrarian change in Vietnam's northern upland region. *Journal of Contemporary Asia* 32 (1): 3–28.

Kemper, N., and R. Klump. 2009. Land reform and the formalization of household credit in rural Vietnam. Paper presented at the annual Poverty Reduction, Equity and Growth Network (PEGNet) conference, September 3–4, in The Hague, Netherlands.

Kerkvliet, B. J. T. 1995. Rural society and state relations. In *Vietnam's rural transformation*, ed. B. J. T. Kerkvliet and D. J. Porter. Boulder, Colo., U.S.A.: Westview.

Kirk, M., and D. A. T. Nguyen. 2009. *Land-tenure policy reforms: Decollectivization and the Doi Moi System in Vietnam*. IFPRI Discussion Paper 927. Washington D.C.: International Food Policy Research Institute.

Minot, N. 2003. Income diversification and poverty reduction in the northern uplands of Vietnam. Paper presented at the American Agricultural Economics Association annual meeting, July 27–30, in Montreal. Washington, D.C.: International Food Policy Research Institute.

Minot, N., and F. Goletti. 2000. *Rice market liberalization and poverty in Vietnam*. Research Report 114. Washington, D.C.: International Food Policy Research Institute.

Neef, A., P. Sirisupluxana, T. Wirth, C. Sangkapitux, F. Heidhues, D. C. Thu, and A. Ganjanapan. 2007. Resource tenure and sustainable land management—Case studies from northern Vietnam and northern Thailand. In *Sustainable land use in mountainous regions of southeast Asia: Meeting the challenges of ecological, socio-economic and cultural diversity*, ed. F. Heidhues, L. Herrmann, A. Neef, S. Neidhart, J. Pape, and V. Zarate. Berlin: Springer.

Nguyen, D. A. T. 2006. *Agricultural surplus and industrialization in Vietnam since the country's reunification.* The Hague, the Netherlands: Shaker Publisher / Institute of Social Studies.

———. 2008. *Relationship between investment and agricultural growth.* Hanoi, Vietnam: Ministry of Agriculture and Rural Development.

Niimi, Y., P. Vasudeva-Dutta, and L. A. Winters. 2004. Rice reform and poverty in Vietnam in the 1990s. *Journal of Asia Pacific Economy* 9 (2): 170–190.

Pingali, P. L., and V.-T. Xuan. 1992. Vietnam: Decollectivization and rice productivity growth. *Economic Development and Cultural Change* 40 (4): 697–718.

Ravallion, M., and D. van de Walle. 2003. Land allocation in Vietnam's agrarian transition. World Bank Policy Research Working Paper 2951. Washington, D.C.: World Bank.

———. 2006. Does rising landlessness signal success or failure for Vietnam's agrarian transition? World Bank Policy Research Working Paper 3871. Washington, D.C.: World Bank.

———. 2008. *Land in transition: Reform and poverty in rural Vietnam.* Washington, D.C. / New York: World Bank / Palgrave Macmillan.

Rozelle, S., and J. F. M. Swinnen. 2004. Success and failure of reform: Insights from the transition of agriculture. *Journal of Economic Literature* 42 (2): 404–456.

Saint-Macary, C., A. Keil, M. Zeller, F. Heidhues, and P. T. M. Dung. 2008. *Land titling policy and soil conservation in the uplands of northern Vietnam.* Research in Development Economics and Policy Discussion Paper 3/2008. Beuren, Stuttgart, Germany: University of Hohenheim.

Sepehri, A., and A. H. Akram-Lodhi. 2002. A crouching tiger? A hidden dragon? Transition, saving and growth in Vietnam, 1975–2000. ISS Working Paper 359. The Hague, the Netherlands: Institute of Social Studies.

Thanh, T. N., and T. Sikor. 2006. From legal acts to actual powers: Devolution and property rights in the central highlands of Vietnam. *Forest Policy and Economics* 8: 397–408.

Tran, T. Q. 2005. Annex: Land and agricultural land management in Vietnam. In *Impact of socio-economic changes on the livelihoods of people living in poverty in Vietnam,* ed. H. H. Thanh and S. Sakata. Chiba, Japan: Institute of Developing Economies, Japan External Trade Organization.

VCP (Vietnamese Communist Party). 1977. *Nghi quyet dai hoi dai bieu toan quoc lan thu IV* (Resolution of the fourth National Congress of the VCP). Hanoi, Vietnam: NXB Su That.

———. 1996. *Van kien dai hoi dai bieu toan quoc lan thu VIII* (Documents of the eighth National Congress of the VCP). Hanoi, Vietnam: NXB Chinh Tri Quoc Gia.

Vo, N. T. 1990. *Vietnam's economic policy since 1975.* Singapore: Institute of Southeast Asian Studies.

Watts, M. 1998. Recombinant capitalism: State, de-collectivization and the agrarian question in Vietnam. In *Theorising transition: The political economy of post-communist transformation,* ed. J. Pickles and A. Smith. London: Routledge.

White, C. 1982. *Debates in Vietnamese development policy.* Discussion Paper 171. Brighton, England, U.K.: University of Sussex, Institute of Development Studies.

World Bank. 1998. Vietnam. Advancing rural development from vision to action. Chapter prepared for the Consultative Group meeting, December 7–8, in Haiphong, Vietnam.

———. 1999. *Vietnam development report 2000: Attacking poverty.* Washington, D.C.

———. 2002. *Land use rights and gender equality in Vietnam.* Promising Approaches to Engendering Development Series 1. Washington, D.C.

———. 2004. *Vietnam development report 2005.* Report 30462-VN. Washington, D.C.

Chapter 20

Improving Diet Quality and Micronutrient Nutrition: Homestead Food Production in Bangladesh

Lora Iannotti, Kenda Cunningham, and Marie T. Ruel

A critical yet often overlooked component of food security is diet quality. Even households that have access to sufficient amounts of food and calories may still lack essential micronutrients, increasing their risk of both short- and long-term health and development consequences. Worldwide, the numbers of food insecure—as measured by insufficient availability of or access to calories—are declining (FAO 2008). These figures, however, fail to capture the even more widespread problem of poor-quality diets and the resulting risks of micronutrient deficiencies—often referred to as "hidden hunger." Recent data estimate that 127 million preschool children are vitamin A deficient and nearly 5 million suffer from xerophthalmia, a condition resulting from this deficiency, which causes irreversible eye damage and blindness in extreme cases. Vitamin A deficiency is a public health problem in nearly 80 developing nations: in these low-income countries, more than 7 million pregnant women suffer from insufficient vitamin A (West 2002; West and Darnton-Hill 2008). More than half of the prevalence of anemia globally is estimated to be due to iron deficiency (Rastogi and Mathers 2002). Similarly, current estimates suggest that one-third of the world's population consumes diets inadequate in zinc (Hess and King 2009). Overall, micronutrient deficiencies raise the risk of mortality from diarrhea, pneumonia, malaria, and measles (Black et al. 2008). These micronutrient deficiencies are responsible for a large proportion of infections, poor physical and mental development, and excess mortality in the developing world. Vitamin A deficiency alone is responsible for 6 percent of all deaths of children under five years old; 4 percent are attributed to zinc deficiency. Iron deficiency increases the risk of maternal mortality by 20 percent and reduces child IQ by 1.73 points for every 10 grams per liter decrease in hemoglobin concentration (Stolzfus, Mullany, and Black 2004).

Several strategies to combat micronutrient deficiencies in Bangladesh have been undertaken in recent decades, including supplementation, fortification, and the promotion of dietary diversification. One such strategy was a homestead food production (HFP) program by Helen Keller International (HKI), specifically introduced to address the issue of vitamin A deficiency. The original HFP model included support for gardening and nutrition education to promote year-round production of vitamin A–rich fruits and vegetables and to increase the availability of and household access to these foods to improve nutrition in vulnerable populations. HFP programming has now evolved and expanded to embody a unique, holistic intervention to increase the availability of micronutrient-rich foods for millions of families while addressing several other aspects of food insecurity, including improved incomes and livelihoods, community development, and the empowerment of women. To date, HFP has directly benefited more than 5 million people in Bangladesh, reaching out to nearly 4 percent of Bangladesh's population and covering more than half of all the country's subdistricts (HKI 2006a, 2006b). The program has been implemented across a range of agroecological zones in the country through various organizations and individuals and with support from a variety of bilateral and multilateral funding sources.

Many of the evaluations of HFP to date suggest improvements in household income and food production, availability, and diversity, but few of these studies have examined the impact of HFP on maternal and child nutritional status. Rigorously designed evaluations—which rely on state-of-the-art impact evaluation techniques such as randomization, appropriate control groups, consistency of intervention design over time, and precise measurement of the program's various impacts and costs (and in particular of maternal and child anthropometry and micronutrient status)—have not been fully applied to HFP interventions in Bangladesh. As a result, there is a paucity of rigorous evidence on the exact extent of the program's impacts, on how the model works to improve nutrition (including vitamin A deficiency), and whether it is cost-effective when compared to other approaches. Therefore, although the evaluations that have been conducted so far do provide consistent and plausible indications that HFP has been successful in improving various aspects of its beneficiaries' well-being, additional studies are required to reliably substantiate these indications, and in particular to establish its impact on the intake of micronutrients.

The HFP Intervention: Program Model and Scale-Up

HFP programming in Bangladesh has spanned two decades and has directly reached about 4 percent of the population in 240 of the 466 subdistricts in the country, covering diverse agroecological zones (see Iannotti, Cunningham, and Ruel 2009 for additional details). The original model focused primarily on vitamin A deficiency, aiming to increase consumption of vitamin A–rich vegetables and fruits available from home gardens, such as sweet gourd, shazna shak, black arum leaves, and bottle gourd leaves (HKI 2003b). More recently, the scope of the HFP model was significantly broadened to address multiple micronutrient deficiencies, including those in iron and zinc. This expansion included incorporating small animal husbandry into the model, because animal-source foods are the best sources of bioavailable (easily absorbed and used) iron and zinc.

The objectives of HFP programs are to

- increase year-round production, varieties, and quantities of vegetables and fruits produced by home gardening;

- increase animal foods through small animal husbandry;

- increase consumption of micronutrient-rich foods through increased household production and income, enhanced by improved knowledge and awareness through nutrition education;

- improve the health and nutritional status of women and children; and

- empower women by giving them control over the resources that ensure better childcare practices.

To implement its HFP programs, HKI works though local nongovernmental organization (NGO) partners at the subdistrict level (see Iannotti, Cunningham, and Ruel 2009 for additional details). Each NGO, in turn, supports approximately 25 to 30 village model farms (VMFs). Typically, there are 2 mothers' groups and 40 households per VMF. HKI may provide the NGOs with inputs (for example, seeds, seedlings, chicks, and the like) and technical assistance (for example, key nutrition messages). The local NGOs then provide similar support to the communities more directly. The mothers' groups are already existing community groups that meet regularly to support women in the community in a variety of ways. The duration of an HFP program is three years with HKI involvement, followed by another two years of ongoing community support from partner NGOs.

HFP Approaches

Several prominent features of HFP have been responsible for its success:

- incorporating into the model nutrition education and behavior-change communication,

- building on local practices and using existing structures and organizations,

- focusing on empowering women,

- fostering income generation,

- including strong technical assistance and capacity-building components, and

- implementing monitoring and evaluation activities.

Home gardens alone do not improve nutrition; nutrition education is necessary to translate food production into improved dietary intakes, particularly for vulnerable household members. Several reviews examining the potential for agricultural programs and food-based strategies to improve nutrition have highlighted the importance of explicit nutritional objectives and nutrition education activities—specifically behavior-change communication—to affect positive nutrition outcomes (Ruel 2001; Berti, Krasevec, and FitzGerald 2004; World Bank 2007). A consistent recommendation from these reviews is that nutrition education or behavior-change communication should include education about appropriate intrahousehold allocation of resources that favor vulnerable household members such as mothers and young children, as well as key messages regarding optimal infant and young child feeding and care practices (World Bank 2007).

In the HFP model, key nutrition messages are usually communicated through group meetings or individual counseling sessions. In the Char project, for example, group leaders conduct meetings and counseling sessions to discuss the importance of regular consumption of foods rich in vitamin A and iron, taking iron tablets during pregnancy, iodized salt consumption, good breastfeeding, and complementary feeding practices (HKI, Bangladesh 2006). Similarly, in the Chittagong Hill Tracts project, partner NGOs train group leaders, who in turn conduct group meetings and counseling sessions with mothers, husbands, and mothers-in-law. Cooking demonstrations are also widely used to illustrate positive food preparation practices, such as washing vegetables before cutting, using oil to cook leafy vegetables, and including more pulses, meat, and eggs in dishes (HKI, Bangladesh 2008).

A recent innovation in the nutrition education component of HFP has been the application of the essential nutrition actions (ENA) framework. ENA represents a comprehensive approach to improving nutrition for children under two years old and reproductive-age women by supporting better practices in the seven key nutrition areas shown to have the greatest impact on improving maternal and child health (USAID 2006). These key nutrition areas include optimal breastfeeding during the first six months of life; optimal complementary feeding from six months of age, with continued breastfeeding until age two and beyond; adequate nutritional care of a sick and malnourished child; optimal maternal nutrition during pregnancy and lactation; and the control of vitamin A deficiency, anemia, and iodine deficiency. In addition to incorporating these messages into group and counseling sessions, HFP programs are now working to create linkages among partner NGOs, beneficiaries, and government health services. The ENA approach was introduced into the Chittagong Hill Tracts project in 2006.

The second prominent feature of HFP is that it builds on local practices and involves existing structures and organizations. This factor helps ensure both acceptance within communities and the sustainability of gardening and positive nutrition behaviors. Local practices include using local cultivation techniques and varieties, understanding and working with traditional customs, and navigating the cultural barriers and opportunities for adopting optimal infant and young child feeding and household dietary practices. To achieve success in this area, it is important to focus on creating community resources and venues, such as VMFs, to provide farmers with ongoing inputs and advice. These local sources and support services allow better market access for participants; for example, women farmers can gain access to inputs, technical information, and better marketing opportunities. In other words, the VMFs provide support to the community, while the HFP systems are owned and operated individually. Finally, local-level ownership is ensured through a cost-sharing requirement: inputs are not provided for free, but rather farmers are required to contribute financially. Thus the project is owned by participants from the outset.

Rooted in local values, customs, and practices, HFP inherently emphasizes community participation at all stages of the program—design, implementation, and monitoring and evaluation. HKI's collaboration with NGOs through particularized local approaches includes strategic planning workshops, proposal and work plan development, program monitoring, financial management, and organized involvement of government and other local authorities (HKI 2003a). An additional strength of the model is that it links agricultural activities to other health and development activities in the community, in recognition of the complex nature of food insecurity and undernutrition.

Third, HFP works to empower women—specifically, poor rural women. In Bangladesh, women are traditionally responsible for managing homestead activities, preparing family meals, and feeding children, among their many responsibilities. The HFP approach supports women in these culturally acceptable roles to upgrade their skills and knowledge to improve food production, income, and practices. The result is often better allocation of household resources, improvements in caring practices, and overall empowerment (HKI 2006a). For instance, as HFP beneficiaries of gardening activities, nutrition education, and income generation, women enhance their bargaining power and become more productive in their traditional roles. Additionally, pregnant women and children younger than three years old are two populations for which inadequate nutrition has the largest impact; therefore, targeting women with nutrition-oriented interventions is also critically important for reducing childhood undernutrition.

Other distinguishing characteristics of the HFP model contribute to its success. The income-generating component for VMF owners and households potentially contributes to its economic viability, reflecting returns to land and labor. At the organizational level, viability and sustainability are enhanced through cost sharing with the many local NGOs. The model also incorporates technical assistance and capacity building at all levels of the program structure. Finally, program monitoring and evaluation are integral aspects of the model, incorporating information systems that provide feedback and enable improvements in HFP interventions.

Key Players

The Government of Bangladesh has played a critical role in HFP programming. Two agencies in particular, the Department of Agricultural Extension (DAE) and the Rural Development Academy (RDA) (under the Ministry of Local Government and Rural Development) have worked with the HFP program. The DAE operates throughout Bangladesh and is an important contact for HKI in all the subdistricts where projects are implemented. For instance, the DAE coordinates HFP programs and incorporates them into its overall district and national plans; provides information on input sources; ensures quality control and coordinates with the private sector for inputs and marketing; facilitates trainings with HKI; and provides updated research findings to enhance programming. It also disseminates information on positive results and findings of the program through its local networks. The role of the other government agency, RDA, has mainly been to provide training facilities and human resources for training and staff development of local NGO partners.

Through the provision of long-term funding, various bilateral and multilateral donors have also played an important role in the HFP program. The Australian Government Overseas Aid Program, Danish International Development Agency, United States Agency for International Development, U.K. Department for International Development, Netherlands Organization for International Development Cooperation (now Oxfam Novib), and World Bank provided funding to supplement the Government of Bangladesh support. These funding sources, combined with the overall program support provided by HKI, ensured that a more diverse set of local NGOs would be able to carry out program operations.

A strong enabling environment was also a prerequisite for the success of HFP throughout Bangladesh. Certain policies, institutional frameworks, and social norms provided the foundation for this type of intervention. First, gardening is an ancient food production method in Bangladesh. Reliance on existing practices, local varieties, and even tools smoothed the introduction of the project (Talukder et al. 2000). Second, women have traditionally been responsible for gardening and the provision of food for the family in Bangladesh. Therefore, targeting women for HFP programs was logical and culturally appropriate (Bushamuka et al. 2005). Third, working with long-established partner NGOs within intervention communities fostered acceptance and facilitated success: NGOs in Bangladesh already know how to mobilize local resources, build community support, and encourage participatory involvement (HKI 2003a).

Scaling Up

A national nutrition survey and a national nutritional blindness survey found that the prevalence of vitamin A–related night blindness in Bangladesh was 3 percent and that more than 1 million children younger than six years of age suffered from some degree of xerophthalmia (Nutrition Survey of Rural Bangladesh 1981–82; National Nutritional Blindness Prevalence Survey 1982–83; Talukder et al. 1993). These findings provided the impetus for what would become a large-scale program throughout Bangladesh in just two decades—a comprehensive HFP intervention promoting the production and consumption of foods rich in vitamin A and other micronutrients.

Beginning in 1990, a pilot HFP program targeted 1,000 households. A midterm evaluation of this project demonstrated that its combined home gardening, nutrition education, and gender interventions could improve vegetable consumption among women and children, and the NGO Gardening and Nutrition Education Surveillance Project (NGNESP) was subsequently launched in 1993.

The NGNESP project combined home gardening, nutrition education, and other community development activities. After seven years of operation, NGNESP covered more than 860,000 households in 210 of the 460 subdistricts of Bangladesh and was deemed successful in achieving its household food security aims (Bushamuka et al. 2005). Whereas the HFP program objectives, basic inputs, and organizational structure remained relatively constant, the flexibility of the model allowed for context-specific adaptations, enhancing its effectiveness and ensuring its continuation in Bangladesh—and its expansion to other parts of the world.

The year 2002 marked another milestone in the history of HFP in Bangladesh: a pilot animal husbandry project was introduced in response to new findings. Endorsed by consensus in the international community, new findings demonstrated that vitamin A from vegetables and fruits was less bioavailable (that is, it took more pro–vitamin A to convert it into usable vitamin A in the body) than previously thought (IOM 2000). (Some of the research contributing to this shift in fact originated from work at HKI on home gardening promotion [de Pee et al. 1998; de Pee, Talukder, and Bloem 2008].) Therefore, a subsequent movement promoted animal-source foods in the diets of vulnerable populations as more efficient and bioavailable sources of essential micronutrients, including vitamin A, iron, and zinc and as the only sources of vitamin B-12 (Randolph et al. 2007; Schroeder 2008).

HKI accordingly carried out a pilot project in Bangladesh, Cambodia, and Nepal to test the feasibility of including animal husbandry in its existing home gardening model. In Bangladesh, the pilot project was carried out with two NGOs—the Social Development Committee and Gono Kallayan Sangstha—and involved 600 households in the Faridpur and Sirajgonji districts of northwest Bangladesh (HKI 2004).

Three more projects were initiated in subsequent years and continue to operate today. In 2003 the Novib Char project (funded by Oxfam Novib) was launched in the chars (islands of silt within rivers) and low-lying flood plains, reaching 10 sub-districts in the north and 10,000 households. The Jibon-O-Jibika project (funded by Save the Children–U.S.A.) in the Barisal division was started in 2004 with the NGO Forum and Cyclone Preparedness Program. And finally, the HFP to Improve Household Food and Nutrition Security in the Chittagong Hill Tracts (funded by Oxfam Novib), was initiated in 2005, targeting 10,000 households. Each year new program areas and partners are added, emphasizing agroecological diversity—for example, tea estates in hilly terrains, flood-prone areas, periurban and urban slums, and areas with high-salinity soil (Talukder et al. 2000).

Scaling up HFP within Bangladesh has been effective and sustainable because of the large number of partner NGOs, with their extensive infrastructure throughout the country and their dedicated focus on working with poorer households (Talukder et al. 2000). Table 20.1 summarizes the scaling up of HFP programs in Bangladesh.

HFP Impacts

Food and Nutrition Security
A variety of assessments and evaluations have been carried out on the Bangladesh HFP program, including small-scale assessments of pilot projects, interim midterm

Table 20.1 Scaling up homestead food production projects in Bangladesh, 1988–2011

Project	Dates	Partner NGOs	Subdistricts	Households	Beneficiaries
Original pilot	1988–90	1	1	150	825
Larger pilot	1990–93	0	2	1,000	5,500
NGNESP	1993–2003	47[a]	210	877,850	4,700,000
ASF pilot	2002–03	2	2	600	3,300
Char I	2003–05	4	10	10,000	70,000
Char II	2005–08	7	10	10,000	65,000
Char III	2008–11	8	10	10,000	64,000
Novib–CHT	2005–08	7	10	10,000	55,000
JOJ–Barisal	2004–10[b]	9	11	22,440	116,688

Sources: Various presentations, publications, and internal reports of Helen Keller International.
Notes: ASF means animal-source foods; CHT means Chittagong Hill Tracks; JOJ means Jibon-O-Jibika; NGNESP means NGO Gardening and Nutrition Education Surveillance Project.
[a] The total number of partners for NGNESP was 49, but 2 of these were governmental organizations.
[b] JOJ–Barisal was originally to end in 2009, but the end date was extended because of a cyclone.

evaluations, monitoring data surveillance, and larger-scale impact evaluations. The first evaluation of the pilot home gardening project was conducted in 1991, the most recent in 2008. Two of the nine evaluations were independent assessments by outside reviewers. Six used a pre–post design to study changes occurring between baseline and endline points of the project; two of these evaluations included a control group to account for external conditions influencing program impact.

Most of the published evidence describing the HFP impact in Bangladesh draws on the cross-sectional evaluation of NGNESP conducted in 2002, comparing three groups: active participants (households receiving assistance for fewer than three years); former participants (households that had completed the program and had still been participating in HFP for three or more years without HKI); and a control group (households in areas without NGNESP activities that are within target subdistricts).

Evaluations of HFP have typically lacked rigor, used poor designs, not been based on program theory and impact pathway frameworks, measured only a subsample of all outcomes that may have been affected by the program, and accounted for only part of program and participants' costs. Some common flaws in evaluation designs are present across several HFP evaluations. For example, many of the evaluations rely on pre–post designs but do not include a control group to account for external factors that may also influence outcomes. Different timing for carrying out baseline and endline surveys (and failure to address seasonal variations) often makes results related to food production, access, and intake difficult to interpret. Another issue is the lack of comparability of the comparison groups. For example, the three

groups included in the NGNESP evaluation had different socioeconomic characteristics at baseline and therefore were not a good choice for comparisons. Important
programmatic differences between the active and completed groups were also likely
to exist, again making the group comparisons difficult to interpret. Finally, many
other pro-poor programs in Bangladesh have been operating simultaneously with
the HFP and targeting similar beneficiary groups, and their impact has not been
rigorously controlled for in any of the analyses. As a result of these deficiencies,
the reported evidence on program impact may be inflated or attenuated in studies
using pre–post designs by nonprogram factors. For example, a drought occurring
in the postsurvey period could attenuate the program's impact, while a drought in
the presurvey period could inflate its impact. The use of a control group in other
studies does not rule out biases, especially if groups have not been randomized and
are not comparable. Thus, although the consistent pattern of positive results in the
various studies is reassuring, the evidence is only suggestive in nature.

With these limitations in mind, the main outcomes in evaluations of HFP can
be classified according to the three components of food security: (1) increases in
food availability—measured by type of garden (improved or developed) and changes
in the amounts and varieties of fruits and vegetables produced, (2) increases in food
access—measured by increased income and expenditures on micronutrient-rich
foods and household-level consumption, and (3) increases in food use—measured
by changes in individual intake of micronutrient-rich foods, as well as indicators of
micronutrient status, anthropometry, or functional outcomes (for example, night
blindness). These are now examined.

Food Availability. To assess the adoption of HFP production interventions, HKI uses a measure of garden type: traditional, improved, and developed.
Traditional gardens involve the production of gourds and traditional vegetables,
are seasonal, and are found in scattered plots. Improved gardens are typically fixed
plots involving the production of a wider variety of vegetables but not year-round.
Developed gardens offer a wider range of vegetables and fruits produced year-round
in fixed plots (HKI, Bangladesh 2003). Changes in the quantity and diversity of
food commodities produced are viewed as an indicator of impacts on household
food availability. As shown in Figure 20.1, the NGNESP evaluation documented a
marked increase in the percentage of households with improved or developed gardens from baseline to postintervention. This was also accompanied by an increase
in the average number of vegetable types grown, from three to six in the same time
period (Taher et al. 2004a).

The NGNESP evaluation (2002) also demonstrated significant increases in the
proportion of households growing fruits and vegetables year-round among active
participants (77.8 percent) as compared to former participants (50.4 percent) and
control households (15.4 percent) (Bushamuka et al. 2005). The quantity and variety

Figure 20.1 Gardening practices among NGNESP households

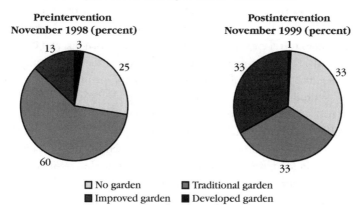

Preintervention
November 1998 (percent)

Postintervention
November 1999 (percent)

☐ No garden   ▨ Traditional garden
■ Improved garden   ■ Developed garden

Sources: Compiled by authors using information from NGNESP evaluations (Talukder et al. 2000) and a Helen Keller International presentation (HKI 2000).
Note: NGNESP means NGO Gardening and Nutrition Education Surveillance Project.

of foods produced in home gardens were also improved through HFP when compared to controls. In a three-month period, households produced a median amount of 135 kilograms and 120 kilograms of vegetables in the active and former groups, respectively, compared to 46 kilograms in the controls. Increases in fruit production were of smaller magnitude. In an evaluation of the animal production pilot in 2003, egg production was considerably higher for the HFP intervention group—200 eggs in the three-month period—than the control group (21 eggs) (HKI 2004).

Later assessments reexamined monitoring data from HFP. An assessment evaluating the impact of the HFP Chars project on mitigating the *monga* (cyclical food insecurity) season in 2007, after the first year and a half, showed that the percentage of households with developed gardens increased from 0 to 49 and the percentage with improved garden types increased from 1 to 18 (HKI 2008b). Similarly, the Jibon-O-Jibika impact assessment found that the percentage of households with developed gardens increased from 0.1 to 60.4 percent and that with improved gardens increased from 1.3 to 31.3 percent (HKI 2008a).

Food Access.   HKI typically analyzes the access component of food security by examining differences in the income earned from home gardens in the last two to three months of the survey, as well as the kinds of expenditures made with the proceeds from the home gardens. This approach does not take into account the complete picture of household income, expenditures, and wealth: information is not collected on how income earned from gardens and small animal production contributes to overall household income, total expenditures, asset acquisition, and savings. Nor does it tell us whether households substitute home production for some food purchases or whether they experience reductions in other forms of household revenue due to redirecting household labor and investments. Nevertheless, this information does reveal spending priorities of target households and demonstrates ways in which HFP engenders broader food security and community development. Figure 20.2 shows the patterns of spending for income generated from the sale of garden produce and poultry products among HFP participating households (based on two separate surveys). It shows a variety of uses of this income, as well as some differences between income earned from garden produce and that from poultry products. Overall, roughly one-third of households report spending some of this income on food, productive assets, and education. Interestingly, 40 percent of households report using additional income earned from poultry products for savings. (No data were available showing income from garden produce used for savings.) Households also report spending some of this additional income on clothes, healthcare, housing, and social activities.

In the NGNESP evaluation, former participants reported the highest income earned from gardening produce (Tk 490), followed by active participants (Tk 347),

Figure 20.2 Uses of additional income earned from HFP garden produce (NGNESP) and poultry products (2004 ASF pilot)

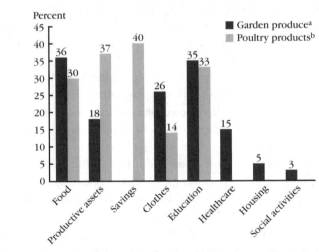

Sources: Compiled by authors using NGNESP evaluation data (Bushamuka et al. 2005, Table 20.3) and ASF pilot project data (HKI 2004).

Notes: ASF means animal-source foods; HFP means homestead food production; NGNESP means NGO Gardening and Nutrition Education Surveillance Project.

[a]Statistics from active participating households in NGNESP evaluation, 2000.

[b]Endline statistics (n = 309) from pilot project of animal husbandry integration, 2003.

then controls (Tk 200). Food was the most frequently reported expenditure (from home gardening proceeds) in all groups; food expenditure was significantly higher in the groups of active and former participants compared to the control groups, as were expenditures on education, clothing, productive assets, and healthcare (Bushamuka et al. 2005).[1] In the animal production pilot evaluation in 2004, income generation from poultry production sales showed a greater impact on earnings than did home gardening—a finding that merits further exploration. A later evaluation of the HFP Chars project showed that 46.3 percent of household income was earned from home gardens compared to 6.7 percent at baseline, and participating households spent significantly more on education, savings, and food (especially fish) (HKI, Bangladesh 2006).

Few studies have examined how homestead gardens have changed total household income. One International Food Policy Research Institute (IFPRI) study examined the profitability of fish and vegetable production in Bangladesh compared with rice, along with concomitant changes in household income; it showed modest increases in income as a result of switching from growing rice to growing foods promoted under HKI's gardening programs (Ruel 2001). HFP program assessments have yet to investigate how the promotion of fruit and vegetable production affects household decisionmaking about other crops produced. As well, there is limited information about market accessibility issues in relation to income earned from garden or animal production. Further, one study showed that households without home gardens primarily depend on the market for their consumption of vegetables (97.5 percent), compared with only 3.2 percent of households with developed gardens (HKI, Bangladesh 2003).

Food Use. By increasing food availability through production and increasing access through higher income and food expenditures, HFP is expected to promote increased intake of higher-quality foods and, ultimately, improved micronutrient status. HFP evaluations have primarily focused on the intake of micronutrient-rich foods by vulnerable groups (mothers and young children). Few studies have examined nutritional impacts using markers such as anthropometry, anemia, or vitamin A (serum retinol or night blindness).

Monitoring data from NGNESP showed an important association between household food availability (in quantity and number of varieties) and vitamin A intake (Bloem et al. 1996). Taher and colleagues concluded that HKI's homestead gardening programs in Bangladesh increased the intake of pro–vitamin A carotenoids, especially among women and children (Taher et al. 2004a): the percentage of mothers and children aged 6–59 months eating dark green leafy vegetables increased from approximately one-third to over three-fourths (Table 20.2).

During the pilot phase of animal production, beneficiary groups produced 200 eggs in three months, whereas control groups produced only 21 eggs. The evalua-

Table 20.2 Consumption patterns among NGNESP households, 1999 and 2000

Indicator	1999	2000
Mothers' intake of dark green leafy vegetables: 3 of 7 days prior (percent)	37	86
Children's (6–59 months) intake of dark green leafy vegetables: 3 of 7 days prior (percent)	28	76
Mothers' intake of retinol equivalent per day from fruits and vegetables	30	230
Children's intake of retinol equivalent per day from fruits and vegetables	10	40

Source: Taher et al. (2004a).
Note: NGNESP means NGO Gardening and Nutrition Education Surveillance Project.

tion found increases in the percentage of households consuming eggs (a 5 percent increase in control groups compared to a 27 percent increase in intervention groups) as well as in the number of eggs consumed in the last 7 days, for children aged 6–59 months and mothers. There was also evidence that money from the sale of poultry was used for the purchase of other foods (HKI 2004).

In sum, the evidence suggests (subject to the methodological qualifications highlighted earlier) that the HFP program in Bangladesh has improved individual intakes of micronutrient-rich foods through gardening and animal production. Although the home gardening and animal production programs initially focused on vitamin A consumption, it is probable that the intake levels of other important nutrients also improved with the increased consumption of vegetables, fruits, and animal-source foods (Ramakrishnan and Huffman 2008). Although HKI has carefully selected regions of Bangladesh vulnerable to undernutrition based on anthropometry and micronutrient status indicators (for example, night blindness), they have yet to evaluate the impact of their programs on these nutritional status outcomes in the country. In Cambodia, however, anthropometric measures and hemoglobin concentrations were collected on mothers and children younger than five years of age at baseline and endline, and no evidence was found that increases in consumption of micronutrient-rich foods led to either improved anthropometry or reduced anemia—a finding that may reflect inadequate evaluation design, problems with the intervention, or both (Olney et al. 2009). Other countries such as Nepal are planning to evaluate the nutritional status impacts of large-scale HFP programs.

Table 20.3 summarizes the findings of the NGNESP evaluation of the HFP program in Bangladesh on food security and other outcomes (discussed below, along with other development outcomes).

The HFP model has been applied quite successfully in various agroecological settings. Since 2002, HKI has implemented HFP in the riverine "chars" region of northern Bangladesh. These chars, or temporary islands, are especially vulnerable to erosion and flooding. Char dwellers experience *monga*, a local term for cyclical poverty and hunger (HKI 2008b). This particular HFP project encouraged home

Table 20.3   Homestead food production food security and other impacts, 1993–2003

Impact category	Description	Example metric
Production	More home gardens	Year-round gardening increased to 33 percent
	Increased varieties of foods	Vegetable varieties increased by more than twofold
	Increased quantities of foods	135 kg instead of 46 kg of vegetables produced in 3 months
Economic status	Improved socioeconomic status through sales	Average extra income: $8 bimonthly
	Future economic benefits	10 percent saved out of income earned
	Employment opportunities	NGNESP alone created more than 60,000 rural jobs
Consumption	Increased consumption of homegrown vitamin A–rich foods	Egg consumption increased by 48 percent
	Increased expenditures on noncereal foods	Lentils and animal products bought with income earned
Intake	Increased intake of vegetables and fruits	Children in households with developed gardens ate 1.6 times more vegetables
Women's status	Garden management	73 percent managed by women
	Income decisionmakers	Women as main decisionmakers
		At least 90 percent of target households represented by women

Sources: Compiled by authors with information drawn from various reports (Sifri 2007; World Bank 2007; Sarkar et al. n.d.).
Notes: The table summarizes changes in relation to the control group or the baseline (NGNESP). NGNESP means NGO Gardening and Nutrition Education Surveillance Project.

gardening and raising livestock through the use of specially adapted low-cost technologies; it also offered nutrition education (HKI 2006b). Project beneficiaries increased their production of micronutrient-rich foods, as well as their dietary diversity and income from the sale of the foods produced, as shown in Table 20.4, although in the absence of a comparator nonparticipant group the exact program impact cannot be ascertained. It was reported that the HFP "activities enhanced [households'] ability to mitigate food insecurity and to cope with flood or *monga*" (HKI 2008b, 4).

The Chittagong Hill Tracts HFP project was also deliberately located in an especially vulnerable region of the country. Chittagong Hill Tracts has suffered long-standing civil conflict. The area is characterized by great cultural diversity (with 13 different tribal groups), poor physical infrastructure, underdeveloped agricultural practices, and isolation from government services (HKI, Bangladesh 2008). Project evaluations show increased production and consumption of nutritionally rich foods

Table 20.4 Food security impacts: Char II, 2006–08

Impact indicator	Target	Baseline (2006)	Endline (2008)
Households established improved or developed home gardens (percent)	70	1	98
Mothers eating dark green leafy vegetables 4 days a week (percent)	40	2	78
Children (12–59 months) eating dark green leafy vegetables 4 days a week (percent)	40	2	67
Mothers eating eggs at least 2 days a week (percent)	30	15	43
Children (12–59 months) eating eggs at least 2 days a week (percent)	30	1	40
Children (12–59 months) eating chicken and other meat at least 1 day a week (percent)	30	17	35
Women participating in decisions on how home gardening income is spent (percent)	90	69	99
Households generating income from homestead food production surplus (percent)	30	27	59
Number of NGOs able to develop a sustainable mechanism for implementing HFP activities	4	3	5

Source: Adapted from HKI, Bangladesh (2008).
Note: HFP means homestead food production; NGOs means nongovernmental organizations.

Table 20.5 Food security impacts: Chittagong Hill Tracts, 2007–08

Impact indicator	Target	Baseline (2006)	Endline (2008)
Households established improved or developed home gardens (percent)	70	4	98
Mothers eating dark green leafy vegetables 4 days a week (percent)	25	5	93
Children (12–59 months) eating dark green leafy vegetables 4 days a week (percent)	21	1	87
Mothers eating red, orange, and yellow fruits 4 days a week (percent)	82	62	93
Children (12–59 months) eating red, orange, and yellow fruits 4 days a week (percent)	84	64	87
Mothers eating eggs 3 days a week (percent)	1	0	43
Children (12–59 months) eating eggs 3 days a week (percent)	1	1	47
Mothers eating chicken and other meat 3 days a week (percent)	2	1	21
Children (12–59 months) eating chicken and other meat 3 days a week (percent)	1	0.4	18
Households generating income from production surplus (percent)	30	12	61
Women participating in decisions on how home gardening income is spent (percent)	70	85	100
Number of NGOs able to develop a sustainable mechanism for implementing HFP activities	4	2	5

Source: Adapted from HKI, Bangladesh (2008).
Note: HFP means homestead food production; NGOs means nongovernmental organizations.

(Table 20.5). Thus, HKI has successfully targeted and scaled up the HFP intervention in regions with particular vulnerabilities.

Through surveys by the Nutritional Surveillance Project of HKI and the Institute of Public Health Nutrition of the Government of Bangladesh, the Barisal division of Bangladesh was identified as an especially vulnerable region based on a high prevalence of underweight and stunting in children and chronic energy deficiencies in nonpregnant women. The Jibon-O-Jibika (Bengali for "Life and Livelihood") project targeted this particularly vulnerable region, focusing on 2,200 "ultra-poor" women for home gardening and goat rearing and some early food security achievements (HKI 2007). Although this particular project is not complete, the provisional data presented in Table 20.6 show that food security is being enhanced throughout the Barisal division (although comparisons to nonparticipant groups are not available, precluding exact assessment of HFP impact).

Other Development Indicators

There have been other important development outcomes of HFP in Bangladesh, most notably the empowerment and improved status of women in households and improvements in the livelihoods of the most vulnerable groups.

Empowerment of women, as demonstrated by greater decisionmaking power within households, is considered an important impact of HFP. Women reported greater contributions to household income because of home gardens. Higher pro-

Table 20.6 Food security impacts: Barisal division Jibon-O-Jibika, 2005–07

Impact indicator	Baseline (2005)	MN Rd-3 (2007)
Households established improved home gardens (percent)	1	31
Households established developed home gardens (percent)	0	60
Households consuming 3 or more eggs per week (percent)	38	50
Mothers eating eggs at least 3 times in last 7 days (percent)	15	22
Children eating eggs at least 3 times in last 7 days percent)	15	24
Mothers eating pulses at least 3 times in last 7 days (percent)	29	47
Children eating pulses at least 3 times in last 7 days (percent)	18	37
Mothers eating vitamin A–rich vegetables at least 3 times in last 7 days (percent)	45	86
Children eating vitamin A–rich vegetables at least 3 times in last 7 days (percent)	27	65
Women decide themselves how/whether to spend money they earned (percent)	28	38
Women decide jointly how/whether to spend money they earned (percent)	8	44
Women have no role in deciding how/whether to spend money they earned (percent)	64	17
Amount of vegetables and fruits produced (kilograms)	3	60
Money earned from vegetables, fruits, and poultry sales (taka)	223	340

Sources: Compiled by authors using information drawn from HKI, Bangladesh (2007); HKI (2008).
Note: MN Rd-3 denotes data from the third round of monitoring.

portions of women reported "full" decisionmaking power on a range of issues among the active and former participant groups as compared to the control group (Bushamuka et al. 2005). Also, HFP programs engendered new employment opportunities for women; on average, at least 70 percent of the households targeted by HKI and partnering NGOs are represented by women. There is an additional food security benefit in that, when programs target women, there is a higher probability that the vegetables will be consumed, particularly by children (Talukder et al. 2000). Table 20.7 illustrates the advances made through NGNESP for women's empowerment.

Vulnerable groups in Bangladesh have been targeted by HKI's programs for several decades. In general, the HFP program targets poor, rural, and landless beneficiaries who are recognized to have limited access to micronutrient-rich foods (Faber et al. 2002). HKI and their partners deliberately locate projects in regions of the country that are especially vulnerable owing to environmental or human-made conditions. Char dwellers in the northern part of the country, residents of the Chittagong Hill Tracts, and households in the Barisal division have all been identified as vulnerable

Table 20.7 Intrahousehold decisionmaking power: Women with full influence before and after NGNESP (percent)

Type of decision	Group	Before NGNESP	2002	Percentage increase
Household land use	FP	10.6	34.5	225
	AP	3.8	26.9	608
	Control	7.0	16.0	129
Group meeting participation	FP	8.6	51.2	495
	AP	2.0	32.8	1,540
	Control	4.0	18.3	358
Making purchases (small household goods)	FP	14.1	49.1	248
	AP	6.7	41.7	522
	Control	7.6	21.8	187
Making purchases (large household goods)	FP	11.1	23.3	110
	AP	5.8	22.7	291
	Control	6.5	12.3	89
Daily workload	FP	25.2	65.0	158
	AP	23.0	64.0	178
	Control	18.2	36.6	101
Vegetable consumption (type and quantity)	FP	34.4	80.5	134
	AP	28.5	77.3	171
	Control	26.7	53.7	101

Source: HKI (2002a).
Notes: *Full influence* is defined as women's making final decisions, either alone or by consulting their husbands. AP means active participants; FP means former participants; NGNESP means NGO Gardening and Nutrition Education Surveillance Project.

populations and have therefore been targeted by HFP projects aiming to increase their food security and improve their quality of life and livelihoods.

Challenges

Overall, the HFP program has been readily accepted and adopted by the stakeholders involved. However, there have been a range of internal and external obstacles to implementation.

Improving Micronutrient Status through Gardening Alone. An initial debate concerns whether households' micronutrient status, including that in vitamin A, iron, and zinc, can be improved through home gardening considering the low bioavailability of these nutrients in fruits and vegetables. HFP evaluations have not adequately measured the programs' impact on micronutrient status to allow for definitive conclusions.

Animal production was added to HFP to better ensure improvements in micronutrient nutrition, but this component has introduced other controversies, mostly related to feasibility (labor and capital inputs are greater for animal production than for home gardens). Also, this component still needs to be studied for its impact on consumption of animal-source foods (especially by young children), the contribution to household income, and again, its effect on micronutrient status. The risk of zoonotic disease associated with animal-source food production has also been debated. Highly pathogenic avian influenza (HPAI), as well as other poultry diseases such as Newcastle disease, threaten poultry stocks and human health in Asia and have hindered the intervention's acceptance, especially by government officials. In response, HFP programming provides vaccines for the animals along with information about prevention of HPAI and other zoonotic diseases.

Cultural and Economic Barriers: Production and Consumption Norms. Bangladeshi farmers traditionally grow rice because of its economic value; they were initially hesitant to switch to fruits and vegetables (Talukder et al. 2000). As well, increasing the production of higher-quality foods through HFP has not always ensured corresponding changes in the diets of vulnerable groups. Some HFP programs —but not all—have conducted preliminary research to provide insight into the barriers to and opportunities for changing behavior and to suggest behavior-change and communication techniques—including negotiation and dialogue—that may be used for positive change.

There have also been internal implementation challenges associated with HFP. Monitoring and evaluation data show that certain inputs, such as seeds, saplings, and seedlings, may be available in the village nurseries without reaching the beneficiaries: the weak linkages between household farmers and village nursery owners have been identified as an implementation constraint (HKI, Bangladesh 2003).

HKI collaborates with many local NGOs throughout Bangladesh. Nevertheless, with program expansion, one challenge has been the lack of NGOs working in more remote areas of the country. Related challenges are problems of coordination among some partner NGOs, limited NGO capacity (especially to implement at scale), and lack of resources. To successfully implement the HFP model, several partner NGOs require more than just technical assistance; they need extensive staff training (capacity building) and additional staff. Overall, the diversity of partner NGOs has allowed for efficient and effective program scale-up, but it has also resulted in variable quality of implementation (Talukder et al. 2000).

Although many of these challenges have been recognized and addressed over the years, new challenges continue to arise. Observers have identified the following remaining tasks: "the development of innovative regional and national marketing systems for garden produce; the establishment of stronger linkages with commercial seed producers; the integration of homestead gardening with other food production schemes; and the opportunities and challenges of using the gardening networks to deliver other services, such as micronutrient supplements" (Talukder et al. 2000, 171).

Civil Conflicts and Environmental Factors.   The Chittagong Hill Tracts program works with 13 diverse tribal groups that are sometimes engaged in low-grade civil conflict that can disrupt the program. Also identified as obstacles have been extreme water conditions—both scarcity and overabundance, depending on the time of year and the location. Natural disasters, including floods, cyclones, and *monga*, also present challenges to food production, delaying program implementation and disrupting previous achievements, and these are particularly prevalent in parts of the country targeted by HFP. In the new HFP char project, Char II, HKI (together with partner NGOs) developed strategies to protect farming assets, such as movable animal shelters and waterproof storage vessels (HKI, Bangladesh 2005), focusing on chars and low-lying flood plains in 10 subdistricts of northern Bangladesh. In 2007, however, flooding disrupted communications, scattered households, and destroyed 238 of the 250 VMFs (Huq 2007).

The Sustainability of HFP

In a typical HFP model, HKI is involved for a period of three years followed by another two years of ongoing partner NGO engagement; following this period, communities are expected to operate HFP independently. It has been estimated that throughout Asia, approximately 95 percent of the households continue to engage in HFP even after their program participation is over. In Bangladesh and Cambodia, fewer than 3 percent of participants drop out of HFP projects annually (HKI, Asia Pacific 2001). The 2002 NGNESP evaluation included a category of former participants—those who had completed the program and operated without

HKI assistance for at least three years—as a comparison group alongside the active participants and control groups (Bushamuka et al. 2005). Although not attaining the levels of the active group, the former participants showed better food security indicators than did the control group, relating to year-round production, crop diversification, production, and consumption. Regarding income earned from gardening products, the former participants actually did better than the active participants, possibly reflecting longer experience and better access to marketing channels (Bushamuka et al. 2005).

HFP sustainability is similarly demonstrated by an ongoing interest in the program and its growing presence throughout the country. The Government of Bangladesh continues to invest and participate in HFP. Each year, hundreds of local NGOs approach HKI and request participation (Talukder et al. 2000). New and continuing HFP programs have by now directly reached more than 5 million Bangladeshi households (HKI 2006a, 2006b).

HFP represents a food-based and household-based strategy that comprehensively addresses availability of and access to higher-quality foods. It addresses some of the underlying structural determinants of undernutrition: poverty, food insecurity, and the low social status of women. HFP thus improves the likelihood of long-term, sustained positive change in nutrition outcomes. HFP can also be a critically important coping strategy for poor households faced with crises such as the recent food and energy price crisis and the unfolding economic recession (HKI 2008c).

The program's most challenging objective is the goal of reducing micronutrient deficiencies in vulnerable groups over the long term. Ultimately, effectively addressing micronutrient deficiencies involves increasing households' intake of bioavailable sources of micronutrients. There are several complementary approaches to achieve this goal: micronutrient supplementation, food fortification, and food-based strategies such as HFP (and biofortification).

Agricultural interventions (like other development interventions) should be financially sustainable. Positive indications of the financial viability of the HFP model are seen in the cost sharing among HKI, local NGOs, and households. Returns on labor and land investments in home gardens can be high compared to field agriculture. Relatively little investment is required, and locally available materials—including indigenous crop varieties, fencing, home-generated manure, and indigenous pest control—may be used (de Pee, Talukder, and Bloem 2008). Moreover, the multiple uses of many of the garden products create a range of opportunities for returns: for example, products from gardens have been used as fodder for animals or for making handicrafts such as baskets.

HFP programming costs have been only minimally examined in relation to benefits. One frequently cited estimate from NGNESP puts costs at $7.66 per household, derived by dividing the project's estimated budget since 1993 (after seven years of operation) by the total number of beneficiaries (Bushamuka et al. 2005). Sometimes cited as the cost per garden, this figure represents the cost of agricultural inputs, training, and technical assistance. It is unclear, however, whether it includes overhead costs or the additional costs borne by other implementing organizations. More rigorous cost–benefit studies of HFP could yield information such as the rate of return on investments and a more comprehensive analysis of costs and outputs throughout the entire program period.

Even in the absence of more thorough analyses, it seems clear that the cost-sharing option of HFP contributes to its financial sustainability. Financial responsibility for the program is shared among HKI, participating households, and partner NGOs (HKI 2003b); this co-financing applies to both program costs and agricultural inputs. Sharing the financial burden of activities throughout a three-year collaboration period reinforces the concept of joint ownership and enhances program sustainability (Talukder et al. 2000).

Homestead gardening is also an environmentally sustainable endeavor. A founding principle integral to the initial program goal of increasing production and consumption of vitamin A–rich foods was the use of ecologically sound methods (HKI 2003b). Some of the environmentally friendly aspects of home gardening practices in general include recycling, safe household waste management and waste water usage, and an increased appreciation of nature among participating families (Landon-Lane 2004). HFP programs, more specifically, embrace additional environmentally friendly agricultural practices: planting trees, using organic fertilizers and pesticides, safe use of pesticides, and using live fencing that enriches the soil with nitrogen. The majority of households involved in HFP programs maintain soil fertility through the use of animal manure and compost as an alternative to chemical fertilizers (HKI 2002b).

Regarding the social and political aspects of sustainability, the active role of the Government of Bangladesh is a vital factor. For instance, HFP has been incorporated into district- and national-level planning, particularly through the DAE. In addition, the involvement and capacity building of partner NGOs have also ensured the longevity of HFP.

Collaboration and capacity strengthening for partner NGOs are regularly practiced through planning workshops and information sharing. Homestead gardening is flexibly integrated into already existing community-based health and development programs (HKI 2003b). Evaluations reveal that partner NGOs have higher skill capacities following project involvement than do nonpartner NGOs. For example, 84 percent of partner NGOs (compared to 70 percent of nonpartner NGOs) have staff with training skills, 100 percent (compared to 90 percent) have gender equality as a component of their programs, 84 percent (compared to 40 percent) have a monitoring system, and 100 percent (compared to 20 percent) conduct nutrition education

(HKI 2003a). Regular reviews of lessons learned have led to modifications of the home gardening model and to collaboration with additional organizations.

Lessons Learned

Two decades of operation in Bangladesh have generated lessons for optimizing HFP implementation and impact. Some of the ideas have already been incorporated into the design and delivery of the program, and others will be applied in future projects, in Bangladesh and elsewhere.

Standard Inputs with Flexible Components

The HFP intervention can be described as a model that is applicable in many diverse settings but has a standard set of inputs and an organizational structure. As such, it allows for centralized quality control, as well as for adjustments by implementing organizations in response to new evidence. For example, when the conversion factor for pro–vitamin A and retinol equivalents was formally increased (IOM 2000), the HFP model was revised to include small animal production and promotion of animal-source foods. Another advantage of standardization is the ease of replication and scaling up, as shown in the reach of HFP both within Bangladesh and elsewhere in South Asia.

A completely static model cannot respond to context and community needs and will ultimately be less effective. HKI therefore included aspects of the HFP intervention model that are dynamic and responsive to local conditions. One mechanism of flexibility is the formation of partnerships with local NGOs: since 1993, HKI has worked with 79 partner organizations throughout Bangladesh, which use varying approaches suitable to the communities where they operate (Talukder et al. 2000). Another element of flexibility is the use of pilot projects to facilitate the early introduction of culturally appropriate and localized components (Sifri 2007). NGNESP was originally a pilot started in 1990, as was the animal husbandry component introduced in 2001.

Local Partnerships

Although HKI has been the primary NGO designing and promoting HFP in Bangladesh and elsewhere, many important stakeholders have also been involved. As mentioned earlier, the Government of Bangladesh and local NGOs are integral to the planning and execution of this program. Local partnerships are vital for program sustainability, because existing structures, institutions, and ways of operating are generally more effective for implementing a program over the long term. Local capacity is also strengthened through partnerships; all parties share information, respond to new challenges, and work together to strengthen HFP programs throughout Bangladesh. More documentation of these processes and partnerships is needed to better understand the challenges and opportunities.

Community Roots

The importance of a community-based approach has long been recognized in the development literature. In the case of HFP, it cannot be overstated. The HFP intervention is built on local farming practices and an understanding of the sociocultural norms of the targeted population (Sifri 2007). The paradox of HFP's apparent success is that it both relies on and transcends local practices in order to increase production and improve consumption and nutrition. HKI and partner NGOs recognize that—like indigenous crops that grow best in local conditions—agricultural practices evolve more smoothly when starting with commonly accepted methods. Village nurseries are often used as sites for demonstrating and experimenting with new techniques, such as the use of botanical pesticides and fertilizers, crop rotation practices, and live fencing (Talukder et al. 2000).

Community participation is important throughout the project cycle, from design through implementation and evaluation. According to one study, "Having a two-way channel for information exchange has been instrumental for achieving sustainable, improved gardening practices" (Talukder et al. 2000, 170). Ultimately, HFP is a community-based intervention; participant ownership is a critical aspect contributing to its success.

Information Systems

Throughout the world, HKI operates and participates in various kinds of surveillance and information systems. The history of HFP in Bangladesh shows a process of active feedback between information collected and programming interventions.

Three forms of information sources have been used. First are the national (or sometimes local) surveys that guide the siting of projects in particular regions. The best example is the 1982–83 Institute of Public Health Nutrition / HKI survey on vitamin A deficiencies that inspired the NGNESP HFP.

Second, monitoring surveys are conducted at program sites every four months to identify problems and compile progress reports. The information is shared and responded to within the community in a problem-solving approach (Talukder et al. 2000). Plans are being discussed to make even better use of this information, to integrate more operations research methodologies, and possibly to use more modern technologies (for example, cell phones) for data collection.

Third, evaluations are used to inform and improve HFP programming. Evaluations are important for motivating donor investments in HFP and fostering commitment on the part of governments and other partners. Stronger evidence of the nutritional impacts of food-based, comprehensive interventions such HFP is needed. HKI is currently working with organizations such as IFPRI to improve its evaluation design, apply program theory, and investigate impact pathways more systematically.

Food-Based Approach

Two important lessons may be drawn regarding the HFP food-based approach to preventing and alleviating micronutrient malnutrition. First, the concept of the best foods to promote for consumption by vulnerable groups has evolved within the HFP program framework, responsive to scientific findings. The shift to animal-source foods as more bioavailable sources of micronutrients is one example, and there have been other changes as well, in messages about foods and the best methods of preparation to enhance absorption of micronutrients from gardening produce.

Second, HFP is among the few programs that provide insights on the contributions of agricultural interventions in improving the quality of diets of vulnerable populations. HFP results show that foods grown in gardens, such as dark green leafy vegetables or eggs from animal production, may be more frequently consumed by children and mothers, especially when a behavior-change communication component is integrated into the agricultural intervention. Whether these changes in diet improve households' micronutrient status is a question that needs further research.

Multidisciplinary Action

HFP programming blends the contributions of several disciplines to address the complex problems of food insecurity and undernutrition in Bangladesh. Inputs are shaped by expertise from the fields of agronomy, nutrition, public health, economics, and environmental science to form an integrated package of interventions. Programs are more likely to have an impact if they simultaneously address both the underlying and the immediate determinants of malnutrition (Ruel 2008). HFP benefits from the synergies among interventions aimed at improving household food security, income, women's empowerment, health, and nutrition. In recent years there has been recognition of the need for linkages with the health sector in particular, in view of the well-known interactions between nutrition and disease. These linkages can help provide both preventive and curative healthcare for mothers and children. HFP is still in the early stages of this effort, and more involvement is needed with ministries of health, primary healthcare systems, and other healthcare stakeholders.

Dialogue, Communication, and Skill Building

Education and technical assistance have long been integral to HFP programming. Capacity building at all levels—for partner NGOs, community members, care-givers, other household members, and so on—is viewed as critical to the success and sustainability of HFP. Education raises awareness and helps ensure better choices and better practices for growing foods year-round that are rich in micronutrients (Sifri 2007). Over the years, the nutrition education component of HFP has evolved and taken different forms to more effectively communicate with caregivers. Behavior-change communication techniques have been shown effective for improving nutrition, especially in young children. HKI is currently working to strengthen and expand the behavior-change communication aspects of HFP programming and to adopt the essential nutrition actions approach.

Conclusion

In Bangladesh HFP has contributed to improved food security for nearly 5 million vulnerable people living in diverse agroecological zones. This holistic, integrated package of home gardening, small livestock production, and nutrition education aims not only to improve household food production and diet quality but also to empower women, households, and communities through economic and social development. It respects local customs and practices and gains longevity in return. HFP leaves a legacy of knowledge, awareness, and understanding with its many partners and beneficiaries. If HFP continues to be responsive to new information and receptive to changes in the environment and the sociopolitical landscape, it will overcome barriers to implementation and effectiveness and continue on its path to improving household food security and diet quality, along with the well-being of many throughout the world.

HFP programming has been replicated in several countries of Asia and the Pacific. In Cambodia, Nepal, and the Philippines, HFP provided support to the food security and livelihoods of more than 1 million households. HFP programming has begun to take hold in Sub-Saharan Africa, with programs operating in Burkina Faso, Mozambique, Niger, Senegal, and Tanzania. Programs in Africa tend to be smaller and more variable than in Asia, ranging from the development of school gardens in Burkina Faso to the promotion of biofortified orange-fleshed sweetpotatoes in Mozambique. Animal production has not yet been introduced in any of these country programs.

HFP programs have not fully met expectations, however, in demonstrating improvements in maternal and child micronutrient status. Although there is suggestive evidence of HFP's impact on household production, improvement of diet quality, and intake of micronutrient-rich foods, its contribution to reducing the prevalence of deficiencies in vitamin A, iron, or zinc has yet to be revealed. It may be that inputs and delivery strategies need to be modified to better reach mothers and young children—or, alternatively, that better impact evaluations need to be designed to capture impacts on anthropometry and markers of micronutrient nutrition. HFP monitoring and evaluation strategies should be grounded in program theory and should clearly identify, measure, and analyze program impact pathways.

Stronger, more rigorous experimental evaluation designs are needed to measure the impact and cost-effectiveness of HFP and to address skepticism about its potential role in addressing micronutrient deficiencies in a sustainable way. Finally, HKI and its partners need to establish and strengthen linkages to the health sector to fulfill its multifaceted approach. HFP's nutritional goals can be fully achieved only by incorporating the bidirectional linkages between health and nutrition to complement its strong agriculture and food security inputs.

Note

1. However, it should be noted that the control group in this study was not fully comparable to program participants.

References

Berti, P. R., J. Krasevec, and S. FitzGerald. 2004. A review of the effectiveness of agriculture interventions in improving nutrition outcomes. *Public Health Nutrition* 7 (5): 599–609.

Black, R., L. Allen, Z. Bhutta, L. Caulfield, M. de Onis, M. Ezzati, C. Mathers, and J. Rivera. 2008. Maternal and child undernutrition: Global and regional exposures and health consequences. *Lancet* 371 (9608): 243–260.

Bloem, M. W., N. Huq, J. Gorstein, S. Burger, T. Kahn, N. Islam, S. Baker, and F. Davidson. 1996. Production of fruits and vegetables at the homestead is an important source of vitamin A among women in rural Bangladesh. *European Journal of Clinical Nutrition* 50 (supp. 3): S62–S67.

Bushamuka, V. N., S. de Pee, A. Talukder, L. Kiess, D. Panagides, A. Taher, and M. Bloem. 2005. Impact of a homestead gardening program on household food security and empowerment of women in Bangladesh. *Food and Nutrition Bulletin* 26 (1): 17–25.

de Pee, S., A. Talukder, and M. Bloem. 2008. Homestead food production for improving nutritional status and health. In *Nutrition and health: Nutrition and health in developing countries,* 2nd ed., ed. R. D. Semba and M. W. Bloem. Totowa, N.J., U.S.A.: Humana Press.

de Pee, S., C. E. West, D. Permaesih, S. Martuti, Muhilal, and J. G. A. J. Hautvast. 1998. Orange fruit is more effective than are dark-green, leafy vegetables in increasing serum concentrations of retinol and β-carotene in schoolchildren in Indonesia. *American Journal of Clinical Nutrition* 68 (5): 1058–1067.

Faber, M., M. A. S. Phungula, S. L. Venter, M. A. Dhansay, and A. J. S. Benade. 2002. Home gardens focusing on the production of yellow and dark-green leafy vegetables increase the serum retinol concentrations of 2–5-year-old children in South Africa. *American Journal of Clinical Nutrition* 76 (5): 1048–1054.

FAO (Food and Agriculture Organization of the United Nations). 2008. *The state of food insecurity in the world.* Rome.

Hess, S. Y., and J. C. King. 2009. Effects of maternal zinc supplementation on pregnancy and lactation outcomes. *Food and Nutrition Bulletin* 30 (1, supp.): S60–S78.

HKI (Helen Keller International). 2000. NGO gardening and nutrition education surveillance project (NGNESP): USAID/BD Mission Meeting. PowerPoint presentation. Helen Keller Worldwide, New York.

———. 2002a. NGO home gardening and nutrition education surveillance project: Economic and social impact evaluation. PowerPoint presentation. USAID–HKI Meeting, October 9, at Helen Keller Worldwide, New York.

———. 2002b. Economic and social impact evaluation: NGO gardening and nutrition education surveillance project (NGNESP). PowerPoint presentation. Helen Keller International, Dahka, Bangladesh.

———. 2003a. *Strengthening the capacity of local NGOs through food production and nutrition programs in Bangladesh, Cambodia, and Nepal.* Asia-Pacific Regional Bulletins, Special Issue. New York: Helen Keller Worldwide.

———. 2003b. *HKI's homestead food production program sustainably improves livelihoods of households in rural Bangladesh.* Homestead Food Production Bulletin 1. New York: Helen Keller Worldwide.

———. 2004. *Homestead food production improves household food and nutrition security.* Homestead Food Production Bulletin 2. New York: Helen Keller Worldwide.

———. 2006a. *Homestead food production: The potential and opportunity to improve the food security and rural livelihood in Barisal division.* Homestead Food Production Bulletin 3. New York: Helen Keller Worldwide.

———. 2006b. *Homestead food production—An effective integrated approach to improve food security among the vulnerable char dwellers in northern Bangladesh.* Homestead Food Production Bulletin 4. New York: Helen Keller Worldwide.

———. 2008a. *Homestead food production—Improving nutrition and food security and empowering women in the rural Barisal Division of Bangladesh.* Homestead Food Production Program Bulletin 5. New York: Helen Keller Worldwide.

———. 2008b. *Homestead food production—An effective strategy in mitigating household food insecurity and coping with disasters (flood and monga).* Homestead Food Production Program Bulletin 6. New York: Helen Keller Worldwide.

———. 2008c. *Global food crisis and HKI's response.* June 3. <http://www.hki.org/research/documents/Food%20Crisis%20and%20HKI.pdf>. Accessed April 23, 2009.

HKI (Helen Keller International), Asia-Pacific. 2001. *Homestead food production—A strategy to combat malnutrition and poverty.* Jakarta, Indonesia.

HKI (Helen Keller International), Bangladesh. 2003. Monitoring of activities in village nurseries and household gardens: A summary report of surveys 14–19 (July 1999–June 2001). NGO

Gardening and Nutrition Education Surveillance Project (NGNESP), Grant HRN-A-0098-00013-00 and BAN-501107-0000434. Helen Keller Worldwide, New York.

———. 2005. Improving nutrition and food security through homestead food production in the riverine islands and floodplains of Bangladesh. Proposal for program cost extension. Helen Keller International, Dhaka, Bangladesh.

———. 2006. Improving nutrition and food security through homestead food production in the riverine islands and floodplains of Bangladesh (HKI Char Project 2). Progress report for project BAN-501107-0006006, July–December 2005. Helen Keller Worldwide, New York.

———. 2007. Homestead food production program in Barisal division: Jibon-O-Jibika project. Program update, June. Helen Keller Worldwide, Dhaka, Bangladesh.

———. 2008. Homestead food production to improve household food and nutrition security in the Chittagong Hill Tracts. End of project report for project BAN-501107-0006005, July 2007–June 2008. Helen Keller Worldwide, New York.

Huq, S. M. 2007. Devastating flood in Bangladesh: HFPP CHAR project. PowerPoint presentation. Helen Keller International, Dhaka, Bangladesh.

Iannotti, L., K. Cunningham, and M. Ruel. 2009. *Improving diet quality and micronutrient nutrition: Homestead food production in Bangladesh.* Discussion Paper 00928. Washington, D.C.: International Food Policy Research Institute.

IOM (Institute of Medicine). 2000. *Dietary reference intakes for vitamin A, vitamin K, arsenic, boron, chromium, copper, iodine, iron, manganese, molybdenum, nickel, silicon, vanadium, and zinc: A report of the panel on micronutrients, food, and nutrition board.* Washington, D.C.: National Academies Press.

Landon-Lane, C. 2004. *Livelihoods grow in gardens: Diversifying rural incomes through home gardens.* Rome: Food and Agriculture Organization of the United Nations.

National Nutritional Blindness Prevalence Survey. 1982–83. Dhaka, Bangladesh: Institute of Public Health Nutrition and Helen Keller International.

Nutrition Survey of Rural Bangladesh. 1981–82. Dhaka, Bangladesh: University of Dhaka, Institute of Nutrition and Food Science.

Olney, D. K., A. Talukder, L. L. Iannotti, M. T. Ruel, and V. Quinn. 2009. Assessing impact and impact pathways of a Homestead Food Production program on household and child nutrition in Cambodia. *Food and Nutrition Bulletin* 30 (4): 355–369.

Ramakrishnan, U., and S. L. Huffman. 2008. Multiple micronutrient malnutrition: What can be done. In *Nutrition and health in developing countries,* ed. R. D. Semba and M. W. Bloem. Totowa, N.J., U.S.A.: Humana Press.

Randolph, T. F., E. Schelling, D. Grace, C. F. Nicholson, J. L. Leroy, D. Peden, D. C. Cole, M. W. Demment, A. Omore, J. Zinsstag, and M. Ruel. 2007. Role of livestock in human nutrition and health for poverty reduction in developing countries. *Journal of Animal Sciences* 85: 2788–2800.

Rastogi, R., and C. D. Mathers. 2002. Global burden of iron deficiency anemia in the year 2000. In *Global burden of disease 2000*. Geneva: World Health Organization. <http://www.who.int/ healthinfo/statistics/bod_irondeficiencyanaemia.pdf>. Accessed August 23, 2010.

Ruel, M. T. 2001. *Can food-based strategies help reduce vitamin A and iron deficiencies? A review of recent evidence*. Washington, D.C.: International Food Policy Research Institute.

————. 2008. Addressing the underlying determinants of undernutrition: Examples of successful integration of nutrition in poverty-reduction and agriculture strategies. *SCN News* 36: 21–29.

Sarkar, N. R., A. Taher, A. Talukder, and A. Hall. n.d. An evaluation of the household food security through nutrition gardening programme. PowerPoint presentation. Helen Keller International, Bangladesh and Indonesia.

Schroeder, D. G. 2008. Malnutrition. In *Nutrition and health in developing countries*, ed. R. D. Semba and M. W. Bloem. Totowa, N.J., U.S.A.: Humana Press.

Sifri, Z. 2007. Large-scale home gardening programs: The Helen Keller International experience in Bangladesh. Report for the International Food Policy Research Institute, Washington, D.C. Photocopy.

Stolzfus, R., L. Mullany, and R. Black. 2004. Iron deficiency anaemia. In *Comparative quantification of health risks: Global and regional burden of disease attributable to selected major risk factors*, ed. M. Ezzati, A. Lopez, A. Rodgers, and C. Murray. Geneva: World Health Organization.

Taher, A., A. Talukder, N. R. Sarkar, V. N. Bushamuka, A. Hall, S. de Pee, R. Moench-Pfanner, L. Kiess, and M. W. Bloem. 2004a. Homestead gardening for combating vitamin A deficiency: The Helen Keller International, Bangladesh, experience. In *Alleviating malnutrition through agriculture in Bangladesh: Biofortification and diversification as sustainable solutions*, ed. N. Roos, H. E. Bouis, N. Hassan, and K. A. Kabir. Washington, D.C.: International Food Policy Research Institute.

Talukder, A., R. K. N. Islam, R. Klemm, and M. Bloem. 1993. *Home gardening in South Asia: The complete handbook*. Dhaka, Bangladesh: Helen Keller International.

Talukder, A., L. Kiess, N. Huq, S. de Pee, I. Darnton-Hill, and M. W. Bloem. 2000. Increasing the production and consumption of vitamin A–rich fruits and vegetables: Lessons learned in taking the Bangladesh homestead gardening programme to a national scale. *Food and Nutrition Bulletin* 21 (2): 165–172.

USAID (U.S. Agency for International Development). 2006. Technical reference materials: Nutrition. PVO Child Survival and Health Grants Program, Washington, D.C.

West, K. P., Jr. 2002. Extent of vitamin A deficiency among preschool children and women of reproductive age. *Journal of Nutrition* 132 (supp. 9): 2857S–2866S.

West, K. P., Jr., and I. Darnton-Hill. 2008. Vitamin A deficiency. In *Nutrition and health in developing countries*, ed. R. D. Semba and M. W. Bloem. Totowa, N.J., U.S.A.: Humana Press.

World Bank. 2007. *From agriculture to nutrition: Pathways, synergies and outcomes*. Washington, D.C.

Chapter 21

Impacts of Agricultural Development on Food Security Goals: Methods, Approaches and Best Practices for Improving the Proof

Mywish K. Maredia

The goal of any development effort is to achieve positive impacts on people's lives. Throughout the past several decades, this has served as a key objective in guiding the planning and implementation of a broad range of global and national development efforts. In the area of agriculture, major attention focuses on achieving people-related impacts, such as reducing poverty, hunger, and food insecurity. Whether, what, and how development efforts have an impact on these goals is at the heart of "impact assessment," which refers to the umbrella concept encompassing all analyses focused on outcomes and impacts. The motivation to assess the impacts of development efforts stems from the need for accountability (whether and what impacts development efforts have on people and their environment) and an interest in institutional learning (how impacts are achieved or not achieved and what lessons can be derived to improve programs). Assessment and documentation of impacts provide the "proof" that development does or does not work and in what contexts.

The motivation for this chapter stems from an interest in better understanding the theory, practice, and use of different impact assessment methods and approaches in agricultural development. Toward this goal, this chapter first presents the con-

Comments and feedback received from the editors and anonymous reviewers on an earlier version of this chapter are greatly appreciated. All the normal disclaimers apply.

ceptual framework that links agricultural development efforts with the impact goal of enhancing food security in the context of broader developmental goals, followed by a review of the evidence of methods of impact assessment used in recent studies and suggestions for best practices in the assessment of developmental impacts related to food security.

Setting the Stage: Defining the Concepts

According to the latest figures available, the total annual donor commitment to agricultural development in the 1990s and early 2000s was in the range of $10–12 billion (FAO 2005).[1] Although a breakdown of these investments by program goals is not available, it can be safely assumed that a majority of these development efforts in agriculture are devoted to achieving goals related to ending poverty and hunger. Enhancing food security is intrinsically linked with the achievement of these developmental goals and is thus considered one of the major focuses of agricultural development interventions. For the purposes of this chapter, development interventions in the category of agriculture are broadly defined to include rural development, infrastructure development that affects agriculture directly (roads, irrigation, drainage systems), policy changes (related to rural land, labor, capital, outputs, inputs, and prices), agricultural services in rural areas encompassing marketing and financial systems, research and provision of improved technologies, practices and other inputs (fertilizer, water, pesticides, seeds), and the provision of extension services to producers, among others.

Because "food security" and "impact assessment" are the main themes of this chapter, it is necessary to look closely at these concepts to establish the explicit or implied definitions.

Food Security

Several hundred definitions of *food security* are found in the literature as a result of the evolution in thinking and understanding of the international community regarding the complexities involved in the technical and policy aspects related to this concept. The official definition of *food security* has evolved from the initial focus in the 1970s on food supply problems—ensuring food availability and to some degree price stability—to the focus in the 1980s and 1990s, which was shifting toward the demand side and issues of consumption and access for vulnerable people at all times (see FAO 2003 for this historical perspective on the evolution of the concept of food security). In this chapter we view food security as a phenomenon relating to individuals. Following FAO (2003), useful working definitions of *food security* and *food insecurity* implied in this chapter are as follows:

- *Food security* exists when all people, at all times, have physical, social, and economic access to sufficient, safe, and nutritious food that meets their dietary needs and food preferences for an active and healthy life. Household food security is the application of this concept to the family level, with individuals in households the focus of concern.

- *Food insecurity* exists when people do not have adequate physical, social, or economic access to food as just defined.

Based on these concepts, the ultimate goal of agricultural development focused on increasing food security (or reducing food insecurity) is enhancing the nutritional status of the individual household member and reducing the risk that that adequate status will not be achieved or become undermined. Over the past four decades, the practical response from the international community to the need to achieve this concept of food security has been to focus on simpler objectives around which to organize international and national public action, such as reducing and eliminating poverty. According to the FAO (2003), the United Nations exemplified the direction of policy for international action on food security by setting the targets for the first Millennium Development Goal as halving the number of hungry and poor people by 2015. In this chapter, therefore, enhancing food security is interpreted as implicitly embedded in the development goals of reducing poverty and hunger. In other words, development efforts targeted at reducing hunger and poverty are viewed as contributing to the goal of enhancing food security. The overview of impact assessments of agricultural development efforts presented in this chapter takes this broader perspective of the goal of enhancing food security.

Impact Assessment

Impact assessment (IA), as used in this chapter, is the systematic analysis of the significant or lasting changes—positive or negative, intended or not—in people's lives brought about by a given action or series of actions in relation to a counterfactual. Figure 21.1 illustrates a simplified and generalized impact pathway (or a results chain) of how actions related to agricultural development (also referred to as an intervention in the form of a program, project, policy change, or activity) affect the goal of enhancing food security. It also introduces the concept of "impact assessment," which is concerned with the evaluation of the final effects (long-term impacts on poverty or hunger, for example) and intermediate effects (medium-term outcomes on production, income, consumption, and prices) *caused* by a given activity (Baker 2000).

Figure 21.1   A generalized impact pathway of agricultural interventions focused on enhancing food security

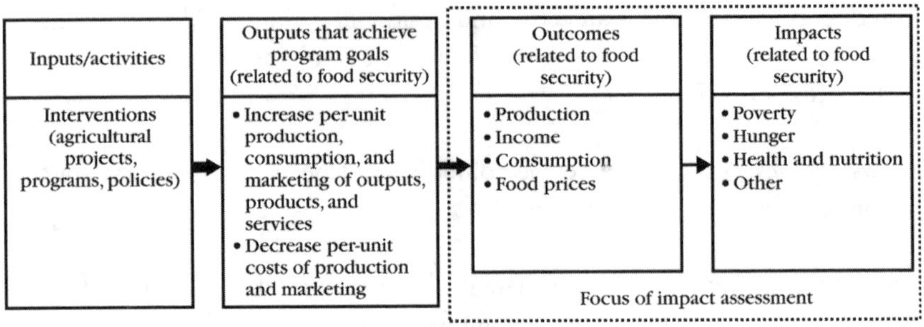

Inputs/activities	Outputs that achieve program goals (related to food security)	Outcomes (related to food security)	Impacts (related to food security)
Interventions (agricultural projects, programs, policies)	• Increase per-unit production, consumption, and marketing of outputs, products, and services • Decrease per-unit costs of production and marketing	• Production • Income • Consumption • Food prices	• Poverty • Hunger • Health and nutrition • Other
		Focus of impact assessment	

Source: Author.

Note: Each arrow in the impact chain indicates the direction of influence, and its thickness indicates the degree of influence on an effect.

The impact chain provides a useful way of conceptualizing the cause-and-effect relationship between different types of changes, with IA focusing mainly on the changes occurring at the outcome and impact levels. However, in practice, the boundaries of the IA focal area are sometimes blurred and may encompass evaluation of program goals. This is especially the case when the interventions being evaluated are pilot-scale programs/projects and IA is integrated into the design of the program or takes place immediately after the intervention is completed.

Several concepts and terminologies found in the literature are closely associated with impact assessment or belong to the family of monitoring, evaluation, and impact assessment. These are introduced in Figure 21.2, which combines the concept of an impact chain with the timeline of a project cycle to provide the classical view of the differences among project appraisal, monitoring, and project evaluation and illustrates where impact assessment fits in this two-dimensional view of the world.[2]

According to this classical view of the project cycle, monitoring, project evaluation, and IA are shown to occur in a specific time frame and focus on specific nodes of the impact chain. However, in practice the time frame of these assessments and the impact chain focal points may not be so well defined. Although retrospective IA occurs ex post an intervention, the planning of such IA as an activity needs to occur concurrently with project planning and implementation to ensure that baseline data are collected and indicators are monitored efficiently. Moreover, although IA focuses mainly on the outcome/impact nodes of the causal chain, it needs to explore the entire chain if reliable conclusions can be drawn about the degree to which any observed change can be attributed to a given intervention. A feedback loop connects all types of assessments (ex ante, monitoring, project/program evaluation,

**Figure 21.2   A generalized view of the types of assessments in a project life
cycle and impact chain framework**

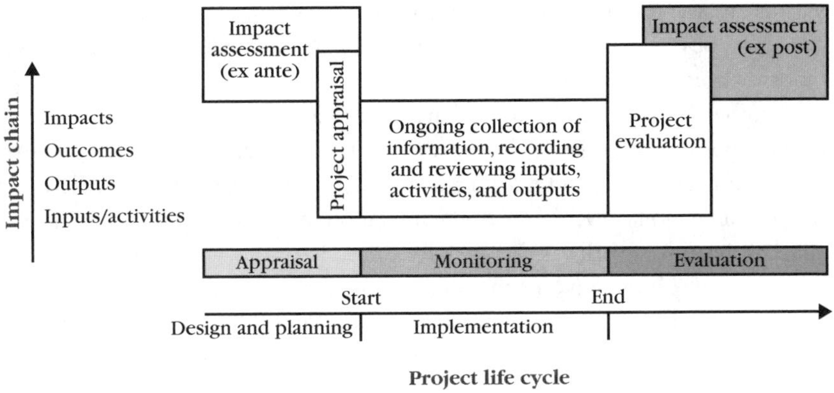

Source: Author.

and ex post IA) with each other, and although the retrospective IA comes closest to
providing the proof, it is the close integration and linkages between the assessments
occurring at different stages of a project cycle that improves project effectiveness.

Thus, although retrospective IAs concerned with providing proof of development effectiveness are the major focus of this chapter, ex ante IAs and other types of
monitoring and evaluation assessments done throughout the continuum of impact
pathway and project life cycles are included in the discussion where appropriate.

Impact Assessment of Food Security Goals

Conceptually, the interventions (agricultural programs, projects, policies, and
activities), as the first node in the results chain, attempt to bring about changes
in the use of farm- and community-level resources and assets (land, labor, capital,
entrepreneurship) to increase per unit production or marketing of outputs, products,
and services or decrease per unit costs at the farm household level (referred to in
Figure 21.1 as program goals resulting from project outputs). These affect four
major indicators that are directly linked with the goal of enhancing food security:
production, income, consumption, and food prices.

In theory, increase in income and consumption and reduction in food prices
serve as impact channels on the demand side (they increase the effective demand
and intake of food), and an increase in agricultural production serves as an impact
channel that affects the supply of food. However, both in theory and in practice,
food production affects the supply- as well as the demand-side indicators related
to income, consumption, and food prices. For example, an increase in production

can increase producers' income, which can increase consumption. However, an increase in production can also reduce price and thus farmers' income; but it can also increase the real wages in the economy and positively affect the consumption of net food buyers. Sorting out the net impact of any outcome on the goal of food security is thus a complex issue.

Another source of complexity is that agricultural development efforts can effect changes (positive or negative, intended or not) in the ultimate impact goals of poverty, hunger, health, and nutrition at an individual level, a community level, or an aggregate national level.[3] One of the limitations of the simple two-dimensional model of the impact pathway depicted in Figure 21.1 is that it does not include the size or the scale dimension for either the development effort or that of the outcome and impact indicators. From the perspective of a single intervention, the channel of impact transmission can be viewed as influencing impact indicators at the micro level (individuals, households, or a community). However, taking the perspective of a development goal at an aggregate level (reflected in macro-level indicators), the generic pathway can be viewed as resulting from collective changes in impact indicators aggregated across a large number of individuals across communities, over a sustained period of time, and resulting from a multitude of interventions (the equilibrium effects). The tapering thickness of the arrows along the results chain in Figure 21.1 indicates this reducing influence of a single intervention on aggregate-level outcomes and impacts.

The conceptual framework depicted in Figure 21.1 also does not reflect the stage at which IA is conducted, which could be at the initial pilot or "proof of concept" stage or at the stage at which project outputs are already scaled up and scaled out. The perspective from which an IA is conducted (from a single project perspective or a development goal perspective), the stage in the impact pathway at which it is conducted (the initial pilot stage or after the program/policy is scaled up), and the individual or group from whose perspective it is assessed (individual beneficiary or the society) have implications for the methods and approaches used to provide the proof of development effectiveness. The overview presented next clarifies and distinguishes among several broad categories of IAs that occur ex post an activity and contribute to providing different types of proofs of development effectiveness.

Overview of Impact Assessment Methods and Approaches

Although there is broad agreement on the definition and general purpose of IA, there is no one way of doing IA. The methods and approaches vary depending on the time, scale, motivation, and types of impacts focused on by the assessment, the level of assessment (beneficiary versus society), and the types of data available or collectible by the impact evaluator (in the context of resource and time constraints). Table 21.1

Table 21.1 Typology of methods and approaches of impact assessment related to agricultural development and their contribution to "improving the proof"

Broad category	Type of approach	Examples of methods	Salient features	Contribution to improving the proof
Macro-level (not specific to an intervention)	Quantitative	Society-level assessments: Econometric and statistical models	Assess causal contribution Use models/methods based on secondary data	Provide estimates of the relative contribution of an aggregated sector-level investment/effort to changes in macro-level impact indicators
Micro-level (intervention specific)	Quantitative	Beneficiary-level assessments: Experimental designs Quasi-experimental and non-experimental designs Meta-analysis	Assess causal attribution Focus on estimating average treatment effects in relation to a counterfactual Use multiple data sources (primary and secondary) for evaluating benefits Use innovative methods of econometric and statistical analysis	Provide estimates of the size and scope of the "effectiveness" and "efficacy" of an intervention from the perspective of impacted units (individuals, households, communities)
		Society-level assessments: Social benefit–cost analysis	Involve first estimating impacts per unit of beneficiary and then the use of models to estimate impacts by aggregating benefits/costs across beneficiaries and over time	Provide measures of relative efficiency from a society's perspective, expressed as a ratio between the total values of the inputs and the total effects/impacts generated from those inputs
	Qualitative	Beneficiary-level assessments: Rapid Rural Appraisal Sustainable Livelihoods Framework	Use nonexperimental and nonstatistical methods Focus on understanding processes, behaviors, and conditions as they are perceived by the individuals or groups being studied Unable to assess causal attribution	Provide an understanding of the processes and changes in conditions, behavior, and perceptions of an intervention ex post from the perspective of impacted units (individuals, households, communities)
	Mixed method	Beneficiary-level assessments: Real World Evaluation	Guided by the principles of practicality and complementarities	Establish the causal link between an intervention and its effect and thus provide proof of project efficacy

Source: Author's compilation.

presents a typology of IA methods and approaches relevant to the development field in general and to agriculture in particular, with the aim of not only clarifying their meanings but also understanding their contributions to "improving the proof."[4]

Two broad categories of IA (relevant to agricultural development) are distinguished in this chapter based on the focus and level of aggregation of impact analysis—micro-level and macro-level (see Table 21.1). These are further subgrouped into three broad types of approaches—quantitative, qualitative, or mixed—and levels (or perspectives) of IA—beneficiary-level or society-level (see Table 21.1). IAs that are micro-level are intervention specific and trace the inputs–outputs–outcomes–impacts relationship along the impact pathway from left to right for a specific activity or group of activities. Micro-level quantitative IAs establish the causal link among inputs–outputs–outcomes–impacts in order to attribute the estimated or observed impacts to a specific intervention (including a natural shock or a policy change). Thus they are best suited to providing proof of the effectiveness of a specific intervention in achieving immediate and medium-term outcomes at the beneficiary level or a long-term goal aggregated across groups of beneficiaries (societal impacts). The methods and approaches used in micro-level IA vary across this spectrum of focus, ranging from short- and medium-term outcomes at the beneficiary level to long-term equilibrium effects at the aggregate society level (see Table 21.1).

IAs that are macro-level studies use quantitative approaches and also focus on long-term equilibrium effects at the aggregate level. But they focus on assessing the contribution of past investments, shocks, or events to some realized macro-level goals aggregated over a category of development effort (for example, investments in infrastructure, research, or extension) or over a subsector (crops, livestock, agroforestry). These types of assessments cannot provide information on the contribution or effectiveness of a specific project, program, or intervention to a developmental goal (see Table 21.1). Classic macro-level IA studies focusing on the economic impacts of sectorwide investments include those of Griliches (1963, 1964) and Evenson and Kislev (1975). The macro-level impact analysis has evolved over the decades to include other dimensions of impacts closer to food security goals such as poverty and nutrition (Evenson and Rosegrant 2003; Fan 2007; Fan et al. 2007). Such studies are useful in showing the cause-and-effect relationship between development efforts aggregated across a sector, type of effort, or spatial and temporal dimension and macro-level indicators of impact goals (such as number of people lifted out of poverty or number of children saved from malnutrition).

Although the micro-level (or intervention-specific) and macro-level IAs (aggregate goal-level assessments that are not linked to a specific intervention) have evolved in tandem over the past five decades, the methods and approaches used for IA are quite different (see Table 21.1).[5] Whereas the evaluation methodology for micro-level analysis is based on assessment of causal attribution, the methodology for macro-level analysis depends on estimates based on causal contribution. Macro-level IAs typically take a statistical approach that relates changes in macro-level indicators (such as total factor productivity or poverty level) to some aggregate-level indicators of "inputs" (usually investments in a type of agricultural development or a subsector). They are based on secondary data and require specialized skills in econometric models and statistical methods, particularly time-series data analyses. In contrast, micro-level studies that focus on one or more well-defined interventions draw on multiple data sources for evaluating benefits, invest in primary data collection, and interact with project teams directly involved in generating the outputs being assessed (see Table 21.1).

Micro-level ex post impact assessment based on a social cost–benefit analysis framework is one of the approaches that has been widely used in evaluating long-term effects of agricultural interventions, especially those related to the adoption of new technologies (that is, interventions related to agricultural research and extension). Studies based on this approach provide evidence of high rates of return to agricultural research investments, which supports expanded public investments in agricultural research (Alston et al. 2000; Evenson 2001). In terms of the food security implications, the evidence of high rates of return from these IAs provides proof of the positive causal link between development interventions that increase productivity and gains in producer and consumer welfare as a result of increased income (for producers) and lower prices (for consumers).

The social cost–benefit framework underlying the quantitative micro-level IAs is designed to estimate relative efficiency, expressed as a ratio between the total values of the inputs and the society-level effects or impacts generated from those inputs. Against this, the beneficiary-level quantitative assessments provide estimates of the size of the "effectiveness" of an intervention from the perspective of impacted units such as individuals, households, and communities (see Table 21.1). The push for learning from evaluations over the past decade or so has seen responses from both quantitative and qualitative approaches of such micro-level assessments focused on beneficiary-level impacts. The qualitative assessments are based on the principles of participation and interpretation of information documented throughout the process of evaluation. Qualitative techniques are used for carrying out IA with the intent to determine impact by relying on something other than the counterfactual (Mohr 1995, 1999). In one sense, these methods use nonexperimental and non-statistical methods. Attribution or contribution is assessed using approaches such as reference to secondary data, program theory (logic models), theory of change, and concept mapping. The focus is on understanding processes, behaviors, and conditions as they are perceived by the individuals or groups being studied (Valadez and Bamberger 1994). One of the qualitative techniques of impact assessment is the use of the sustainable livelihoods framework as the underlying theory of change and a tool for evaluation (Adato and Meinzen-Dick 2007; Mancini, Van Bruggen, and Jiggins 2007; La Rovere et al. 2008) (see Table 21.1). These types of assessments are multidimensional and qualitative, and they present results in terms of indicators that measure capabilities and assets of participants or beneficiaries of an intervention.

The benefits of qualitative assessments are that they are flexible, can be specifically tailored to the needs of the evaluation using open-ended approaches, can be carried out quickly using rapid techniques, and can greatly enhance the findings of an IA by providing a better understanding of stakeholders' perceptions and priorities and the conditions and processes that may have affected program impact. Thus, in the context of food security goals they are good at asking the "why" and "how" questions related to intrahousehold patterns of food consumption, nutrition, and distribution. Among the main drawbacks are the subjectivity involved in data collection, the lack of a comparison group, and the lack of statistical robustness given the mainly small sample sizes, all of which make it difficult to generalize the results to a larger, representative population.

In recent years there has also been an increased emphasis on learning from quantitative methods built on the movement toward "evidence-based policy" (Center for Global Development 2006). As a result, rigorous quantitative methods for estimating counterfactuals such as experimental (E) designs (based on the principles of random assignment) and quasi-experimental (QE) designs using statistical techniques (such as propensity score matching, instrumental variable, difference-in-difference, and discontinuity regression) have gained importance for estimating beneficiary-level impacts. These methods seek to improve the quality of the control or comparison group for estimating the size and magnitude of average effects of an intervention on beneficiaries. In general, impact evaluations (IEs) based on social experimental research design are considered more robust because they provide a high level of internal validity for the causal link between an intervention and the social outcomes by overcoming the problem of selection bias that is endemic to nonrandomized evaluations. Due to this advantage, social experiment has been advocated as the main tool for studying development effectiveness by an influential group of academic economists (called the "randomistas") led by MIT's Poverty Action Lab and promoters of evidence-based policymaking (Duflo 2006; Banerjee et al. 2007; Duflo, Glennerster, and Kremer 2007).[6]

Some recent examples of IEs based on experimental designs in the area of food, agriculture, and rural development include the evaluation of the productivity impacts of a cashless microcredit program (DrumNet) in Kenya (Ashraf, Giné, and Karlan 2008), social experiments to assess whether the provision of insurance against a major source of production risk induces demand for credit to adopt new crop technologies (Giné and Yang 2009), ongoing experiments to evaluate the effectiveness of Integrated Agricultural Research for Development as a new approach for conducting agricultural research and extension in Africa (FARA 2009), and the IE of the promotion of fortified orange-fleshed sweetpotatoes on health and nutrition indicators among children and women in rural communities in Mozambique (Arimond et al. 2007).[7]

The E and QE methods represent the emerging trends in IE and have the potential for providing rigorous proofs of the development effectiveness of a publicly supported program or policy on indicators related to poverty and food security. However, lack of external validity, ethical concerns, high costs, a sole focus on average impact parameters, a focus on short-term effects, and the impracticality of maintaining the treatment and control groups in a real world setting are often cited by critiques as limitations of social experiments. These drawbacks make experimental designs unsuitable and less than a gold standard in the practice of evaluations of many types of development programs, projects, and policies (Deaton 2007, 2009; Rodrik 2008; Ravallion 2009).

Systematic syntheses and reviews (for example, those undertaken by What Works Clearinghouse in the area of education and by the Campbell Collaboration in the areas of education, crime and justice, and social welfare), meta-cost-benefit analysis (for example, Maredia and Raitzer 2006; Raitzer and Kelley 2008), and statistical meta-analysis (Alston et al. 2000) also seem to be new emerging trends in IA of agricultural development interventions. These types of meta-syntheses not only help take stock of what is known in IA in a given focused area of inquiry but also help systematically build a body of evidence and identify gaps in knowledge about what works in development (and what does not). By systematically combining studies, one attempts to overcome limits of size or scope in individual studies to obtain more reliable information about the impact of a "treatment." Thus these types of analyses have the potential for not only providing the proof but also improving the proof about what works and what does not in development at the micro level.

Among the different methods of synthesizing documented evidence from past evaluations, statistical meta-analysis has become increasingly popular in recent decades. Despite the controversy about its validity (see, for example, Thompson and Pocock 1991 and Bailar 1997, cited in Berman and Parker 2002), it has been applied with increasing frequency in the areas of education, health, and psychology (Glass 1976) and to some extent in the field of economic research (Espey, Espey, and Shaw 1994; Görg and Strobl 2001; Thiam, Bravo-Ureta, and Rivas 2001). But its application to evaluations of agricultural and rural development interventions is a recent phenomenon. With the advent of the application of experimental and quasi-experimental impact evaluation methods, there is scope to expand the use of meta-analysis techniques to improving the proof. This is because such quantitative methods provide a common methodological framework with which to synthesize findings across impact studies.

The growing debate between the randomistas and their critics in recent years has seen the rise of mixed-methods approaches such as the Real World Evaluation (RWE) designs (see Table 21.1). RWE is guided by the following principles to strengthen practical IE designs: (1) basing the evaluation design on a program theory model, (2) complementing the quantitative (summative) evaluation with process evaluation, (3) incorporating contextual analysis (because many external factors along the impact pathway can influence the outcome and impact of a development effort), (4) reconstructing baseline conditions, (5) using mixed-methods approaches to strengthen the validity of indicators and to improve interpretation of findings, and (6) adjusting for differences between the project and comparison groups (Bamberger, Rugh, and Mabry 2006). Like the quantitative methods for IEs (E and QE), RWE is concerned with micro-level impact assessments focused on establishing the causal link between an intervention and its effect and thus with providing the proof of project efficacy.

Assessments of Food Security Impacts: Past Evidence and Potential for the Future Based on Best Practices

Past Evidence

Although food security assessments (and famine analysis) have long been a topic of research and inquiry, the goals of achieving food security and ending poverty and hunger did not receive explicit attention in IAs of development interventions even though they were becoming explicit goals of development efforts in the 1970s and 1980s. Even today, it is rare for IAs to include an explicit indicator called "food security" to measure the impact on this goal. Most studies measure food security impacts through changes in outcome indicators related to consumption or imply this impact through changes in outcomes related to production, income, and prices. Table 21.2 identifies some of the common indicators under these outcomes that are found in the IA literature and lists examples of studies (focusing on the past 10 years) related to those indicators.[8] These include micro-level (Low et al. 2007; Duflo, Kremer, and Robinson 2008) as well as macro-level IAs (Evenson and Rosegrant 2003; Fan et al. 2007), and IEs of project-specific interventions (Gilligan, Hoddinott, and Taffesse 2008), as well as IAs conducted ex post the scaling-up of program outputs (Evenson and Gollin 2003).

The impacts on food security and related goals (such as poverty) have been assessed at both the household or farm level and the aggregate societal level. The indicators specifically used by an impact study are a function of the objectives and program goals of an intervention (see Table 21.2). Thus, for interventions specifically designed to address the issue of food security (such as food aid, social safety nets, school feeding programs, and cash transfers), the indicators commonly used relate to consumption. These include food expenditures and the food gap (Gilligan, Hoddinott, and Taffesse 2008), per capita food, and calorie and nutritional intake (Low et al. 2007; Dillon 2008). For development activities designed to address the issue of productivity, the indicators commonly used relate to yield, production, and profitability.

IA questions are good indicators of the objectives and scope of the development activity (or intervention) being evaluated. As exemplified in Table 21.2, these can be diverse depending on the size and scale of an intervention and the timing and motivation for conducting an IA. The method or approaches used for IA are functions of both the IA questions and the type of data available. The growing practice in micro-level analysis (that focuses on specific interventions) is to use experimental or quasi-experimental designs to identify a counterfactual and attribute the change in impact indicators to the intervention (for example, Hoddinott and Skoufias 2004; Gilligan and Hoddinnott 2007; Low et al. 2007; Dillon 2008). Such IEs require specific data collection through surveys across intervention and nonintervention sites and perhaps also across multiple time periods. These types of studies have been implemented to address evaluation questions related to food security impacts at the household level.

At the aggregate level, many studies use econometric models to assess long-term equilibrium effects at the macro level of a change in past investments, shocks, or outcomes to some realized macro-level goals.[9] For example, Evenson and Rosegrant (2003) look at the impacts of increased productivity as a result of crop germplasm improvement research on aggregate-level welfare indicators related to poverty and food security. After taking into account the general equilibrium effects of increased production and reduced prices on demand and supply of food, they report that in the counterfactual scenario (the absence of improved crop varietal technologies), world food production would have been 4–5 percent lower in developing countries, and 13–15 million more children would have suffered from hunger and malnourishment.

IAs based on existing sources of data (secondary data or primary data collected for other purposes) usually use quasi-experimental or nonexperimental designs and require the use of sophisticated econometric techniques to establish causal attribution of an intervention (a policy change, program implementation) to the observed changes in welfare indicators derived from pre-existing surveys. Usually such evaluations are conducted at an aggregate program level and are not specific to time- and/or space-bound interventions (for example, Evenson and Gollin 2003; Dercon et al. 2008). The Living Standards Measurement Studies sponsored by the World Bank in many developing countries have served as a rich source of data with which to assess the macro-level impacts of major policy changes or develop-

Table 21.2 Common indicators related to food security found in the impact assessment (IA) literature and illustrative examples of methods and types of IAs

Outcomes related to food security goals	Common indicators found in the IA literature	Notable observations from the literature	Illustrative examples of IA studies (since 2000) that measure some of the given outcomes and indicators		Methods and approaches used in the studies cited
			Study citation	Impact question related to food security outcomes	
Production	Farm level: Yield Unit cost of production Profitability	One of the most common outcomes found in the assessments of many types of agricultural interventions. The impact channels between these indicators and food security are implied and indirect and are not always investigated.	Evenson and Gollin (2003); Duflo, Kremer, and Robinson (2008)	What are the productivity impacts of improved crop germplasm research? Do fertilizer and hybrid seed increase yield and profitability on small farms?	Mixed methods, mostly natural experiment (impacts derived from long-term agricultural experiments and statistical analysis of secondary data and adoption surveys) Experimental design applied to a community-level intervention
	Aggregate level: Total factor productivity (TFP)		Thirtle, Lin, and Piesse (2003)	What is the impact of research-led agricultural productivity growth on poverty?	Econometric analysis and multiequation models using time-series data
Consumption	Household level: Food expenditure Food gap Total or per capita consumption Calorie intake Nutritional status measured by height and weight	These are the most direct indicators related to food security and nutritional security. These are used to assess impacts of different types of interventions (including nonagricultural poverty	Hoddinott and Skoufias (2004); Gilligan and Hoddinott (2007); Low et al. (2007); Dillon (2008)	What are the impacts of PROGRESA, an anti-poverty program, on the quantity and quality of food intake? What is the impact of the social safety-net program on food security, consumption	Experimental design based on pipeline analysis Quasi-experimental design using propensity score matching Quasi-experimental design using double difference and

	Indicators	Description	References	Research questions	Methods
	Aggregate level: Consumption growth Consumer surplus Food expenditures	reduction programs) targeted to increase household or individual income and consumption. Micro-level IAs based on these indicators are more common.	Dercon et al. (2008)	levels, and use of income? What is the effect of increased agricultural production induced by irrigation on household consumption? What are the nutritional impacts of orange-fleshed sweetpotatoes on young children? What are the impacts of public investments in extension and roads on consumption growth and poverty in rural areas?	propensity score matching Quasi-experimental design using fixed effects regression model Econometric analysis based on Instrumental Variables model using Generalized Methods of Moments
Income	Household level: Farm income Nonfarm income Total income Asset holdings Gross margins	These are most directly linked to the poverty goal. The impact channels between these indicators and food security are implied rather than explicitly stated.	Cocchi and Bravo-Ureta (2007); Rutherford (2008)	What is the relationship among farm income, adoption of conservation technologies, and output? What is the impact of broad bed maker technology adoption on gross margins?	Quasi-experimental design based on matching techniques Natural experiment design based on comparison of with/without farmer household surveys
	Aggregate level: Producer surplus Number of people living below $x per day		Fan et al. (2007); Dercon et al. (2008)	See the description above. What is the impact of productivity-enhancing rice research on rural poverty in India and China?	Econometric analysis based on Instrumental Variables model using Generalized Methods of Moments Econometric analysis and models using time series data

(continued)

Table 21.2 Continued

Outcomes related to food security goals	Common indicators found in the IA literature	Notable observations from the literature	Illustrative examples of IA studies (since 2000) that measure some of the given outcomes and indicators		
			Study citation	Impact question related to food security outcomes	Methods and approaches used in the studies cited
Food prices	Aggregate level: Market price of food Price stability Producer surplus Consumer surplus	The impact channels between these indicators and food security depend on the types of groups impacted (net food consumers vs. producers) and the type of commodity (food vs. nonfood). The evidence on the relationship between these indicators and food security is therefore ambiguous and not generalizable.	Evenson and Rosegrant (2003); van den Berg and Ruben (2006)	How would food prices, food production, and food consumption have differed if there were no investments in crop genetic improvement research in developing countries? What are the food-price impacts of irrigation development?	Econometric analysis and general equilibrium multicountry models to estimate the counterfactual Statistical analysis of before and after intervention data

Source: Author's compilation.
Note: PROGRESA is a successful conditional cash transfer program in Mexico.

ment programs implemented over a large scale (see Jacoby, Li, and Rozelle 2002 for China; Ravallion and van de Walle 2004 and CIEM 2008 for Vietnam). These types of rich data offer great opportunities to conduct IAs and explore causal links and associations between interventions and outcome variables.

Looking toward the Future: Best Practices

IAs play a vital role in generating information or knowledge about what works and what does not. However, sorting out what works and what does not in achieving outcomes related to food security and poverty goals is complex. Rarely will it be possible for a single study to trace these complex relationships in a comprehensive manner. Moreover, the strategy to answer these questions may vary depending on the time frame—what works in the short term versus the long term. Generation of robust knowledge that informs developmental policies and investment decisions requires a hierarchical and incremental approach to "improving the proof" through a variety of rigorous IA methods applied incrementally at the project, program, and system levels. The principles of the hierarchical approach are suggested in the following four steps:

1. Improve the proof between project inputs, outputs, and potential outcomes. As a first step toward the hierarchical approach to improving the proof, it is essential that (a) *all* investment decisions be based on a strong proposal, a conceptual framework, and a planning document that clearly lays out the impact pathway with clearly identified links between activities or inputs, outputs, and projected outcomes; and (b) *all* projects have an integrated monitoring and evaluation (M&E) plan to monitor outputs and progress toward outcomes. Whether such an M&E plan is based on classical tools of log frame analysis or the more recent participatory methods should be left to project planners and implementers. The important thing is that any method that is used should provide information to establish the link between inputs and outputs and expected outcomes.

2. Improve the proof of development outcomes realized at the beneficiary level as a result of outputs generated by development efforts. Unlike the M&E assessments proposed for all interventions in step 1, this type of assessment is suggested for a subset of development activities with common project outputs that collectively contribute to a common indicator related to food security. Thus, for a selected (randomly, if feasible) and possibly representative sample of development activities that lead to similar types of project outputs, there should be periodic documentation of development outcomes at the beneficiary level using rigorous methods. Ideally, such IAs should be planned and coordinated across major development investors to ensure a common assessment framework and

methodology to derive generalizable results. The literature provides guidelines for good practices in assessing impacts of outcomes at the beneficiary level. These can be summarized under "methods" and "robustness of results":

Best practices for methodology:

- Estimating a counterfactual that minimizes selection bias

- Controlling for pre- and postprogram differences in participants

- Collecting policy-relevant data at baseline and follow-up to estimate program impacts

- Allowing sufficient time for program impacts

- Incorporating qualitative techniques to allow for the triangulation of findings

- Assessing impacts and reporting results beyond the "mean outcomes"

Best practices for robustness of results:

- Using more than one technique to infer patterns of impact from the data collected

- Ensuring that the treatment and comparison groups are of sufficient sizes to establish statistical inferences with minimal attrition

Despite the claims of advocates of different methods (qualitative, E, or QE, for example) discussed in the previous section, no single method dominates across all the criteria of best practices. As suggested by Ravallion (2008), rigorous, policy-relevant evaluations should be open-minded about methodology, adapting to the problem setting and data constraints. Depending on the resources available, a two-step approach can also be used toward building the evidence of development effectiveness at the beneficiary level, where, as a first step, low-cost, less rigorous studies of a wide range of interventions are conducted to identify areas in which an additional research investment using more rigorous methods is warranted. The value of this filtering step would be to generate hypotheses about what works that merit confirmation in more rigorous studies (Ravallion 2008).

3. Periodically synthesize results across step 2 studies and assessments. Such syntheses should establish the link between generic categories of project or program outputs and common shorter-term developmental outcomes (for example, increased production, income, consumption, food prices). A recent attempt to conduct statistical meta-analyses of IEs that assess agricultural productivity impacts is one of the few examples of a step 3–type assessment (IEG 2009). However, one of the limitations faced by this study in applying the methodology of statistical MA, which is commonly used in other fields, was an insufficient number of observations with common, comparable measure of effect to enable the derivation of statistically meaningful results. Concerted and coordinated efforts to conduct IAs in step 2 studies that adhere to best practices in methods and robustness of results and that use common indicators of program effectiveness can potentially lead to a large pool of studies over time to conduct statistical meta-analysis from which to derive generalizable patterns of what works and what does not in development.[10] As indicated before, such meta-analyses are standard practice in other fields of public-sector investments (education, health, social welfare) in which the application of rigorous quantitative methods based on best practices to establish the causal link between an intervention and its outcome(s) are common in the IE of social programs. Also, protocols exist on documenting the evidence of such IEs from published and unpublished sources, undertaking systematic reviews, and synthesizing the results across all the documented evidence (for instance, those of What Works Clearinghouse and Campbell Collaboration). The experience gained in other fields in meta-analysis and other types of systematic syntheses can serve as a source of learning about the best practices for the agricultural development community concerned with the question of what works and what does not.

4. Document longer-term developmental impacts realized across the beneficiaries (and nonbeneficiaries) as a result of scaling up generic types of developmental outputs (such as yield-enhancing technologies and long-term impacts on level and distribution of income and food consumption effects). Strategic ex post IAs of program-level outputs and macro-level assessments across programs and sectors should be conducted periodically to gain insights on the longer-term impacts of broader developmental goals. However, because of limited resources, impact assessments focused on longer-term impacts are conducted with the motivation of proving successes rather than learning from failures. Given the stage at which such assessments are conducted (many years after the program or project has ended) and their motivation (to provide proof), step 4 studies can best serve the purpose of deriving lessons on what worked, why, and how.

The vision for the hierarchical approach suggested is that the knowledge gained from proofs of development effectiveness from steps 2, 3, and 4 studies will contribute to the development of a "theory of change" that can feed into the development of future interventions and the M&E plan in step 1.

Conclusions

Researchers have always conducted research on the impacts of development programs, projects, and policies. Assessing, understanding, analyzing, and evaluating impacts of public programs, projects, policies, and activities has captured the interest and attention of all the stakeholders involved. *Impact assessment* as a distinct field of inquiry on which many of these impact stories are based is defined in this chapter as an assessment of changes in indicators of program goal(s) that can be attributed to a particular intervention. Depending on their timing and the level of aggregation when they are conducted, IAs can serve as vital tools in determining how investments in a development activity can be most efficiently and equitably made to achieve program goals such as enhancing food security, reducing poverty, or stimulating economic growth.

As shown in this chapter, the theory and practice of IA as applied to agricultural development (including research, extension, infrastructure, marketing, policy, and other types of interventions in the area of agriculture) varies from intervention-specific micro-level assessments to macro-level assessments aggregated over a category of interventions. The methods and approaches of IAs also depend on whether they are beneficiary-level assessments or aggregate society-level impacts. The traditional approach to accounting for impacts of a new technology that increases agricultural productivity and availability of food has been to use a cost–benefit analysis framework to assess social benefits or costs emanating from the adoption of a new technology. In this method, benefits are estimated based on primary and secondary data spanning many time periods and geographic areas, assumptions about the underlying relationships between model parameters, and then an estimation of economic rates of return on project- or program-specific investments. Also, IA based on success stories has been a common practice, especially in the agricultural research IA literature. These approaches and methods played an important role in meeting the demand for accountability and "strategic validation" from donor communities. However, with the increasing emphasis on learning from IA so as to improve the design and implementation of agricultural development programs, the value of "impact assessment" occurring several years after the end of an intervention and only focusing on success stories is questionable. A hierarchical and incremental approach to IA that integrates M&E, IE, meta-analysis, and macro-level IA in hierarchical steps is suggested as a best practice for making evidence-based policies and investment decisions.

Subjecting as many development interventions as resources allow to rigorous IA based on a common IA framework (such as a common structure and model relationships, indicators of measured impacts, method of analysis) can help build a body of evidence on the impacts of development interventions. After a critical mass of evidence is accumulated, these can be subjected to meta-analyses to help assimilate results across different IAs and build a knowledge base of what works, what does not, and why in different contexts. This can then serve as the foundation for building more efficient and effective programs. Such a common framework is urgently needed in the field of agricultural development and could be organized around broad development goals such as poverty, food security, and environmental sustainability. What such a common framework will look like and the organizational and coordination challenges of implementing it across development interventions funded by different donor agencies and ministries need to be further explored.

Notes

1. This figure is an underestimate of the total global effort on agricultural development because it does not include developing countries' own support and commitment to agriculture.

2. The term *project evaluation* as used in this chapter is distinct from impact evaluation. It is also referred to by some as "performance evaluation" (Frey and Osterloh 2006, for example).

3. The intended positive effects for the beneficiaries are the goals of the intervention. The unintended effects are positive or negative side effects. These welfare effects can be directly caused by the intervention or indirectly linked to the intervention through changes in the environment or the resource base for societies, institutions, groups, or individuals. All these types of effects fall within the scope of IA.

4. See Maredia (2009) for an overview of the historical evolution of different methods and approaches of impact assessment over the past six decades.

5. Alston, Norton, and Pardey (1998) provide a comprehensive overview of methodologies applied to both macro- and micro-level quantitative methods of assessing society-level impacts as they relate to agricultural research, which is one of the important contributors to indicators of food security impacts.

6. See the Evidence-Based Policy website (www.evidencebasedpolicy.org) for resources and tools for social experimental designs.

7. The website of the Millennium Challenge Corporation (http://www.mcc.gov/mcc/panda/activities/impactevaluation/ie-sectors/sector-agriculture.shtml) also lists several ongoing IEs in the area of agricultural development using randomized control trials.

8. The examples and observations noted in Table 21.2 are samples for illustrative purposes based on the author's familiarity with the literature. It is not based on a comprehensive review of the literature on impact assessments that focus on food security goals. The purpose is to highlight the diversity in methods or approaches and impact assessment questions found in the development literature related to agriculture.

9. The modeling approach to assess changes in macro-level indicators of development impact is commonly used as a tool for ex ante analysis. However, given the focus on ex post IAs, these types of models and ex ante IAs are not discussed in this chapter.

10. To ensure that step 3 analyses lead to such knowledge and information requires that evaluations in step 2 not be self-selected (in other words, only those interventions that have high chances of success be subjected to IE) and that all the evaluation results (positive, neutral, or negative) be in the public domain (published in one form or another to minimize publication bias in meta-syntheses). In theory, the way to minimize the self-selection bias would be to subject all projects or programs of a certain size and scale to step 2 types of IA. Alternatively, if the resources available to conduct evaluation are limited, a better approach would be to randomize such IEs at a program/system/investor level (for example, across all the World Bank- or OECD-funded projects). In other words, development programs to be evaluated in any given time period would be randomly picked from a pool of all eligible candidates meeting the minimum requirements in terms of size and scale. There are certainly challenges, but more thought needs to be devoted to how to make this suggestion practically and politically feasible to implement.

References

Adato, M., and R. Meinzen-Dick, eds. 2007. *Agricultural research, livelihoods and poverty: Studies of economic and social impact in six countries.* Washington, D.C., and Baltimore: International Food Policy Research Institute and Johns Hopkins University Press.

Alston, J. M., G. W. Norton, and P. G. Pardey. 1998. *Science under scarcity: Principles and practice for agricultural research evaluation and priority setting.* Wallingford, U.K.: CAB International.

Alston, J. M., C. Chan-Kang, M. C. Marra, P. G. Pardey, and T. J. Wyatt. 2000. *A meta-analysis of rates of return to agricultural R&D: Ex pede Herculem?* IFPRI Research Report 113. Washington, D.C.: International Food Policy Research Institute.

Arimond, M., A. de Brauw, D. Gilligan, C. Hotz, R. Labarta, C. Loechl, and J. V. Meenakshi. 2007. *Baseline survey report of OFSP evaluation.* HarvestPlus Challenge Program, International Food Policy Research Institute, Washington, D.C. Photocopy.

Ashraf, N., X. Giné, and D. Karlan. 2008. Finding missing markets (and a disturbing epilogue): Evidence from an export crop adoption and marketing intervention in Kenya. Working Paper 08-065. Boston: Harvard Business School. Also published as World Bank Policy Research Working Paper 4477, Washington, D.C.

Bailar, J. C. 1997. The promise and problems of meta-analysis. *New England Journal of Medicine* 337 (8): 559–561.

Baker, J. L. 2000. *Evaluating the impact of development projects on poverty: A handbook for practitioners.* Washington, D.C.: World Bank.

Bamberger, M., J. Rugh, and L. Mabry. 2006. *Real World evaluation: Working under budget, time, data, and political constraints.* Newbury Park, Calif., U.S.A.: Sage.

Banerjee, A. V., A. H. Amsdan, R. H. Bates, J. Bhagwati, A. Deaton, and N. Stern. 2007. *Making aid work.* Cambridge, Mass., U.S.A.: MIT Press.

Berman, N. G., and R. A. Parker. 2002. Meta-analysis: Neither quick nor easy. *BMC Medical Research Methodology* 2: 10.

CGD (Center for Global Development). 2006. *When will we ever learn? Improving lives through impact evaluation.* Washington, D.C.

CIEM (Central Institute for Economic Management). 2008. Final report of the impact evaluation of the Northern Mountains Poverty Reduction Project. Submitted to the World Bank, Hanoi, Vietnam, June.

Cocchi, H., and B. E. Bravo-Ureta. 2007. On-site costs and benefits of soil conservation among hillside farmers in El Salvador. Working Paper 4/07, Office of Evaluation and Oversight. Washington, D.C.: Inter-American Development Bank.

Deaton, A. 2007. Evidence-based aid must not become the latest in a long string of development fads. In *Making aid work,* ed. A. V. Banerjee, A. H. Amsden, R. H. Bates, J. Bhagwati, A. Deaton, and N. Stern. Cambridge, Mass., U.S.A.: MIT Press.

———. 2009. Instruments of development: Randomization in the tropics, and the search for the elusive keys to economic development. NBER Working Paper 14690. Cambridge, Mass., U.S.A.: National Bureau for Economic Research.

Dercon, S., D. O. Gilligan, J. Hoddinott, and T. Woldehanna. 2008. *The impact of agricultural extension and roads on poverty and consumption growth in fifteen Ethiopian villages.* Discussion Paper 840. Washington, D.C.: International Food Policy Research Institute.

Dillon, A. 2008. *Access to irrigation and the escape from poverty: Evidence from northern Mali.* Discussion Paper 782. Washington, D.C.: International Food Policy Research Institute.

Duflo, E. 2006. Field experiments in development economics. Prepared for the World Congress of the Econometric Society, Department of Economics and Abdul Latif Jameel Poverty Action Lab. Massachusetts Institute of Technology, Cambridge, Mass., U.S.A.

Duflo, E., R. Glennerster, and M. Kremer. 2007. *Using randomization in development economics research: A toolkit.* CEPR Discussion Paper 6059. London: Center for Economic Policy Research.

Duflo, E., M. Kremer, and J. Robinson. 2008. How high are rates of return to fertilizer? Evidence from field experiments in Kenya. *American Economics Review* 98 (2): 482–488.

Espey, M., J. Espey, and W. D. Shaw. 1994. Price elasticity of residential demand for water: A meta-analysis. *Water Resources* 33: 1369–1374.

Evenson, R. E. 2001. Economic impacts of agricultural research and extension. In *Handbook of agricultural economics,* vol. 1A, ed. B. L. Gardner and G. C. Rausser. Amsterdam, The Netherlands: Elsevier.

Evenson, R. E., and D. Gollin, eds. 2003. *Crop variety improvement and its effect on productivity: The impact of international agricultural research.* Wallingford, U.K.: CAB International.

Evenson, R. E., and Y. Kislev. 1975. *Agricultural research and productivity.* New Haven, Conn., U.S.A.: Yale University Press.

Evenson, R. E., and M. W. Rosegrant. 2003. The economic consequences of CGI programs. In *Crop variety improvement and its effect on productivity: The impact of international agricultural research,* ed. R. E. Evenson and D. Gollin. Wallingford, U.K.: CAB International.

Fan, S. 2007. Agricultural research and urban poverty in China and India. In *Agricultural research, livelihoods, and poverty: Studies of economic and social impacts in six countries,* ed. M. Adato and R. Meinzen-Dick. Washington, D.C. and Baltimore: International Food Policy Research Institute and Johns Hopkins University Press.

Fan, S., C. Chan-Kang, K. Qian, and K. Krishnaiah. 2007. National and international agricultural research and rural poverty: The case of rice research in India and China. In *Agricultural research, livelihoods, and poverty: Studies of economic and social impact in six countries,* ed. M. Adato and R. Meinzen-Dick. Washington, D.C. and Baltimore: International Food Policy Research Institute and Johns Hopkins University Press.

FAO (Food and Agriculture Organization of the United Nations). 2003. *Trade reforms and food security: Conceptualizing the linkages.* Rome: Commodity Policy and Projections Service, Commodities and Trade Division, FAO.

———. 2005. Summary of world food and agricultural statistics. Rome. <http://www.fao.org/es/ess/sumfas/sumfas_en_web.pdf>. Accessed June 2009.

FARA (Forum for Agricultural Research in Africa). 2009. *Sub-Saharan Africa challenge programme: Research plan and programme for impact assessment.* Accra, Ghana.

Frey, B. S., and M. Osterloh. 2006. Evaluations: Hidden costs, questionable benefits, and superior alternatives. Working Paper 2006-23. Basel, Switzerland: CREMA Gellertstrasse.

Gilligan, D., and J. Hoddinott. 2007. Is there persistence in the impact of emergency food aid? Evidence on consumption, food security, and assets in rural Ethiopia. *American Journal of Agricultural Economics* 89 (2): 225–242.

Gilligan, D., J. Hoddinott, and A. Taffesse. 2008. The impact of Ethiopia's productive safety net programme and its linkages. International Food Policy Research Institute, Washington, D.C. Photocopy.

Giné, X., and D. Yang. 2009. Insurance, credit, and technology adoption: Field experimental evidence from Malawi. *Journal of Development Economics* 89 (1): 1–11.

Glass, G. V. 1976. Primary, secondary, and meta-analysis of research. *Educational Researcher* 5: 3–8.

Görg, H., and E. Strobl. 2001. Multinational companies and productivity spillovers: A meta-analysis. *Royal Economic Society* 111 (475): F723–F739.

Griliches, Z. 1963. The sources of measured productivity growth: U.S. agriculture 1940–1960. *Journal of Political Economy* 74 (4): 331–346.

———. 1964. Research expenditures, education, and the aggregate agricultural production function. *American Economic Review* 54 (6): 961–974.

Hoddinott, J., and E. Skoufias. 2004. The impact of PROGRESA on food consumption. *Economic Development and Cultural Change* 53 (1): 37–61.

IEG (Independent Evaluation Group). 2009. Measuring the impacts of agricultural projects: A meta-analysis and three impact evaluations. IEG of the World Bank, Washington, D.C. Photocopy.

Jacoby, H., G. Li, and S. Rozelle. 2002. Hazards of expropriation: Tenure insecurity and investment in rural China. *American Economic Review* 92 (5): 1420–1447.

La Rovere, R., D. Flores, P. Aquino, and J. Dixon. 2008. Through the livelihoods lens: Lessons from Mexico and Nepal on integrating approaches and metrics for impact assessment. Paper presented at the workshop Rethinking Impact: Understanding the Complexity of Poverty and Change, March 26–28, in Cali, Colombia.

Low, J., M. Arimond, N. Osman, B. Cunguara, F. Zano, and D. Tschirley. 2007. A food-based approach introducing orange fleshed sweet potatoes, increased vitamin A intake and serum retinol concentrations in young children in rural Mozambique. *Journal of Nutrition* 137 (5): 1320–1327.

Mancini, F., A. H. Van Bruggen, and J. L. S. Jiggins. 2007. Evaluating cotton integrated pest management farmer field school outcomes using the sustainable livelihoods approach in India. *Experimental Agriculture* 43: 97–112.

Maredia, M. 2009. *Improving the proof: Evolution of and emerging trends in impact assessment methods and approaches in agricultural development.* Discussion Paper 929. Washington, D.C.: International Food Policy Research Institute.

Maredia, M. K., and D. A. Raitzer. 2006. *CGIAR and NARS partner research in Sub-Saharan Africa: Evidence of impact to date.* Rome: Consultative Group on International Agricultural Research Science Council Secretariat.

Mohr, L. B. 1995. *Impact analysis for program evaluation,* 2nd ed. Thousand Oaks, Calif., U.S.A.: Sage.

————. 1999. The qualitative method of impact analysis. *American Journal of Evaluation* 20: 69–84.

Raitzer, D., and T. G. Kelley. 2008. Benefit–cost meta-analysis of investment in the international agricultural research centers of the CGIAR. *Agricultural Systems* 96: 108–123.

Ravallion, M. 2008. Evaluating anti-poverty programs. In *Handbook of development economics,* vol. 4, ed. H. Chenery and T. N. Srinivan. Amsterdam, the Netherlands: Elsevier.

————. 2009. Should the randomistas rule? *The Economists' Voice* 6 (2): Article 6.

Ravallion, M., and D. van de Walle. 2004. Breaking up the collective farms: Welfare outcomes of Vietnam's massive land privatization. *Economics of Transition* 12 (2): 201–236.

Rodrik, D. 2008. The new development economics: We shall experiment, but how shall we learn? In *What works in development? Thinking big vs. thinking small,* ed. J. Cohen and W. Easterly. Washington, D.C.: Brookings Institution Press.

Rutherford, A. S. 2008. *Broad bed maker technology package innovations in Ethiopian farming systems: An ex post impact assessment.* Research Report 20. Nairobi, Kenya: International Livestock Research Institute.

Thiam, A., B. E. Bravo-Ureta, and T. E. Rivas. 2001. Technical efficiency in developing country agriculture: A meta-analysis. *Agricultural Economics* 25 (2/3): 235–243.

Thirtle, C., L. Lin, and J. Piesse. 2003. The impact of research-led agricultural productivity growth on poverty reduction in Africa, Asia, and Latin America. *World Development* 31 (12): 1959–1975.

Thompson, S. G., and S. J. Pocock. 1991. Can meta-analyses be trusted? *Lancet* 338: 1127–1130.

Valadez, J., and M. Bamberger, eds. 1994. *Monitoring and evaluating social programs in developing countries.* Economic Development Institute of the World Bank Series. Washington, D.C.: World Bank.

van den Berg, M., and R. Ruben. 2006. Small-scale irrigation and income distribution in Ethiopia. *Journal of Development Studies* 42 (5): 868–880.

Advisory Committee Members and IFPRI Project Team

Advisory Committee Members

As of November 2009

Harris Mule, Committee Co-Chair, Former Permanent Secretary, Ministry of Finance, Kenya

Raul Montemayor, Committee Co-Chair, General Secretary, Federation of Free Farmers Cooperatives Inc., Philippines

Chris Dowswell, Executive Director, Sasakawa Africa Association, Mexico

Mahabub Hossain, Executive Director, Bangladesh Rural Advancement Committee, Bangladesh

Isatou Jallow, Chief, Women, Children, and Gender Policy, World Food Programme, Italy

Marina Joubert, Science Communication Editor, Southern Science, South Africa

Ruth Levine, Vice President, Programs and Operations, Center for Global Development, U.S.A.

Xiaopeng Luo, Professor, China Academy for Rural Development, Zhejiang University, China

Stephen Muliokela, Executive Director, Golden Valley Agricultural Research Trust, Zambia

Raj Paroda, Executive Secretary, Asia-Pacific Association of Agricultural Research Institutions, Thailand

Christie Peacock, Chief Executive, Farm Africa, United Kingdom

Prabhu Pingali, Deputy Director, Agricultural Development, The Bill & Melinda Gates Foundation, U.S.A.

Martín Piñeiro, Director, GrupoCeo, Argentina

Papa Seck, Director General, Africa Rice Center, Benin

Camila Toulmin, Director, International Institute for Environment and Development, United Kingdom

Ajay Vashee, President, International Federation of Agricultural Producers, Zambia

Joachim von Braun, Ex-Officio Member, Director General, International Food Policy Research Institute, U.S.A.

IFPRI Project Team
As of November 2009

Rajul Pandya-Lorch, Project Leader–Chief of Staff, Director General's Office, and Head, 2020 Vision Initiative

David J. Spielman, Project Research Lead–Research Fellow, Knowledge, Capacity, and Innovation Division

Klaus von Grebmer, Project Communications Lead–Director, Communications Division

Kenda Cunningham, Research Support–Senior Research Assistant, Director General's Office

Sivan Yosef, Research Support–Senior Research Assistant, Director General's Office

Contributors

Joshua Ariga (arigajos@msu.edu) is a research fellow with Tegemeo Institute of Agricultural Policy and Development, Egerton University, Kenya, and is currently a Ph.D. candidate at Michigan State University, U.S.A.

Tanguy Bernard (bernardt@afd.fr) is a research fellow with Agence Française de Développement, France.

John P. Brennan (jpjkbrennan@gmail.com) is an economist at Coolamaine Economic Research, Australia.

John W. Bruce (jwbruce@ladsiinc.com) is president of Land and Development Solutions International Inc., U.S.A.

Eugenio J. Cap (ecap@correo.inta.gov.ar) is director of the Institute of Economics and Sociology (INTA), Argentina.

Ashwini Chhatre (achhatre@illinois.edu) is an assistant professor in the Department of Geography at the University of Illinois at Urbana–Champaign, U.S.A.

Kenda Cunningham (k.cunningham@cgiar.org) is a senior research assistant in the Director General's Office at the International Food Policy Research Institute, U.S.A.

H. J. Dubin (hjdubin@comcast.net) is a consultant and was formerly the associate director of the Wheat Program, International Maize and Wheat Improvement Center (CIMMYT), Mexico.

Olaf Erenstein (o.erenstein@cgiar.org) is a senior scientist with the International Maize and Wheat Improvement Center, Mexico, and is currently outposted in Ethiopia.

Peter B. R. Hazell (p.hazell@cgiar.org) was formerly director of the Development Strategy and Governance Division at the International Food Policy Research Institute, U.S.A., and is now retired and living in England.

Derek Headey (D.Headey@cgiar.org) is a research fellow in the Development Strategy and Governance Division at the International Food Policy Research Institute, U.S.A.

Mahabub Hossain (hossain.mahabub@brac.net) is the executive director of the Bangladesh Rural Advancement Committee, Dhaka, and the former head of the Social Sciences Division at the International Rice Research Institute, the Philippines.

Jacqueline d'Arros Hughes (jackie.hughes@worldveg.org) is deputy director general of research at the AVRDC–The World Vegetable Center, Taiwan.

Lora Iannotti (liannotti@wustl.edu) is an assistant professor at Washington University in St. Louis, U.S.A., and was formerly a postdoctoral research fellow in the Poverty, Health, and Nutrition Division at the International Food Policy Research Institute, U.S.A.

T. S. Jayne (jayne@msu.edu) is a professor of international development in the Department of Agricultural, Food, and Resource Economics at Michigan State University, U.S.A.

Jonathan Kaminski (kaminski.jonathan@gmail.com) is a research fellow at the Hebrew University of Jerusalem, Israel.

J. D. H. Keatinge (dyno.keatinge@worldveg.org) is director general of the AVRDC–The World Vegetable Center, Taiwan.

Michael Kirk (kirk@wiwi.uni-marburg.de) is a professor of development economics at Marburg University, Germany.

Jiming Li (jiming.li@pioneer.com) is a senior research manager at Pioneer Hi-Bred International, the Philippines.

Zongmin Li (zli@ladsiinc.com) is vice president of Land and Development Solutions International Inc., U.S.A.

Valeria N. Malach (vmalach@correo.inta.gov.ar) is a researcher at the Institute of Economics and Sociology (INTA), Argentina.

Mywish K. Maredia (maredia@msu.edu) is an associate professor in the Department of Agricultural, Food, and Resource Economics at Michigan State University, U.S.A.

Latha Nagarajan (nagarajan@aesop.rutgers.edu) is a research associate in the Department of Agriculture, Food, and Resource Economics at Rutgers, The State University of New Jersey, U.S.A.

Nguyen Do Anh Tuan (ndatuan@gmail.com) is director of the Southern Office of the Institute of Policy and Strategy for Agriculture and Rural Development, Vietnam.

Felix I. Nweke (nwekefel@yahoo.com) is a visiting professor at Michigan State University, U.S.A.

Hemant R. Ojha (ojhahemant1@gmail.com) is editor of the *Journal of Forest and Livelihoods* and a natural resource governance specialist at ForestAction Nepal.

Rajul Pandya-Lorch (r.pandya-lorch@cgiar.org) is head of the 2020 Vision Initiative and chief of staff in the Director General's Office at the International Food Policy Research Institute, U.S.A.

Lauren Persha (lpersha@umich.edu) is a research fellow in the School of Natural Resources and Environment at the University of Michigan, U.S.A.

Carl E. Pray (pray@aesop.rutgers.edu) is a professor in the Department of Agriculture, Food, and Resource Economics at Rutgers, The State University of New Jersey, U.S.A.

Chris Reij (c.reij@chello.nl) is a natural resource management specialist with the Center for International Cooperation of Vrije Universiteit Amsterdam, the Netherlands.

Karl M. Rich (kr@nupi.no) is a senior research fellow at the Norwegian Institute of International Affairs in Oslo, Norway.

Peter Roeder (peter.roeder@taurusah.com) is an independent veterinary consultant specializing in control of transboundary animal diseases, United Kingdom.

Marie T. Ruel (m.ruel@cgiar.org) is director of the Poverty, Health, and Nutrition Division at the International Food Policy Research Institute, U.S.A.

Subramanyam Shanmugasundaram (sundar19392004@yahoo.com) was formerly deputy director general of research, AVRDC–The World Vegetable Center, and is currently an agricultural consultant, U.S.A.

Melinda Smale (msmale@oxfamamerica.org) is a senior researcher, agriculture and trade, Oxfam America, and was formerly a senior research fellow in the Environment and Production Technology Division at the International Food Policy Research Institute, U.S.A.

David J. Spielman (d.spielman@cgiar.org) is a research fellow with the Knowledge, Capacity, and Innovation Division of the International Food Policy Research Institute, U.S.A., and is currently outposted in Ethiopia.

Gray Tappan (tappan@usgs.gov) is a physical geographer with the U.S. Geological Survey at the EROS Center in Sioux Falls, U.S.A.

Eduardo J. Trigo (etrigo@grupoceo.com.ar) is director of Grupo CEO, Argentina.

Federico Villarreal (fedevillarreal78@gmail.com) is a research associate at Grupo CEO, Argentina.

Sivan Yosef (s.yosef@cgiar.org) is a senior research assistant in the Director General's Office at the International Food Policy Research Institute, U.S.A.

Yeyun Xin (xinyeyun@hotmail.com) is a research professor at the China National Hybrid Rice Research and Development Center, China.

Longping Yuan (lpyuan@hhrrc.ac.cn) is the Director General of the China National Hybrid Rice Research and Development Center, China.

Index

Page numbers for entries occurring in figures are suffixed by *f*, those for entries occurring in notes by *n*, and those for entries occurring in tables by *t*.